WILLIAM M. MAIESE

The Discovery of Natural Products with Therapeutic Potential

BIOTECHNOLOGY

JULIAN E. DAVIES, *Editor*
Pasteur Institute
Paris, France

Editorial Board

BIOTECHNOLOGY SERIES

1. R. Saliwanchik — *Legal Protection for Microbiological and Genetic Engineering Inventions*

2. L. Vining (editor) — *Biochemistry and Genetic Regulation of Commercially Important Antibiotics*

3. K. Herrmann and R. Somerville (editors) — *Amino Acids: Biosynthesis and Genetic Regulation*

4. D. Wise (editor) — *Organic Chemicals from Biomass*

5. A. Laskin (editor) — *Enzymes and Immobilized Cells in Biotechnology*

6. A. Demain and N. Solomon (editors) — *Biology of Industrial Microorganisms*

7. Z. Vaněk and Z. Hošťálek (editors) — *Overproduction of Microbial Metabolites: Strain Improvement and Process Control Strategies*

8. W. Reznikoff and L. Gold (editors) — *Maximizing Gene Expression*

9. W. Thilly (editor) — *Mammalian Cell Technology*

10. R. Rodriguez and D. Denhardt (editors) — *Vectors: A Survey of Molecular Cloning Vectors and Their Uses*

11. S.-D. Kung and C. Arntzen (editors) — *Plant Biotechnology*

12. D. Wise (editor) — *Applied Biosensors*

13. P. Barr, A. Brake, and P. Valenzuela (editors) — *Yeast Genetic Engineering*

14. S. Narang (editor) — *Protein Engineering: Approaches to the Manipulation of Protein Folding*

15. L. Ginzburg (editor) — *Assessing Ecological Risks of Biotechnology*

16. N. First and F. Haseltine (editors) — *Transgenic Animals*

17. C. Ho and D. Wang (editors) — *Animal Cell Bioreactors*

18. I. Goldberg and J.S. Rokem (editors) — *Biology of Methylotrophs*

19. J. Goldstein (editor) — *Biotechnology of Blood*

20. R. Ellis (editor) — *Vaccines: New Approaches to Immunological Problems*

21. D. Finkelstein and C. Ball (editors) — *Biotechnology of Filamentous Fungi*

22. R. Doi and M. McGloughlin (editors) — *Biology of Bacilli: Applications to Industry*

23. J. Bennett and M. Klich (editors) — Aspergillus: *Biology and Industrial Applications*

24. J. Davies and W. Reznikoff (editors) — *Milestones in Biotechnology: Classic Papers on Genetic Engineering*

25. D. Woods (editor) — *The Clostridia and Biotechnology*

26. V. Gullo (editor) — *The Discovery of Natural Products with Therapeutic Potential*

The Discovery of Natural Products with Therapeutic Potential

Edited by

Vincent P. Gullo
Schering-Plough Research Institute
Kenilworth, New Jersey

Butterworth–Heinemann
Boston London Oxford Singapore Sydney Toronto Wellington

Recognizing the importance of preserving what has been written, it is the policy of Butterworth–Heinemann to have the books it publishes printed on acid-free paper, and we exert our best efforts to that end.

Library of Congress Cataloging-in-Publication Data
The Discovery of natural products with therapeutic potential / edited by Vincent P. Gullo.
p. cm.—(Biotechnology series ; 26)
Includes bibliographical references and index.
ISBN 0-7506-9003-8 (acid-free paper)
1. Pharmacognosy. 2. Natural products—Therapeutic use.
I. Gullo, Vincent Philip, 1950– . II. Series: Biotechnology series (Reading, Mass.) ; 26.
RS160.D58 1994
615′.3—dc20 93-23111
CIP

British Library Cataloguing-in-Publication Data
A catalogue record for this book is available from the British Library.

Butterworth–Heinemann
80 Montvale Avenue
Stoneham, MA 02180

10 9 8 7 6 5 4 3 2 1

Printed in the United States of America

R. Albin
Antiviral Chemotherapy
Schering-Plough Research Institute
Kenilworth, New Jersey

Angela Belt
The Biotic Network
Sonora, California

Anna M. Casazza
Department of Experimental
  Therapeutics
Bristol Myers Squibb
  Pharmaceutical Research
  Institute
Princeton, New Jersey

I.H. Chapela
Preclinical Research
Sandoz Pharmaceuticals Ltd.
Basel, Switzerland

Beth J. DiDomenico
Molecular Genetics/Chemotherapy
Schering-Plough Research Institute
Kenilworth, New Jersey

M.M. Dreyfuss
Preclinical Research
Sandoz Pharmaceuticals Ltd.
Basel, Switzerland

Akira Endo
Department of Applied Biological
  Science
Tokyo Noko University
Tokyo, Japan

Otto D. Hensens
Merck Research Laboratories
Rahway, New Jersey

Jill E. Hochlowski
Bioactive Microbial Metabolites
  Project
Pharmaceutical Products
  Discovery Research and
  Development
Abbott Laboratories
Abbot Park, Illinois

Ann C. Horan
Microbial Products
Schering-Plough Research Institute
Kenilworth, New Jersey

J.C. Hunter-Cevera
The Biotic Network
Sonora, California

Sunil Kadam
Anti-infective Research Division
Pharmaceutical Products
  Discovery Research and
  Development
Abbott Laboratories
Abbott Park, Illinois

A. Douglas Kinghorn
Program for Collaborative
  Research in the Pharmaceutical
  Sciences and Department of
  Medicinal Chemistry and
  Pharmacognosy
College of Pharmacy
University of Illinois at Chicago
Chicago, Illinois

Toru Kino
Exploratory Research
  Laboratories
Fujisawa Pharmaceutical Co., Ltd.
Tsukuba, Japan

Donald R. Kirsch
Agricultural Research Division
American Cyanamid
Princeton, New Jersey

Frank E. Koehn
Division of Biomedical Marine
    Research
Harbor Branch Oceanographic
    Institution
Fort Pierce, Florida

Byron H. Long
Department of Experimental
    Therapeutics
Bristol Myers Squibb
    Pharmaceutical Research
    Institute
Princeton, New Jersey

Ross E. Longley
Division of Biomedical Marine
    Research
Harbor Branch Oceanographic
    Institution
Fort Pierce, Florida

James B. McAlpine
Bioactive Microbial Metabolites
    Project
Pharmaceutical Products
    Discovery Research and
    Development
Abbott Laboratories
Abbott Park, Illinois

Oliver J. McConnell
Division of Biomedical Marine
    Research
Harbor Branch Oceanographic
    Institution
Fort Pierce, Florida

Masakuni Okuhara
Exploratory Research
    Laboratories
Fujisawa Pharmaceutical Co., Ltd.
Tsukuba, Japan

E. Rozhon
Antiviral Chemotherapy
Schering-Plough Research Institute
Kenilworth, New Jersey

J. Schwartz
Antiviral Chemotherapy and
    Molecular Pharmacology
Schering-Plough Research Institute
Kenilworth, New Jersey

Kazuo Umezawa
Department of Applied Chemistry
Faculty of Science and Technology
Keio University
Yokohama, Japan

CONTENTS

**PART III.   ISOLATION AND STRUCTURE DETERMINATION
OF NATURAL PRODUCTS**

Although effective drugs for many of the diseases that afflict humankind have been discovered, many health problems remain untreatable. These problems include various types of cancer; viral infections such as HIV; severe fungal infections, particularly in immunocompromised patients; cardiovascular diseases; and inflammatory and allergic disorders. Even when significant progress is made, as in the treatment of bacterial infections, resistant, highly pathogenic organisms can appear. However, as knowledge of various disease processes expand, new targets for intervention are discovered. Therefore, the search for novel therapeutic agents continues.

From where are drugs derived? There are only two sources of drugs, natural products and synthetic chemicals. The trend in recent years has been to focus attention on rational drug design, highlighting the synthetic chemical approach. However, rational drug design has proven to be effective in refining structural leads that interact with a particular target, for example, an enzyme or receptor. The need still remains to uncover the initial structural lead that interacts with the therapeutic target. The myriad of structurally diverse compounds found in nature offers a unique source for drug discovery. It is in this scenario that natural products play an important role.

The authors of each chapter are a diverse group of experts, who, together, present the knowledge base required to discover potential therapeutic agents from natural sources. The subject matter included in each chapter is intended to be useful for scientists currently working in this broad field, as well as to enlighten those interested in the current status and complexity of natural product research. The authors have highlighted successes, but, for the most part, they have concentrated on the strategies employed. I am deeply indebted to them for their efforts.

The purpose of this book is to capture what we perceive as the rational process of natural product drug discovery. The book has been divided into three major sections that represent this process. Chapters 1–5 describe some of the diverse sources of natural products. Both terrestrial and marine environments represent ecological niches where therapeutically active, complex molecules can be discovered. Chapters 6–12 describe how molecular biological and biochemical research have increased knowledge of biological systems and human disease. This knowledge has led to the design of targeted assays, amenable to high-volume screening. Chapters 13 and 14 describe the current methods that natural product chemists employ to isolate and identify compounds with biological activity from natural product extracts.

Vincent P. Gullo

The Discovery of Natural Products with Therapeutic Potential

PART
I

The Source of Chemical Diversity

Aerobic Actinomycetes: A Continuing Source of Novel Natural Products

Ann C. Horan

A Natural Products Program begins at the source, whether it be extracts of plants and/or marine organisms or fermentation products of actinomycetes, fungi, and bacteria. All have yielded abundant diverse compounds, many with therapeutic utility. Ten of the top selling 30 branded prescription products for 1990 (*Scripts Review Issue 1990*, p. 21) were either natural products or their derivatives and accounted for U.S. $8.2 billion in sales.

Methods for detecting biological activity with therapeutic potential have shifted from simple determinations of antibacterial and antifungal activity to assays using genetically engineered eukaryotic cells that can detect regulators of specific gene expression, inhibitors of signal transduction, small molecular weight agonists/antagonists of cytokines, and/or specific receptors and receptor subtypes. The shift in emphasis away from antibacterials is reflected in the reported activities of novel secondary metabolites found in *The Journal of Antibiotics*. During the years from 1986 to 1991 (Table 1–1), the total number of novel compounds increased 20%; anti-infectives were reduced by 24%, and pharmacologically active compounds increased 87%. When new targets for drug intervention are defined and new assays developed to exploit

TABLE 1-1 Novel Secondary Metabolites Reported in *The Journal of Antibiotics* by Therapeutic Area, 1986 versus 1991

Therapeutic Area	1986	1991	Percent Change
Anti-infectives	25	19	−24
Tumor biology	14	18	+28
Pharmacology	15	28	+87
Total	54	65	+20

the target, chemical leads are not available as starting points for potent new drugs. "Random" screening will be initiated and should include diverse natural products as well as synthetic compounds.

Aerobic, saprophytic actinomycetes remain the major source of novel, fermentation derived, secondary metabolites (Table 1–2) and are well suited to the production of secondary metabolites. Their minimum nutritional requirements seem to leave large segments of the genome available for functions other than growth. This may not be true for organisms whose survival requires complex growth conditions, such as anaerobes and nitrogen fixers. The natural reservoir of the aerobic actinomycetes is soil, where their primary ecological role appears to be decomposing organic matter (Lechevalier and Lechevalier 1981). Labeda and Shearer (1990) reported that the number of recoverable actinomycetes per gram of agricultural soil is in excess of 10^6 colony forming units (cfu) per gram.

Our Natural Products Program has focused on isolating aerobic actinomycetes that contain meso-diaminopimelic acid in their cell walls (Lechevalier and Lechevalier 1981, 1985). We have developed a testing paradigm that evaluates the effect of soil type, media constituents, and selective pressure. Physiological data used for taxonomic analysis of the various genera of actinomycetes formed the basis for our initial choices of selective pressures (antibiotics) and media constituents. The purpose of these investigations has been to discover appropriate conditions for isolating various groups of actinomycetes, thereby enhancing the ability to discover novel natural products. Numerous reviews (Cross 1981a, 1982; Goodfellow and Williams 1983; Williams and Wellington 1981, 1982; Williams and Vickers 1988; Williams et al. 1984) have detailed procedures for the isolation of a wide variety of actinomycetes from soil. A recent review by Labeda and Shearer (1990) is highly

TABLE 1-2 Novel Secondary Metabolites Reported in *The Journal of Antibiotics* by Producing Organism, 1986 versus 1991

Producing Organism	1986	Percent of Total	1991	Percent of Total
Actinomycetes	47	87	51	78
Fungi	7	13	14	22
Total	54	100	65	100

recommended. This chapter will detail experiments performed in our laboratory directed toward the isolation of micromonosporae, actinomadurae, and nocardioform actinomycetes.

1.1 MICROMONOSPORAE

The micromonosporae (Kawamoto 1989) have been crucial to the success of the Natural Products Program at Schering-Plough. The exploitation of these bacteria proved that non-streptomycetes were valuable sources of novel natural products. For a detailed history of the micromonosporae's role in Schering-Plough's Natural Products Program, the reader is referred to Dr. George Luedemann's recent publication, *Free Spirit of Inquiry, The Uncommon Common Man in Research and Discovery, The Gentamicin Story* (Luedemann 1991).

Micromonosporae are the second largest group of actinomycetes in soil, surpassed only by the streptomycetes. They have been isolated from a variety of habitats. Jensen (1932) isolated micromonosporae from dry, neutral, and alkaline soils. Solovieva and Singal (1972) found them to be widely distributed, occurring preferentially in moist Russian soils. Shearer (1987) reported that two or three micromonosporae can be isolated from most soil types. Cross (1981b) places the number at 10×10^6 cfu per gram of soil. Micromonosporae have been isolated from freshwater lakes and lake mud, where they are the most frequently isolated mesophilic actinomycetes (Willoughby 1969; Johnston and Cross 1976). The surface of lake mud was reported to contain 2.4×10^6 micromonosporae cfu per gram, which represented 50% of the total population; deep mud contained 4×10^5 cfu per gram, and in which the micromonosporae predominated. Micromonosporae have also been isolated from the marine environment, from a salt marsh by Hunter et al. (1981), from sea water and sand by Watson and Williams (1974), and antibiotic producers have been isolated from marine sediments (Okazaki and Okami 1976).

Viable micromonosporae spores have been found from 100-year-old sediments (Cross and Attwell 1974), and there has been much discussion over the ecological role these organisms play in mud and marine sediments. We have isolated numerous strains of micromonosporae from lake mud. Macroscopic and microscopic evaluation of the isolates did not reveal a wide range of diversity nor were these organisms prolific producers of secondary metabolites.

Lake mud isolates and type species were tested for their ability to utilize glucose fermentatively using the Hugh and Leifson test as described by Gordon and Horan (1968). None of the strains tested gave a positive fermentative response, nor were they able to grow along the stab line in the aerobic tube. Growth was limited to the surface of the agar, indicating that all the strains were strict aerobes. The results suggest that micromonosporae are not phys-

TABLE 1–3 Soil Isolation Media for Micromonosporae

Media Constituent[1]	Medium (g)			
	M-1	M-2	M-3	Chitin
Carbon source				
Soluble starch	1.0	2.0	1.0	—
Chitin	—	—	—	2.0–2.5 colloidal chitin[2]
Nitrogen source				
Yeast extract	1.0	—	—	—
Casein (vitamin free)	—	—	1.0	—
NaNO$_3$	—	2.0	—	—

[1] All media contain CaCO$_3$, 1.0 g; Difco agar, 15.0 g; tap water, 1000 ml; nystatin, 75 μg/ml.
[2] Lingappa and Lockwood 1962.

TABLE 1–4 Ability of *Micromonospora* Species to Grow in the Presence of Antibiotics

Organism	No. of Strains	Genta-micin[1]	Neo-mycin	Erythro-mycin	Rosara-micin	Rifa-mycin
Type species						
M. chalcea	(1)	+ —	+ —	+ —	+ —	+ —
Aminoglycoside producers						
M. chalcea	(2)	-- —	+ +	+ —	+ +	+ +
M. echinospora	(16)	+ +	— —	— —	+ —	— —
M. grisea	(2)	+ +	— —	— —	— —	+ +
M. inyoensis	(14)	+ +	— —	— —	+ —	— —
M. olivoasterospora	(3)	+ +	— —	— —	— —	+ —
M. zionensis	(2)	+ +	— —	— —	— —	— —
M. purpurea	(2)	+ +	— —	— —	+ —	— —
Macrolide producers						
M. polytrota	(2)	+ +	— —	+ ±	+ —	— —
M. rosaria	(11)	+ +	— —	+ +	+ +	— —
Others						
M. carbonacea	(16)	— —	— —	— —	+ +	+ ±
M. halophytica	(7)	— —	— —	— —	+ ±	+ +

[1] The left-hand column under each type of antibiotic represents a 10 μg/ml concentration, the right represents a 50 μg/ml concentration.

iologically active in lake mud or sediments. Their presence may be the result of the spores' physical properties. Micromonosporae spores are hydrophilic, wettable, and easily removed from soil by water (Ruddick and Williams 1972). They withstand ultrasonication, moist heat treatment for 20 minutes at 60°C, dry heat up to 75°C, and are resistant to various chemical solutions (Vobis 1992). Their physical nature suggests that micromonosporae spores that occur in soil along rivers, streams, and lakes are easily washed into the water, eventually falling to the bottom and accumulating in the mud and sediments.

In addition to being more resistant than most actinomycete spores, micromonosporae spores may also have a different density. Karwowski (1986) was able to separate micromonosporae spores from streptomycete and fungal spores using $CsCl_2$ density gradient centrifugation of saline extracted soil. W. Reiblein (unpublished observations), from our laboratory, experimented with the use of renographin-76 density gradients in order to isolate micromonosporae from soils that were heavily contaminated with bacteria. He was able to isolate 3×10^3 cfu per gram of soil in specific regions of the gradient, which was free from bacteria, fungi, and streptomycetes.

Micromonosporae hydrolyze complex carbons (Luedemann 1971), degrade biopolymers (Erikson 1941), attack lignin complexes (McCarthy and Broda 1984), and are proteolytic (Umbreit and McCoy 1941), preferring hydrolyzed vegetable protein over meat protein. Jensen (1932) emended soil

Everninomicin	Novobiocin	Nalidixic Acid	Penicillin G	Tetracycline	Antibiotic Production
+ +	+ +	+ +	+ +	— —	— None detected
+ —	+ +	+ +	+ ±	— —	+ Neomycin
+ —	+ ±	+ —	— —	— —	+ Gentamicin
— —	+ —	+ +	— —	— —	+ Verdamicin
V —	+ —	+ +	— —	— —	+ Sisomicin
— —	+ —	+ +	— —	— —	+ Fortimicin
+ +	+ ±	+ +	— —	— —	+ G-52
V —	+ ±	+ —	— —	— —	+ Gentamicin
+ —	+ ±	+ +	— —	— —	+ Mycinamycins
+ +	+ —	+ +	+ +	— —	+ Rosaramicin
+ +	+ +	+ +	— —	— —	+ Everninomicins
+ —	+ +	+ +	— —	— —	+ Halomicins

with cellulose and lignic acid to enhance the isolation of micromonosporae. Soil samples were plated on dextrose-casein medium (Jensen 1930) and incubated at 30 to 32°C. Cross (1981a) used colloidal chitin agar (Lingappa and Lockwood 1962) for the selective isolation of micromonosporae while Sandrak (1977) isolated strains from the rhizosphere of winter wheat and corn using Kadoto's medium emended with 2% sodium benzoate. Species of *Micromonospora* are sensitive to acid conditions (Luedemann 1971) and NaCl, rarely growing at saline levels greater than 3% (Luedemann 1971; A.C. Horan, unpublished observations). Kawamoto et al. (1983) found inorganic nitrogen to be a poor source of nitrogen; one-half of the strains tested were unable to grow in the presence of sodium nitrate or potassium nitrate, but all the strains tested grew to varying degrees in the presence of ammonium nitrate. We have used soil isolation media that exploit the physiological properties of the micromonosporae. Sodium nitrate has been used successfully in isolation media and results in a significant reduction of the bacterial population. The media constituents are presented in Table 1–3.

The use of antibiotics as selective pressures that enhance the isolation of actinomycetes from soil has been extensively reported and reviewed (Cross 1982; Labeda and Shearer 1990). Early studies in our laboratories, confirmed by Sveshnikova et al. (1976), identified novobiocin as an effective selective pressure in isolation media, enhancing the growth of micromonosporae while suppressing streptomycetes. Prior to the reports of antibiotic gene clustering and the presence of resistance genes within the cluster, we had devised a specific screen to search for novel aminoglycoside antibiotics by plating soil samples on M1 and M2 agars containing 50 to 100 µg/ml gentamicin. Interestingly, as high as 90% of the micromonosporae, the predominating organisms under these conditions, were aminoglycoside producers. Ivanitskaya et al. (1978), using 1, 5, and 10 µg/ml of gentamicin, showed a marked increase in the frequency of micromonosporae detected, 1 µg/ml being optimal. Antibiotic producers were three times higher than on media without gentamicin. Tunicamycin at 25 to 50 µg/ml resulted in the isolation of four different micromonosporae from one soil sample, on average (Wakisaka et al. 1982). Data, generated in our laboratory, on the ability of secondary metabolite producing species of *Micromonospora* to grow in the presence of antibiotics, are presented in Table 1–4.

Micromonospora have proven to be prolific producers of a wide variety of antibacterial antibiotics. However, certain chemical classes produced by other actinomycetes, including polyenes, β-lactams, and tetracycline, have never been detected or isolated. The micromonosporae-produced compounds isolated at Schering-Plough are presented in Table 1–5.

1.2 NON-MICROMONOSPORAE–NON-STREPTOMYCETES

The micromonosporae's inability to produce certain chemical classes of secondary metabolites has led to a systematic examination of isolation conditions that favor other actinomycetes. Although no specific group was targeted, we

TABLE 1-5 Micromonosporae Isolated at Schering-Plough and Associated Products

Antibiotic Class and Producing Organism	Fermentation Product	Selective Pressure	Reference
Aminoglycosides			
M. echinospora	Gentamicin	None	Weinstein et al. 1964
M. purpurea	Gentamicin	None	Weinstein et al. 1964
M. inyoensis	Sisomicin	Gentamicin	Weinstein et al. 1970
M. chalcea	Neomycin	None	Wagman et al. 1973
M. rhodorangia	G-418	Gentamicin	Wagman et al. 1974
M. grisea	Verdamicin	Gentamicin	Weinstein et al. 1975
M. zionensis	G-52	Gentamicin	Marquez et al. 1976b
M. sp.	Fortimicin	Gentamicin	A.C. Horan, unpublished observations
Macrolides			
M. megalomicea	Megalomicin	None	Weinstein et al. 1969
M. rosaria	Rosaramicin	Novobiocin	Wagman et al. 1972
M. sp. (SCC 1880)	Erythromycin	Novobiocin	Marquez et al. 1976a
M. polytrota	Mycinamycin	Gentamicin	Waitz et al. 1983

(continued)

TABLE 1-5 (continued)

Antibiotic Class and Producing Organism	Fermentation Product	Selective Pressure	Reference
Orthosomycins			
M. carbonacea	Everninomicins	None	Weinstein et al. 1965
M. carbonacea var. *africana*	Novel everninomicins	Everninomicin	Waitz and Horan 1988
Ansamycins			
M. halophytica	Halomicins	None	Weinstein et al. 1968
M. ellipsospora	Rifamycin	Novobiocin	Wagman et al. 1975
Actinomycins			
M. floridensis	Actinomycins	None	Wagman et al. 1976
Peptide like			
M. arborensis	Thiostrepton	None	Weinstein et al. 1978
M. carbonacea var. *africana*	Thiostrepton	Everninomicin	Waitz and Horan 1988
M. sp. (SCC 1792)	Peptide	Novobiocin	Cooper et al. 1988
Isonitrile			
M. echinospora var. *chalsenensis*	Hazimicin	Gentamicin	Marquez et al. 1983

TABLE 1–6   Ability of *Actinomadura* Species to Grow in the Presence of Antibiotics[1]

Organism	Genta-micin[2]		Neo-mycin		Erythro-mycin		Rosara-micin		Rifa-mycin		Evernino-micin		Novo-biocin		Nalidixic Acid		Penicillin G		Tetra-cycline		Antibiotic Production
A. kijaniata (ATCC 31588)	+	+	+	+	+	+	+	−	+	+	+	+	±	−	+	+	+	+	+	+	+
A. madura (SCC 908)	+	−	−	−	−	−	−	−	+	+	−	−	−	−	+	+	−	−	−	−	+
A. pelletieri (SCC 918)	−	−	+	+	+	−	+	−	+	+	+	−	−	−	+	+	+	+	−	−	−
A. citrea (ATCC 27887)	+	+	−	−	+	−	+	−	+	+	+	+	−	−	+	−	+	+	−	−	−
A. flava (ATCC 29533)	−	−	+	−	+	−	−	−	+	+	−	−	−	−	+	+	+	+	−	−	+
A. helvata (ATCC 27295)	+	+	+	+	+	+	+	+	+	+	+	−	+	+	+	−	+	+	+	−	−
A. malachitica (ATCC 27888)	+	+	+	−	+	+	+	+	+	+	+	+	−	−	+	+	+	+	−	−	−
A. pusilla (ATCC 27296)	+	+	+	+	+	+	+	+	+	+	+	+	+	+	+	+	+	+	+	−	+
A. roseoviolacea (ATCC 27297)	+	+	+	+	+	+	+	+	+	+	+	+	−	−	+	+	+	+	+	−	−
A. spadix (ATCC 27298)	−	−	−	−	−	−	−	−	+	+	+	−	−	−	−	−	+	+	−	−	−
A. verrucosospora (ATCC 27299)	+	+	+	+	−	−	−	−	+	+	−	−	+	−	+	−	+	+	−	−	−

[1] See Horan and Brodsky (1982) for methods.
[2] The left-hand column under each type of antibiotic represents a 10 µg/ml concentration, the right represents a 50 µg/ml concentration.

began by incorporating antibiotics into soil isolation agar that did not inhibit actinomadurae (Table 1–6). The resulting actinomycetes were isolated, examined microscopically, and identified as either micromonosporae or non-micromonosporae. Beneficial results were those antibiotics (rifamycin, mefoxitin, lincomycin, rosaramicin, gentamicin, everninomicin, neomycin, streptomycin) resulting in greater than 50% non-micromonosporae. The data are presented in Figure 1–1. Although some treatments such as neomycin, strep-

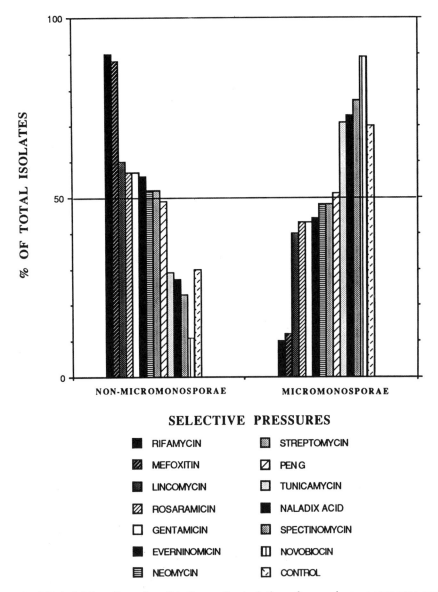

FIGURE 1–1 The effect of antibiotics on the isolation of non-micromonosporae and micromonosporae.

tomycin, and lincomycin, resulted in a high percentage of non-micromono-sporae, the yield of isolates was extremely low (Figure 1–2). Rifamycin resulted in the greatest percentage of non-micromonosporae having a moderate yield, confirming the early results of Gordon and Barnett (1977). Novobiocin and spectinomycin selected for and enhanced the micromonosporae; 54% of the total isolates were from these treatments, while the other 12 antibiotics accounted for the remaining isolates. The large numbers of micromonosporae isolated in the presence of these antibiotics obscured the effects of soil type and media constituents.

Experiments to further evaluate selective pressures and to determine whether media constituents and soil type affected the isolation of actino-mycetes were initiated. Eight soil types, four media constituents, and six

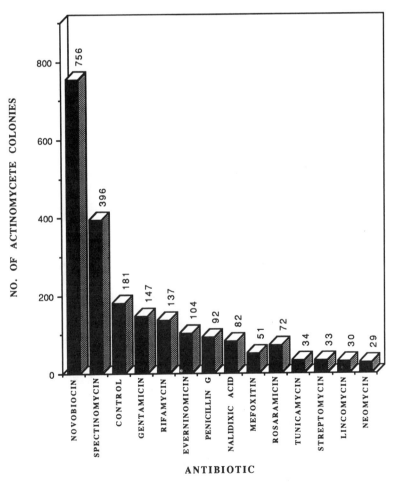

FIGURE 1–2 Effect of antibiotics on the yield of actinomycetes.

TABLE 1–7   An Examination of Environmental Variables

Soil Type	Selective Pressure	Media Constituents[1]
Water logged	Mefoxitin (10 µg/ml)	C source
Sandy	Rifamycin (15 µg/ml)	Soluble starch or glucose
Organic	Everninomicin (10 µg/ml)	
Clay	Rosaramicin (10 µg/ml)	N source
Dry	Neomycin (10 µg/ml)	Yeast extract or NaNO₃
Wood	Gentamicin (5 µg/ml)	
Cultivated		
Mud		

[1] All ingredients at 1.0 g. All media contained Difco agar, 15.0 g; $CaCO_3$, 1.0 g; tap water, 1000/ml, pH 7.0.

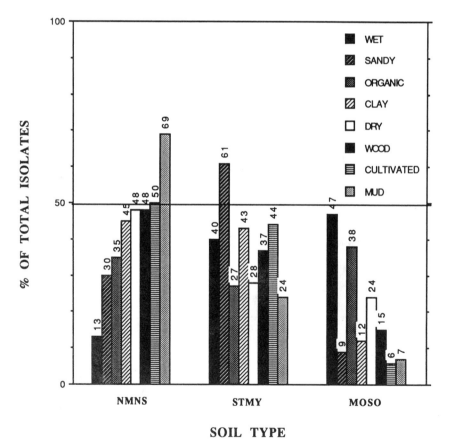

FIGURE 1–3 Effect of soil type on the isolation of actinomycetes.

selective pressures were included in the analysis (Table 1–7). Selective pressures (antibiotic additions) were chosen from previous studies that resulted in greater than 50% non-micromonosporae–non-streptomycetes (NMNS). Soils were air-dried for 24 to 48 hours in a fume hood, 1.0 g samples were suspended in 10 ml of sterile distilled water, serially diluted (1:10, 1:100, 1:1000), and 0.1 ml of all dilutions streaked onto the agar surface using glass rods or sterile cotton swabs presoaked in sterile water. All media contained 75 μg/ml nystatin. Plates were incubated for 3 to 4 weeks at 30°C. Resulting actinomycete colonies were removed to ATCC medium 172 agar (Gherna et al. 1989), incubated 7 to 14 days at 30°C, and transferred to ATCC medium 172 broth, 10 ml/25 mm tubes. The broth was incubated, shaken at 30°C, and shaken at 250 rpm for 3 to 5 days, and the resulting biomass restreaked

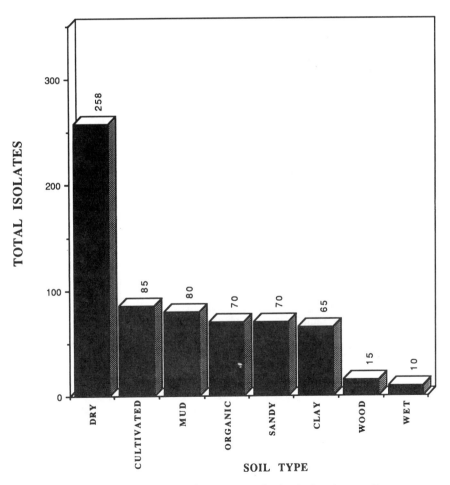

FIGURE 1–4 Number of actinomycete colonies isolated per soil type.

onto ATCC medium 172 and water agars (Sigma agar, 15 g; tap water 1000 ml). Plates were incubated for 14 to 21 days at 30°C and examined microscopically. The organisms were classified as NMNS, streptomycetes (STMY), or micromonosporae (MOSO). Approximately 2000 actinomycetes were isolated and identified using this technique.

The elimination of novobiocin and spectinomycin permitted the effect of soil type to be seen (Figure 1–3). Cultivated soils and mud resulted in greater than 50% NMNS. Sandy soils were the richest in streptomycetes; wet and organic soils were the richest in micromonosporae. The largest number of isolates per soil sample (Figure 1–4) was found in dry soils; wet, waterlogged soils were the least productive. The addition of the antibiotics gentamicin, neomycin, rosaramicin, everninomicin, and rifamycin to the isolation agar resulted in greater than 50% NMNS (Figure 1–5). Rifamycin resulted in the most isolates, neomycin the fewest (Figure 1–6). A simple carbon source in

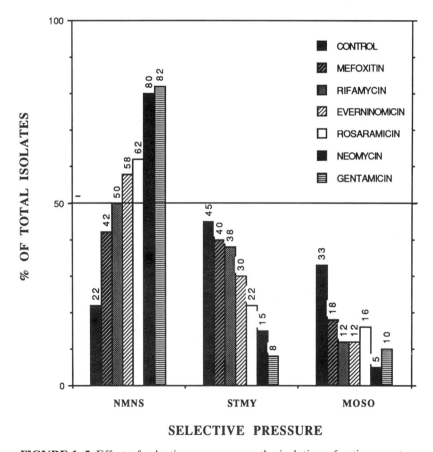

FIGURE 1–5 Effect of selective pressures on the isolation of actinomycetes.

combination with organic nitrogen favored NMNS (Figure 1–7). This effect was negated in the presence of antibiotics, as illustrated in Figure 1–8. Without antibiotic additions, none of the media resulted in greater than 50% NMNS. Soil type, media constituents, and selective pressures all affected the isolation of NMNS. Antibiotic additions to the agar exerted the greatest effect.

The procedures outlined have led to the isolation of a wide variety of secondary metabolite-producing maduramycetes (Williams and Wellington 1981; Goodfellow and Cross 1984). Their high level of resistance to rifamycin (Gordon and Barnett 1977; Goodfellow and Orchard 1974; Athalye et al. 1981; see Table 1–6) and their relative numbers in soil (Goodfellow 1992;

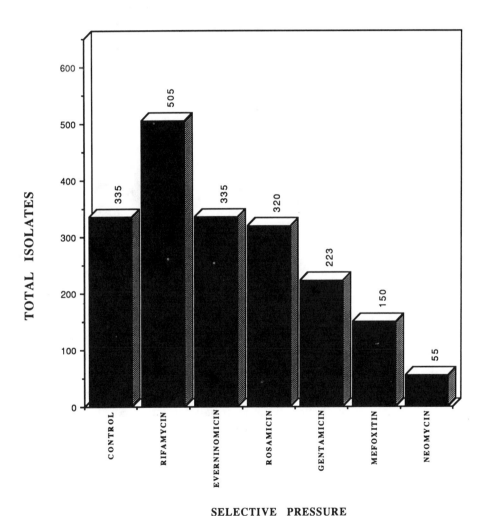

FIGURE 1–6 Effect of selective pressures on the yield of actinomycetes.

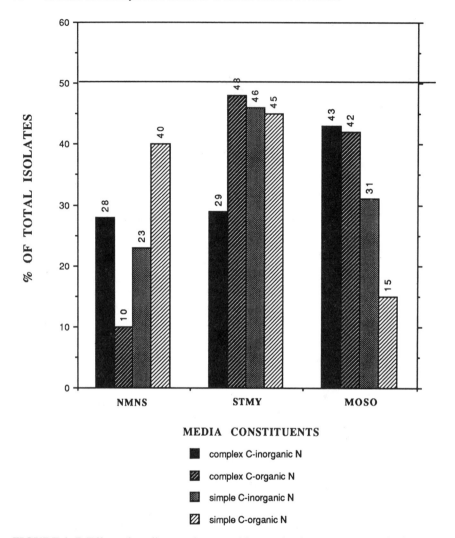

FIGURE 1–7 Effect of media constituents without selective pressures on the isolation of actinomycetes.

Kroppenstedt and Goodfellow 1992) contributed to the ease with which mad-uramycetes were isolated. The cultures and their products are presented in Table 1–8.

1.3 NOCARDIOFORM ACTINOMYCETES

After establishing a protocol capable of evaluating some of the variables involved in isolating non-streptomycete actinomycetes from natural habitats,

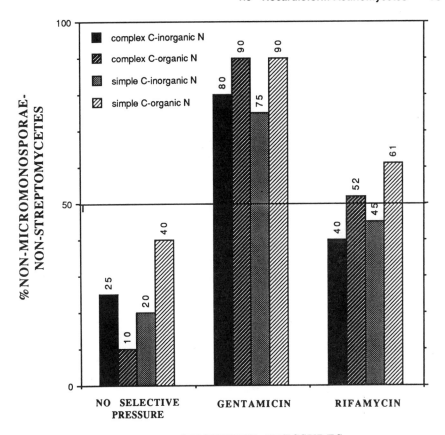

FIGURE 1–8 The effect of selective pressures in various media on the isolation of NMNS.

we applied the same methods to isolating nocardioform actinomycetes. No-cardioform is a general term used to describe actinomycetes that reproduce by fragmentation of the vegetative and aerial mycelium into irregular to rod-like to coccoid elements that give rise to new mycelia (Figures 1–9 and 1–10) (Prauser 1976). We were interested in the aerobic, saprophytic nocardi-oforms that have a cell wall composition of Type III or IV (Lechevalier and Lechevalier 1985).

The following experiments employed 1 g soil samples that were air-dried for 48 hours, suspended in 10 ml of sterile distilled water, and sonicated for 1 minute in an ice bath to break up clumps. Soil suspensions were serially diluted (1:10, 1:100, 1:1000), plated on agar media containing 75 μg/ml nystatin, incubated 14 to 21 days at 28°C, and the resulting colonies removed to ATCC medium 172 agar. The cultures were evaluated morphologically

TABLE 1–8 Maduramycetes Isolated at Schering-Plough and Their Products

Producing Organism	Fermentation Product (Class)	Activity	Reference
Actinomadura			
A. kijaniata	Kijanimicins (macrolides)	Antitumor	Waitz et al. 1981
A. sp.	Oxanthromicin (anthrone peroxide)	Phosphodiesterase inhibitor	Patel et al. 1984
A. brunnea	2'-N-methyl-8-methoxy-tetracycline (tetracycline)	Antibacterial	Patel et al. 1987
A. melliaura	AT 2433 (indolocarbazole)	Antitumor	Horan et al. 1989
A. sp.	Sch 40873 (guanidino)	Mycelial phase of *C. albicans*	Gullo et al. 1991
Microtetraspora			
M. fulva	Sch 38512, 38513, 38511 (macrocyclic lactams)	Antifungal	Cooper et al. 1992
M. vulgaris	Sch 38518, 39185, 38516 (macrocyclic lactams)	Antifungal	Hegde et al. 1992
Microbispora			
M. sp.	Nebularin (nucleoside)	Antifungal	Cooper et al. 1986
M. sp.	EV-22 (triacetylene)	Antifungal	Patel et al. 1988
Thermomonospora			
T. sp.	Sch 38519 (chromanoquinone)	Platelet aggregation inhibitor	Patel et al. 1989

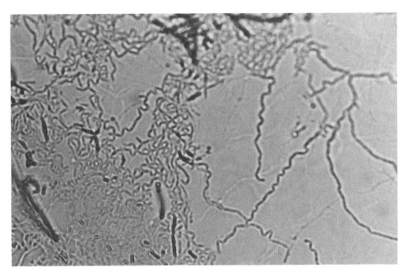

FIGURE 1–9 Typical fragmenting zig-zag vegetative mycelium of a nocardioform actinomycetes isolated from soil. Culture was grown on water agar for 21 days at 28°C.

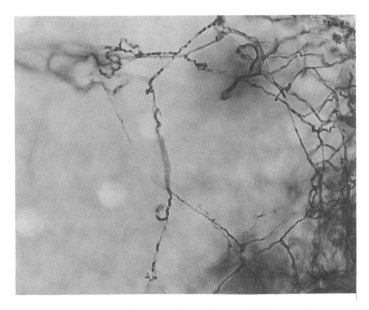

FIGURE 1–10 Fragmenting aerial mycelium of a nocardioform actinomycete isolated from soil. Culture was grown on water agar for 21 days at 28°C.

after growth on water agar for 14 to 21 days following the procedures outlined previously. The data on number of nocardioforms isolated were calculated as the mean number per variable tested.

The effect of soil type is illustrated in Figure 1–11. One hundred and forty-four soils, plated on all media and under all conditions, were evaluated. Rain forest and conifer forest soils were the most productive, yielding between three and four isolates per soil type. Like the NMNS, wet, water-logged soils yielded the fewest nocardioforms as well as the fewest total isolates. Five random soils were chosen to evaluate the effect of air-drying and heating soils to 120°C for 1 hour (Table 1–9). Heating reduced the numbers of nocardioforms whereas air-drying reduced the growth of spreading bacteria.

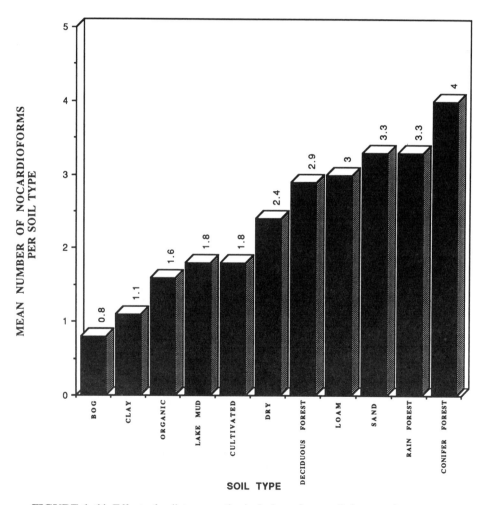

FIGURE 1–11 Effect of soil type on the isolation of nocardioform actinomycetes.

TABLE 1–9 **Effect of Pretreatments on the Isolation of Nocardioform Actinomycetes[1]**

Treatment	Mean, Nocardioforms/Soil
Control (no treatment)	0[2]
Air-dry, 48 hours	1.8
Air-dry, 48 hours and heat, 1 hour at 120°C	0.8

[1] Five soils/all media/all additions.
[2] Plates overgrown with bacteria.

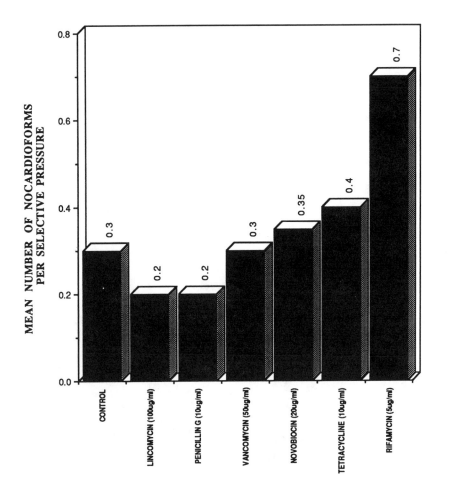

FIGURE 1–12 Effect of selective pressures on the isolation of nocardioform actinomycetes.

TABLE 1–10   Effect of Media on the Isolation of Nocardioform Actinomycetes[1]

Agar Medium[2]	No. of Soils	Mean, Nocardioforms/Soil
AV	168	0.9
SCN	73	1.0
GA	77	1.1

[1] All soil types/soil air-dried/all additions.
[2] AV, arginine B vitamin (Nonomura and Ohara 1969); SCN, starch casein nitrate (Küster and Williams 1964); and GA, glycerol asparagine (Waksman 1961).

Antibiotic additions were based on the taxonomic data generated by Goodfellow and Orchard (1974) and on our previous results in reducing streptomycetes with minimum effect on other actinomycetes. Of the selective pressures evaluated, rifamycin was the most effective (Figure 1–12). Three media frequently used to isolate actinomycetes (Shearer 1987), AV-agar (Non-

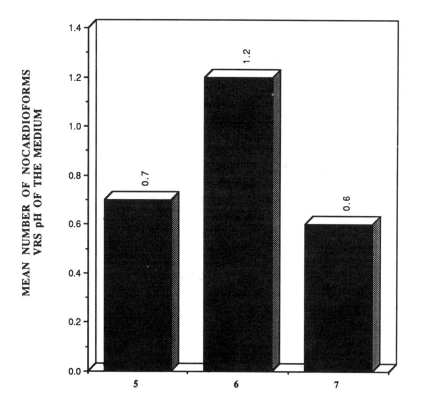

pH

FIGURE 1–13 Effect of pH on the isolation of nocardioform actinomycetes.

TABLE 1–11   Optimization of Variables for the Isolation of Nocardioform Actinomycetes[1]

Variable	Mean, Nocardioforms/Soil
Medium	
GA	1.0
SCN	4.2
pH	
7.0	2.4
6.0	2.8
Antibiotic	
None	2.1
Rifamycin (5 µg/ml)	3.1
Soil	
Conifer forest	5.8
Dry tropical	4.5

[1] Twelve soils/all soils air-dried.

omura and Ohara 1969), starch-casein-nitrate agar (Kuster and Williams 1964), and glycerol-asparagine agar (Waksman 1961) were used. The three media were equally effective in supporting the growth of isolated nocardioforms (see Table 1–10). However, adjusting the media to pH 6.0 (Figure 1–13) increased nocardioform isolations.

An experiment to optimize isolation conditions consisted of examining 12 soils, six conifer forest and six dry tropical soils, plated on glycerol-asparagine and starch-casein-nitrate agars at pH 6.0 and 7.0, comparing no selective pressure to rifamycin at 5 µg/ml. These data are presented in Table 1–11. Conifer forest soils plated on starch-casein-nitrate agar at pH 6.0 in the presence of rifamycin resulted in 5.8 nocardioform isolates per soil plated.

We have isolated two novel secondary metabolite-producing nocardioform actinomycetes using these techniques, and both have been identified as members of the genus *Saccharothrix* (Labeda 1992). One strain produces a series of 14 novel indolocarbazole metabolites, which are potent inhibitors of protein kinase C (Brodsky et al. 1991). The second series of novel metabolites are currently being evaluated.

1.4 CONCLUSIONS

Procedures can be defined that will successfully enhance the isolation of specific groups of actinomycetes from natural habitats. The most effective method appears to be the incorporation of antibiotics into isolation media comprised of nutrients that favor the group being pursued. However, the soil harbors a myriad of organisms that have eluded isolation. Scanning electron micrographs of soil samples clearly show that only a fraction of the organisms present have been cultured.

If an organism has not been isolated, its nutritional requirements cannot be defined; therefore determining appropriate conditions for its isolation is impossible. Yet, these are exactly the organisms we want to evaluate for the production of novel secondary metabolites. When faced with the dilemma of isolating organisms that have not been characterized, it will be the microbiologist skilled in the isolating, identifying, and classifying the actinomycetes who will be able to recognize rare and unusual strains. We tried to measure the effect of a microbiologist's training for successfully isolating NMNS. Two groups of microbiologists were given the same soil samples, media, and selective pressures, and were asked to isolate non-streptomycete actinomycetes. All isolates were identified to their major group by an actinomycete taxonomist (A.C. Horan) not directly involved in culture isolation. The results are presented in Figure 1–14. Group 1 consisted of microbiologists who were

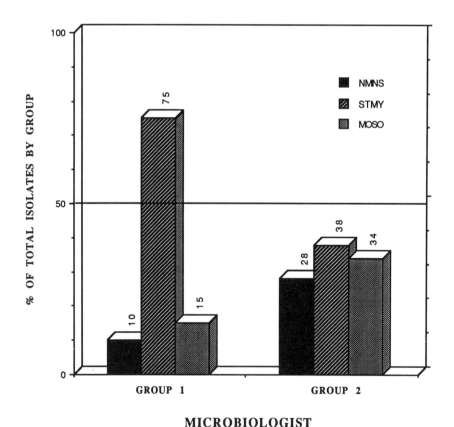

MICROBIOLOGIST

FIGURE 1–14 Effect of microbiologist training on the isolation of NMNS.

familiar with some genera of actinomycetes but were not trained in the macroscopic and microscopic identification of a broad range of soil isolates. They isolated predominantly streptomycetes. Group 2, however, had extensive experience in isolating and identifying actinomycetes. The results clearly show that they were better able to distinguish NMNS. The extent of the participating microbiologist's experience and training in actinomycete isolation and identification is a key element in a successful Natural Products Program.

Actinomycetes remain a major source of novel, therapeutically relevant natural products. To best exploit this group, a continuing source of geographically and ecologically diverse soil samples along with experienced microbiologists trained in the identification, classification, and preservation of actinomycetes are required. Taxonomic data on newly described genera should be evaluated in order to define media constituents and selective pressures that favor their growth over other actinomycetes, particularly streptomycetes. New materials and methods should then be incorporated into an isolation protocol and monitored for effectiveness in enhancing the isolation of novel, secondary metabolite-producing actinomycetes.

REFERENCES

Athalye, M., Lacey, J., and Goodfellow, M. (1981) *J. Appl. Bacteriol.* 51, 281–298.

Brodsky, B.C., Cardaci, K., Shearer, M.C., Barrabee, E., and Horan, A.C. (1991) in *8th Int. Symp. Biol. Actinomycetes*, Abstract no. P2-032, University of Wisconsin, Madison.

Cooper, R., Horan, A.C., Gentile, F., et al. (1988) *J. Antibiot.* 41, 13–19.

Cooper, R., Horan, A.C., Gunnarsson, I., Patel, M., and Truumees, I. (1986) *J. Indus. Microbiol.* 1, 275–276.

Cooper, R., Truumees, I., Yarborough, R., et al. (1992) *J. Antibiot.* 45, 517–522.

Cross, T. (1981a) in *The Procaryotes, A Handbook on Habitats, Isolation, and Identification of Bacteria*, vol. II, 1st ed. (Starr, M.P., Stolp, H., Trüper, H.G., Balows, A., and Shlegel, H.G., eds.), pp. 2091–2102, Springer-Verlag, New York.

Cross, T. (1981b) *J. Appl. Bacteriol.* 50, 397–423.

Cross, T. (1982) *Dev. Ind. Microbiol.* 23, 1–18.

Cross, T., and Attwell, R.W. (1974) in *Spore Research 1973* (Barker, A.N., Gould, G.W., and Wolf, J., eds.), pp. 11–20, Academic Press, London.

Erikson, D. (1941) *J. Bacteriol.* 41, 277–300.

Gherna, R., Pienta, P., and Cote, R. (eds.) (1989) *American Type Culture Collection Catalogue of Bacteria and Phages*, 17th ed., p. 298, American Type Culture Collection, Rockville, MD.

Goodfellow, M. (1992) in *The Procaryotes, A Handbook on the Biology of Bacteria: Ecophysiology, Isolation, Identification, Applications* (Balows, A., Trüper, H.G., Dworkin, M., Harder, W., Schleifer, K.-H., eds.), pp. 1115–1138, Springer-Verlag, New York.

Goodfellow, M., and Cross, T. (1984) in *The Biology of the Actinomycetes* (Goodfellow, M., Mordarski, M., and Williams, S.T., eds.), pp. 7–164, Academic Press, London.

Goodfellow, M., and Orchard, V. (1974) *J. Gen. Microbiol.* 83, 375–387.
Goodfellow, M., and Williams, S.T. (1983) *Annu. Rev. Microbiol.* 37, 189–216.
Gordon, R.E., and Barnett, D.A. (1977) *Int. J. Syst. Bacteriol.* 27, 176–178.
Gordon, R.E., and Horan, A.C. (1968) *J. Gen. Microbiol.* 50, 223–233.
Gullo, V., Gunnarsson, I., Horan, A., et al. (1991) *J. Indus. Microbiol.* 8, 65–68.
Hegde, V., Patel, M., Horan, A.C., et al. (1992) *J. Antibiot.* 45, 507–515.
Horan, A.C., and Brodsky, B.C. (1982) *Int. J. Syst. Bacteriol.* 32, 195–200.
Horan, A.C., Patel, M., Matson, J.M., et al. (1989) *J. Antibiot.* 42, 1547–1555.
Hunter, J.C., Eveleigh, D.E., and Cassala, G. (1981) in *Actinomycetes Zentralbl. Bakteriol. Microbial. Hyg.*, Suppl. 11 (Schaal, K.P., and Pulverer, G., eds.), pp. 195–200, Fisher-Verlag, Stuttgart, Germany.
Ivanitskaya, L.P., Singal, S.M., Bibikova, M.V., and Vostrov, S.N. (1978) *Antibiotiki* 8, 690–693.
Jensen, H.L. (1930) *Proc. Linn. Soc. New So. Wales* 55, 231–249.
Jensen, H.L. (1932) *Proc. Linn. Soc. New So. Wales* 57, 173–180.
Johnston, D.W., and Cross, T. (1976) *Freshwater Biol.* 6, 457–463.
Karwowski, J.P. (1986) *J. Indus. Microbiol.* 1, 181–186.
Kawamoto, I. (1989) in *Bergey's Manual of Systematic Bacteriology*, vol. 4 (Williams, S.T., ed.), pp. 2442–2450, Williams and Wilkins, Baltimore.
Kawamoto, I., Oka, T., and Nara, T. (1983) *Agric. Biol. Chem.* 47, 203–215.
Kroppenstedt, R.M., and Goodfellow, M. (1992) in *The Procaryotes, A Handbook on the Biology of Bacteria: Ecophysiology, Isolation, Identification, Applications*, vol. II, 2nd ed. (Balows, A., Trüper, H.G., Dworkin, M., Harder, W., Schleifer, K.-H., eds.), pp. 1085–1114, Springer-Verlag, New York.
Küster, E., and Williams, S.T. (1964) *Nature* 202, 928–929.
Labeda, D.P. (1992) in *The Prokaryotes, A Handbook on the Biology of Bacteria: Ecophysiology, Isolation, Identification, Applications*, vol. II, 2nd ed. (Balows, A., Trüper, H.G., Dworkin, M., Harder, W., Schleifer, K.-H., eds.), pp. 1061–1068, Springer-Verlag, New York.
Labeda, D.P., and Shearer, M.C. (1990) in *Isolation of Biotechnological Organisms from Nature* (Labeda, D.P., ed.), pp. 1–19, McGraw-Hill, New York.
Lechevalier, H.L., and Lechevalier, M.P. (1981) in *The Procaryotes, A Handbook on Habitats, Isolation, and Identification of Bacteria*, vol. II, 1st ed. (Starr, M.P., Stolp, H., Trüper, H.G., Balows, A., and Schlegel, H.G., eds.), pp. 1915–1922, Springer-Verlag, New York.
Lechevalier, M.P., and Lechevalier, H.A. (1985) in *Biology of Industrial Microorganisms* (Demain, A.L., and Solomon, N.A., eds.), pp. 315–358, Benjamin/Cummins, Menlo Park, CA.
Lingappa, Y., and Lockwood, J.L. (1962) *Phytopathology* 52, 317–323.
Luedemann, G.M. (1971) *Trans. N.Y. Acad. Sci.* 33, 207–218.
Luedemann, G.M. (1991) *Actinomycetes* 2(Suppl. 1), 1–149.
Marquez, J.A., Horan, A.C., Kalyanpur, M., et al. (1983) *J. Antibiot.* 36, 1101–1108.
Marquez, J.A., Kirschner, A., Truumees, I., et al. (1976a) 172nd Annual Meeting of the American Chemical Society, Abstract no. MICR 55, New Orleans, LA.
Marquez, J.A., Wagman, G.H., Testa, R.T., Waitz, J.A., and Weinstein, M.J. (1976b) *J. Antibiot.* 29, 483–487.
McCarthy, A.J., and Broda, P. (1984) *J. Gen. Microbiol.* 130, 2905–2913.
Nonomura, H., and Ohara, Y. (1969) *J. Ferment. Technol.* 47, 463–469.

Okazaki, T., and Okami, Y. (1976) in *Actinomycetes: The Boundary Microorganisms* (Arai, T., ed.), pp. 123–161, Toppan Co. Limited, Tokyo.

Patel, M., Conover, M., Horan, A.C., et al. (1988) *J. Antibiot.* 40, 794–797.

Patel, M., Gullo, V.P., Hegde, V.R., et al. (1987) *J. Antibiot.* 40, 1408–1413.

Patel, M., Hegde, V., Horan, A.C., et al. (1989) *J. Antibiot.* 42, 1063–1069.

Patel, M.P., Horan, A.C., Gullo, V.P., et al. (1984) *J. Antibiot.* 37, 413–415.

Prauser, H. (1976) in *Actinomycetes: The Boundary Organisms* (Arai, T., ed.), pp. 193–207, Toppan Co. Limited, Tokyo.

Ruddick, S.M., and Williams, S.T. (1972) *Soil Biol. Biochem.* 4, 93–103.

Sandrak, N.A. (1977) *Mikrobiologiya* 46, 478–481.

Shearer, M.C. (1987) *Dev. Ind. Microbiol.* 28, 91–97.

Solovieva, N.K., and Singal, E.M. (1972) *Antibiotiki* 17, 778–781.

Sveshnikova, M.A., Chormonova, N.T., Laurova, N.V., Terekhova, L.P., and Preobrazhenskaya, T.P. (1976) *Antibiotiki* 9, 784–787.

Umbreit, W.W., and McCoy, E. (1941) in *Symposium on Hydrobiology—1940*, pp. 106–114, University of Wisconsin, Madison.

Vobis, G. (1992) in *The Prokaryotes, A Handbook on the Biology of Bacteria: Eco-physiology, Isolation, Identification, Applications*, vol. II, 2nd ed. (Balows, A., Trüper, H.G., Dworkin, M., Harder, W., Schleifer, K.-H., eds.), pp. 1029–1060, Springer-Verlag, New York.

Wagman, G.H., Marquez, J.A., Watkins, P.D., et al. (1973) *J. Antibiot.* 26, 732–736.

Wagman, G.H., Marquez, J.A., Watkins, P.D., et al. (1976) *Antimicrob. Agents Chemother.* 9, 465–469.

Wagman, G.H., Patel, M., Testa, R.T., et al. (1975) in *15th Int. Conf. Antimicrob. Agents Chemother.*, Abstract no. 420.

Wagman, G.H., Testa, R.T., Marquez, J.A., and Weinstein, M.J. (1974) *Antimicrob. Agents Chemother.* 6, 144–149.

Wagman, G.H., Waitz, J.A., Marquez, J.A., et al. (1972) *J. Antibiot.* 25, 641–646.

Waitz, J.A., and Horan, A.C. (1988) U.S. patent no. 4,735,903.

Waitz, J.A., Horan, A.C., Kalyanpur, M., et al. (1981) *J. Antibiot.* 34, 1101–1106.

Waitz, J.A., Reiblein, W., and Truumees, I. (1983) U.S. patent no. 4,367,287.

Wakisaka, Y., Kawamura, Y., Yasuda, Y., Koizumi, K., and Nishimoto, Y. (1982) *J. Antibiot.* 35, 822–836.

Waksman, S.A. (1961) *The Actinomycetes*, vol. 2, Williams and Wilkins Co., Baltimore.

Watson, E.T., and Williams, S.T. (1974) *Soil Biol. Biochem.* 6, 43–52.

Weinstein, M.J., Luedemann, G.M., Oden, E.M., and Wagman, G.H. (1964) *Antimicrob. Agents Chemother.* 1963, 1–7.

Weinstein, M.J., Marquez, J.A., Testa, R.T., et al. (1970) *J. Antibiot.* 23, 551–554.

Weinstein, M.J., Luedemann, G.M., Oden, E.M., and Wagman, G.H. (1968) *Antimicrob. Agents Chemother.* 1967, 435–441.

Weinstein, M.J., Luedemann, G.M., Oden, E.M., and Wagman, G.H. (1965) *Antimicrob. Agents Chemother.* 1964, 24–32.

Weinstein, M.J., Wagman, G.H., Marquez, J.A., and Testa, R.T. (1978) U.S. patent no. 4,078,056.

Weinstein, M.J., Wagman, G.H., Marquez, J.A., Testa, R.T., and Waitz, J.A. (1975) *Antimicrob. Agents Chemother.* 7, 246–249.

Weinstein, M.J., Wagman, G.H., Marquez, J.A., et al. (1969) *J. Antibiot.* 22, 253–258.

Williams, S.T., and Wellington, E.M.H. (1981) in *The Procaryotes, A Handbook on Habitats, Isolation, and Identification of Bacteria*, vol. II, 1st ed. (Storr, M.P., Stolp, H., Trüper, H.G., Balows, A., and Schlegel, H.G., eds.), pp. 2103–2117, Springer-Verlag, New York.

Williams, S.T., and Wellington, E.M.H. (1982) in *Bioactive Microbial Products: Search and Discovery* (Bu'lock, J.D., Nisbet, L.J., and Winstanley, D.J., eds.), pp. 9–26, Academic Press, London.

Williams, S.T., Lanning, S., and Wellington, E.M.H. (1984) in *The Biology of the Actinomycetes* (Goodfellow, M., Mordarski, M., and Williams, S.T., eds.), pp. 481–528, Academic Press, London.

Williams, S.T., and Vickers, J.C. (1988) in *Biology of Actinomycetes '88* (Okami, Y., Beppu, T., and Ogawara, H., eds.), pp. 265–270, Japan Scientific Society Press, Tokyo.

Willoughby, L.G. (1969) *Hydrobiologia* 34, 465–483.

2

Bacteria As a Source of Novel Therapeutics

J.C. Hunter-Cevera
Angela Belt

A naturalist once stated that the most important discoveries of the laws, methods and progress of nature have nearly always sprung from the examination of the smallest objects which she contains (Slater et al. 1983). Bacteria, being small objects in nature, come in a variety of sizes and shapes, and so do their secondary metabolites, some of which have proven to be useful as therapeutic agents for humans and animals.

Babes, a biochemist, in 1885 was the first to realize that "microbial antibiosis" was due to the production of an inhibitory chemical substance by the antagonistic organism (Vandamme 1984). Years later, microbial physiologists, Bouchard, Emmerich, and Loew made the extract "pyocyanase" from *Pseudomonas aeruginosa*, which was active against *Corynebacterium diphteriae*, *Salmonella typhi*, *Pasteurella pestis*, and other pathogenic cocci (Vandamme 1984). Pyocyanase was used for two decades as a treatment for a variety of infectious diseases. However, due to its toxicity, further research on this compound as an antimicrobial agent was not continued.

Interest in examining bacteria for antimicrobial activity was rekindled in 1939 when Dubos discovered tyrothricin, produced by *Bacillus brevis* (Vandamme 1984). This work marked the beginning of the "Golden Era of Antibiotics," during which in 1945, bacitracin, produced by *Bacillus subtilis* and *B. licheniformis*, was discovered along with polymyxins (*B. polymyxa*) and the gramicidins A, C, D, and S (*B. brevis*) (Berdy 1974, 1980).

Other economically important antibiotic compounds discovered since the 1940s are butirosin, produced by *B. circulans*; 6 aminopenicillanic acid (6-APA), produced by *Escherichia coli* and *B. megaterium*; 6-azauridine, produced by *E. coli* from 6-azauracil; asparaginase, produced by *Erwinia* sp. and *E. coli* (cytotoxic and antitumor activity); tyrocidine, produced by *B. brevis* (Vandamme 1984); myxin, produced by *Lysobacter antibioticus* (Weigele and Leimgruber 1967; Christensen and Cook 1978); and ambruticin produced by *Sorangium cellulosum* (Ringel et al. 1977).

The development of microbial resistance to marketed drugs necessitated the discovery of new antibiotics. In the late 1970s and early 1980s there was renewed interest in screening bacteria for novel metabolites. Some of the resulting discoveries included the monobactams, quinones and bulgecins (O-sulfated glycopeptides) (Sykes et al. 1981a, 1981b; Wells et al. 1982a, 1982b, 1982c; Cooper and Unger 1985, 1986; Parker and Rathnum 1982; Meyers et al. 1983; Cooper et al. 1983; Singh et al. 1983; Cooper et al. 1985; Katayama et al. 1985; Kato et al. 1985; Box et al. 1988; Gwynn et al. 1988; Asai et al. 1981; Imada et al. 1981, 1982; Kintaka et al. 1981a, 1981b), the beta-lactones (Parker et al. 1982a; Wells et al. 1982c), oxazinomycin (Tymiak et al. 1984), the carbapenems (Parker et al. 1982b), the catacandins (Meyers et al. 1985), and the cephabacins (Ono et al. 1984).

Other compounds demonstrating antibiotic activity but not used for human applications are bioinsecticides, produced by *B. thuringiensis* (Vandamme 1984), and nisin (Rayman and Hurst 1984), a polypeptide used in the food industry, produced by *Streptococcus lactis* and *S. cremoris*. In addition, sorangicin, a broad-spectrum nontoxic antibacterial compound produced by *Sorangium cellulosum* (myxobacteria) (Reichenbach et al. 1988; Ringel et al. 1977) is used to treat animal infections. Table 2–1 lists antimicrobial compounds produced by bacteria. It is obvious from the above examples that bacteria can produce novel metabolites of interest and commercial value.

Bacteria provide several advantages as a source of novel metabolites. The metabolite production time is shorter than that required for actinomycetes and/or fungi. In addition, genes responsible for novel metabolite production and regulation can usually be more easily cloned into other bacteria, or improved through random or specific site mutagenesis, or protein engineering for greater stability. Bacteria may also be an untapped source of other therapeutic activity for cardiovascular disease, inflammatory arthritis, cancers, and neurological disorders.

2.1 SOURCES OF BACTERIA FOR SCREENS

To initiate a program that examines bacteria found in nature as a source of novel metabolites, one must first address several questions. What are the in-

TABLE 2–1 Antibiotics Produced by Bacteria

Organism	Antibiotic/Therapeutic	Reference
Achromobacter sp.	Xerosin	Korzybski et al. 1978
Agrobacterium radiobacter	Monobactam	Parker and Rathnum 1982
Alteromonas aurantia	Unnamed	Truper et al. 1992
A. citrea	Unnamed	Truper et al. 1992
A. denitrificans	Prodigiosin	Truper et al. 1992
A. luteoviolacea	Violacein	Truper et al. 1992
A. rubra	Prodigiosin	Truper et al. 1992
Angiococcus disciformis	Myxothiazol	Reichenbach et al. 1988
Arthrobacter oxamiceticus	Oxamicetin	Korzybski et al. 1978
Arthrobacter sp.	Beta lactone	Wells et al. 1982c
B. alvei	Alvein	Korzybski et al. 1978
B. aurantinus	Antibiotic KM214	Korzybski et al. 1978
Bacillus badius	Thiocillins	Korzybski et al. 1978
B. brevis	Colisan, tyrocidine	Korzybski et al. 1978
	Edeine, gramicidin	Truper et al. 1992
	S. tyrothricin, brevolin, brevisitin	Neidleman 1988
B. cereus	Cerein, cerexins, biocerin	Korzybski et al. 1978
B. circulans	Butirosins, circulin, polpeptin, antibiotic B43	Korzybski et al. 1978
B. cocovenaus	Toxoflavin	Korzybski et al. 1978
B. colistinus	Colistin	Korzybski et al. 1978; Neidleman 1988
B. laterosporus	Laterosporamine, spergualin, laterosporin	Korzybski et al. 1978
B. licheniformus	Bacitracin, proticin, licheniforminB	Korzybski et al. 1978
B. licheniformus var. *mesentericus*	Proticin	Korzybski et al. 1978
B. megaterium	Bacimethrin, thiocollins, megacine	Korzybski et al. 1978
B. mesentericus	Esperin	Korzybski et al. 1978
B. natto	Unnamed	Korzybski et al. 1978
B. niger	Nigrin	Korzybski et al. 1978
B. polymyxa	Polymyxins	Korzybski et al. 1978; Neidleman 1988
B. polymyxa var. *colistinus*	Gatavalin, jolipeptin	Korzybski et al. 1978
B. pumilus	Amicoumacin A, tetaine, pumilin	Korzybski et al. 1978; Truper et al. 1992
B. simplex	Simplexin	Korzybski et al. 1978
Bacillus sp.	Beta lactones, xylostasine, galantins, petrin	Parker et al. 1982a; Korzybski et al. 1978

(*continued*)

TABLE 2–1 (continued)

Organism	Antibiotic/Therapeutic	Reference
B. subtilis	Bacitracin, subtilin, bacilysin, bacilipns A and B, bacillin, globicin, antibiotic TL-119, balliomycin, fungomycin, eumycin, toximycin, bacillomycin, Rhizoctonia and *Aspergillus* factors, fungistatin, rhizobacidin, mycosubtilin, fluvomycin, subtenolin, polychlorosubtilin, xanthellin, iturin, neocidin, obutin, endosubtilysine, subtilisine, analysine, trypanotoxin, subsporin	Korzybski et al. 1978 Froyshov 1984
B. thiaminolyticus	Baciphelacin, octopytin	Korzybski et al. 1978
Bacterium antimyceticum	Comirin	Korzybski et al. 1978
Brevibacterium ammoniagenes	Brevimycins A and B	Korzybski et al. 1978
B. crystalloiodinum	Iodin	Korzybski et al. 1978
Chromobacterium violaceum	Glycopeptides, beta lactamase Arphamenines A and B, Aerocyanidin, aerocavin, Factor Y, monobactam, violacein	Sykes et al. 1981b Korzybski et al. 1978 Truper et al. 1992
C. iodinum	Iodinin	Korzybski et al. 1978
Corallococcus coralloides	Myxothiazol	Reichenbach et al. 1988
Cytophaga johnsonae	Monobactams	Truper et al. 1992; Reichenbach et al. 1988
Cy. sp.	Cephabacins	Ono et al. 1984
Cy. uliginosa	Marinactan	Truper et al. 1992
Erwinia sp.	Carbapenems	Parker et al. 1982b
Escherichia coli	Coliformin, colicins	Korzybski et al. 1978
Felxibacter sp.	Beta lactams	Truper et al. 1992
Gluconobacter sp.	Monobactam	Wells et al. 1982c
Lactobacillus acidophilus	Acidolin, lactacin B	Korzybski et al. 1978 Truper et al. 1992
L. helveticum	Helveticin J, lactocin 27	Korzybski et al. 1978; Truper et al. 1992
L. plantarum	Plantaricin A	Korzybski et al. 1978; Truper et al. 1992

TABLE 2-1 (continued)

Organism	Antibiotic/Therapeutic	Reference
L. sake	Sakecin A	Korzybski et al. 1978; Truper et al. 1992
Lysobacter antibioticus	Myxin	Truper et al. 1992
L. gummosous	Catacandin A and B	Meyers et al. 1985
L. sp.	Myxocidin A and B, quinolines, myxin, cephabacins, lactivicin, lysobactin, decapeptide peptolides, nyxosidins A and B, acyltetramic acids	Truper et al. 1992; Reichenbach et al. 1988
Micrococcus sp.	Micrococcin	Korzybski et al. 1978
Mycoplasma sp.	Unnamed	Korzybski et al. 1978
Myxococcus fulvus	Unnamed	Korzybski et al. 1978
	Myxothiazol	Reichenbach et al. 1988
Mx. vinescens	Unnamed	Korzybski et al. 1978
Mx. virescens	Antibiotic TA	Korzybski et al. 1978; Truper et al. 1992
Mx. xanthus	Xanthacin, antibiotic TA	Truper et al. 1992
Nanocystis (?)	Gliobactins	Truper et al. 1992
Pseudomonas acidophila	Sulfazecin	Kintaka et al. 1981a, 1981b
P. aeruginosa	Sulfazecin, isosulfazecin, pyocyanine, hemipyocyanine, antibiotics PX, YC73, 173t, pyoluteorin, butanoic acid, aeruginic acid	Imada et al. 1981; Korzybski et al. 1978
P. chlororaphis	Chlororaphin	Korzybski et al. 1978
P. cocovenenans	Monobactams	Box et al. 1988; Gwynn et al. 1988
P. fluorescens	Safracin	Meyers et al. 1983
P. magnesiorubra	Magnesidin	Korzybski et al. 1978
P. viscosa	Viscosin	Korzybski et al. 1978
P. pyrrocinia	Pyrrolnitrin, pyolipic acid, pioklastin	Korzybski et al. 1978
Pseudomonas sp.	Oxazinomycin	Tymiak et al. 1984
	Bactobolin, DB-2073, dopastin, cycloserine, aminoglycosidase	Korzybski et al. 1978 Bushell 1981
Proteus vulgaris	Protaptin	Truper et al. 1992
Serratia marcescens	Prodigiosin, cabapenems	Korzybski et al. 1978 Parker et al. 1982b
Sorangium cellulosum	Ambruticin	Ringel et al. 1977
Sorangium sp.	Myxin	Korzybski et al. 1978

(*continued*)

TABLE 2-1   (continued)

Organism	Antibiotic/Therapeutic	Reference
Staphylococcus epidermis	Mycobactocidin, epidermidins	Korzybski et al. 1978
S. haemolyticus	Antigonnococcal peptides	Truper et al. 1992
Staphylococcus sp.	Lantibiotics, staphylococcin	Truper et al. 1992
S. staphylolyticus	Lysostaphin	Korzybski et al. 1978
Stigmatella aurantiaca	Myxothiazol, auracins, stigmatellin	Reichenbach et al. 1988
Streptococcus lactis	Nisin, diplococcin	Rayman and Hurst 1984
Waksmania aerata	Iodinin	Korzybski et al. 1978
Xenorhabdus luminescens	Unnamed	Truper et al. 1992

house capabilities for isolating and fermenting certain types of bacteria? What are the assay and purification capabilities? After answering these basic questions, additional decisions need to be made. For instance, what is the molecule of interest and which bacteria would probably be capable of producing this molecule under both natural and artificial conditions? If one is looking for a molecule that is stable at high pH, one would not usually isolate or examine species of acid-producing or acidophilic bacteria such as *Gluconobacter* or *Acetobacter*. On the other hand, there are genetic modifications through random or specific site mutagenesis that could induce these species to produce molecules at high pH. In this case, a classic geneticist or molecular biologist could work with the microbial physiologist to accomplish this task. Another example is examining anaerobes for novel antibiotics. They are interesting to work with but special equipment and training in media preparation are needed to examine strict or obligate anaerobes. The acids they produce can exhibit antimicrobial activity and would have to be rejected as positives in a screen for novel antibiotics produced by anaerobic growth in submerged culture.

2.1.1   Sources Found in the Literature

It is interesting to note that when authors describe the discovery of a new compound, the source and isolation of the organism and isolation method used is often the least described event and yet one of the first steps to success! Without the sample and isolation methods employed, there would be no microorganisms to screen or to produce the desired novel metabolite. Isolation of microorganisms from nature is both an art and science. Dixon (1983) predicted that biotechnology companies that employ a microbial ecologist will be the most successful in finding novel compounds in the next decade. However, the "isolation scientist" is often indeed isolated within his or her own department, and the importance of microbial ecology to a natural product screening program is often not recognized.

A few articles have described and even correlated the isolation of known antibiotic producing bacteria to sample source (Wells et al. 1982a) (Table 2–2). The examples described below illustrate the logic that soil microbiologists have used to combine ecological parameters, isolation techniques, and screen design in the discovery of active molecules (Hunter et al. 1984).

The first monobactam discovered was isolated from an acidic water and soil sample taken from the Pine Barrens in southern New Jersey. Using a rice-baiting technique for *Chromobacterium violaceum* (Corpe 1951) (the producing strain), many *Chromobacterium* isolates were obtained from a variety of sample types. However, only those strains isolated from low pH samples produced the novel monobactam structure; when type strains from ATCC and other isolates obtained from neutral samples were examined, they did not exhibit activity (Wells et al. 1982c).

Gluconobacter and *Acetobacter* species isolated from 41 samples were the two genera most frequently found to produce monobactams and were therefore positive in the Colworth agar droplet/B-lactamase screen for related beta lactams (Sykes et al. 1981a, 1981b).

The isolation methods that incorporated natural extracts and low pH isolation media did favor development of these colony types (Hunter, J.C., unpublished observations). These strains produced acid and still maintained metabolic activity at low pH. The fermentation media was not buffered, thus one can hypothesize that low pH may trigger the formation of monobactams. It is also interesting that Takeda's monobactam-producing (sulfazecin and isosulfazecin) strains were probably isolated from a low pH sample, since the strains were identified as *Pseudomonas acidophila* and *P. mesoacidophila*, both of which grow better at low pH compared with other members of this genus (Imada et al. 1981).

Pseudomonas strains, which produced the beta-lactones, are another example where microbial ecology and metabolite production are correlated. Most isolates originated from a variety of old or decaying mushrooms (Wells et al. 1982c). See Table 2–2 for a description of these samples. This correlation asks what association do these beta-lactone producing bacteria have with basidiomycetes? Is the beta-lactone produced by bacteria a response to a change in the physiology of the mushroom? By making such observations on which sample types yielded "positives," one was able to choose samples from different habitats that had a high probability of yielding bacteria that was positive in the beta-lactam/lactone screen.

2.2 METHODS FOR ISOLATING BACTERIA

No single isolation method reveals the total number and types of microorganisms present in a "niche" (Hungate 1962; Slater et al. 1983). Support for

TABLE 2–2   Sources of Some Antibiotics Produced by Bacteria

Organism	Isolation Source	Metabolite	Reference
Agrobacterium radiobacter	Plant material	Monobactams	Wells et al. 1982c
Alteromonas sp.	Stones, sand, fish, algae, marine sediment, seawater	Prodigiosin, violacein, unnamed compound	Truper et al. 1992
Bacillus licheniformis	Wound	Bacitracin	Korzybski et al. 1978
B. subtilis	Contaminated actinomycete culture	Bacillomycin	Korzybski et al. 1978
Chromobacterium violaceum	Soil, spruce needles, compost bog sediment, decaying roots, forest soil, oak leaf litter	Monobactams	Wells et al. 1982c
Gluconobacter sp.	Moss, soil, plants, water	Monobactams	Wells et al. 1982c
Micrococcus sp.	Sewage	Micrococcin	Korzybski et al. 1978
Mycoplasma sp.	Rat hypophyses	Factors I, II, III	Korzybski et al. 1978
Pseudomonas acidophila	Soil	Sulfazecin	Kintaka et al. 1981a, 1981b
P. mesoacidophila	Soil under a *Diospynoskaki* tree	Sulfazecin	Imada et al. 1981
P. cocovenenans	Soil	Monobactams	Gwynn et al. 1988
P. magnesiorubra	Marine alga	Magnesidin	Korzybski et al. 1978
Pseudomonas sp.	Decaying mushrooms, leaf litter, soil, water, beach sand, algae	Beta-lactone	Wells et al. 1982c

this statement lies in the fact that there are over 300 genera listed in *Bergy's Manual of Systematic Bacteriology* (Holt 1989) and four volumes of *The Prokaryotes: A Handbook on the Biology of Bacteria: Ecophysiology, Isolation, Identification, Applications* (Truper et al. 1992), both of which describe the huge biodiversity of bacteria that exists in nature. The types and numbers described are representative of the successful isolation techniques employed.

The numbers and types of bacteria one can isolate from nature depends upon (1) the ecosystem's limiting parameters, such as pH, redox potential, temperature, salinity, and available substrates, which determine the bacterial types best suited for survival and proliferation; and (2) the method, including media and incubation conditions, used to isolate any one specific or general group of bacteria (Hunter et al. 1984).

2.2.1 Sampling within Ecosystems

Considerable detail and attention should be given to sampling when trying to isolate the natural biota from a given ecosystem; especially to species occurring in unique micro-environmental niches (Hunter et al. 1984; Hunter-Cevera et al. 1986). One can find the same genus or perhaps the same species occupying a number of different niches. However, there are often genetic and phenotypic differences among strains of the same species that can influence secondary metabolite production (Wells et al. 1982c). The main steps involved in examining an ecosystem for microorganisms to be screened for natural products are summarized below. Using ecological approaches for the isolation of bacteria from nature can actually increase the number and types of bacteria isolated.

1. List the groups of microorganisms to be isolated; aerobic Gram-negative rods, blue-green algae, dematiaceous hyphomycetes, etc.
2. Describe the ecosystem or habitat from which the samples are to be collected; e.g., hardwood forest, tropical forest, tundra, desert, etc.
3. Groups samples into types; e.g., plants and plant parts (leaves, flowers, stems, roots), soils and horizons (detritus, first and second year, etc.), rocks surfaces (top and bottom), water (column depth), insects and parts (legs, mandibles, wings, abdomen, etc.).
4. List the environmental parameters to be considered and measured, such as pH, salinity, and temperature.
5. List the special natural substrates in the ecosystem, e.g., chitin and cellulose in forest soils, metals in mining sites.
6. Design isolation experiments around data obtained from steps 1 through 5, i.e., diluents, agars, natural extracts, and incubation temperature.
7. Do not be limited to any "standard" procedure for isolation of bacteria.
8. Use standard methods as controls to evaluate "ecological isolation methods" and enrichments.
9. Modify known procedures to the extent required by ecological parameters of material to be examined.
10. Try enrichment procedures and compare with results obtained by direct examination.

It is clearly not the diversity of bacteria present in a sample that limits the success in recovery of many different isolates but rather one's own imagination and application of these ideas (Hunter-Cevera and Belt 1992).

2.2.2 Sample Collection

Samples should be representative of a site; i.e., if certain types of plants are present within the site, then each plant and plant part should be sampled because different bacterial populations occur on different parts of the plants. For example, bacteria isolated from leaf surfaces are frequently pigmented and represent different genera than bacteria isolated from the roots where a rhizosphere population would exist (Hunter et al. 1984).

It is important to document the collection site of the sample, the time, the date, and biophysical parameters, when applicable, such as redox potential, temperature, salinity, and pH. These parameters may be useful for designing fermentation conditions as well as isolation media. For example, bacteria isolated from a salt marsh where a salt gradient exists may produce different secondary metabolites in the presence of halide ions; or knowing the redox potential of the sample would aid in determining whether to isolate for faculative anaerobes or strict anaerobes. In addition, there would be different metabolic pathways used in broths that have a high redox potential. For future reference, it is helpful to assign a number to each sample collected. Once collected, the samples should be examined immediately upon return to the laboratory or stored at 5°C overnight, separate from the plating and/or enrichment area. It is advisable to collect more than 1 g of sample material; e.g., at least 5 to 10 g of a soil sample, 200 ml of water samples, or 2 to 5 g of plant material, so more than one isolation method can be used and the sample can be examined again for further documentation. The U.S. Department of Agriculture regulates inter- and intrastate transfer of soils, plant materials, and certain microbes, as well as the importation of foreign samples. Check with the county, state and/or federal government agencies to determine what regulations and permits may be needed for the handling and containment of soil, plant, and water samples.

2.2.3 Materials Used to Collect Samples

Sterile equipment such as gloves, spatulas, spoons, paper towels, scissors, bags, bottles, etc., should be used at all times to avoid any misinterpretation of results due to cross-contamination from sampling. Collection vessels should be leakproof, especially when sampling water.

2.3 SAMPLE PROCESSING

Samples should be processed under a laminar flow hood that has been on for at least 20 minutes. All polluted, contaminated, or foreign samples should be handled with gloves. Media and diluents should be made at least 2 days prior to use and examined for any laboratory contaminants. In addition, all equipment should be prelabeled before beginning the isolation experiments. It is also recommended that sample pH be recorded by use of a glass electrode for solid and/or liquid samples before processing. One method that gives accurate pH readings is to mix a one-to-one ratio of soil or other materials and distilled water and let stand for 30 minutes with intermittent stirring (Pramer and Schmidt 1964). The following section describes a few different methods that have been used successfully to isolate bacteria from soils, plants, and waters that were subsequently screened for novel metabolites in which "positives" were obtained.

2.3.1 Direct Isolation Techniques

It is usually considered standard procedure to dilute samples in distilled water, vortex each dilution for a few seconds, and then plate them onto the surface of isolation media. However, it has been observed that the incorporation of natural extracts and pH or salinity adjustments made to media, as well as the diluent employed in plating out the sample, can affect the number and types of bacteria isolated by direct examination (Straka and Stokes 1957; Hunter et al. 1984). Butterfield (1932) observed that the number of bacteria present in natural waters can decrease 60% in 30 minutes when diluted in distilled water. If one does not have time to make diluents incorporated with 25% of a natural soil or plant extract, then the use of 0.1% peptone in distilled water is recommended as a diluent (Straka and Stokes 1957).

2.3.1.1 Soils. Mix a 5-gram sample (wet weight) of soil with 99 ml of sterile distilled water contained in a 250 ml flask with foam or cotton plug and shake at 150 rpm for 20 minutes at room temperature. Serially dilute the soil-water suspension according to turbidity in the appropriate diluent and transfer 0.1 ml of at least three suspensions on to four different isolation agars. Choosing the right dilution to obtain representative bacterial numbers and types often correlates with soil type; e.g., soils with high clay and sand content can usually be plated at 10^{-3} to 10^{-5}, whereas garden soils are plated at higher dilutions of 10^{-7} to 10^{-9}. Spread the suspension over the surface of the agar plate with a bent L-shape glass rod that has been surface sterilized in alcohol and flamed briefly. Inverted plates are taped and incubated between 20 to 25°C for mesophiles for 5 to 7 days.

2.3.1.2 Plants. Cut individual plant parts with sterile scissors and transfer approximately 2 g to a flask containing 99 ml of quarter-strength plant extract and glass beads (optional). Shake at 175 rpm for 20 minutes to loosen the attached surface microbiota. Withdraw a 1-ml sample and serially dilute in quarter-strength plant extract and plate in the same manner as described for soils. Inverted plates are taped and incubated at 25 to 28°C for 5 to 10 days.

Another method for direct examination of plant parts is to swab the surfaces with a sterile swab dipped in plant extract and zig-zag over the agar surface. Or, gently roll a piece of cut plant material over the surface of the plant with sterile forceps. Implants of plant pieces into the agar can also be used to isolate bacteria that will grow out from the implant material (Hunter et al. 1984).

2.3.1.3 Waters. Filter-sterilize a 100-ml water sample through a 0.45 μm filter. When the sediment settles on the filter membrane, add 1 or 2 ml of the filtrate. With a flat, wide-mouthed 2-ml pipette, gently scrape to uplift the sediment. Transfer the suspension to a 9-ml peptone water dilution bank, vortex for a few minutes, and serially dilute and plate as described above. Inverted plates are taped and incubated at 20 to 25°C for 7 days.

2.3.2 Enrichment and Selective Isolation

The bacteria present in any sample represent a group of competing species of different metabolic types. To promote dominance of one or two specific types, the physical and/or chemical environment should be altered through use of physical treatments, antibiotics, and other chemical agents.

2.3.2.1 Physical Methods. Air drying of samples will often increase the percentage of spore-forming bacteria present in a population by eliminating some of the competition from nonsporeformers that cannot survive under dry conditions. By lowering the media pH to below 5.0, one can select for acidophilic bacteria. The medium (without agar) is adjusted to the pH of the particular soil or water sample using a 5% KH_2PO_4 solution, adjusted to a pH of 2.8, using 6 N HCL. The agar solution of 30.0 g per 500 ml of distilled water is autoclaved separately at pH 7.0. After sterilization the medium and agar are separately cooled to 65°C and combined. The 65°C temperature is needed to achieve complete mixing of the agar. The medium plus agar is then readjusted to the appropriate pH value by use of sterile KH_2PO_4, HCl, or NaOH. Changes in incubation temperature can enrich for thermophiles (>55°C) or psychrophiles (0–5°C).

2.3.2.2 Chemical Methods. Use of inorganic nitrogen and sulfur sources can result in the isolation of nitrogen-fixing bacteria and autotrophic bacteria, such as the sulfur oxidizing bacteria, respectively. Thallium selects for *Mycoplasma* species while use of selenite will increase the number of *Salmonella* in a sample. Other chemicals used to alter the microbial population are triphenylmethane dyes (malachite green, brilliant green, crystal violet, basic fushin), which inhibit Gram-positive bacteria. Use of phenylethyl alcohol inhibits Gram-negative bacteria. Incorporation of nystatin, cyclohexamide, vancomycin, and colistin (50–75 μ/ml each) inhibits fungi from appearing on the isolation plates. There are many chemical enrichment/selection techniques used for specific groups of bacteria, all of which are described elsewhere in greater detail (Truper et al. 1992).

2.3.2.3 Substrate Additions in Minienvironments.

The use of minienvironment enrichment procedures can dramatically influence the isolation of specific bacterial populations. Enrichment procedures that result in a greater number of monobactam-producing bacteria from soils and plants than by direct examination are described below. Essentially, the incorporation of 0.10 to 0.50 g of a complex substrate suspended in 5 to 10 ml of water such as colloidal chitin, swollen cellulose, pectin, corn meal, peptone, and/or amino acids, were mixed with 1 g of soil or chopped plants in a sterile 50-ml beaker. The beaker and contents were then covered with sterile brown paper and incubated for 7 to 10 days at room temperature in a sealed vacuum jar in which 200 ml of water was placed in the bottom. A moisture chamber was thus created once the lid was sealed. A 1 ml sample was pipetted from the minienvironment and serially diluted and plated in the same manner as described for direct examination for soil samples. The recorded pH of the mixture was usually lower after a 10-day incubation, especially when polysaccharides were used as the enriching substrate.

Use of the minienvironment enrichment method resulted in the isolation of 47 out of 82 bacterial "monobactam positives" from a total of 546 soils, 102 plant part samples, and 28 water samples, representative of 26 different habitats and processed over an eight month period (J.C. Hunter, unpublished observations).

2.3.3 Media Design

There are several "Betty Crocker"-type media recipe cookbooks available. In fact, every pharmaceutical house has a collection of favorite recipes for the isolation and fermentation of microorganisms. However, there are several factors that should be considered when designing isolation media. Some of these factors are the ecological parameters of the sample niche, broth and/or solid media, and use of synthetic and/or natural extracts. The media used to isolate bacteria from nature should be modified to suit the habitat being examined. For example, when examining salt marsh samples, add sea water at the level of salinity present in the marsh salt gradient. Lower pH to match the pH of the acidic samples, such as that collected from a forest or bog. Substrate concentration should be suited to the content present in the sample. Frequently, mixed complex carbon, organic and inorganic nitrogen sources in small amounts, and trace metals are more representative of what is really present in a soil sample rather than 10 g of dextrose and 5 g of yeast extract!

The use of several different agars that range from lean to rich is recommended for the general isolation of bacteria from soils, plants, and water.

Many of the prepared agars sold by Difco (Detroit, MI) and BBL (Baltimore, MD) diluted either quarter- or half-strength work very well for the isolation of bacteria from nature (Hunter et al. 1984; Hunter-Cevera et al. 1986). Incorporation of a small amount of natural extracts made via the steeping tea bag method (Hunter-Cevera et al. 1986; Hunter-Cevera and Belt 1992) will enhance more of the representative biota present in the samples being examined (Hunter-Cevera et al. 1986).

2.4 SUBCULTURING AND PURIFICATION

After isolation, cultures must be further purified from other contaminating cells. Colonies may be picked with a needle or loop and plated onto media using the isolation streak method. The media used for this process should be similar to the isolation media, with the ability to support growth and yet "lean" enough to minimize the occurrence of variants. Do not include antibiotics or other such selective compounds in the purification media; doing so may mask the presence of low levels of resistant contaminating cells. When the culture is believed pure, one may streak this onto a richer media that is capable of supporting growth of a wider range of potential contaminants, in order to verify that it is axenic.

2.5 PRESERVATION AND MAINTENANCE

Maintaining growth of cells does not necessarily mean maintaining productivity of cells, thus one must consider both qualities carefully when determining how to preserve the culture. It is important to preserve the bacterium so that it has maximum viability, while retaining the characteristics of the original isolate; i.e., the ability to produce the metabolite of interest.

There is no excuse for trying to maintain bacterial cultures by continuous subculture or by freezing at $-20°C$. These methods may have been acceptable 30 years ago; however, results have been very poor and current technology provides more satisfactory options. Among the more successful methods are lyophilization and cryopreservation at $-80°C$ or in liquid nitrogen ($-196°C$). Results with different strains vary. For example, although lyophilization is an effective method of preserving viability of some bacteria, mutations and loss of physiological characteristics of the preserved cells have been recorded (Ashwood-Smith 1965; Ashwood-Smith and Grant 1976; Dietz 1975). Although many cells exhibit good viability at $-80°C$, there is evidence that liquid nitrogen at $-196°C$ is more reliable over long time periods (Moore and Carlson 1975; Clark et al. 1962; Chang and Elander 1986). In addition to the final storage temperature, one must also consider the sample preparation method. Reports of poor results after freezing strains at low temperatures can be explained by the fact that the poor results are caused by the lack of cryoprotectant, an uncontrolled freezing rate, temperature fluctuations from excessive handling and transferring in frozen state, and slow thawing of frozen cells rather than the final storage temperature. It is advisable to carefully work out specific protocols for each culture being preserved, with routine viability and productivity checks (Lapage et al. 1970; Shannon et al. 1975; Choate 1973).

2.6 FUTURE PROJECTIONS

The resurgence of interest in natural products for use as therapeutics has forced us to return to nature as a source for novel metabolites. It has been estimated that we have isolated only about one-tenth of all existing microorganisms. As our understanding of microbial ecology and physiology improves, our isolation techniques will also improve and result in the discovery of new and diverse species that could produce novel structures of therapeutic value. The increased use of molecular biology methods as a screening tool has helped to elucidate new targets and receptors. Protein engineering and specific-site mutagenesis have served as tools for improving upon what nature has already provided in the form of compounds yet to be discovered. If we work together, combining biochemistry, organic chemistry, microbial physiology, ecology, and genetic and protein engineering, we could expect a flood of new products by the year 2000. Nature has provided the raw materials and to a certain extent has already done the experiments for us. We can uncover these hidden treasures if we apply our understanding of microbial ecology and physiology in a rational, yet creative manner.

REFERENCES

Asai, M., Haibara, M., Muroi, M., Kintaka, K., and Kishi, T. (1981) *J. Antibiot.* 34(6), 621–627.

Ashwood-Smith, M. (1965) *Cryobiology* 2, 39–43.

Ashwood-Smith, M., and Grant, E. (1976) *Cryobiology* 13, 206–213.

Berdy, J. (1974) *Adv. Appl. Microbiol.* 14, 309–406.

Berdy, J. (1980) *Proc. Biochem.* 15, 28–35.

Box, S., Brown, A., Gilpin, M., Gwynn, M., and Spear, S. (1988) *J. Antibiot.* 41(1), 7–12.

Bushell, M. (1981) in *Topics in Enzyme and Fermentation Biotechnology* 6, (Wiseman, A., ed.), pp. 33–67, Ellis Horwood Ltd., London.

Butterfield, C. (1932) *J. Bacteriol.* 23, 355–368.

Chang, L., and Elander, R. (1986) in *Manual of Industrial Microbiology and Biotechnology* (Demain, A., and Solomon, N., eds.), pp. 4–55, American Society for Microbiology, Washington, DC.

Choate, R. (1973) *Sci. Teacher* 40, 33–35.

Christensen, P., and Cook, F. (1978) *J. Syst. Bacteriol.* 28, 367–393.

Clark, W., Horneland, W., and Klein, A. (1962) *Appl. Microbiol.* 10, 463–465.

Cooper, R., Bush, K., Principie, P., et al. (1983) *J. Antibiot.* 36, 1252–1257.

Cooper, R., Wells, J., and R. Sykes (1985) *J. Antibiot.* 38(4), 449–454.

Cooper, R., and Unger, S. (1985) *J. Antibiot.* 38(1), 24–30.

Cooper, R., and Unger, S. (1986) *J. Org. Chem.* 51, 3942–3945.

Corpe, W. (1951) *J. Bacteriol.* 62, 515–517.

Dietz, A. (1975) in *Round Table Conference on Cryogenic Preservation of Cell Cultures* (Riefret, A., and LaSalle, B., eds.), National Academy of Science, Washington, DC.

Dixon, B. (1983) *Trends Biotechnol.* March, 45, 131.

Froyshov, O. (1984) in *Biotechnology of Industrial Antibiotics* (Vandamme, E., ed.), pp. 665–694, Marcel Dekker, New York.

Gwynn, M., Box, S., Brown, A., and Gilpin, M. (1988) *J. Antibiot.* 41(1), 1–6.

Holt, J., ed. (1989) *Bergy's Manual of Systematic Bacteriology*, 2nd ed., Williams and Wilkins, Baltimore.

Hungate, R. (1962) in *The Bacteria, Vol. IV: The Physiology of Growth* (Gunsalas, C., and R. Stainer, eds.), pp. 95–119, Academic Press, New York.

Hunter, J., Fonda, M., Sotos, L., Toso, B., and Belt, A. (1984) in *Developments in Industrial Microbiology* (Underkoffler, L., ed.), pp. 247–266, Society for Industrial Microbiology, Arlington, VA.

Hunter-Cevera, J., Fonda, M., and Belt, A. (1986) in *Manual of Industrial Microbiology and Biotechnology* (Demain, A., and Solomon, N., eds.), pp. 3–31, American Society for Microbiology, Washington, DC.

Hunter-Cevera, J., and Belt, A. (1992) in *Encyclopedia of Microbiology* (Lederberg, J., ed.), pp. 561–570, Academic Press, New York.

Imada, A., Kitano, K., Kintaka, K., Muroi, M., and Asai, M. (1981) *Nature* 289, 590–591.

Imada, A., Kintaka, K., Nakao, M., and Shinagawa, S. (1982) *J. Antibiot.* 35, 1400–1403.

Katayama, N., Nozaki, Y., Okonogi, K., et al. (1985) *J. Antibiot.* 38, 1117–1127.

Kato, T., Hinoo, H., Terui, Y., et al. (1985) *J. Antibiot.* 40, 139–144.

Kintaka, K., Kazuaki, K., Uukimasa, N., et al. (1981a) *J. Ferment. Technol.* 59(4), 263–268.

Kintaka, K., Kitano, K., Nozaki, Y., et al. (1981b) *J. Antibiot.* 34, 1081–1089.

Korzybski, T., Kowszyk-Gindifer, Z., and Kurylowicz, W. (1978) *Antibiotics: Origin, Nature and Properties.* American Society for Microbiology, Washington, DC.

Lapage, S., Shelton, J., Michell, T., and Mackenzie, A. (1970) in *Methods in Microbiology* (Norris, J., and Ribbons, D., eds.), pp. 135–228, Academic Press, London.

Meyers, E., Cooper, R., Trejo, W., Georgopapakakou, N., and R. Sykes (1983) *J. Antibiot.* 36(2), 190–193.

Meyers, E., Cooper, R., Dean, L., et al. (1985) *J. Antibiot.* 38(12), 1642–1648.

Moore, L., and Carlson, R. (1975) *Phytopathology* 65, 246–250.

Neidleman, S. (1988) in *CRC Handbook of Microbiology, 2nd Edition* (Laskin, A., and Lechevalier, H., eds.), pp. 57–65, CRC Press, Inc., Boca Raton, FL.

Ono, H., Nozaki, Y., Katayama, N., and Okazaki, H. (1984) *J. Antibiot.* 37, 1528–1535.

Parker, W., and Rathnum, M. (1982) *J. Antibiot.* 35(3), 300–305.

Parker, W., Rathnum, L., and Liu, W.-C. (1982a) *J. Antibiotics* 35(7), 900–902.

Parker, W., Rathnum, M., Wells, J., et al. (1982b) *J. Antibiot.* 35, 653–660.

Pramer, D., and Schmidt, E. (1964) *Experimental Soil Microbiology*, Burgess Publishing Co., Minneapolis.

Rayman, K., and Hurst, A. (1984) in *Biotechnology of Industrial Antibiotics* (Vandamme, E., ed.), pp. 607–628, Marcel Dekker, New York.

Reichenbach, H., Gerth, K., Irschik, H., Kunze, B., and Höfle, G. (1988) *Trends Biotechnol.* 6, 115–121.

Ringel, S., Greenough, R., Roemer, S., et al. (1977) *J. Antibiot.* 30, 371–375.

Shannon, J., Gherna, R., and Jong, S. (1975) in *Round Table Conference on Cryogenic Preservation of Cell Cultures* (Riefret, A., and LaSalle, B., eds.), National Academy of Science, Washington, DC.

Singh, P., Johnson, J., Ward, P., et al. (1983) *J. Antibiot.* 36, 1245–1251.

Slater, J., Whittenbury, R., and Wimpenny, M., eds. (1983) *Microbes in their Natural Environment*, Cambridge University Press, Cambridge.

Straka, R., and J. Stokes (1957) *Appl. Microbiol.* 5, 21–25.

Sykes, R., Cimarusti, C., Bonner, D., et al. (1981a) *Nature* 291, 489–491.

Sykes, R., Bonner, D., Bush, K., Georgopapadakou, N., and Wells, J. (1981b) *J. Antimicrob. Chemother.* 8(Suppl. E), 1–16.

Truper, H., Balows, A., Dworkin, M., Jarder, W., and Schleifer, K.-H. (1992) *The Prokaryotes: A Handbook on the Biology of Bacteria: Ecophysiological, Isolation, Identification, Applications*. vol. I–IV, 2nd ed., Springer-Verlag, New York.

Tymiak, A., Culver, C., Goodman, J., Seiner, S., and R. Sykes (1984) *J. Antibiot.* 37, 416–418.

Vandamme, E. (1984) in *Biotechnology of Industrial Antibiotics* (Vandamme, E., ed.), pp. 3–31, Marcel Dekker, New York.

Weigele, M., and W. Leimgruber (1967) *Tetrahedron Lett.*, pp. 715–718.

Wells, J., Trejo, W., Principie, P., et al. (1982a) *J. Antibiot.* 35(2), 184–188.

Wells, J., Trejo, W., Principie, P., et al. (1982b) *J. Antibiot.* 35(3), 295–299.

Wells, J., Hunter, J., Astle, et al. (1982c) *J. Antibiot.* 35(7), 814–821.

Potential of Fungi in the Discovery of Novel, Low-Molecular Weight Pharmaceuticals

M.M. Dreyfuss
I.H. Chapela

A completely new era in medicine and biotechnology began in the mid-1940s with the discovery and subsequent broad therapeutic use of penicillin from the filamentous fungus *Penicillium chrysogenum*. The discovery of such an effective compound was proof and motivation enough for others to join a new gold rush in the search for other potent antibacterials. From the mid-1950s to the early 1960s and on, practically all major pharmaceutical companies had initiated a microbiological screening program.

Although attention soon drifted away from fungi, focusing more and more on actinomycetes, it was again a filamentous fungus, *Tolypocladium inflatum* and its selectively immunosuppressive metabolite cyclosporin, that created another quantum leap in medicine in the 1980s.

Microorganisms had by now proven not only to be capable producers of antibacterials but also of compounds for other therapeutic applications. As a consequence of this and the discovery of several other useful compounds or important lead structures (e.g., polyoxins, bialaphos, avermectins, acarbose, mevinolin, asperlicin, FK-506, rapamycin) on the one hand and the extremely high standard of natural and semisynthetic antibiotics on the other hand, the 1980s may be seen as the decade in which microbiological screening programs continued, focused on new antibiotics, and to a greater extent, on compounds useful in other therapeutic areas as well as in agriculture.

Looking at the rich harvests of the past from microbiological screening programs and the many exciting discoveries and developments currently taking place, we can foresee an even stronger role of this discovery strategy in the future. Our recent and growing understanding of microbial abundance and diversity, in particular that of fungi (Hawksworth 1991), shows that the fungal world has until now been, at best, superficially scratched. Coupled with this knowledge, new technological developments in areas such as robotics, ligand analysis, informatics, chemical analysis, and synthesis should pave the road to a dynamic, discovery-led growth in the pharmaceutical industry. Not only should the mycologist be amazed by the beauty, diversity,

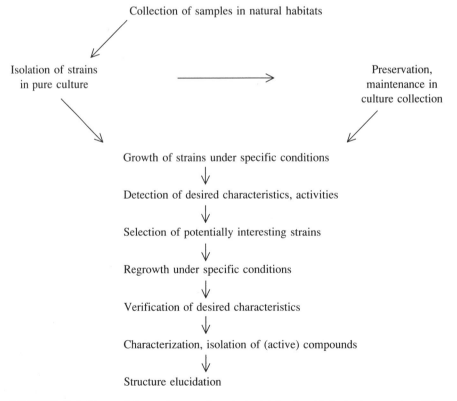

FIGURE 3–1 General flow-scheme of an industrial microbiological screening (For further details, see Berdy 1989).

and creativity of fungi, industrial managers and marketing experts should be in awe over the economic value of certain fungi, in particular the producers of penicillin, cephalosporin, mevinolin, and cyclosporin. The existence of a huge, still unknown and untapped microbial pool justifies hopes of discovering a rich menu of novel, useful, and economically profitable microbial compounds.

In this chapter we will not discuss secondary metabolism (for this topic see Barabás 1980; Zähner et al. 1983; Bennett 1983; Campbell 1984; Vining 1986; Davies 1990; Luckner 1990) but will concentrate on relating our experience in the pharmaceutical industry on fungal strains and the habitats from where they originate. Strain selection represents the very beginning, but also a crucial step in the discovery undertaking—the so-called screening program (Figure 3–1). At one time, all strains used in industry originated from natural habitats, and future searches for novel microbial strains will continue to start exactly there: in nature. Conducting a microbiological discovery program without considering taxonomy, ecology, or going out to nature might be compared to gold mining or oil drilling without considering geology and physical prospecting in the field. Here we summarize some empirical knowledge and provide a blueprint of a more rational approach to fungal strain isolation and selection, based on ecological and taxonomic criteria. Some useful practical tips are provided in Appendix A.

3.1   CREATIVITY, TAXONOMY, AND ECOLOGY

3.1.1   Creativity: Creative and Uncreative Fungal Groups
When we search for novel fungal metabolites we are forced to ask an initial key question: which are the worthwhile, creative organisms to search and where will we find them? However, we must first define *creative organism.* An initial approach for defining creativity of a fungus might be to look at its ability to produce compounds of interest for human activities. Creative fungi in this sense are summarized in Table 3–1.

It is evident from Table 3–1 that several taxa can produce the same metabolites, even though the producing fungi might not be taxonomically closely related. This is often true for the comparatively few available commercial compounds as well as for the plethora of noncommercial metabolites, documented in numerous examples (e.g., Turner 1971; Purchase 1974; Floss and Anderson 1980; Cole and Cox 1981; Mislivec 1981; Turner and Aldridge 1983; Frisvad and Filtenborg 1989; Samson and Pitt 1990; Bhatnagar et al. 1992). We believe metabolites that are produced exclusively by only one taxon or a single strain are the exception rather than the rule. In addition to examples in the literature (e.g., Loeffler 1984), we have various examples from our own screening of novel metabolites or an unusual metabolite pattern that occurs only in single strains. Examples are the novel cyclosporin-like peptolide SDZ 214-103 from a single *Cylindrotichum oligospermum* strain (Lawen et

TABLE 3–1 Fungal Taxa Able to Produce Secondary Metabolites of Past and Present Commercial Relevance as Pharmaceuticals or for Other Uses[1]

Fungal Taxa	Metabolite/Product	Reference
Penicillium chrysogenum Penicillium spp. **(Eupenicillium, Talaromyces)** Aspergillus **(Emericella)** Epidermophyton (Gymnoascus) Trichophyton Polypaecilum (Thermoascus) Malbranchea	Penicillins	Rehm 1980 Kitano 1983
Cephalosporium acremonium Cephalosporium/Acremonium spp. **(Nigrosabulum, Emericellopsis)** Pleurophomopsis Spiroidium **(Arachnomyces)** Scopulariopsis **(Anixiopsis)** Paecilomyces Diheterospora	Cephalosporins	Higgins et al. 1974 Rehm 1980 Kitano 1983 Dreyfuss 1986
Penicillium griseofulvum Penicillium patulum Penicillium nigricans Penicillium coprophilum Penicillium canescens Penicillium janczewskii Penicillium jensenii Penicillium lanosum Penicillium raistricki Penicillium sclerotigenum Penicillium aethiopicum	Griseofulvin	Rehm 1980 Frisvad and Filtenborg 1990a, 1990b Neidleman 1991

Claviceps purpurea
Clavicpes paspali
"Grass endophytes"
	Ergot alkaloids	Rehm 1980
		Kobel and Sanglier
		  1986
		Bacon and De Battista
		  1991

Tolypocladium inflatum Tolypocladium geodes Tolypocladium nubicola Tolypocladium terricola Tolypocladium cylindrosporum Tolypocladium tundrense Chaunopycnis alba[2] Aphanocladium sp.[2] Beauveria bassiana Beauveria brongniarti Acremonium spp. Paecilomyces spp. Verticillium spp. Isaria felina Fusarium spp. Trichoderma viride **Neocosmospora vasinfecta**	Cyclosporins	Sawai et al. 1981 Dreyfuss 1986 Nakajima et al. 1989 Jegorov et al. 1990 Weiser et al. 1991

TABLE 3–1   (continued)

Fungal Taxa	Metabolite/Product	Reference
Penicillium brevicompactum Penicillium citrinum Penicillium solitum Penicillium canescens Penicillium lanosum Penicillium hirsutum Hypomyces chrysospermus Paecilomyces sp. **Eupenicillium sp.** Trichoderma longibrachiatum Trichoderma pseudokoningii	Compactin	Endo et al. 1986 Stolk et al. 1990 Frisvad and Filtenborg 1990a, 1990b
Aspergillus terreus Monascus ruber Phoma sp. Aspergillus niger **Gymnoascus umbrinus** Doratomyces nanus Several sterile endophytic fungi[2]	Mevinolin	Turner 1971 Endo et al. 1986 Buckland et al. 1989
Pleurotus mutilis Pleurotus passeckerianus Drosophila subatrata Clitopilus pseudopinsitus	Pleuromutilin	Anke 1986
Fusidium coccineum Cephalosporium spp. Epidermophyton floccosum Microsporum gypseum Keratinomyces longifuscus Calcarisporium antibioticum Chrysosporium sp. **Isaria kogane** Phycomyces spp. Mucor remannianus	Fusidic acid	Loeffler 1984 Bérdy 1986
Fusarium roseum **(Giberella)** Fusarium tricinctum Fusarium oxysporum	Zearalenone	Rehm 1980 Turner and Aldridge 1983

[1] Italicized taxa, used for industrial production; boldface: teleomorphs.
[2] Unpublished data.

al. 1991) or the Sporiofungin family from a *Penicillium arenicola* strain (Dreyfuss 1986). Inability to produce a compound cannot be easily concluded from negative data, such as the failure to detect that compound without information on the metabolic machinery necessary to produce it. However, different fungal classes or subdivisions seldom produce the same secondary metabolites; e.g., beauvericin, occurring in fungi imperfecti and basidiomycetes (Gupta et al. 1991) and Oosporein in fungi imperfecti, ascomycetes, and basidiomycetes

(Lechevalier 1975). Production of a given secondary metabolite by eukaryotes (fungi) and prokaryotes (streptomycetes, bacteria) have been reported in several cases; e.g., β-lactam antibiotics (Kitano 1983), nebularin and questiomycin A (Lechevalier 1975; Cooper et al. 1986) and terferol (H.U. Nägeli, personal communication). In still rarer cases, higher plants were demonstrated to be producers of metabolite classes usually found in fungi; e.g., citrinin (Lechevalier 1975), tricothecenes (Jarvis et al. 1988), vioxantin (Provost and Garcia 1990), and ergot alkaloids (Floss and Anderson 1980).

To return to our attempt to define fungal creativity, the relatively few commercially relevant compounds (see Table 3–1) represent only a minute fraction of the known fungal secondary metabolites, making our first attempt rather unreliable. Commercially, relevant products might not be discovered in their most useful form, but rather as chemical "leads," which attain their highest biological activity only after derivatization or other forms of chemical optimization. To define creativity, therefore, we require more comprehensive data, including metabolites that initially may not display any obvious biological activity. The total number of characterized fungal secondary metabolites, including antibiotics, mycotoxins, pharmacologically active compounds, and those without known biological activities, are not easy to find. Estimates range between 3000 and 4000 compounds (Turner 1971; Turner and Aldridge 1983; Loeffler 1984; Hawksworth and Kirsop 1988; Berdy 1989). These compounds are not randomly produced by different fungi, but appear to occur in particularly creative fungal groups. Loeffler (1984) recognized "highly active" and "inactive" areas within the fungal system. Highly active areas correspond to fungal groups potentially able to produce a high diversity of secondary metabolites, independent of their commercial usefulness. Genera such as *Aspergillus, Penicillium, Acremonium,* and *Fusarium* (teleomorphs in the Eurotiales and Sphaeriales) clearly belong to these areas and account for over 40% of the antibiotics known in 1984. In contrast, fungi belonging to the Zygomycetes, Yeasts, and Ascomycetes of the order Pezizales seem to represent inactive areas.

By using information from our database on fungal metabolites and their producers, we have attempted to quantify the degree of activity or, as we prefer to call it, *creativity* (Table 3–2) of various fungal groups. The number of secondary metabolites recorded in the database for a selected group of fungi (e.g., genus, family or an ad hoc group such as "yeasts"), divided by the number of species in that particular group (for taxonomic information we used Gams 1971; von Arx 1981; Barnett et al. 1990) gives a "Creativity Index" (CI) or measurement of the activity in that group. As shown in Table 3–2, there are wide variations in CI between various groups, coinciding with the observation of Zähner (1982) and Loeffler (1984) and the data in Table 3–1.

In producing such CI, however, we are faced with various limitations. We do not have absolutely accurate, up-to-date figures on metabolites and taxa, many metabolites and taxa have not been reviewed critically, and/or many secondary metabolites might be minor variants of a single chemical

TABLE 3–2   Creativity Indexes of Arbritrarily Selected Fungal Taxa and Groups (Boldface: Teleomorphic States)

Fungal Taxa, Fungal Group	Estimated No. of Species	Approximate No. of Known Metabolites	Creativity Index
Aspergillus[1], **Eurotium**, **Emericella**	200	525	2.6
Penicillium, **Talaromyces**, **Eupenicillium**	200	380	1.9
Fusarium, **Giberella**, **Nectria**, **Calonectria**	70	200	2.8
Trichoderma, **Hypocrea**	20	54	2.7
"Cephalosporium-like hyphomycetes", Acremonium, Tolypocladium, Verticillium, Monocillium, **Emericellopsis**	140	116	0.8
Mucor, Rhizopus, Phycomyces	70	26	0.4
Oomycetales, Chytridiales	450	3	0.007
"Yeasts"	600	50	0.08
Basidiomycetes[2]	30,000	300	0.01
Fungal species in culture[3]	7000	4000	0.6
Fungi total[4]	$1.5 \text{ times } 10^6$	?	?

[1] From various strains of one species, *Aspergillus terreus*, over 20 different metabolites have been reported (Cane et al. 1987).

[2] Anke 1978, 1989; Anke and Steglich 1988.

[3] Hawksworth and Kirsop 1988.

[4] Hawksworth 1991.

class. We are also aware that comparing specific genera with more loosely defined groups can be problematic, and given the available data, it is not possible to recognize whether creative and noncreative fungal groups really exist on the grounds of phylogeny, i.e., genetic information.

Nevertheless, differences in CI are so large that we expect tendencies to be maintained. Data in Table 3–2 thus reinforce our view that secondary metabolite production is not homogenously distributed in nature. On the contrary, there seems to be highly creative and less creative fungal groups. Creative fungal groups have proven their potential. Groups classed as uncreative, however, might yet provide useful compounds. Why they should appear as uncreative could be related to one or several of the following considerations.

1. In the past a large proportion of known metabolites were detected and subsequently characterized according to their biological activity (antibiotic, toxicity), thus it cannot be excluded that apparently uncreative fungi produce metabolites in abundance that display neither antibiotic nor toxic activities.
2. Strains of fungi belonging to apparently uncreative groups might simply not have been cultivated appropriately. They might require radically different, yet unknown, conditions in order to produce secondary metabolites.
3. Uncreative groups might have been simply quantitatively underexamined (as is certainly the case for Basidiomycetes, and generally for all fungal groups that are not readily isolated and grown in axenic culture).

For now these questions must remain unanswered. Further experimentation will shed more light on these problems but it will also be a challenge for those involved in drug discovery to take a risk and change their "winning horses." Possibly, the apparently uncreative fungal groups will yet provide us with completely novel chemical structures.

3.1.2 Taxonomy: Taxon Specificity of Secondary Metabolites

Chemotaxonomy is based on the empirical observation that phylogenetically related organisms share common physiological and biochemical characteristics. The potential and actual production of secondary metabolites is a part of this physiological/biochemical identity.

Some secondary metabolites are produced by constitutive metabolic machineries, so that specific taxa will always have the potential to produce such "marker" metabolites. Pigments, toxins, and other secondary metabolites from mushrooms (basidiomycetes), lichens, and other large, fleshy, or sclerotial fungi are widely accepted as constant characteristics and therefore reliable markers for identification. Data on the consistence of secondary metabolites in microfungi are, by contrast, rather scanty; therefore, their use for identification has been discussed and proposed only in exceptional cases (see Pitt and Samson 1990; Samson and Pitt 1990; Thrane 1990). It is clear that there are taxa of microfungi for which specific metabolites or metabolite patterns are characteristic and therefore good identity markers. The taxonomic identity of a strain can indicate the presence of the corresponding marker metabolite; alternatively, a single metabolite or metabolite pattern frequently can point to a specific fungal taxon or series of taxa. This simple concept can be of great help in a fungal screening program, since it opens the possibility of predicting secondary metabolite production on the basis of taxonomic information. Table 3–3 gives several examples of fungal species for which a metabolite or metabolite pattern was shown to be consistent.

TABLE 3–3 Examples of Fungal Taxa that Consistently Produce Mycotoxins or Other Biologically Active "Marker" Metabolites[1]

Fungal Species	*Typical "Marker" Metabolite(s) for All or Majority of Strains Investigated*	*Reference*
Tolypocladium inflatum *Tolypocladium geodes*	Cyclosporins	Dreyfuss 1986
Emericella rugulosa	Echinocandins	Dreyfuss 1986
Trichothecium roseum	Trichothecin	Loeffler 1984
Myrothecium verrucaria		Loeffler 1984
Myrothecium roridum	Verrucarins, roridins	Loeffler 1984
Myrothecium leucotrichum		Loeffler 1984
Calcarisporium antibioticum	Verrucarins, roridins, fusidic acid	Loeffler 1984
Epidermophyton floccosum	Fusidic acid, floccosin, floccosinic acid	Loeffler 1984
Malbranchea sulfurea	Penicillin, viomellein	Loeffler 1984
Fusarium moniliforme	Fumonisin B 1	Nelson et al. 1991
Penicillium coprophilum	Griseofulvin, dechlorogriseofulvin, roquefortine C, meleagrin, oxaline	Frisvad and Filtenborg 1990b
Penicillium griseofulvum	Griseofulvin, cyclopiazonic acid, patulin, roquefortin C	Frisvad and Filtenborg 1990b
Penicillium brevicompactum	Mycophenolic acid, raistric phenols	Frisvad and Filtenborg 1990b
Emericella javanicum *(= Penicillium janthinellum)*	Xanthomegnin	Frisvad et al. 1990

[1] For further examples in the terverticillate group of *Penicillium*, see Frisvad and Filtenborg 1989.

Although the examples shown in Table 3–3, particularly those concerning the genus *Penicillium,* suggest that marker metabolites are generally found in fungi, it would be unwise to indiscriminately apply this concept. There also appear to be cases, as discussed above, in which a metabolite is known from only one strain of a given taxon, although such negative results must be accepted with some reservations since the apparent absence of a metabolite from strains expected to be producers might be due to inadequate culture and detection techniques. Nevertheless, day-to-day experience instructs us that taxon-specific "marker metabolites" occur far more frequently in fungi than is reported in the literature. An indication of this is the nuisance of repeatedly rediscovering known metabolites. A closer look at the strains producing such redundant activities usually turn out to be all too-well-known producers of the rediscovered compound. As an example, a characteristic pattern in our antifungal screen is regularly caused by a "green *Trichoderma* sp.," a known producer of the steroid-type antibiotic recently published as ergokonin (Augustiniak et al. 1991). However, such observations are not sufficient evidence to claim that *all* strains of the particular species are capable of producing the known and expected compounds. Moreover, such strains might also coproduce unknown metabolites.

3.1.3 Ecology: Habitat Fidelity

We need not be biologists or ecologists to recognize typical plant and animal communities and their differences according to habitat. Mushroom collectors know precisely the types of habitats, among which plant communities, and when to search for and expect to find their favorite species. We take this habitat-community fidelity for granted, but what seems a trivial undertaking for macroscopic organisms becomes a true challenge when we attempt to describe and distinguish communities of microfungi.

Numerous studies and reviews have contributed to our understanding of fungal ecology, especially in soil (e.g., Bisset and Parkinson 1979a, 1979b, 1979c; Widden 1979, 1986a, 1986b, 1986c, 1987; Pugh 1980; Parkinson 1981; Wicklow 1981a; Bartoli and Massari 1985; Christensen 1981, 1989; Frankland et al. 1990). Fungal communities in other habitats, such as dung, water, wood, leaf litter, living plants, lichens, etc., have also attracted considerable interest. References for such habitats are listed in Table 3–4. Certainly, one of the main difficulties in understanding and describing fungal communities in natural habitats lies in the inadequacy of any one single method to quantify diversity, population numbers, and activities of fungi. This is primarily due to their filamentous structure, making the fungi inherently indeterminate and inhomogeneous, and growing in an inhomogeneous, continuously changing environment where they are constantly forced to interact with other organisms. The second difficulty is that a description of a community can only be as good as our understanding of the taxonomy of its members, which in fungi is often very unsatisfactory.

TABLE 3–4 Some Diverse, Underexploited Fungal Groups of Ecosystems

Fungal Groups	References[1]
Endophytic fungi	Carroll 1986, 1991 Clay 1986, 1988 Hijwegen 1988 Bacon and De Battista 1991 Petrini 1986, 1991 Andrews and Hirano 1991 Siegel and Schardl 1991
Phylloplane fungi	Dickinson 1976 Macauley and Waid 1981 Cooke and Rayner 1984 Marakis and Diamantoglou 1990 Andrews and Hirano 1991 Fokkema 1991 Kinkel 1991
Xylotropic endophytes, wood decay fungi	Cooke and Rayner 1984 Chapela and Boddy 1988a, 1988b Chapela 1989 Rayner and Boddy 1988 Boddy 1991
Mycorrhizal fungi (ecto-, endo and vesicular arbuscular)	Cooke 1977 Mosse et al. 1981 Cooke and Rayner 1984 Currah et al. 1987 Currah et al. 1988 Tylka et al. 1991 Warcup 1991 Bagyaraj 1991 Paulitz and Linderman 1991 Suvercha et al. 1991 Barrett 1991 Norris et al. 1991, 1992
Other specific plant-associated fungi	Starmer 1981 Phaff 1990 Starmer et al. 1991
Plant litter fungi	Dickinson and Pugh 1974 Pugh 1974 Jones 1974 Cooke and Rayner 1984 Gamundi et al. 1987 Kjøller and Struwe 1987 Frankland et al. 1990

(continued)

TABLE 3–4   (continued)

Fungal Groups	References[1]
Moss-symbiontic fungi	Boullard 1988 Felix 1988
Lichen-parasymbiontic fungi and mycobionts	Cooke 1977 Cooke and Rayner 1984 Hawksworth 1988 Petrini et al. 1990 Honegger and Bartnicki-Garcia 1991 Crittenden and Porter 1991
Coprophilous fungi	Wicklow 1981b Cooke and Rayner 1984
Fungal parasites and associates of insects and other anthropods	Cooke 1977 Roberts and Humber 1981 Lichtwardt 1986 Evans 1988 Glare and Milner 1991
Fungal parasites of nematodes and rotifers	Cooke 1977 Barron 1981 Carris et al. 1989 Saxena et al. 1991
Fungicolous fungi, mycoparasites	Cooke 1977 Hawksworth 1981 Lumsden 1981 Samuels 1988
Cryptoendolitic microbial communities	Taylor-George et al. 1983 Johnston and Vestal 1991
Aquatic hyphomycetes, aero-aquatic fungi	Ingold 1975, 1976 Jones 1974 Webster 1981 Bandoni 1981 Fisher and Webster 1981 Jones 1981 Webster and Descals 1981 Bärlocher and Kendrick 1981 Bärlocher and Rosset 1981 Wood-Eggenschwiler and Bärlocher 1983 Descals and Webster 1982 Cooke and Rayner 1984 Ando and Tubaki 1984a, 1984b

(*continued*)

TABLE 3–4 (continued)

Fungal Groups	References[1]
	Suberkropp and Kluge 1981
	Suberkropp 1984, 1991
	Hasija and Singhal 1991
Marine fungi	Jones 1974
	Kohlmeyer and Kohlmeyer 1979
	Kohlmeyer 1981
	Newman et al. 1989
Thermophilic fungi	Sharma and Johri 1992
	Satyanaryana et al. 1992
Psychrophilic environment and psychrophilic filamentous fungi	Wharton et al. 1985
	Petrini et al. 1992

[1] Literature relating to the ecology and methodology for studying specific fungal or microbial groups. This is not intended to be a complete literature review, in most examples we attempted to include some useful specific references as well as more recent literature from which older publications can be traced.

It is not the scope of this section to fully discuss and review all problems of fungal ecology and taxonomy; thus abstracting ourselves from the above problems, we find it useful to quote two basic conclusions from Christensen (1981): (1) "It is apparent that there is an extremely high species diversity among the fungi in any given ecosystem," and (2) "Habitat specificity for individual species and guilds of species appears to be the rule."

Analogous conclusions have been made for aquatic habitats (Bärlocher and Rosset 1981) and for endophytic fungi (Petrini 1991). Over the years we have collected a vast amount of data that particularly reflects and explains the second quote, and from our pragmatic viewpoint is significant in the context of redundancy during metabolite discovery.

The significance of habitat and substrate for the production of some metabolites that we have found during screening can be illustrated by a few examples collected in Table 3–5.

Habitat fidelity or specificity allows us to predict with some confidence that different habitats will yield different fungal groups. With this information, we see possibilities to surmount some of the stumbling blocks of redundancy and take a more direct approach toward creative and novel fungal groups. In similar terms and also with respect to other microbial groups such as actinomycetes and myxobacteria, Porter (1985), Cheetham (1987), Omura (1988), Williams and Vickers (1988), Reichenbach and Höfle (1989), and Steele and Stowers (1991) come to the conclusion that novel organisms will yield novel products. This is confirmed constantly in our own experience, as shown by some targeted isolation work involving endophytic fungi (see Appendix B), parasymbionts of lichens (M.M. Dreyfuss et al., unpublished

TABLE 3–5 Reoccurring Metabolites and their Frequencies in Relation to Habitat and Substrate

Type of Habitat	Metabolites			
	Cyclosporin A	Echinocandin B	Papulacandins	Verrucarins
Soil: arctic/alpine/temperate	++	(+)	++	++
Soil: hot desert/warm tropical	0	++	+	++
Leaf litter: alpine/arctic/temperate	+	0	++	++
Leaf litter: warm tropical	(+)[1]	++	++	++
Mosses: alpine/temperate	++	0	0	+
Mosses: warm tropical	0	0	0	+
Lichens:[2] arctic/alpine/temperate	0	0	0	0
Lichens:[2] hot desert/warm tropical	0	0	++	0
Living plants:[3] alpine/temperate	0	0[4]	0	0
Living plants:[3] warm tropical	0	0	0	0

++, Frequently found; +, sporadically found; (+), rarely found, single records; 0, no records.

[1] Very recently several cyclosporin A producing *Chaunopycnis* strains have been isolated from leaf litter samples collected at an altitude of 2200 m in Malaysia; these strains represent the only secured records of cyclosporin producers from the tropics.

[2] Refers to the highly diverse parasymbiontic filamentous fungi within the lichen thallus (Petrini et al. 1990).

[3] Endophytic fungi isolated from healthy plant tissues (for references see Table 3–4).

[4] Other metabolites of this class of lipopeptides, sporiofungins A, B, and C, 41075 F-1, and L-671, 329 were isolated from endophytic strains of *Cryptosporiopsis* sp. (Dreyfuss 1986; Noble et al. 1991).

observations), as well as from recent literature describing novel metabolites from fungi that were either isolated from plants or can usually be expected to be plant-associated (Shibata et al. 1988; Hedge et al. 1989; Koshino et al. 1989; Sugawara et al. 1991; Mikawa et al. 1988; Satoshi et al. 1988). In addition, aquatic and marine fungi are beginning to be discovered as producers of novel metabolites (Poch and Gloer 1989; Schwartz et al. 1989; Sugita et al. 1990).

Finally, we should not overlook the possibility that in analogy to taxonomic marker metabolites the same could apply with respect to ecology. For example, botrydial derivatives were found to be produced by plant-associated fungi, including some xylotropic endophytes (M.M. Dreyfuss, I.H. Chapela, and L. Hagiman, unpublished observations, see Appendix B), the discomycete *Hymenoscyphus epiphyllus* (Anke et al. 1991) and *Botrytis squamosa* (Kimura et al. 1988). As in the case of taxonomical marker compounds, this information can be used to specifically search for this family of compounds or, alternatively, to avoid their redundant appearance in screening.

3.2 THE UNTAPPED POOL

Although the estimated 3000 to 4000 known fungal secondary metabolites have been isolated after screening hundreds of thousands of fungal cultures, possibly not more than 5000 to 7000 taxonomic species have been studied in this respect, the latter being the number of fungal species maintained in culture collections throughout the world (Hawksworth and Kirsop 1988; Hawksworth 1991) and representing, therefore, fungal species that can be readily grown in axenic culture. The untapped pool of fungal diversity is, however, tremendous. Cheetham (1987) speculates that less than 1% of all microorganisms have been identified and characterized. Some recent evidence provided by DNA and RNA probing seems to indicate that only a minor proportion of prokaryotes in natural environments are being isolated (Giovannoni et al. 1990; Olsen 1990; Ward et al. 1990) and an analogous situation is thought to prevail with eukaryotes such as fungi.

Approximately 70,000 species of fungi have been given a valid name, whereas Hawksworth (1991) concludes that at least 1.5×10^6 species of fungi must exist. From our own experience with endophytic fungi of plants, we suggest that any vascular plant species can host somewhere between 10 and 100 different fungal species. Two to five of them could well be host-species specific. Given the 270,000 species of vascular plants thought to exist on this planet, the endophytic fungi alone could account for up to 1.3×10^6 fungal species. This figure does not include the largely unknown endophytes or parasymbionts of lichens, mosses, and the fungal associates of the 6×10^6 species of insects and other arthropods believed to exist. In addition, it is becoming increasingly clear that fungal strains assigned to the same taxon might be physiologically variable (Petrini 1991).

There are well-known fungal groups that resist artificial cultivation and have been largely excluded from industrial screenings. These include obligate biotrophic fungi (rusts, uredinales, vesicular arbuscular (VA) mycorrhiza, as well as arthropod-associated groups such as the whole class of the Laboulbeniales with 1730 *known* species), but there are also many saprobes that will not grow in the laboratory, probably due to unknown but essential triggers of spore germination. Conversely, there are examples of predominant and culturable fungal groups, their ubiquitous occurrence and diversity being recognized astoundingly late: Ingoldian aquatic hyphomycetes in the 1940s (Ingold 1975), endophytic fungi in the late 1970s (Bernstein and Carroll 1977), and parasymbionts of lichens (Dreyfuss 1986; Petrini et al. 1990). Investigation of underexploited, possibly extreme habitats (see, e.g., Edwards 1990) in conjunction with specific isolation techniques should provide one of the key answers to industrial exploitation of fungal diversity, which might be tantamount to chemical diversity. Strains found in outstanding habitats or missed in traditional isolation procedures could well be producers of metabolites that have been unknown thus far.

Table 3–4 lists some fungal groups we would expect to be diverse and probably largely underexploited in the context of screening and drug discovery.

3.3 ECO-TAXONOMIC LANDSCAPE: A WORKING MODEL

A central theme in the preceding section relates taxonomy and ecology of a fungal group to its creativity in terms of secondary metabolite production.

In the following section we rework this theme to attempt to provide a conceptual working model to help direct our search for new fungal secondary metabolites. The concept is simple: if ecology and taxonomy are the main determinants of the potential to produce secondary metabolites, then ecological and taxonomical principles should direct our search for creative fungi. Giving a formal body to this concept requires, by contrast, complex mathematics or a semi-intuitive, graphical approach. The second of these possibilities is explored in the following discussion.

Ecology is a major determinant of secondary metabolite production. Brock (1966) stated that "the search for new antibiotics is a problem of microbial ecology." We have seen how some ecological niches harbor particularly creative fungi, such as xylotropic endophytes (see Appendix B) and soil fungi. The reasons for this are unknown, but it seems fair to assume that, given a genetic background, certain physical and biological situations in the natural environment favor the production of a diverse range of secondary metabolites.

This hypothesis suggests that the secondary metabolites having relevance to our pharmaceutical screening have some direct, selectable relevance for the survival (or at least the lifestyle) of the fungi concerned (Note: This is a topic under strong discussion, where experimental evidence is sorely missing. For an insight into the discussion, see Vining 1990). If this assumption is true, some ecological guidelines could be derived to direct our search toward highly creative fungal groups. For example, environmental stress and the intensity and frequency of interactions between a given fungus and its biological environment may act as promoters of metabolic diversity: overcoming stress and communicating with other organisms in any kind of symbiosis will favor the production of finely designed, highly selected low-molecular compounds. Examples of generic groups of these molecules include neurotransmitters, pheromones, antibiotics, venoms, interspecific attractants or repellents, morphogens, etc. All these compound classes have the required communication function in nature; given the known parsimony in evolution, it is then possible that the secondary metabolites used by fungi in nature could also be biologically active in other organisms, such as humans, agricultural plants, or their pathogens. It might not be a coincidence that many cyclosporin producers are opportunistically associated with insects or plants (Dreyfuss 1986; Jegorov et al. 1990).

How does one locate ecological niches where highly creative fungi live? The first, necessary step toward this goal is to describe the niches where we might search. To precisely describe an ecological niche is practically an impossible task, since there are innumerable factors that conform the physical and biological environment of each species and of each individual within a species.

Nevertheless, some major factors can be defined that determine the distribution of a given organism. Temperature, O_2/CO_2 ratios, humidity, carbon, and nitrogen sources are examples of common determinants of fungal ecological niches. Each of these factors can be represented graphically as a section on an axis: for example, penicillia exhibits rapid growth from 5 to 37°C (Pitt 1979), occupying only the 5 to 37°C section of the temperature axis. By introducing a second factor, e.g., humidity, a "surface" can be found, which is limited by temperature and water availability and where penicillia can thrive. A third factor defines a three-dimensional "volume" where penicilli can grow, whereas adding a fourth factor results in a four-dimensional "hypervolume" (Figure 3–2). The ecological niche of a given organism is thus defined as an n-dimensional hypervolume in which that organism can thrive (for a further explanation of this familiar ecological concept, see Begon et al. 1986).

The other major determinant of fungal potential to produce secondary metabolites is taxonomy. We have seen that certain taxonomic groups are proven wizards of secondary metabolite production while others appear to be very uncreative (see Sections 3.1.1 and 3.1.2). This implies that through evolution certain groups have been selected that display a higher diversity and abundance of secondary metabolites. It is of great practical importance, therefore, to locate taxa exhibiting creativity. But first, one must identify and describe the taxa to be studied. Again, the number of factors to be considered is enormous. Entire research groups have been dedicated to this task, i.e., taxonomy, for generations. We can find similarities between the identification of a fungal taxon and the definition of its ecological niche, as explained above. Characteristics such as spore size, DNA C-G content, growth rates on specific media, and utilization of various carbon sources are routinely used to define

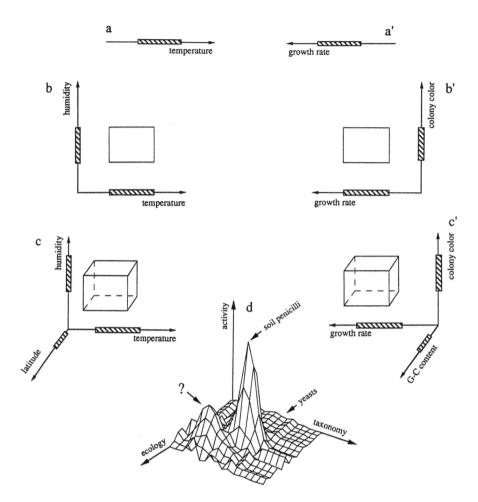

FIGURE 3–2 Construction of a multidimensional "hyperspace" from basic ecological and taxonomical characters. **a–c,** Addition of ecological characters to produce an ecological hyperspace, the so-called ecological niche of a species. The species under consideration can thrive only within limited ranges along the axis representing ecological variables; to define the ecological niche of that species, a large number of variables should be considered, but only three variables are represented here for the sake of illustration. **a'–c',** Addition of taxonomic characters to produce a hyperspace defining a taxonomic species. Although a large number of variables (here depicted as axes) should be considered to define a fungal species accurately, three variables are depicted here for the sake of illustration. **d,** all ecological characters are brought into one "ecology" complex axis, all taxonomic characters are condensed into one "taxonomy" axis, and a third "activity" axis is added, indicating the activity of an ecologically and taxonomically defined fungal group in a screening. The activity axis could be replaced, for example, by a CI axis (CI, as defined in Section 3.1.1).

In this fictitious landscape, we have labeled three regions: one of low activity, mapping to the yeasts, one of high creativity corresponding to soil penicillia, and a yet undiscovered peak of high activity. The task of a directed screening would be to discover and exploit such undiscovered or understudied fungal groups with high creativity.

a taxon. As an analogy to the representation of an ecological niche, these characters can be graphically expressed as sections along various axes. For example, penicillia can be found with growth rates between 0 and 10 mm/d (Pitt 1979), defining the section between 0 and 10 mm/d on the size axis. If DNA G-C content is given as a second axis, a "surface" will be found where spore sizes and G-C contents of penicillia are contained. Adding factors in this way we approach the identification of a taxon, which, by analogy, could be defined as an n-dimensional hypervolume of characters containing all individuals of the taxon studied.

However, ecology and taxonomy are only two sides of the same biological coin, with fungal taxa occurring over diverse ecological niches. Consequently, both ecological and taxonomical principles should dictate the criteria to follow in our search for metabolically richer fungal groups. To this purpose, we can expand the graphical approach already applied to bring ecology, taxonomy, and secondary metabolite production into a unified conceptual framework. An "eco-taxonomic hyperspace" can be constructed by projecting all ecological determinants onto one axis and all taxonomic characters onto a second axis. On this plane, defined groups of fungi can be located; for example, soil aspergilli, wood-inhibiting basidiomycetes, and mycorrhizal phycomycetes would represent three distinct areas on this surface. On this hyperspace, a new axis could be added to represent the ability of each particular fungal group to produce secondary metabolites (i.e., CIs, as defined in Section 3.1.1). The result is a landscape with peaks and troughs of CIs (see Figure 3–2). This graphical display shows that there are well defined ecological niches and taxonomical groups with a higher proportion of creative fungal species (see above), i.e., with higher CIs. From a practical point of view, this means that we should be able to find areas within the vast landscape of fungal diversity where our chances of finding new (or avoiding known) secondary metabolites are greatly increased.

A good example is provided by fungi in the family Clavicipitaceae: while grass-associated, endophytic and pathogenic clavicipitaceous strains are common producers of ergot-type alkaloids (Bacon and De Battista 1991), to our knowledge, none of the taxonomically related, insect-associated *Cordyceps* species have ever been shown to produce these compounds.

Confronted with a given eco-taxonomic landscape, how does one steer the search specifically toward enclosures where particularly creative fungi gather; i.e., how does one find peaks of high CIs that promise multiple new secondary metabolites? Of course, outside the relatively well-explored peak corresponding to temperate soil-inhabiting fungi, our knowledge of the fungal eco-taxonomic landscape as related to secondary metabolite production is small.

We must by necessity assume that a fungal group with high CIs *is* essentially a creative producer of secondary metabolites, that it is worth looking for and exploring peaks of high creativity in the eco-taxonomic landscape irrespective of the specific biological activity that forms the goal of a screening. The fact that soil deuteromycetes continue to provide new compounds of industrial interest despite the radically new goals of screening programs tends to reinforce this assumption. Here, however, we find ourselves in a dilemma: in order to achieve high success rates we should screen fungi belonging to taxa having high CIs. However, we simultaneously encounter the problem of rediscovering the same metabolites over and over again (redundancy). It is at this point when the knowledge of other peaks of creativity might encourage us to establish new isolation and culturing techniques in order to climb novel peaks within the eco-taxonomic landscape.

When (if ever) we should move away from a proven peak of high creativity is a complex question, but a systematic approach can help in making that decision. By carefully registering newly encountered species or compounds we find as we proceed in the investigation of a given area of the fungal eco-taxonomic landscape, we can establish a point where an increased search effort is no longer matched by a reasonable gain in novel species or compounds. An example of this analysis is provided in Figure 3–3, where the number of fungal species found during the study of two ecological niches is plotted against a measure of the sampling effort invested. Under similar conditions, and given that CIs in the two groups are similar, a reasonable gain in new species would be found at approximately 400 isolates from soil or approximately 10 branch samples. Gain/investment ratios could thus be approximately calculated for the investigation of novel ecological or taxonomic groups.

In summary, it seems appropriate to see a microbiological screening as consisting of two main phases. In the first, the exploratory phase, ecologic and taxonomic criteria are applied to search for an activity peak of fungi with high CIs. Once such a group has been identified, a second phase of consolidation strives to optimize sampling, handling, and fermentation techniques

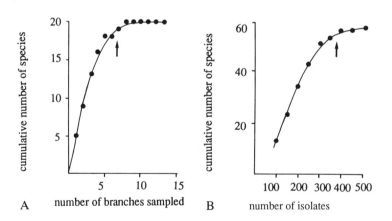

A   number of branches sampled   B   number of isolates

FIGURE 3–3 Incremental gain in new species found as sampling effort was increased in two ecological groups of fungi. (A) Fungi in attached, decaying beech branches (from Chapela 1987); (B) soil fungi (data from States 1981, recalculated by Chapela 1987).

to exploit the potential of the newly found creativity pool. These two phases, whether consciously or not, normally occur in parallel in most large-scale microbiological screenings.

3.4 CONCLUSIONS

1. We find creative and uncreative fungal groups. Creativity of a given group seems determined by its taxonomic situation (evolutionary history, phylogeny) as well as by its expression in its immediate environment (ecology). Within the creative groups various taxa may produce the same metabolites or the same metabolite patterns.
2. There are taxa-specific metabolites and metabolite patterns. This might be more widely true than is documented.
3. There are habitat-specific fungi. By investigating novel ecological niches, novel fungal species and strains should be expected.
4. Through the application of ecological and taxonomical criteria, a practically infinite expansion of our menu of metabolic diversity could be achieved. The search for highly creative fungal groups can be rationally steered by such criteria.

APPENDIX A.   USEFUL TIPS FROM AN INDUSTRIAL SCREENING PROGRAM

A number of books and papers have been published (Cole and Kendrick 1981; Hawksworth and Kirshop 1988; Gams et al. 1988; Labeda 1990; Arora et al. 1991a, 1991b, 1991c, 1992; Jones et al. 1991; Kirsop and Doyle 1991; Bull et al. 1992) that provide useful information and ideas on collecting, strain isolation, strain preservation, organization of culture collections, etc. We do not wish to duplicate this information here but intend to communicate some thoughts and specific methods developed in our laboratories.

Collecting and Treatment of Material

It is general practice to process samples that have been removed from nature. The removal of samples from nature is a starting point of imbalance and might be crucial for the development of "weedy" fungi at the expense of more delicate ones. Moist soil or plant litter samples, for example, are often shipped around the world and stored under adverse conditions. In our view it makes little sense to investigate such samples, since they will probably be dominated by a few ubiquitous molds. In contrast, isolation of endophytes from plant leaves relies on freshly collected plant material not older than a few hours or days. Removal of substrata from nature is, however, not always undesirable. Xylotropic endophytes, for example, can only be isolated in reasonable numbers after samples have been incubated for some weeks in the laboratory (see Appendix B).

Central to this theme is the human factor involved in the sampling and processing of these materials. The direct involvement of trained biologists, microbiologists, or mycologists is required in order to design protocols and conduct sampling since uninformed correspondents tend to confuse the rational evaluation of exploratory efforts. Mobile isolation stations have been described and successfully used. Such efforts and the associated costs are generally thought to be unnecessary, although they can be essential for isolating delicate groups of fungi.

The decision of where to go, which substrata to collect, and what kind of fungal groups to be sought will directly influence which isolation methods to follow, media preparation, incubation regimes, etc.

Isolation Media and Techniques

Although many so-called specific media or selective agents useful for the isolation of certain fungal genera or groups have been published (e.g., Jong 1981; Bååth 1991; Wildman 1991; Worrall 1991), true specificity can only be fully satisfied in very few, exceptional cases. In this respect, fungi provide a contrast with actinomycetes and eubacteria, for which media can be designed and chemical agents incorporated for the enrichment and selection of certain taxa (Labeda and Shearer 1990; Bull et al. 1992). We have been using 1 to 2 mg cyclosporin/l of agar as a general growth retardant in conjunction with a sharply pointed welding rod to eliminate undesired, rapidly growing "weedy" species that would otherwise overgrow the isolation plates. This allows the development and isolation of slow-growing fungi (Dreyfuss 1986).

Cultivation and Stress

There is no such thing as an ideal culture condition. Ideal conditions for growth and fermentation must be found for every strain. Any manipulation of a fungal culture is stressful, as soon as it is removed from its natural habitat. There is no such thing as a nonstressed condition in the laboratory. In relation to natural conditions, cultures are probably more stressed in pure culture on an agar surface and in an "optimally" designed high nutrient medium that is aerated, stirred, and depleted, than in a more oligotrophic environment with nutrient limitation.

APPENDIX B. ECOLOGICALLY DIRECTED SEARCH FOR NEW FUNGI AND THEIR PRODUCTS: A STUDY ON XYLOTROPIC ENDOPHYTES

In an effort to move away from trivial fungal groups, novel ecological niches can be specifically studied. The following is an example of such a search, where we compared a well-defined ecological group, xylotropic endophytes (XE), and other fungi (control group, CG) being tested in a high throughput screening.

Study Design

We chose XE, a group of fungi particularly adapted to live in association with woody plant organs, as the experimental group. XE inhabit healthy stems and branches of various trees, where they live in an inconspicuous form until the host tissues are stressed or die (Chapela and Boddy 1988a, 1988b). This particular way of life makes it possible to selectively isolate these fungi by artificially stressing tree samples, whereupon XE develop and can be re-covered from the wood using appropriate culture media (Chapela and Boddy 1988b). XE were included in the Sandoz High Throughput Screening for a period of 6 months. As a control group (CG), we used other fungi included in the Sandoz Screening during the same 6 months. A total of 1164 strains were included in this study, 400 XE and 764 CG fungi.

A measure of performance in the screening was obtained from the per-centage of strains from each of the two groups that was considered to be active in each of 14 tests by an independent, unbiased researcher. Tests were coded from 1 to 14 (Figure B–1) since their exact nature cannot be disclosed for proprietary reasons. Two main groups of tests can be distinguished: those related to antimicrobial activity (tests 2–5) and those of nonantimicrobial nature (tests 1 and 6–14).

The ratio of percentages XE/CG was calculated and, for the sake of comparison, the $\log_2$ of this ratio was plotted for each of the 14 tests. In this way, $\log_2$ XE/CG values close to 0 indicated similar performance rates for the two groups, while a doubling in relative activity is shown as an increase of one unit on the $\log_2$ scale (see Figure B–1).

Findings (see Figure B–1)

XE showed a clearly higher performance rate over the control group in general and in independent assays. Taking all tests together, XE showed a 73% higher performance than controls (49% versus 28% positive activities). If individual tests are considered, however, a distinct profile can be observed for XE. This fungal group performed up to six times better than the control group in some tests (test codes 1–5) but in others no difference in performance or even a reduction of 1/3 was observed. Importantly, all but one of the tests in which a statistically significant better performance of XE was found were related to antimicrobial activities (tests 2–5). This might be explained by the specialized life strategy of XE, which are believed to provide some kind of protection to their host plant against harmful organisms. Similarly, two of the tests where XE showed a statistically significant lower activity than the control group (tests 12 and 14) are related to herbicidal activity. Since XE live a good part

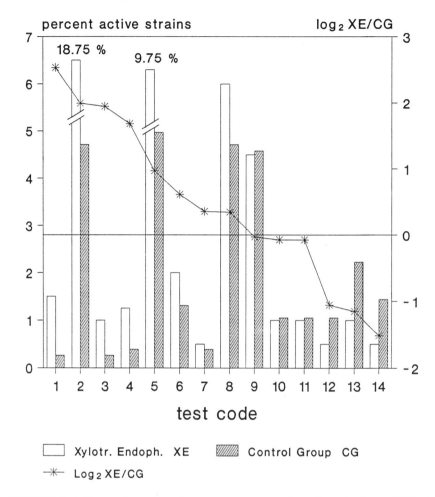

FIGURE B–1 Comparison of performance between xylotropic endophytes (XE) and a control group (CG) of "other" fungi in 14 biological screening assays (For details, see text in Appendix B).

of their life within plants, it might not be surprising that they produce few phytotoxic compounds.

Early follow-up of these results produced at least six new chemical structures from only 13 XE strains chemically investigated.

Conclusions

It can be said from the above results that XE represent one of the many expected peaks of creativity in the landscape of fungal diversity (see Section 3.3). As a *different* peak on that landscape, however, novel activity profiles and secondary metabolites could be found in XE that were not present in other fungal groups taken as control.

At least in this ecological group, there appears to be some correlation between the activities displayed by secondary metabolites from XE and their way of life. Ecology provides clues about the nature of such activities.

REFERENCES

Ando, K., and Tubaki, K. (1984a) *Trans. Mycol. Soc. Jpn.* 25, 21–37.

Ando, K., and Tubaki, K. (1984b) *Trans. Mycol. Soc. Jpn.* 25, 39–47.

Andrews, J.H., and Hirano, S.S., eds. (1991) *Microbial Ecology of Leaves,* Springer, New York.

Anke, H., Helfer, A., Arendholz, W.R., Casser, I. and Steglich, W. (1991) *Planta Medica* 57(Suppl. 2), A34.

Anke, T. (1978) *Z. Mykol.* 44(1), 131–141.

Anke, T. (1986) in *Biotechnology,* vol. 4 (Rehm, H.-J., and Reed, G., eds.), pp. 611–628, VCH, Weinheim, Germany.

Anke, T. (1989) in *Bioactive Metabolites from Microorganisms; Progress in Industrial Microbiology,* vol. 27 (Bushell, M.E., and Gräfe, U., eds.), pp. 51–66, Elsevier, Amsterdam.

Anke, T., and Steglich, W. (1988) *Forum Mikrobiologie,* 1-2/1988, 21–25.

Arora, D.K., Ajello, L., and Mukerji, K.G., eds. (1991a) *Handbook of Applied Mycology,* vol. 2, Marcel Dekker, Inc., New York.

Arora, D.K., Elander, R.P., and Mukerji, K.G., eds. (1992) *Handbook of Applied Mycology,* vol. 4, Marcel Dekker, Inc., New York.

Arora, D.K., Mukerji, K.G., and Marth, E.H., eds. (1991b) *Handbook of Applied Mycology,* vol. 3, Marcel Dekker, Inc., New York.

Arora, D.K., Rai, B., Mukerji, K.G., and Knudsen, G.E., eds. (1991c) *Handbook of Applied Mycology,* vol. 1, Marcel Dekker, Inc., New York.

Augustiniak, H., Forche, E., Reichenbach, H., et al. (1991) *Leibigs Ann. Chem.* 4, 361–366.

Bååth, E. (1991) *Mycol. Res.* 95(9), 1140–1152.

Bacon, Ch.W., and De Battista, J. (1991) in *Handbook of Applied Mycology,* vol. 1 (Arora, D.K. Rai, B., Mukerji, K.G., and Knudsen, G.R., eds.), pp. 231–256, Marcel Dekker, Inc., New York.

Bagyaraj, D.J. (1991) in *Handbook of Applied Mycology,* vol. 1 (Arora, D.K., Rai, B., Mukerji, K.G., and Knudsen, G.R., eds.), pp. 3–34, Marcel Dekker, Inc., New York.

Bandoni, R.J. (1981) in *The Fungal Community* (Wicklow, D.T., and Carroll, G.C., eds.), pp. 693–708, Marcel Dekker Inc., New York.

Barabás, G. (1980) *Folia Microbiol.* 25, 270–277.

Bärlocher, F., and Kendrick, B. (1981) in *The Fungal Community* (Wicklow, D.T., and Carroll, G.C., eds.), pp. 743–760, Marcel Dekker Inc., New York.

Bärlocher, F., and Rosset, J. (1981) *Trans. Br. Mycol. Soc.* 76(3), 479–483.

Barnett, J.A., Payne, R.W., and Yarrow, D. (1990) *Yeasts: Characteristics and Identification,* 2nd ed., Cambridge University Press, Cambridge, MA.

Barrett, V. (1991) in *Handbook of Applied Mycology,* vol. 1 (Arora, D.K., Rai, B.,

Mukerji, K.G., and Knudsen, G.R., eds.), pp. 217–229, Marcel Dekker, Inc., New York.

Barron, G.L. (1981) in *Biology of Conidial Fungi,* vol. 2 (Cole, G.T., and Kendrick, B., eds.), pp. 167–200, Academic Press, New York.

Bartoli, A., and Massari, G. (1985) *Ecol. Mediterr.* 11(2/3), 73–86.

Begon, M., Harper, J.L., and Townsend C.R. (1986) *Ecology. Individuals, Populations and Communities,* pp. 72–73, Blackwell Scientific Publications, Oxford.

Bennet, J.W. (1983) in *Secondary Metabolism and Differentiation in Fungi* (Bennett, J.W., and Ciegler, A., eds.), pp. 1–32, Marcel Dekker, Inc., New York.

Berdy, J. (1986) in *Biotechnology,* vol. 4, (Rehm, H.-J., and Reed, G., eds.), pp. 465–507, VCH, Weinheim, Germany.

Berdy, J. (1989) in *Bioactive Metabolites from Microorganisms; Progress in Industrial Microbiology,* vol. 27 (Bushell, M.E., and Gräfe, U., eds.), pp. 3–26, Elsevier, Amsterdam.

Bernstein, M.E., and Carroll, G.C. (1977) *Can J. Bot.* 55, 644–653.

Bhatnagar, D., Lillehoj, E.B., and Arora, D.K., eds. (1992) *Handbook of Applied Mycology,* vol. 5, Marcel Dekker, Inc., New York.

Bissett, J., and Parkinson, D. (1979a) *Can. J. Bot.* 57, 1609–1629.

Bissett, J., and Parkinson, D. (1979b) *Can. J. Bot.* 57, 1630–1641.

Bissett, J., and Parkinson, D. (1979c) *Can. J. Bot.* 57, 1642–1659.

Boddy, L. (1991) in *Handbook of Applied Mycology,* vol. 1 (Arora, D.K., Rai, B., Mukerji, K.G., and Knudsen, G.R., eds.), pp. 507–540, Marcel Dekker, Inc., New York.

Boullard, B. (1988) in *Coevolution of Fungi with Plants and Animals* (Pirozynski, K.A., and Hawksworth, D.L., eds.), pp. 107–124, Academic Press, London.

Brock. T.D. (1966) *Adv. Appl. Microbiol.* 8, 61–75.

Buckland, B., Gbewonyo, K., Hallada, T., Kaplan L., and Masurekar, P. (1989) in *Novel Microbial Products for Medicine and Agriculture* (Demain, A.L., Somkuti, G.A., Hunter-Cevera, J.C., and Rossmoore, H.W., eds.), pp. 161–169, Elsevier, Amsterdam.

Bull, A.T., Goodfellow, M., and Slater, J.H. (1992) *Annu. Rev. Microbiol.* 46, 219–252.

Campbell, I.M. (1984) *Advances in Microbial Physiology* 25, pp. 1–60, Academic Press, London.

Cane, D.E., Rawlings, B.J. and Yang, C.-C. (1987) *J. Antibiot.* 40(9), 1331–1334.

Carris, L.M., Glave, D.A., Smyth, C.A., and Edwards, D.I. (1989) *Mycologia* 81(1) 66–75.

Carroll, G.C. (1986) in *Microbiology of the Phyllosphere* (Fokkema, N.J., and Van Den Heuvel, J., eds.) pp. 205–222, Cambridge University Press, Cambridge, MA.

Carroll, G.C. (1991) in *Microbial Ecology of Leaves* (Andrews, J.H., and Hirano, S.S., eds.), pp. 358–375, Springer-Verlag, New York.

Chapela, I.H. (1987) Ph.D. thesis, University of Wales, Cardiff, United Kingdom.

Chapela, I.H. (1989) *New Phytologist,* 113, 65–75.

Chapela, I.H., and Boddy, L. (1988a) *New Phytologist* 110, 39–45.

Chapela, I.H., and Boddy, L. (1988b) *New Phytologist* 110, 47–57.

Cheetham, P.S.J. (1987) *Enzyme Microb. Technol.* 9, 194–213.

Christensen, M. (1981) in *The Fungal Community* (Wicklow, D.T., and Carroll, G.C., eds.) pp. 201–232, Marcel Dekker Inc., New York.

Christensen, M. (1989) *Mycologia* 81(1), 1–19.

Clay, K. (1986) in *Microbiology of the Phyllosphere* (Fokkema, N.J., and Van Den Heuvel, J., eds.) pp. 188–204, Cambridge University Press, Cambridge, MA.

Clay, K. (1988) in *Coevolution of Fungi with Plants and Animals* (Pirozynski, K.A., and Hawksworth, D.L., eds.), pp. 79–105, Academic Press, London.

Cole, G.T., and Kendrick, B. (1981) *Biology of Conidial Fungi,* vol. 1 and 2, Academic Press, New York.

Cole, R.J., and Cox, R.H. (1981) *Handbook of Toxic Fungal Metabolites,* Academic Press, New York.

Cooke, R. (1977) in *The Biology of Symbiotic Fungi,* John Wiley & Sons, London.

Cooke, R.C., and Rayner, A.D.M. (1984) *Ecology of Saprotrophic Fungi,* Longman, London.

Cooper, R., Horan, A.C., Gunnarsson, I., Patel, M., and Truumees, I. (1986) *J. Ind. Microbiol.* 1, 275–276.

Crittenden, P.D., and Porter, N. (1991) *Trends Biotechnol.* 9(12) 409–414.

Currah, R.S., Hambleton, S., and Smreciu, A. (1988) *Am. J. Bot.* 75(5), 739–752.

Currah, R.S., Sigler, L., and Hambleton, S. (1987) *Can. J. Bot.* 65, 2473–2482.

Davies, J. (1990) *Mol. Microbiol.* 4(8), 1227–1232.

Descals, E., and Webster, J. (1982) *Trans. Br. Mycol. Soc.* 79(1), 45–64.

Dickinson, C.H. (1976) in *Microbiology of Aerial Plant Surfaces* (Dickinson, D.H., and Preece, T.F., eds.), pp. 293–324, Academic Press, London.

Dickinson, C.H., and Pugh, G.J.F., eds. (1974) *Biology of Plant Litter Decomposition,* vol. 1 and 2, Academic Press, London.

Dreyfuss, M. (1986) *Sydowia* 39, 22–36.

Edwards, C., ed. (1990) *Microbiology of Extreme Environments,* Open University Press, Milton Keynes, United Kingdom.

Endo, A., Hasumi, K., Yamada, A., Shimoda, R., and Takeshima, H. (1986) *J. Antibiot.* 39(11), 1609–1610.

Evans, H.C. (1988) in *Coevolution of Fungi with Plants and Animals* (Pirozynski, K.A., and Hawksworth, D.L., eds.), pp. 149–171, Academic Press, London.

Felix, H. (1988) *Bot. Helvet.* 98(2), 239–269.

Fisher, P.J., and Webster, J. (1981) in *The Fungal Community* (Wicklow, D.T., and Carroll, G.C., eds.), pp. 709–730, Marcel Dekker Inc., New York.

Floss, H.G., and Anderson, J.A. (1980) in *The Biosynthesis of Mycotoxins* (Steyn, P.S., ed.), pp. 17–67, Academic Press, New York.

Fokkema, N.J. (1991) in *Microbial Ecology of Leaves* (Andrews, J.H., and Hirano, S.S., eds.), pp. 3–18, Springer-Verlag, New York.

Frankland, J.C., Dighton, J., and Boddy, L. (1990) in *Methods in Microbiology,* vol. 22 (Grigorova R., and Norris, J.R., eds.), pp. 343–404, Academic Press, London.

Frisvad, J.C., and Filtenborg, O. (1989) *Mycologia* 81(6), 837–861.

Frisvad, J.C., and Filtenborg, O. (1990a) in *Modern Concepts in Penicillium and Aspergillus Systematics* (Samson, R.A., and Pitt, J.I., eds.), pp. 159–172, Plenum Press, New York.

Frisvad, J.C., and Filtenborg, O. (1990b) in *Modern Concepts in Penicillium and Aspergillus Systematics* (Samson, R.A., and Pitt, J.I., eds.), pp. 373–384, Plenum Press, New York.

Frisvad, J.C., Samson, R.A., and Stolk, A.C. (1990) in *Modern Concepts in Penicillium and Aspergillus Systematics* (Samson, R.A., and Pitt, J.I., eds.), pp. 445–454, Plenum Press, New York.

Gams, W. (1971) *Cephalosporium-artige Schimmelpilze (Hyphomycetes),* Gustav Fischer Verlag, Stuttgart, Germany.

Gams, W., Hennebert, G.L., Stalpers, J.A., et al. (1988) *J. Gen. Microbiol.* 134, 1667–1689.

Gamundi, I.J., Arambarri, A.M., and Spinedi, H.A. (1987) *Rev. Mus. La Plata* 14(92), 89–116.

Giovannoni, S.J., Britschgi, T.B., Moyer, C.L., and Field, K.G. (1990) *Nature* 345, 60–63.

Glare, T.R., and Milner, R.J. (1991) in *Handbook of Applied Mycology,* vol. 2, (Arora, D.K., Ajello, L., and Mukerji, K.G., eds.), pp. 547–612, Marcel Dekker, Inc., New York.

Gupta, S., Krasnoff, S.B., Underwood, N.L., Renwick, J.A.A., and Roberts, D.W. (1991) *Mycopathologia* 115, 185–189.

Hasija, S.K., and Singhal, P.K. (1991) in *Handbook of Applied Mycology,* vol. 1 (Arora, D.K., Rai, B., Mukerji, K.G., and Knudsen, G.R., eds.), pp. 481–505, Marcel Dekker, Inc., New York.

Hawksworth, D.L. (1981) in *Biology of Conidial Fungi,* vol. 1 (Cole, G.T., and Kendrick, B., eds.), pp. 171–244, Academic Press, New York.

Hawksworth, D.L. (1988) in *Coevolution of Fungi with Plants and Animals* (Pirozynski, K.A., and Hawksworth, D.L., eds.), pp. 125–148, Academic Press, London.

Hawksworth, D.L. (1991) *Mycol. Res.* 95(6), 641-655.

Hawksworth, D.L., and Kirsop, B.E. (1988) *Living Resources for Biotechnology, Filamentous Fungi,* Cambridge University Press, Cambridge, MA.

Hedge, V.R., Wittreich, H., Patel, M.G., et al. (1989) *J. Ind. Microbiol.* 4, 209–214.

Higgens, C.E., Hamill, R.L., Sands, T.H., et al. (1974) *J. Antibiot.* 27(4), 298–300.

Hijwegen, T. (1988) in *Coevolution of Fungi with Plants and Animals* (Pirozynski, K.A., and Hawksworth, D.L., eds.), pp. 63–77, Academic Press, London.

Honegger, R., and Bartnicki-Garcia, S. (1991) *Mycol. Res.* 95(8), 905–914.

Ingold, C.T. (1975) *An Illustrated Guide to Aquatic and Water-Borne Hyphomycetes,* Freshwater Biological Association, Scientific Publication no. 30, The Ferry House, Ambleside, Cambria.

Ingold, C.T. (1976) in *Recent Advances in Aquatic Mycology* (Jones, E.B.G., ed.), pp. 335–357, Elek Science, London.

Jarvis, B.B., Midiwo, J.O., Bean, G.A., Bassam Aboul-Nasr, M., and Barros, C.S. (1988) *J. Nat. Prod.* 51(4), 736–744.

Jegorov, A., Matha, V., and Weiser, J. (1990) *Microb. Lett.* 45, 65–69.

Johnston, C.G., and Vestal, J.R. (1991) *Appl. Environ. Microbiol.* 57(8), 2308–2311.

Jones, E.B.G. (1974) in *Biology of Plant Litter Decomposition,* vol. 2 (Dickinson, C.H., and Pugh, G.J.F., eds.), pp. 337–383, Academic Press, London.

Jones, E.B.G. (1981) in *The Fungal Community* (Wicklow, D.T., and Carroll, G.C., eds.), pp. 731–743, Marcel Dekker Inc., New York.

Jones, R.J., Sizmur, K.J., and Wildman, H.G. (1991) *The Mycologist* 5(4), 184–185.

Jong, S.C. (1981) in *Biology of Conidial Fungi,* vol. 2 (Cole, G.T., and Kendrick, B., eds.), pp. 551–575, Academic Press, New York.

Kimura, Y., Fujioka, H., Nakajima, H., Hamasaki, T., and Isogai, A. (1988) *Agric. Biol. Chem.* 52, 1845–1847.

Kinkel, L. (1991) in *Microbial Ecology of Leaves* (Andrews, J.H., and Hirano, S.S., eds.), pp. 253–270, Springer-Verlag, New York.

Kirsop, B.E., and Doyle, A., eds. (1991) *Maintenance of Microorganisms and Cultured Cells,* 2nd ed., Academic Press, London.

Kitano, K. (1983) *Prog. Ind. Microbiol.* 17, 37–69.

Kjøller, A., and Struwe, S. (1987) *Pedobiologia* 30, 151–159.

Kobel, H., and Sanglier, J.-J. (1986) in *Biotechnology,* vol. 4 (Rehm, H.-J., and Reed, G., eds.), pp. 569–609, VCH, Weinheim, Germany.

Kohlmeyer, J. (1981) in *Biology of Conidial Fungi,* vol. 1 (Cole, G.T., and Kendrick, B., eds.), pp. 357–372, Academic Press, New York.

Kohlmeyer, J., and Kohlmeyer, E. (1979) *Marine Mycology,* Academic Press, New York.

Koshino, H., Togiya, S., Terada, S., et al. (1989) *Agric. Biol. Chem.* 53(3), 789–796.

Labeda, D.P., ed. (1990) *Isolation of Biotechnological Organisms from Nature,* McGraw-Hill Publishing Company, New York.

Labeda, D.P., and Shearer, M.C., (1990) in *Isolation of Biotechnological Organisms from Nature* (Labeda, D.P., ed.), pp. 1–19, McGraw-Hill Publishing Company, New York.

Lawen, A., Traber, R., and Geyl, D. (1991) *J. Biol. Chem.* 266(24), 15567–15570.

Lechevalier, H.A. (1975) *Adv. Appl. Microbiol.* 19, 25–45.

Lichtwardt, R.W. (1986) *The Trichomycetes,* pp. 1–360, Springer-Verlag, New York.

Loeffler, W. (1984) *Forum Mikrobiol.* 7, 219–229.

Luckner, M. (1990) *Secondary Metabolism in Microorganisms, Plants, and Animals,* 3rd ed., Springer-Verlag, Berlin.

Lumsden, R.D. (1981) in *The Fungal Community* (Wicklow, D.T., and Carroll, G.C., eds.), pp. 295–318, Marcel Dekker Inc., New York.

Macauley, B.J., and Waid, J.S. (1981) in *The Fungal Community* (Wicklow, D.T., and Carroll, G.C., eds.), pp. 501–531, Marcel Dekker Inc., New York.

Marakis, S., and Diamantoglou, S. (1990) *Cryptogamie, Mycol.* 11(4), 243–254.

Mikawa, T., Takahashi, N., Ohkishi, H., et al. (1988) European Patent Application No. 88101493.0.

Mislivec, P.B. (1981) in *Biology of Conidial Fungi,* vol. 2 (Cole, G.T., and Kendrick, B., eds.), pp. 38–74, Academic Press, New York.

Mosse, B., Stribley, D.P., and LeTacon, F. (1981) in *Advances in Microbial Ecology,* vol. 5 (Alexander, M., ed.), pp. 137–210, Plenum Press, New York.

Nakajima, H., Hamasaki, T., Tanaka, K., et al. (1989) *J. Agric. Biol. Chem.* 53(8), 2291–2292.

Neidleman, S. (1991) in *Biotechnology* (Moses, V., and Cape, R.E., eds.), pp. 297–310, Harwood Academic Publishers, Chur, Switzerland.

Nelson, P.E., Plattner, R.D., Shackelford, D.D., and Desjardins, A.E. (1991) *Appl. Environ. Microbiol.* 57(8), 2410–2412.

Newman, D.J., Jensen, P.R., Clement, J.J., and Acebal, C. (1989) in *Novel Microbial Products for Medicine and Agriculture* (Demain, A.L., Somkuti, G.A., Hunter-Cevera, J.C., and Rossmoore, H.W., eds.), pp. 239–251, Elsevier, Amsterdam.

Noble, H.M., Langley, D., Sidebottom, P.J., Lane, S.J., and Fisher, P.J. (1991) *Mycol. Res.* 95(12), 1439–1440.

Norris, J.R., Read, D.J., and Varma, A.K., eds. (1991) *Methods in Microbiology,* vol. 23, pp. 1–480, Academic Press, London.

Norris, J.R., Read, D.J., and Varma, A.K., eds. (1992) *Methods in Microbiology,* vol. 24, Academic Press, London.

Olsen, G.J. (1990) *Nature* 345, 20.

Omura, S. (1988) in *Biology of Actinomycetes '88* (Okami, O., Beppu, T., and Ogawara, H., eds.), pp. 26–31, Japan Scientific Societies Press, Tokyo.

Parkinson, D. (1981) in *Biology of Conidial Fungi,* vol. 1 (Cole, G.T., and Kendrick, B., eds.), pp. 277–294, Academic Press, New York.

Paulitz, T.C., and Linderman, R.G. (1991) in *Handbook of Applied Mycology,* vol. 1 (Arora, D.K., Rai, B., Mukerji, K.G., and Knudsen, G.R., eds.), pp. 77–129, Marcel Dekker, Inc., New York.

Petrini, O. (1986) in *Microbiology of the Phyllosphere* (Fokkema, N.J., and Van Den Heuvel, J., eds.) pp. 175–187, Cambridge University Press, Cambridge.

Petrini, O. (1991) in *Microbial Ecology of Leaves* (Andrews, J.H., and Hirano, S.S., eds.), pp. 179–197, Springer-Verlag, New York.

Petrini, O., Hake, U., and Dreyfuss, M.M. (1990) *Mycologia* 82(4), 444–451.

Petrini, O., Petrini, L.E., and Dreyfuss, M.M. (1992) *Mycologia Helvetica* 5, 9–20.

Phaff, H.J. (1990) in *Isolation of Biotechnological Organisms from Nature* (Labeda, D.P., ed.), pp. 53–79, McGraw-Hill Publishing Company, New York.

Pitt, J.I. (1979) *The Genus Penicillium and its Teleomorphic States Eupenicillium and Talaromyces.* Academic Press, London.

Pitt, J.I., and Samson, R.A. (1990) *Stud. Mycol. (Baarn)* 32, 77–91.

Poch, G.K., and Gloer, J.B. (1989) *J. Nat. Prod.* 52(2), 257–260.

Porter, N. (1985) *Pestic. Sci.* 16, 422.

Provost, J., and Garcia, M. (1990) *Planta Med.* 56, 647.

Pugh, G.J.F. (1974) in *Biology of Plant Litter Decomposition,* vol. 2 (Dickinson, C.H., and Pugh, G.J.F., eds.), pp. 303–336, Academic Press, London.

Pugh, G.J.F. (1980) *Trans. Br. Mycol. Soc.* 75(1), 1–14.

Purchase, I.F.H. (ed.) (1974) *Mycotoxins,* Elsevier, Amsterdam.

Rayner, A.D.M., and Boddy, L. (1988) *Fungal Decomposition of Wood,* John Wiley & Sons, Chichester.

Rehm, H.-J. (1980) *Industrielle Mikrobiologie,* 2nd ed., Springer-Verlag, Berlin.

Reichenbach, H., and Höfle, G. (1989) in *Bioactive Metabolites from Microorganisms; Progress in Industrial Microbiology,* vol. 27 (Bushell, M.E., and Gräfe, U., eds.), pp. 79–100, Elsevier, Amsterdam.

Roberts, D.W., and Humber, R.A. (1981) in *Biology of Conidial Fungi,* vol. 2 (Cole, G.T., and Kendrick, B., eds.), pp. 201–236, Academic Press, New York.

Samuels, G.J. (1988) *Mem. N.Y. Bot. Gard.* 48, 1–78.

Samson, R.A., and Pitt, J.I., eds. (1990) *Modern Concepts in Penicillium and Aspergillus Systematics,* Plenum Press, New York.

Satoshi, N., Koji, Y., Katsuhiko, A. et al. (1988) European Patent Application No. 88105712.9.

Satyanarayana, T., Johri, B.N., and Klein, J. (1992) in *Handbook of Applied Mycology,* vol. 4 (Arora, D.K., Elander, R.P., and Mukerji, K.G., eds.), pp. 729–761, Marcel Dekker, Inc., New York.

Sawai, K., Okuno, T., Terada, Y., et al. (1981) *Agric. Biol. Chem.* 45(5), 1223–1228.

Saxena, G., Mittal, N., Mukerji, K.G., and Arora, D.K. (1991) in *Handbook of Applied Mycology,* vol. 2 (Arora, D.K., Ajello, L., and Mukerji, K.G., eds.), pp. 707–733, Marcel Dekker, Inc., New York.

Schwartz, R.E., Giacobbe, R.A., Bland, J.A., and Monaghan, L. (1989) *J. Antibiot.* 42(2), 163–167.

Sharma, H.S.S., and Johri, B.N. (1992) in *Handbook of Applied Mycology,* vol. 4

(Arora, D.K., Elander, R.P., and Mukerji, K.G., eds.), pp. 707–728, Marcel Dekker, Inc., New York.

Shibata, T., Nakayama, O., Okuhara, M., Tsurumi, Y., Terano, H., and Kohsaka, M. (1988) *J. Antibiot.* 41(9), 1163–1169.

Siegel, M.R., and Schardl, C.L. (1991) in *Microbial Ecology of Leaves* (Andrews, J.H., and Hirano, S.S., eds.), pp. 198–221, Springer-Verlag, New York.

Starmer, W.T. (1981) in *The Fungal Community* (Wicklow, D.T., and Carroll, G.C., eds.), pp. 129–156, Marcel Dekker Inc., New York.

Starmer, W.T., Fogleman, J.C., and Lachance, M-A. (1991) in *Microbial Ecology of Leaves* (Andrews, J.H., and Hirano, S.S., eds.), pp. 158–178, Springer-Verlag, New York.

States, J.S. (1981) in *The Fungal Community* (Wicklow, D.T., and Carroll, G.C., eds.), pp. 185–199, Marcel Dekker, Inc., New York.

Steele, D.B., and Stowers, M.D. (1991) *Annu. Rev. Microbiol.* 45, 89–106.

Stolk, A.C., Samson, R.A., Frisvad, J.C., and Filtenborg, O. (1990) in *Modern Concepts in Penicillium and Aspergillus Systematics* (Samson, R.A., and Pitt, J.I., eds.), pp. 121–137, Plenum Press, New York.

Suberkropp, K. (1984) *Trans. Br. Mycol. Soc.* 82(1), 53–62.

Suberkropp, K. (1991) *Mycol. Res.* 95(7), 843–850.

Suberkropp, K., and Klug, M.J. (1981) in *The Fungal Community* (Wicklow, D.T., and Carroll, G.C., eds.) pp. 761–776, Marcel Dekker Inc., New York.

Sugawara, F., Strobel, S., and Strobel, G., et al. (1991) *J. Org. Chem.* 56, 909–910.

Sugita, K., Itazaki, H., Matsumoto, K., and Kawamura, Y. (1990) European Patent Application No. 90112472.7.

Suvercha, F., Mukerji, K.G., and Arora, D.K. (1991) in *Handbook of Applied Mycology,* vol. 1 (Arora, D.K., Rai, B., Mukerji, K.G., and Knudsen, G.R., eds.), pp. 187–215, Marcel Dekker, Inc., New York.

Taylor-George, S., Palmer, F., Staley, J.T., et al. (1983) *Microb. Ecol.* 9, 227–245.

Thrane, U. (1990) *J. Microb. Methods* 12, 23–39.

Turner, W.B. (1971) *Fungal Metabolites,* Academic Press, London.

Turner, W.B., and Aldridge, D.C. (1983) *Fungal Metabolites II,* Academic Press, London.

Tylka, G.L., Hussey, R.S., and Roncadori, R.W. (1991) *Phytopathology* 81, 754–759.

Vining, L.C. (1986) in *Biotechnology,* vol. 4 (Rehm, H.-J., and Reed, G., eds.), pp. 19–38, VCH, Weinheim, Germany.

Vining, L.C. (1990) *Annu. Rev. Microbiol.* 44, 395–427.

Von Arx, J.A. (1981) *The Genera of Fungi Sporulating in Pure Culture,* J. Cramer, Vaduz, FL.

Warcup, J.H. (1991) *Mycol. Res.* 95(6), 656–659.

Ward, D.M., Weller, R., and Bateson, M.M. (1990) *Nature* 345, 63–65.

Webster, J., and Descals, E. (1981) in *Biology of Conidial Fungi,* vol. 1 (Cole, G.T. and Kendrick, B., eds.), pp. 295–355, Academic Press, New York.

Webster, J. (1981) in *The Fungal Community* (Wicklow, D.T., and Carroll, G.C., eds.) pp. 681–691, Marcel Dekker, Inc., New York.

Weiser, J., Matha, V., and Jegorov, A. (1991) *Folia Parasitol.* 38, 363–369.

Wharton, R.A., Jr., McKay, C.P., Simmons, G.M., Jr., and Parker, B.C. (1985) *BioScience* 35(8), 499–503.

Wicklow, D.T. (1981a) in *Biology of Conidial Fungi,* vol. 1 (Cole, G.T. and Kendrick, B., eds.), pp. 417–447, Academic Press, New York.

Wicklow, D.T. (1981b) in *The Fungal Community* (Wicklow, D.T., and Carroll, G.C., eds.), pp. 47–76, Marcel Dekker Inc., New York.

Widden, P. (1979) *Can. J. Bot.* 57, 1324–1331.

Widden, P. (1986a) *Can. J. Bot.* 64, 1402–1412.

Widden, P. (1986b) *Can. J. Bot.* 64, 1413–1423.

Widden, P. (1986c) *Can. J. Bot.* 64, 1424–1432.

Widden, P. (1987) *Mycologia* 79(2), 298–309.

Wildman, H.G. (1991) *Mycol. Res.* 95(12), 1364–1368.

Williams, S.T., and Vickers, J.C. (1988) in *Biology of Actinomycetes '88* (Okami, O., Beppu, T., and Ogawara, H., eds.), pp. 265–270, Japan Scientific Societies Press, Tokyo.

Wood-Eggenschwiler, S., and Bärlocher, F. (1983) *Trans. Br. Mycol. Soc.* 81(2), 371–379.

Worrall, J.J. (1991) *Mycologia* 83(3), 296–302.

Zähner, H. (1982) in *Handbuch der Biotechnologie* (Präve, P., Faust, U., Sittig, W., and Sukatsch, D.A., eds.), pp. 83–103, Akademische Verlagsgesellschaft, Wiesbaden, Germany.

Zähner, H., Anke, H., and Anke, T. (1983) in *Secondary Metabolism and Differentiation in Fungi* (Bennett, J.W., and Ciegler, A., eds.), pp. 153–171, Marcel Dekker, Inc., New York.

The Discovery of Drugs from Higher Plants

A. Douglas Kinghorn

Although naturally occurring drugs may be obtained from plants, animals, or microorganisms, the natural products with the broadest range of therapeutic application are currently obtained from the plant kingdom. The isolation of the first plant drug dates back almost to the beginning of organic chemistry itself, when the potent analgesic morphine was isolated from the dried latex of the Opium poppy (*Papaver somniferum*) by Serturner in 1805 (Hite 1989). Other important drugs obtained from plants in the nineteenth century were atropine, cocaine, codeine, digitoxin, papaverine, and pilocarpine, and in the twentieth century have been digoxin, ergotamine, ergometrine, reserpine, and vincristine (Baerheim Svendsen and Scheffer 1982). Efforts to discover and develop further important drugs from plant sources have continued up to the present day, as evidenced by the current intense interest in producing larger amounts of the diterpene alkaloid taxol for more extensive clinical trials. Taxol, found as a minor constituent of the bark of the Pacific or Western Yew tree (*Taxus brevifolia*), has shown excellent results in clinical trials on patients with ovarian and breast cancer (Borman 1991).

In this chapter, the importance of plant-derived drugs in western industrialized countries will be briefly described, as will the potential for wider use of compounds obtained from Chinese herbal medicine and other well-developed systems of traditional medicine. In addition, mention will be made of the role that plant secondary metabolites have played in the past as lead compounds for drug synthesis. The approaches that can be taken to discover new prototype biologically active molecules from plants will then be discussed. It is apparent that drug discovery from plants is hindered by number of problems common to all such programs dealing with organisms, including the often restricted supply of resource material, the biological variation of different batches of collected plant material, and the need to rapidly dereplicate active compounds of known structure. Two current concerns particularly relevant to plant drug discovery efforts, however, are the alarming depletion of the world's tropical rain forests and a diminishing knowledge-base about medicinal plants among indigenous peoples in developing countries. Recent progress will be pointed out for some promising drug candidates of plant origin, as exemplified by compounds with antineoplastic, antimalarial, and anti-human immunodeficiency virus (HIV) activities. Finally, future prospects for the discovery of new plant drugs will be discussed.

4.1   IMPORTANCE OF PLANT-DERIVED DRUGS

4.1.1   Plant-Derived Drugs in Western Medicine

Plant secondary metabolites, unlike their precursor primary metabolites, such as fatty acids and amino acids, do not have any apparent metabolic functions, and are of restricted taxonomic distribution. The major role for plant secondary metabolites appears to be ecological, and therefore they are produced in the correct chiral form to serve such functions as chemical defenses against insects, microorganisms, and other predators, or as pollinator attractants. Secondary metabolites tend to be biosynthesized in specialized cell types, at only some of the life stages of the plant, and usually accumulate in much lower quantities than primary metabolites (Balandrin et al. 1985). Among the plant secondary metabolites used as therapeutic agents are alkaloids, anthraquinone glycosides, cardiac glycosides, and lignans, whose biosynthetic precursors are, respectively, amino acids, polyketides, isoprenoids, and shikimate-derived compounds (Balandrin and Klocke 1988; Tyler et al. 1988; Kinghorn 1992). Table 4–1 lists a selection of plant-derived drugs currently of use in the United States.

Farnsworth and Morris (1976) analyzed National Prescription Audit data over the 15-year period 1959 through 1973, and determined that prescriptions containing natural products represented a rather stable market in the United States. Thus, for the year 1973, it was calculated that over 25% of U.S. prescriptions dispensed contained one or more active constituents derived from higher plants, either in the form of pure compounds, such as the alkaloids, atropine, codeine, hyoscyamine, and pilocarpine, and the cardiac glycoside, digoxin, or as "extractives," which contain crude mixtures of compounds, such as the alkaloid-containing tinctures of Belladonna, Ipecac,

TABLE 4–1   Examples of Drugs Used in the United States that Are Obtained from Higher Plants[1]

Drug Name	Compound Class	Clinical Use
Atropine	Tropane alkaloid	Anticholinergic
Cocaine	Tropane alkaloid	Local anesthetic
Codeine	Phenanthrene alkaloid	Analgesic, antitussive
Digoxin	Steroidal glycoside	Cardiotonic
Emetine	Isoquinoline alkaloid	Antiamebic agent
Ephedrine	Phenethylamine derivative	Bronchodilator
Morphine	Phenanthrene alkaloid	Analgesic
Papaverine	Isoquinoline alkaloid	Smooth muscle relaxant
Physostigmine	Indole alkaloid	Ophthalmic cholinergic
Pilocarpine	Imidazole alkaloid	Ophthalmic cholinergic
Quinidine	Quinoline alkaloid	Cardiac depressant
Quinine	Quinoline alkaloid	Antimalarial
Reserpine	Indole alkaloid	Antihypertensive
Sennosides A and B	Anthraquinone glycosides	Laxative
Tubocurarine	Isoquinoline alkaloid	Skeletal muscle relaxant
Vinblastine	Indole alkaloid	Antineoplastic
Vincristine	Indole alkaloid	Antineoplastic

[1] Information taken from Tyler et al. 1988.

Opium, and Rauwolfia. In contrast, U.S. prescription products derived from microorganisms and animals constituted 13.3% and 2.7% of the total, respectively (Farnsworth and Morris 1976). More recently, Principe (1989) has estimated that for 1980, the total value of prescriptions containing products derived from higher plants was $8 billion in the United States.

In addition to the use of established plant-derived drugs, like those mentioned thus far, there is currently a large phytomedicine market in Western European countries, such as Germany and France, inclusive of extracts prepared from *Crataegus oxycantha* (cardiotonic), *Silybum marianum* (antihepatotoxic), and *Valeriana mexicana* (sedative) (Tyler 1986; Tyler et al. 1988). In 1988, the turnover of this European phytomedicine market was estimated at U.S. $2.2 billion, or about 5% of the total pharmaceutical market, with the largest share being taken by *Ginkgo biloba* leaf extracts (about $500 million in value; used for circulatory problems) (O. Sticher, personal communication, 1991). Extracts of ginkgo leaves contain a series of diterpenes called ginkgolides, which have been found to be specific inhibitors of platelet aggregation induced by platelet-activating factor (PAF) (Hamburger et al. 1991). It has been pointed out that official validation of the effectiveness and safety of plant drugs is far easier to obtain in Germany than in the United States, and, for example, does not require the conduct of extensive clinical trials (Tyler 1986; Abelson 1990).

4.1.2 Plant-Derived Drugs from Chinese Herbal Remedies and Other Systems of Traditional Medicine

For thousands of years, plant-derived (herbal) remedies have remained a vital part of traditional Chinese medicine, and even today constitute about a 30% to 50% proportion of the total drug therapy for a fifth of the world's population who live in the People's Republic of China (PRC) (Xiao and Chen 1987). Of about 5500 medicinal plants used in traditional Chinese medicine, between 300 and 500 are commonly used in regular prescriptions (Han et al. 1988). One drug that has been in use in China for at least 5000 years is *Ephedra sinica* (*Ma huang*), from which the potent sympathomimetic amine ephedrine was isolated and pharmacologically tested in the early years of this century, and is now used in western medicine in the form of various salts to combat bronchial asthma (Wang and Liu 1985; Tyler et al. 1988). Recent phytochemical investigations on plants used in Chinese traditional medicine, both in the PRC and elsewhere, have led to the discovery of several hundred pharmacologically active substances, with about 60 new drugs being derived from such compounds (Xiao and Fu 1987).

Examples of plant drugs that are of use in the PRC include the tropane alkaloids anisodamine (Figure 4–1A) and anisodine (Figure 4–1B) from

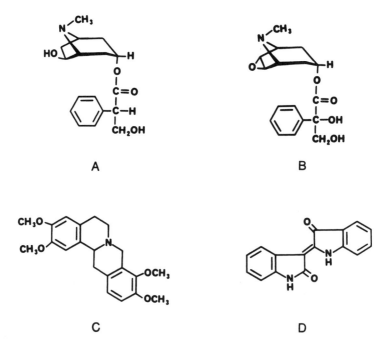

FIGURE 4–1 Structures of four drugs isolated from plants used in traditional Chinese medicine: (A) anisodamine, (B) anisodine, (C) *dl*-tetrahydropalmatine, and (D) indirubrin.

Scopolia tangutica, which are employed as a mild, centrally acting anticholinergic agent for septic shock in cases of bacillary dysentery, and as a migraine treatment, respectively. An isoquinoline alkaloid, racemic tetrahydropalmatine (Figure 4–1C), from *Corydalis ambigua* is used as an analgesic and tranquilizer, and indirubrin (Figure 4–1D), a nitrogen-containing metabolite produced by *Indigofera tinctoria*, is effective in the treatment of chronic myelocytic leukemia (Wang and Liu 1985; Xiao and Chen 1987, 1988; Xiao and Fu 1987; Han et al. 1988; Han 1988). Additional examples of very promising drug candidates obtained from plants used in Chinese traditional medicine will be provided in subsequent sections of this chapter.

Well-developed systems of traditional medicine also exist in India and Indonesia, as exemplified by "Ayurveda" and "Jamu," respectively. An excellent example of how a useful western drug came to be derived from a plant used indigenously in India is reserpine, an indole alkaloid constituent of *Rauwolfia serpentina*, with antihypertensive and tranquilizing properties. This plant has been used by healers in India for centuries to treat many disease conditions, including insanity (Tyler et al. 1988). Forskolin and swietemahonin D are examples of two interesting biologically active compounds that were recently isolated and characterized from Indian and Indonesian medicinal plants, respectively. Forskolin (Figure 4–2A), is a highly oxygenated labdane diterpenoid that was isolated from the roots of the Indian plant, *Coleus forskohlii*, and, as an adenylate cyclase activating agent, has been found to exhibit antihypertensive, antithrombotic, bronchospasmolytic, and positive inotropic activity (de Souza and Shah 1988). A series of tetranortriterpenoids obtained from *Swietenia mahagoni* collected in Sumatra, Indonesia, has been demonstrated as having potent inhibitory activity against PAF-induced aggregation in both in vitro and in vivo bioassays. One of the active compounds obtained was swietemahonin D (Figure 4–2B), which showed roughly comparable activity to a standard drug, verapamil (Ekimoto et al. 1991).

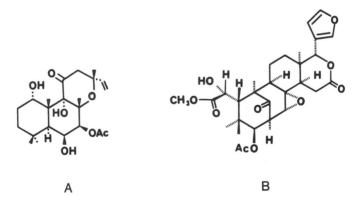

A   B

FIGURE 4–2 Structures of (A) forskolin and (B) swietemahonin D, obtained from an Indian and an Indonesian medicinal plant, respectively.

4.1.3 Plant Constituents as Lead Compounds for Synthetic Modification

Plant secondary metabolites are not only useful as drugs or potential drugs in their natural unmodified form, but are also suitable as synthetic intermediate substances for the production of useful drugs. For example, the readily available plant steroid, diosgenin from *Dioscorea* spp. (Mexican yams), may be synthetically converted to steroids with anabolic, anti-inflammatory, and oral contraceptive activities (Briggs 1990). Other plant sterols that can be used in this manner are hecogenin, stigmasterol, and β-sitosterol, as well as the steroidal alkaloid, solasodine (Galeffi and Marini Bettolo 1988).

Whereas only the more structurally simple plant drugs such as ephedrine and pseudoephedrine are prepared commercially by total synthesis (Farnsworth and Morris 1976; Balandrin and Klocke 1988), plant constituents, along with other natural products, are often extremely useful as "lead" compounds for the synthetic design of analogues with either improved therapeutic activity or reduced toxicity. Of a total of about 1300 therapeutic agents listed in the British National Formulary of 1988, plants, animals, and microorganisms, respectively, accounted for 21, 34, and 29 of a total of 151 lead sources (Midgley 1988).

Of all plant secondary metabolites, probably the one that has been used as a lead compound to the greatest extent is morphine, the principal narcotic analgesic alkaloid of the Opium poppy, *Papaver somniferum*. Modification of the morphine molecule, in the attempt to produce derivatives that have a similar level of analgesic potency but not the same addiction tendency, has led to a plethora of clinically used analgesics, such as levophanol and pentazocine (Hite 1989). Other examples of plant compounds that have served as lead compounds for synthetic drugs are the tropane alkaloid, cocaine, the prototype local anesthetic that led to the development of the more effective procaine (Sneader 1985), and the isoquinoline alkaloid, petaline, the lability of which was utilized in the design of the neuromuscular blocking agent, atracurium besylate, which under physiological conditions rapidly degrades to nontoxic compounds (Sneader 1985; Midgley 1988). An example of a nonalkaloidal plant-derived lead compound is podophyllotoxin (Figure 4–3A), which is a lignan from the American mandrake (*Podophyllum peltatum*) and Indian podophyllum (*P. hexandrum*), and has served as a prototype antineoplastic agent for the development of the less toxic and more water-soluble compound, etoposide (Figure 4–3B; 4'-demethylepipodophyllotoxin-β-D-ethylidine glucoside). In the United States, etoposide is used clinically for the treatment of small-cell lung cancer and testicular cancer (Graham and Chandler 1990).

Occasionally, the analogs of a lead compound may exhibit unexpected biological activity, and then can be modified synthetically even further. For example, the synthetic compound meperidine (pethidine) was found to have analgesic activity, which was discovered during screening for a series of compounds intended as anticholinergic agents based on the plant-derived tropane

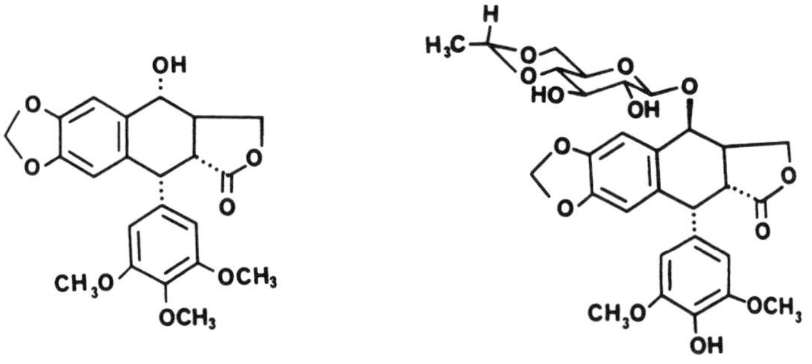

FIGURE 4–3 Structures of the plant-derived lead compound (A) podophyllotoxin and its semisynthetic analog (B) etoposide.

alkaloid, atropine (Midgley 1988; Hite 1989). Structural modification of the meperidine molecule led in turn to the haloperidol series of neuroleptics (Midgley 1988; Soudijn 1991).

4.2 APPROACHES TO THE DISCOVERY OF DRUGS FROM PLANT SOURCES

4.2.1 General Considerations for Selection of Plant Material

Although serendipity has played a distinct role in drug discovery from plants, probably the most crucial aspect in this type of endeavor is the selection of plant material. Theoretically, there would seem to be plenty of scope, since it has been estimated that no scientific studies whatsoever have been performed on 90% of the world's angiosperms (flowering plants), which together with the gymnosperms constitute the higher plants (Spermatophytes) (Schultes 1972). It is accepted that there are at least 250,000 angiosperms and about 700 gymnosperm species (Schultes 1972). However, improper selection of source material can be not only costly but also deleterious to the morale of people involved in plant drug screening efforts. It is pertinent here to review the general approaches that have been used historically for candidate plant selection, namely, general phytochemical screening, random plant selection followed by bioassay, and the follow-up of existing literature reports of the biological activity of plant extracts. In addition, selection of plants using ethnopharmacological leads has come into prominence in the last few years, and will be dealt with separately. To conclude this section of the chapter, some practical aspects involved in a contemporary plant drug discovery program will be reviewed.

Plant extracts containing compounds of a desired structural type (e.g., alkaloids) may be targeted by the application of simple color or chromatographic tests (Farnsworth 1968). However, it does not seem as if this type of approach has ever yielded any benefit in terms of drug discovery (Farnsworth 1984). An analogous approach is chemotaxonomy, which relies upon the fact that taxonomically related plants often biosynthesize chemically similar secondary metabolites. The discovery of digoxin from *Digitalis lanata* is a good example of how a chemotaxonomic approach can lead to a new drug, since this species was investigated phytochemically, based on the known existence of cardiac glycosides, such as digitoxin, in another member of the same genus, *D. purpurea* (Farnsworth et al. 1985). The chemotaxonomic approach is potentially useful in that a higher yield of a given compound of interest can often be found in, for example, other members of the same genus of a species under consideration (Anonymous 1989).

Plant-derived drug discovery programs have been set up in the past, wherein plants have been randomly collected and priorities have then been set after monitoring in one or more biological assays. Undoubtedly, the most extensive program of this type was that organized by the National Cancer Institute (NCI, formerly the Cancer Chemotherapy National Service Center), Bethesda, MD, which ran from 1955 through 1981. In this program, about 114,000 plant extracts were screened using various in vitro cytotoxicity and in vivo antineoplastic assays, with about 4.3% (~1500 genera, ~3400 species) regarded as "confirmed actives." Toward the end of this program, plant taxa that had not previously been examined were collected selectively, and the P-388 murine lymphocytic leukemia assay in vivo was the primary bioassay employed (Suffness and Douros 1982). Among the many structurally diverse antineoplastic compounds discovered as a result of this screening program was taxol, which is still in clinical development and offers much promise against certain types of cancer (Suffness and Douros 1982; Farnsworth 1990; Borman 1991).

There are a large number of literature reports on the biological screening of plant extracts, in which the nature of the active compound or compounds responsible for the observed activity is (are) never determined subsequently by activity-guided fractionation and spectroscopic and/or chemical characterization (Anonymous 1989). Assuming that the taxonomic and biological aspects of such preliminary studies are conducted in a scientific manner, then this type of lead can be very valuable for use by others in later drug discovery programs.

4.2.2 Ethnopharmacological Approach to Plant Selection

Schultes has pointed out that after the Second World War, certain scientists in western countries developed a somewhat cynical attitude to both folk remedies and drugs of plant origin (Schultes 1972). The attitude of some medicinal chemists to the use of folklore data on plants in drug discovery efforts is still not entirely cordial (Korolkovas 1988). However, in recent years, there has been what can only be described as a dramatic upsurge in interest in *ethnopharmacology*. This term refers to the interdisciplinary scientific observation, description, and experimental investigation of indigenous drugs and their biological activities (Rivier and Bruhn 1979). Thus, ethnobotanical information about taxa used as medicinal plants in developing countries, in particular, may be factored into the plant selection process. According to Cox, three features are desirable for the useful ethnopharmacological selection of plants; namely, the local population under study should (1) have a stable history of use of the same medicinal plants for many generations, (2) reside in an area with a diverse flora, and (3) have lived in the same geographical area for many years (Cox 1990). Clearly, the chances of success in advantageously using this type of information is greatly augmented if the same medicinal plant is used for the same purpose by two or more indigenous cultures separated from one another geographically. Examples of important plant-derived drugs that have a folkloric origin are digitoxin (Aronson 1987) and tubocurarine (Locock 1988).

Recent interest in the use of ethnopharmacological information in plant drug discovery has greatly increased, for several reasons. First, at the beginning of the 1990s, there has been greater cross-cultural interest in the practices of native peoples than was formerly the case (de Smet and Rivier 1989), especially with the continuity of these practices currently being threatened. Second, there have been at least two key publications pointing to the effectiveness of an ethnopharmacological approach in plant selection in drug discovery programs. Thus, Farnsworth and colleagues showed that of 119 important plant-derived drugs used in one or more countries, 88 (77%) were regarded as having been discovered as a result of being derived from a plant used in traditional medicine (Farnsworth et al. 1985). Another significant paper came from Spjut and Perdue, who concluded that there is a greater chance of finding extracts with inhibitory activity against experimental tumor systems from plants with reputed medicinal or toxic attributes, rather than from plants collected at random (Spjut and Perdue 1976). In addition, several additional contributions have appeared in the literature recently that report on the value of using an ethnopharmacological approach to the discovery of natural product drugs from plant sources (e.g., Farnsworth and Kaas 1981; Labadie 1986; Steiner 1986; Hosler and Mikita 1987; Phillipson and Anderson 1989; Kinghorn 1992). In 1990, the International Society of Ethnopharmacology was founded at a meeting in Strasbourg, France, which also featured a series of excellent papers dealing with the role of ethnopharmacology and the development of various categories of drugs, inclusive of anticancer agents (Cordell et al. 1991), antimalarial agents (Phillipson and Wright 1991), antiviral agents (Vlietinck and Vanden Berghe 1991), and natural PAF antagonists (Braquet and Hosford 1991). Two international scientific journals were instituted recently that focus entirely or in great part on ethnopharmacology, namely, the *Journal of Ethnopharmacology* (founded in 1979) and *Phytotherapy Research* (founded in 1987). Consequently, through these journals and via several more established journals that publish research articles on natural products chemistry, pharmacognosy, and economic botany, there is an ample scientific forum for publishing research on ethnopharmacological observations and conclusions.

It is probably too early to say how this increased awareness of ethnopharmacology has affected plant drug discovery programs in terms of generating lead molecules. Certainly, this is not a fail-safe method for finding positive leads, since uses in traditional medicine for a given plant part may be many and varied. Also, a species of interest may not be administered to patients as a single therapeutic agent. It has been pointed out that knowledge of the local language used by a healer is necessary to prevent misunderstandings about the uses of plant drug remedies (Cox 1990). However, two examples of biologically active compounds reported as a result of recent collaborative ethnopharmacological studies may be presented. Thus, Lloyd, Fales, and colleagues isolated pyrrole-3-carboxamidine (Figure 4–4A), from the Amazonian plant, *Brunfelsia grandiflora* ssp. *schultesii*, which was shown to exhibit convulsant effects in mice. Water extracts of this species are employed as a medicine (for treating rheumatism and fevers), as well as a narcotic and fish poison (Lloyd et al. 1985). In our laboratory, a novel highly sweet labdane diterpenoid, gaudichaudioside A (Figure 4–4B), was identified as a constituent of *Baccharis gaudichaudiana*, a plant sold as an herbal remedy in a medicinal plants market in Asuncíon, Paraguay, under the name *chilca melosa*, which is suggestive of its sweetness (Fullas et al. 1991). Highly sweet compounds like gaudichaudioside A have potential use as formulation excipients in medicines to mask the taste of bitter drugs (Kinghorn and Compadre 1985).

In recent years, one aspect of considerable practical and ethical importance to emerge concerning ethnopharmacological explorations is that both indigenous peoples and their governments should be duly recompensed in the form of royalties, should any commercial benefit result from the discovery of a drug or lead substance from one of their native plants (Cox 1990; Prance 1991). The NCI now has a formal policy to return a share of royalties to the

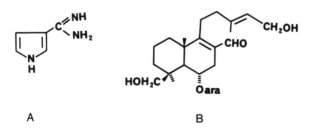

A   B

FIGURE 4–4 Structures of two biologically active compounds obtained as a result of ethnopharmacological investigations: (A) pyrrole-3-carboxamidine and (B) gaudichaudioside A (ara, arabinose).

country of origin of the source plant material from the profits of any plant-derived drug so obtained (Farnsworth 1990).

4.2.3 Practical Aspects of Drug Discovery Efforts from Plant Sources

When conducting a plant screening program, it is obvious that the more samples screened, the higher the chances will be that useful lead molecules will result. In the interests of saving time and money, therefore, it is preferable if only a relatively small amount (approximately 100 g) of plant material is extracted for initial biological screening. It is often profitable to collect more than one part of the same plant, since quite different secondary metabolite profiles may result. The plant samples should be taxonomically identified as far as possible prior to starting the phytochemical portion of the investigation, not only to enable a literature search of the previously established constituents of the plant to be carried out and thus assist in the dereplication of active compounds, but also to aid in subsequent compound identification. In addition, at the time of collection, a voucher specimen of the plant material (preferably in the flowering or fruiting stage) should be deposited in one or more herbaria, both as a necessary step in taxonomic identification, and in case of future need as reference material. As discussed in Sections 4.2.1 and 4.2.2, there are many considerations to be made in candidate plant selection, but a good rule of thumb is to try and obtain previously unstudied species that are endemic to the country where collected. Plants that have a folkloric reputation should be obtained whenever possible, and detailed records should be kept of the medicinal use(s), plant part(s) used, and method of preparation and mode of administration in such cases. It should also be noted that the secondary metabolites may vary considerably in the same plant part at different times of the year, so the date of collection must also be documented. Plant selection in drug discovery programs may be assisted by database surveillance. For example, the NAPRALERT® database at the University of Illinois at Chicago has been used to select and prioritize plants exhibiting potential fertility-regulating activity (Fong et al. 1989) and is currently being employed to select plants for potential antineoplastic activity (Cordell et al. 1991). The use of this database for plant selection involves, for a given species, a consideration of known traditional uses, and of existing literature reports of phytochemical constituents and biological activities (Fong et al. 1989; Cordell et al. 1991).

Initial extraction on air-dried, milled plant material can be carried out by percolation or maceration on a batch-wise basis, using three or four changes of solvent, to provide effective though not quantitative extraction. Extraction at room temperature rather than soxhlet extraction is preferred because of the possible lability of the potentially active compound or compounds present. It is not advisable to add extraneous agents to break emulsions or aid in solvent evaporation, because of the possibility of contaminating or otherwise changing the composition of the resultant plant extract. It is recommended that solvent extraction schemes be kept relatively simple, in order to increase reproducibility in case it is necessary to reisolate an active constituent from recollected plant material. In addition, it is normally advisable to subject to testing only a limited number of dried nonpolar and polar extracts, for reasons of economy and logistics. For example, the extraction procedure used in our laboratory to isolate the cytotoxic principles of the Indonesian medicinal plant, *Plumeria rubra*, is shown in Figure 4–5. Thus, the dried, comminuted plant material was extracted with three changes of petroleum ether, and the marc was then extracted with four aliquots of methanol, with the solvent removed from these two extractives on a rotary evaporator under reduced pressure, at a temperature not exceeding 40°C, in order to prevent compound breakdown. Next, the dried methanol extract was partitioned between chloroform and water to afford, on drying, chloroform and water extracts in addition to

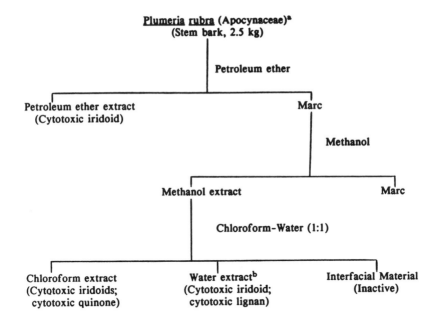

---

[a]Taken from Kardono et al. (1990). Altogether six significantly cytotoxic compounds of known structure were obtained, representing three compound classes (iridoid, lignan, quinone). Each active compound was evaluated against a panel of human and murine cancer cell lines.

[b]Although extract regarded as non-cytotoxic, small quantities of active compounds were obtained after chromatographic fractionation.

FIGURE 4–5 Solvent extraction scheme used for the isolation and identification of the cytotoxic constituents of the Indonesian medicinal plant *Plumeria rubra*.

interfacial material. The petroleum ether, chloroform, and water extracts, and the interfacial material were then tested for cytotoxic activity against P-388 (murine lymphocytic leukemia) cells, with significant activity found only in the petroleum ether and chloroform extracts (Kardono et al. 1990).

A variety of in vitro mechanism-based enzyme inhibition assays, receptor-binding assays, and cytotoxicity screens with high throughput are available today for the initial detection of biological activity of plant extracts (Farnsworth et al. 1985; Freter 1987; Cassady et al. 1990; Cordell et al. 1991; Hamburger et al. 1991). In certain cases, it is possible to use mammalian enzymes produced by recombinant DNA technology, and to have bioassay test data obtained by robots and handled by computers (Freter 1987). Once significant biological activity is detected and confirmed for a given plant extract, then the plant material should be recollected, ideally under the same conditions (collection location, time of year, and plant part) as encountered initially. Quantities of 2 to 5 kg are sufficient for most activity-guided fractionation procedures that lead to the purification and structural characterization of the active constituent(s) present in the initial plant extract. Frequently, significant information on the chemical class of organic compound responsible for the observed biological activity may be gained from residual active extract left from the initial plant material collection. However, for particularly interesting compounds, where evaluation with laboratory animals is needed, it may prove necessary to collect 10 kg or considerably more plant material. Activity-guided fractionation is carried out using a combination of bulk solvent extraction and chromatographic procedures, which vary from laboratory to laboratory, depending upon individual preference. Compound characterization and structural determination is performed routinely by the interpretation of physical and spectroscopic data, with compound degradations or total or partial synthesis carried out as deemed necessary. The following scheme represents a summary of the stages involved in the development of a pure natural product drug candidate from a plant source, and is modified from schemes that have appeared in the literature previously (e.g., Suffness and Douros 1982; Boyd 1989; Hamburger et al. 1991):

1. Collect and identify plant material and deposit voucher samples in local and major herbaria.
2. Extract the dried, milled plant acquisition with solvent, and prepare nonpolar and polar extracts for initial biological testing.
3. Evaluate such plant extracts against a panel of biological test methods, as exemplified by receptor binding, enzyme inhibition, and/or cytotoxicity bioassays.
4. Confirm initial biological activity on larger quantities of recollected plant material.
5. Perform literature surveillance on the plant species selected for isolation studies to aid in the dereplication of known active compounds.
6. Conduct activity-guided fractionation on the extract showing activity, by

monitoring each chromatographic fraction with a bioassay chosen from the panel available to the investigation.

7. Determine structure of pure active isolate(s) using spectroscopic techniques and chemical methods, if necessary.
8. Test each active compound (whether of novel or known chemical structure) in all in vitro and in vivo biological test methods available, in order to determine potency and selectivity.
9. Perform molecular modeling studies and prepare derivatives of the active compound of interest.
10. When total synthesis is not practical, carry out large-scale reisolation of interesting active compounds for toxicological, pharmacological, and formulation studies.
11. Clinical trials (phases I–III).

It should be pointed out that this type of research and development work requires the close and equitable collaboration of researchers in several disciplines, including botanists, pharmacognosists, biologists, chemists, and clinicians.

4.3 PROBLEMS AND CHALLENGES

4.3.1 Supply of Source Material

One of the major barriers to the successful outcome of drug discovery programs from plants is the periodic need to obtain recollections of plant material, whether for structural confirmation of active constituents found in very low abundance in the plant part under study, or for the generation of larger compound quantities for biological or clinical testing. A topical example wherein the generation of continued supplies of an active plant secondary metabolite is both crucial and uncertain is that of taxol, a candidate antitumor drug obtained from the stem bark of *Taxus brevifolia* (Pacific or Western Yew), currently being developed under the auspices of the NCI. Concern over the need to obtain an adequate supply of taxol from its plant of origin for clinical trials, while preserving native stands of *T. brevifolia* in states in the Pacific Northwest, has even been expressed in recent editorial columns of local and national newspapers in the United States. It has been estimated that 25,000 lb (11,364 kg) of bark from 2500 trees are needed to supply 1 kg of taxol, since one *T. brevifolia* tree can provide 10 lb (4.5 kg) of bark, using current isolation methods. Since ultimately there may be a total need to procure 200 to 300 kg of taxol annually, the NCI has taken several initiatives to attempt to increase the supply of this drug, and a major pharmaceutical company (Bristol-Myers Squibb) also has a program to this same end (Cragg and Snader 1991). Taxol has so far proven to be refractory to total synthesis, which is not surprising since it has 11 stereocenters, although the partial synthesis of this compound has been accomplished from several precursors

(Borman 1991; Cragg and Snader 1991). Taxol is found not only in the stem bark of *T. brevifolia*, but also in the roots, needles, twigs, and seedlings of this species and other representatives of the genus *Taxus*, which therefore represent alternative plant sources for this drug (Vidensek et al. 1990; Witherup et al. 1990). Other potential approaches towards solving the supply problem of taxol are the selection and propagation of high-taxol yielding *T. brevifolia* phenotypes and plant tissue culture (Cragg and Snader 1991). In addition, it is possible that less structurally complex taxol analogs will be discovered having the same type and potency of antineoplastic activity as the parent compound and can then be synthesized (Cragg and Snader 1991).

4.3.2 Diminishing Resources

There is great concern about threats to the survival of the world's tropical rain forests due to increasing deforestation. Although the tropical forests cover less than 10% of the surface of the world, they are estimated to contain 50% of the earth's species, with perhaps 150,000 higher plants alone (Plotkin 1988; Soejarto and Farnsworth 1989). Soejarto and Farnsworth have listed 47 plant drugs of major importance that are obtained commercially from plants native to tropical regions of the world, inclusive of atropine, emetine, hyoscyamine, ouabain, physostigmine, pilocarpine, quinine, quinidine, reserpine, scopolamine, tubocurarine, vinblastine, and vincristine (Soejarto and Farnsworth 1989). New drug products from the rain forest continue to be exploited, and in the case of Madagascar, four new plant species have only been exported since 1985, with the primary species being *Areca madagascariensis*, which is used as a veterinary anthelmintic (Rasoanaivo 1990).

However, it has been estimated that each year, 245,000 km^2 of tropical forest are destroyed or degraded, and that all tropical forests outside protected areas will disappear in the next 30 years, and 40,000 species will become extinct (Elliot and Brimacombe 1986). Accordingly, in order to give scientists the time to properly study the bioactive constituents of the tropical flora, several authors have called for stringent conservation efforts (e.g., Elliot and Brimacombe 1986; Plotkin 1988; Balandrin and Klocke 1988; Soejarto and Farnsworth 1989; Principe 1989; Balick 1990; Bentz 1990; Eisner 1990; Prance 1991). At the 1989 Annual Meeting of the International Society of Chemical Ecology held in Göteborg, Sweden, the following resolution was unanimously adopted by the membership present:

> Natural products constitute a treasury of immense value to mankind. The current alarming rate of species extinction is rapidly depleting this treasury, with potentially disastrous consequences. The International Society of Chemical Ecology urges that conservation measures be mounted worldwide to stem the tide of species extinction, and that vastly increased biorational studies be undertaken aimed at discovering new chemicals of use in medicine, agriculture and industry. These exploratory efforts should be pursued by a partnership of developing and devel-

oped nations, in such fashion that the financial benefits flow in fair measure to all participants (Eisner and Meinwald 1990).

Concomitant with the loss of the flora and fauna in the tropical forest regions is the loss of indigenous ethnomedical traditions, due in part to the encroachment of western medicine to once inaccessible locations (Plotkin 1988). To preserve and access indigenous medicinal discoveries, Cox has called for the close collaboration between ethnobotanists, pharmacognosists, natural product chemists, and pharmacologists (Cox 1990).

4.3.3 Additional Problems Encountered During the Search for Bioactive Compounds from Plants

There are many other potential difficulties that can be experienced in the laboratory search for new prototype active compounds from plant sources. For example, the investigation of biologically active natural compounds is often limited by their occurrence as complex mixtures of analogs at low concentration levels in the plant; also, such substances may be thermally or hydrolytically unstable. Occasionally, due to biological variation, recollected plant materials do not exhibit the same biological activity observed in the original extract, so reisolation efforts obviously would prove fruitless. It is important that the natural environment of a plant of interest is not unduly disturbed during collection, a consideration that can greatly limit the amount of material available at a given location. Political unrest in the country of origin may affect further collection efforts of a plant of interest, and hence impede the development of the therapeutic agent. For example, homoharringtonine (described in greater detail in Section 4.4.1) was originally discovered as an antileukemic agent in the United States (Powell et al. 1972). However, this drug candidate could not be developed clinically in this country for a number of years because supplies of the substance were not obtainable from the PRC (Suffness and Douros 1982).

There are a number of problems inherent in performing bioassays on crude plant extracts. Such assays should be highly sensitive and selective, as well as rapid, reproducible, and inexpensive (Suffness and Douros 1982; Boyd 1989; Hamburger et al. 1991). Typically, crude plant extracts are not water soluble and may have a gummy consistency, but organic solvents that are soluble in water, such as dimethylsulfoxide can be added to materials to be bioassayed in order to enhance solubility (Arisawa et al. 1984; Jayasuriya et al. 1989). Also, quite frequently, crude botanical extracts will contain pigments that interfere with spectrophotometric determinations (Cassady et al. 1990). In order to avoid "false positives," bioassays should be insensitive to ubiquitous compounds that occur in crude extracts. In our laboratory, for example, tannins and other polyphenols were found to be responsible for inhibitory effects exhibited by a majority of crude plant extracts tested against HIV type 1 (HIV-1) reverse transcriptase. It was necessary to adopt a tannin removal procedure, involving passage of each extract over a polyamide column (Tan et al. 1991). Along the same lines, it can be pointed out that the Walker 256 assay was dropped from the NCI plant program because it was sensitive to tannins (Suffness and Douros 1982).

4.4 SOME PROMISING DRUG CANDIDATES OF PLANT ORIGIN

4.4.1 Antineoplastic Agents

Despite impressive advances in therapy, cancer continues to be a major threat to human health in all countries in the world. Therefore, of all disease conditions, the search for anticancer agents has been the most intense as far as plant-derived therapeutic agents are concerned. The utility of plant secondary metabolites as clinically active anticancer agents is undisputed, as a result of nearly 30 years of therapeutic use in the United States of the indole alkaloids, vincristine, and vinblastine, which are extracted from the Madagascan periwinkle (*Catharanthus roseus*). However, most progress in developing cancer chemotherapeutic agents has been against highly proliferative tumors; consequently, there is a great need to discover agents that are active against breast, colorectal, lung, and ovarian cancers, among others (Krakoff 1986).

As mentioned earlier in this chapter, the NCI has been at the forefront in the search for new antineoplastic agents from plants for many years. Most recently, the NCI initiated another natural products acquisition program, which will entail the collection of some 20,000 rain forest plants from three continents (Africa inclusive of Madagascar, southeast Asia, and South and Central America) as part of this overall effort (Cassady et al. 1990). The emphasis in the primary biological evaluation of plant-derived and other test materials by the NCI has shifted from the use of the L1210 or P-388 in vivo murine leukemia model to a panel of in vitro screens employing a series of human tumor cell lines (Boyd 1989). It is intended to have as many as 120 cell lines in the panel to test each compound, with the number of individual assays performed daily at NCI being over 50,000 (Boyd 1989). This high-volume screening program is being accomplished using a microculture tetrazolium assay, employing cell lines derived from over ten categories of human cancer (e.g., breast, colon, head, neck, kidney, leukemia, lung, melanoma, ovary, and sarcoma) (Alley et al. 1988). Promising in vitro active isolates in this "disease oriented" approach are then subjected to in vivo testing in selected sensitive human tumor cell lines borne by athymic nude mice, before further consideration for drug development (Boyd 1989).

There are a number of "mechanism based" bioassays that are suitable for screening plant extracts, fractions and isolates, and can be used to complement cytotoxicity data. Included in this category are bioassays involving enzymes such as DNA topoisomerase I and II, protein kinase C, aromatase, and tyrosine kinase, as well as tubulin binding and DNA nicking assays (Cordell et al. 1991). Not all laboratories, where phytochemical research is directed towards the isolation and identification of plant-derived antitumor agents, have sophisticated in-house bioassay facilities. However, McLaughlin and co-workers have developed a series of simple bench-top bioassays that determine brine shrimp lethality, *Lemna* frond proliferation, and the inhibition of crown galls on potato disks (Anderson et al. 1991). The predictive value of these assays for the detection of known P-388 (murine lymphocytic leukemia) in vivo active compounds was compared with several human tumor cell lines, and the potato disk and brine shrimp assays have been found to correlate in a statistically significant manner. These simple prescreens appear to have great potential value for scientists in developing nations, to whom an abundance of unexplored flora is locally available for biological screening (Anderson et al. 1991).

Several plant-derived compounds have attracted particular attention for ultimate development as cancer chemotherapeutic agents, including taxol (Figure 4–6A), homoharringtonine (Figure 4–6B), and several derivatives of camptothecin (Figure 4–6C). Taxol (see Figure 4–6A) was first isolated from the stem bark of *Taxus brevifolia* and structurally characterized in the laboratory of Monroe Wall at Research Triangle Institute in North Carolina, and is a nitrogen-containing diterpenoid based on the taxane skeleton (Wani et al. 1970). Unlike previously known antimicrotubule agents like colchicine and vinblastine that induce depolymerization of microtubules, taxol was found to exhibit an unprecedented mechanism of action in effecting tubule polymerization and thereby producing stable nonfunctional microtubules (Rowinsky et al. 1990). In a very encouraging phase-II clinical trial, taxol was determined to be an active agent for 12 of 47 patients suffering from drug-refractory ovarian cancer (McGuire et al. 1989). This compound has also proven to be somewhat active for patients suffering from breast cancer, metastatic melanoma, and non-small-cell lung cancer, although symptoms of toxicity, such as myelosuppression and peripheral neuropathy, have now been identified as a result of the clinical use of taxol (Rowinsky et al. 1990; Slichenmyer and Von Hoff 1990; Borman 1991). Homoharringtonine (see Figure 4–6B) was first isolated and characterized in the laboratories of the U.S. Department of Agriculture in Peoria, IL, with the source of this compound being the seeds of the Chinese plant, *Cephalotaxus harringtonii* var. *harringtonii* (Powell et al. 1972). This cephalotaxine alkaloid has shown promise in phase-II clinical trials in the United States for patients with acute myelogenous leukemia, but it produces hypotensive and myelosuppressive effects (Slichenmyer and Von Hoff 1990). Along with other cephalotaxine alkaloids from *C. harringtonii*, homoharringtonine has been used in the treatment of leukemia patients in the PRC for several years (Han 1988; Xiao and Chen 1988). The quinoline alkaloid, camptothecin (see Figure 4–6C), was isolated from the stem wood of a tree (*Camptotheca acuminata*) native to mainland China, and found initially to significantly prolong the lifespan of mice implanted with L1210 leukemia (Wall et al. 1966). Early clinical trials on the sodium salt of camp-

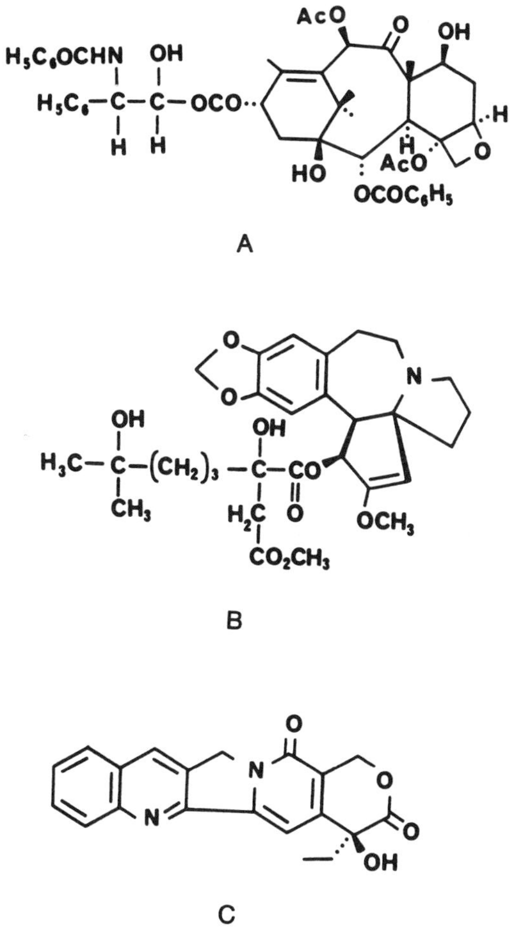

A

B

C

FIGURE 4–6 Structures of the plant-derived antineoplastic agents (A) taxol, (B) homoharringtonine, and (C) camptothecin.

tothecin in the United States were hindered by manifestations of excessive leucopenia and hemorrhagic cystitis (Giovanella et al. 1989). However, there has been a renewed interest in camptothecin, both because it has been shown to be highly unusual in interacting with the enzyme DNA topoisomerase I and because several analogs have shown promising antineoplastic activities. For example, the 20(RS)-9-amino derivative of camptothecin at nontoxic doses was effective for treating immunodeficient mice carrying three different lines of human colon cancer (Giovanella et al. 1989). Clinical trials have been organized for two additional semisynthetic derivatives of camptothecin, CPT-11 and SKF 104864, although their clinical activity profiles have not yet been determined (Slichenmyer and Von Hoff 1990). In the PRC, the 10-hydroxy derivative of camptothecin was shown to afford promising results for patients in stage-II clinical trials suffering from liver cancer and cancer of the head and neck (Xiao and Chen 1988).

4.4.2 Antimalarial Agents

Malaria remains the most prevalent tropical disease for humans, and is effected by several species in the genus *Plasmodium*, including *P. falcaparum*, a parasite of the *Anopheles* mosquito. Despite eradication attempts to wipe out the insect vector using insecticides, in addition to the development of several types of synthetic antimalarials, transmission of malaria still occurs in over 100 countries in the world, affecting approximately 100 million people each year (Trigg 1989). The first antimalarial drug was the plant constituent, quinine, a quinoline alkaloid from the bark of *Cinchona ledgeriana* and other *Cinchona* species. Quinine was the structural model for the design of the 7-chloro-4-aminoquinoline and 8-aminoquinoline classes of synthetic antimalarials and is still used for treating severe episodes of falciparum malaria (Fullerton 1991).

A very promising plant compound has come to the fore as an antimalarial agent in recent years, the sesquiterpene lactone endoperoxide, qinghaosu (artemisinin; Figure 4–7A). This compound is derived from *Artemisia annua*, a plant used for many years in the PRC for the treatment of malaria and fevers (Klayman 1985; Trigg 1989). Qinghaosu was isolated and characterized by Chinese scientists in 1972, and was shown to be useful for the treatment of drug-resistant *Plasmodium* species and has also demonstrated activity against cerebral malaria. This drug has been used to treat thousands of patients in certain southern provinces of the PRC and has given good cure rates, although its lack of aqueous solubility and recurrences of the disease condition have caused problems (Trigg 1989; Woerdenbag et al. 1990). Two derivatives of qinghaosu, sodium artesunate (Figure 4–7B) and artemether (Figure 4–

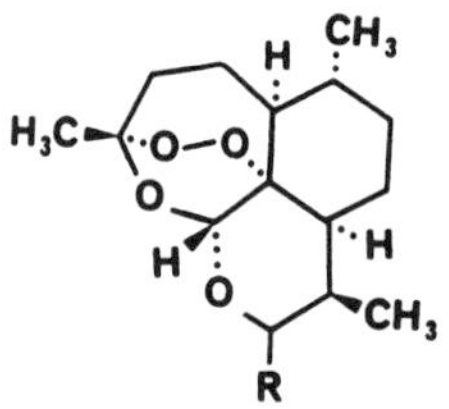

FIGURE 4–7 Structures of the plant-derived antimalarial agent qinghaosu (A) artemisinin, and its semisynthetic analogs (B) sodium artesunate, (C) artemether, and (D) arteether.

A   (R = O)

B   (R = β-OCOCH$_2$CH$_2$CO$_2$Na)

C   (R = β-OCH$_3$)

D   (R = β-OCH$_2$CH$_3$)

7C), are now being manufactured as antimalarial drugs in mainland China. The latter compound has been subjected to successful clinical trials for patients suffering from complicated falciparum malaria in Burma (Trigg 1989). A further derivative of qinghaosu, arteether (Figure 4–7D), which is a stable, crystalline compound, is being used in clinical trials organized by the World Health Organization (Brossi et al. 1988).

In addition to the two plant constituents mentioned above, a number of other plant secondary metabolites have been found that exhibit antimalarial activity, including alkaloids, limonoids, and quassinoids (O'Neill and Phillipson 1989). Effective in vitro and in vivo bioassays are available for the antimalarial evaluation of plant crude extracts and pure isolates (O'Neill and Phillipson 1989; Kardono et al. 1991).

4.4.3 Anti-HIV Agents

Acquired immunodeficiency syndrome (AIDS) emerged as a severe threat to public health in the 1980s and has been traced to a retrovirus, HIV (Piot et al. 1988). There is an urgent need to discover new drugs to prolong the lifespan of patients suffering from HIV infections (De Clercq 1989).

While the number of plant-derived secondary metabolites with demonstrated anti-HIV activity is not yet large, a substantial literature exists on plant constituents known to possess other types of antiviral activity. Thus, in about 25 screening studies on nearly 1000 plants in 150 families, ~30% of the extracts tested were found to have either in vivo or in vitro activity against one or more animal or plant viruses (Vanden Berghe et al. 1986). Although most of these leads were not followed up by activity-guided fractionation, it has been ascertained that the structural classes of plant constituents having antiviral activity against at least one animal virus test system include alkaloids (especially of the acronycine, β-carboline, indole, indolizidine, isoquinoline, purine, quinoline, and tropane types), coumarins , flavonoids, phenols, quinones, sterols, tannins, terpenoids, and xanthones (Vanden Berghe et al. 1986).

Recently, a number of structurally diverse potential anti-AIDS agents of plant origin have been reported. Among the pure substances of plant origin found to inhibit HIV-1 are the indolizidine alkaloid, castanospermine (Figure 4–8A), which was first isolated from the seeds of *Castanospermum australe* (Hohenschutz et al. 1981) and is known to inhibit plant and animal α- and β-glucosidases in vitro (Saul et al. 1983). This compound affects the growth and infectivity of HIV-1 at least in part because it produces changes in the sugar side chains of the gp120 glycoprotein (Tyms et al. 1987). The oleanane glucuronide, glycyrrhizin (Figure 4–8B), obtained in high yield from various *Glycyrrhiza* species, was found to inhibit HIV-induced plaque formation in MT-4 cells at a concentration of 0.06 mM (Ito et al. 1987). The mechanism of action of glycyrrhizin on HIV may result from either inhibition of the phosphorylation enzyme, protein kinase C, or from interference with virus

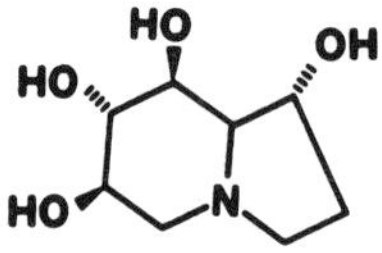

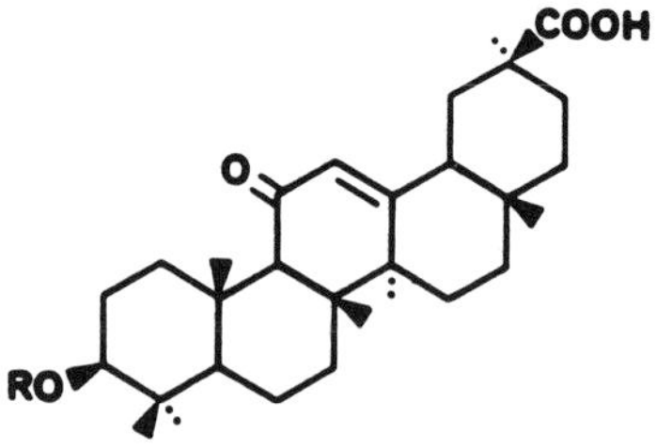

FIGURE 4–8 Structures of some plant-derived compounds with in vitro anti-HIV activity: (A) catanospermine, (B) glycyrrhizin, (C) hypericin, and (D) pseudohypericin (GlcA, glucuronic acid).

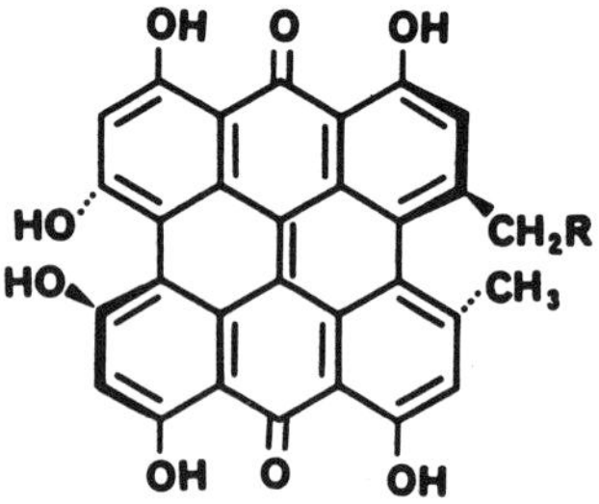

cell binding (Ito et al. 1988). Intravenous administration (400–1600 mg/day) of a glycyrrhizin-containing preparation to a small number of hemophiliacs with AIDS in Japan has given some evidence of the clinical efficacy of this substance (Hattori et al. 1989). Hypericin (Figure 4–8C) and pseudohypericin (Figure 4–8D), aromatic polycyclic dione constituents of *Hypericum perforatum* (St. Johnswort) and other species in this same genus, have proven to inactivate HIV and other human and murine retroviruses directly (Meruelo et al. 1988; Lavie et al. 1989). It has been suggested that a possible mode of action of these compounds is due to the inhibition of protein kinase C during HIV-induced CD4 phosphorylation (Takahashi et al. 1989). Recently, a clinical trial on hypericin was begun on over 20 patients at New York University Medical Center (Holden 1991). A final example of a plant-derived anti-HIV agent is the protein GLQ223, which is obtained from the roots of the Chinese medicinal plant *Trichosanthes kirilowii*. GLQ223 is the purified form of the ribosome-inactivating protein trichosanthin and has a molecular weight of about 26 kDa. Treatment of HIV-infected T-lymphoblastoid cells with GLQ223 appeared to reduce levels of viral protein and RNA synthesis (McGrath et al. 1989).

Although it is by no means certain that the five plant constituents mentioned above will eventually enter the market as anti-AIDS drugs, it is worth noting that they all appear to have a different mechanism of action than that of AZT and ddI, two nucleoside drugs currently approved for the treatment of HIV infections, which inhibit HIV reverse transcriptase (Holden 1991). While none of these plant derivatives with anti-HIV activity appears to have been isolated by activity-guided fractionation, rapid assays exist for the determination of HIV cytopathic effects that are appropriate for this type of endeavor, such as the tetrazolium-based colorimetric assays developed at the NCI (Weislow et al. 1989) and the Rega Institute for Medical Research, Leuven, Belgium (Pauwels et al. 1988). Use of the former assay system recently led to the activity-guided purification of several HIV-active diterpenes from the tropical trees *Homalanthus acuminatus* and *Chrysobalanus icaco* at the Laboratory for Drug Discovery and Development, NCI, Frederick, MD (Gustafson et al. 1991). It has been suggested that plants used in traditional medicine in developing countries may be valuable for the search for novel anti-HIV agents (Anonymous 1989), and it is accordingly worth noting that *H. acuminatus* is employed as a medicinal plant in Samoa to treat viral diseases (Gustafson et al. 1991).

4.5 FUTURE PROSPECTS

The few examples of drug candidates chosen in the preceding section of this chapter clearly show that plant constituents either already have an effective role or else have high promise for treating some of the most serious diseases known to humans. There is a growing interest in plant drugs, and, as an example of this, the consumption of medicinal plants has doubled in the last 10 years in Western Europe (Hamburger et al. 1991). It seems certain that the continued scientific study of medicinal plants will afford a plethora of novel, structurally diverse, bioactive compounds. The current success of taxol, in particular, should encourage additional industrial, academic, and governmental laboratories to engage in research on the discovery and development of plant-derived drugs. Multidisciplinary research on plants should lead to many new drug candidates, as well as to prototype biologically active molecules and biological tools. The suggestion by Eisner (1990), that efforts should be made to transfer the chemical screening aspects of plant drug discovery to laboratories in developing countries where the local flora will be examined, should be encouraged. However, even if such efforts do not become well established, it can be confidently expected that in the future significant lead compounds will be isolated from plants in laboratories outside North America,

Western Europe, and Japan, especially in countries such as the PRC, where such a strong tradition in the use of plant drugs exists. While the vast majority of the plant-derived drugs on the market in the United States today are either glycosides or alkaloids, and thus are either water soluble or easily rendered so by salt formation (Kinghorn 1992), it may be envisioned that a broader structural range of plant constituents, somewhat more difficult to deal with in the laboratory, will become future drug candidates, such as aromatic substances, terpenoids, oligopeptides, and even complex carbohydrates and proteins. According to Tyler (1986), the range of diseases that could be successfully treated with plant products will expand to include conditions like arthritis and parkinsonism, genetic diseases like cystic fibrosis, and self-induced diseases, including alcoholism and stress.

Several technological developments in the last few years offer additional possibilities for the phytochemical aspects of the discovery of drugs from plants. For example, efficient extraction of plant material may now be undertaken using supercritical fluids such as carbon dioxide (Bicchi et al. 1991). Also, there has been a continual improvement in the sensitivity and capability of analytical instruments used to evaluate and elucidate the structure of natural products, whether present in chromatographic fractions or in pure form. The most significant development in recent years for the structure elucidation of plant secondary metabolites is the refinement of one- and two-dimensional nuclear magnetic resonance spectroscopic techniques, which now permit complete determinations of structure with only a few milligrams of compound (e.g., Cordell and Kinghorn 1991; Yoshida et al. 1991).

While there are some who have taken a pessimistic view on the likelihood of ever again discovering useful therapeutic agents from plants (Vane and Cuatrecasas 1984), this viewpoint is now much less widely expressed than formerly. However, formidable challenges remain, as indicated earlier in this chapter. The problem of obtaining a large-scale supply of promising plant-derived compounds may be overcome by plant-tissue culture (Verpoorte et al. 1987), although this type of technology is limited by slow growth rates and insufficient understanding of plant secondary metabolite regulatory processes (Hamburger et al. 1991). The loss of the tropical rain forests and ethnomedical information about plant uses is acute, and conservation efforts are urgently needed so that future generations of scientists can investigate endangered genetic resources more thoroughly than is now possible (Balandrin and Klocke 1988). These types of difficulties notwithstanding, prospects for the future success of multidisciplinary research teams in discovering new drugs from higher plants remains high, provided that imaginative and flexible strategies are employed.

REFERENCES

Abelson, P.H. (1990) *Science* 247, 513.
Alley, M.C., Scudiero, D.A., Monks, A., et al. (1988) *Cancer Res.* 48, 589–601.

Anderson, J.E., Goetz, C.M., McLaughlin, J.L., and Suffness, M. (1991) *Phytochem. Anal.* 2, 107–111.

Anonymous (1989) *Bull. WHO* 67, 613–618.

Aronson, J.K. (1987) *Chem. Br.* 23, 33–36.

Arisawa, M., Pezzuto, J.M., Bevelle, C., and Cordell, G.A. (1984) *J. Nat. Prod.* 47, 453–458.

Baerheim Svendsen, A., and Scheffer, J.J.C. (1982) *Pharmaceutisch Weekblad, Scientific Edition* 4, 93–103.

Balandrin, M.F., and Klocke, J.A. (1988) in *Biotechnology in Agriculture and Forestry. 4. Medicinal and Aromatic Plants. I.* (Bajaj, Y.P.S., ed.), pp. 3–36, Springer Verlag, Berlin.

Balandrin, M.F., Klocke, J.A., Wurtele, E.S., and Bolinger, W.H. (1985) *Science* 228, 1154–1160.

Balick, M.J. (1990) in *Bioactive Compounds from Plants*, Ciba Foundation Symposium 154, pp. 22–39, John Wiley & Sons, Chichester, UK.

Bentz, G.D. (1990) *South. Med. J.* 83, 491–492.

Bicchi, C., Rubiolo, P., Frattani, C., Sandra, P., and David, F. (1991) *J. Nat. Prod.* 54, 941–945.

Borman, S. (1991) *Chem. Eng. News* September 2, pp. 11–18.

Boyd, M.R. (1989) *Principles Practice Oncology Updates* 3(10), 1–12.

Braquet, P., and Hosford, D. (1991) *J. Ethnopharmacol.* 32, 135–139.

Briggs, C.J. (1990) *Can. Pharm. J.* 123, 413–415.

Brossi, A., Venugopalan, B., Dominguez Gerpe, L., et al. (1988) *J. Med. Chem.* 31, 645–650.

Cassady, J.M., Baird, W.M., and Chang, C.-J. (1990) *J. Nat. Prod.* 53, 23–41.

Cordell, G.A., and Kinghorn, A.D. (1991) *Tetrahedron* 47, 3521–3534.

Cordell, G.A., Beecher, C.W.W., and Pezzuto, J.M. (1991) *J. Ethnopharmacol.* 32, 117–133.

Cox, P.A. (1990) in *Bioactive Compounds from Plants*, Ciba Foundation Symposium 154, pp. 40–55, John Wiley & Sons, Chichester, U.K.

Cragg, G.M., and Snader, K.M. (1991) *Cancer Cells* 3, 233–235.

De Clercq, E. (1989) *Antiviral Res.* 12, 1–19.

de Smet, P.G.A.M., and Rivier, L. (1989) *J. Ethnopharmacol.* 25, 127–138.

de Souza, N.J., and Shah, V. (1988) in *Economic and Medicinal Plant Research*, vol. 2 (Wagner, H., Hikino, H., and Farnsworth, N.R., eds.), pp. 1–16, Academic Press, London.

Eisner, T. (1990) *Chemoecology* 1, 38–40.

Eisner, T., and Meinwald, J. (1990) *Chemoecology* 1, 38.

Elliot, S. and Brimacombe, J. (1986) *Manufac. Chem.* October, pp. 31, 33–34.

Ekimoto, H., Irie, Y., Araki, Y., et al. (1991) *Planta Med.* 57, 56–58.

Farnsworth, N.R. (1968) *J. Pharm. Sci.* 55, 255–276.

Farnsworth, N.R. (1984) in *Natural Products and Drug Development*, Alfred Benzon Symp. 20, (Krogsgaard-Larsen, P., Brogger Christensen, S., and Kofod, H., eds.), pp. 17–30, Munksgaard, Copenhagen.

Farnsworth, N.R. (1990) in *Bioactive Compounds from Plants*, Ciba Foundation Symposium 154, pp. 1–21, John Wiley & Sons, Chichester, UK.

Farnsworth, N.R., Akerele, O., Bingel, A.S., Soejarto, D.D., and Guo, Z. (1985) *Bull. WHO* 63, 965–981.

Farnsworth, N.R., and Kaas, C.J. (1981) *J. Ethnopharmacol.* 3, 85–89.

Farnsworth, N.R., and Morris, R.W. (1976) *Am. J. Pharm.* 147, 46–52.

Fong, H.H.S., Farnsworth, N.R., and Beecher, C.W.W. (1989) in *Proc. Int. Symp. on East-West Medicine* (Paik, Y.-H., ed.), Seoul, Korea, October 1989, pp. 255–265.

Freter, K.R. (1987) *Pharm. Res.* 5, 397–400.

Fullas, F., Hussain, R.A., Bordas, E., et al. (1991) *Tetrahedron* 47, 8515–8522.

Fullerton, D.S. (1991) in *Wilson and Grisvold's Textbook of Organic Medicinal and Pharmaceutical Chemistry* (Delgardo, J.N., and Remers, W.A., eds.), pp. 205–225, J.B. Lippincott, Philadelphia.

Galeffi, C., and Marini Bettolo, G.B. (1988) *Fitoterapia* 59, 179–205.

Giovanella, B.C., Stehlin, J.S., Wall, M.E., et al. (1989) *Science* 246, 1046–1048.

Graham, N.A., and Chandler, R.F. (1990) *Can. Pharm. J.* 123, 330–333.

Gustafson, K.R., Munro, M.H.G., Blunt, J.W., et al. (1991) *Tetrahedron* 47, 4547–4554.

Hamburger, M., Marston, A., and Hostettmann, K. (1991) in *Advances in Drug Research*, vol. 20 (Testa, B., ed.), pp. 167–215, Academic Press, London.

Han, G.-Q., Chen, Y.-Y., and Liang, X.-T. (1988) in *The Alkaloids: Chemistry and Pharmacology*, vol. 32 (Brossi, A., ed.), pp. 241–270, Academic Press, San Diego, CA.

Han, J. (1988) *J. Ethnopharmacol.* 24, 1–17.

Hattori, T., Ikematsu, S., Koito, A., et al. (1989) *Antiviral Res.* 11, 255–262.

Hite, G. (1989) in *Principles of Medicinal Chemistry*, 3rd ed. (Foye, W.O., ed.), pp. 239–275, Lea & Febiger, Philadelphia.

Hohenschutz, L.D., Bell, E.A., Jewess, P.J., et al. (1981) *Phytochemistry* 20, 811–814.

Holden, C. (1991) *Science* 254, 522.

Hosler, D.M., and Mikita, M.A. (1987) *J. Chem. Ed.* 64, 328–332.

Ito, M., Nakashima, H., Baba, M., et al. (1987) *Antiviral Res.* 7, 127–137.

Ito, M., Sato, A., Hirabayashi, K., et al. (1988) *Antiviral Res.* 10, 289–298.

Jayasuriya, H., McChesney, J.D., Swanson, S.M., and Pezzuto, J.M. (1989) *J. Nat. Prod.* 52, 325–331.

Kardono, L.B.S., Angerhofer, C.K., Tsauri, S., et al. (1991) *J. Nat. Prod.* 54, 1360–1367.

Kardono, L.B.S., Tsauri, S., Padmawinata, K., Pezzuto, J.M., and Kinghorn, A.D. (1990) *J. Nat. Prod.* 53, 1447–1455.

Kinghorn, A.D. (1992) in *Phytochemical Resources for Medicine and Agriculture*, (Nigg, H.N., and Seigler, D., eds.), pp. 75–95, Plenum, New York.

Kinghorn, A.D., and Compadre, C.M. (1985) *Pharm. Int.* 6, 201–204.

Klayman, D.L. (1985) *Science* 228, 1049–1055.

Korolkovas, A. (1988) *Essentials of Medicinal Chemistry*, 2nd ed., John Wiley & Sons, New York.

Krakoff, I.H. (1986) in *Cancer Chemotherapy: Challenges for the Future*, (Kimura, K., Yamida, K., Krakoff, I.H., and Carter, S.K., eds.), pp. 17–23, Excerpta Medica, Amsterdam.

Labadie, R.P. (1986) *J. Ethnopharmacol.* 15, 221–230.

Lavie, G., Valentine, F., Levin, B., et al. (1989) *Proc. Natl. Acad. Sci. USA* 86, 5963–5967.

Lloyd, H.A., Fales, H.M., Goldman, M.E., et al. (1985) *Tetrahedron Lett.* 26, 2623–2624.

Locock, R.A. (1988) *Can. Pharm. J.* 121, 577–578, 581.

McGrath, M., Hwang, K.M., Caldwell, S.E., et al. (1989) *Proc. Natl. Acad. Sci. USA* 86, 2844–2848.

McGuire, W.P., Rowinsky, E.K., Rosenshein, N.B., et al. (1989) *Ann. Intern. Med.* 111, 273–279.

Meruelo, D., Lavie, G., and Lavie, D. (1988) *Proc. Natl. Acad. Sci. USA* 85, 5230–5234.

Midgley, J.M. (1988) *Pharm. J.* 241, 358–365.

O'Neill, M.J., and Phillipson, J.D. (1989) *Rev. Latinoamer. Quím.* 20-3, 111–118.

Pauwels, R., Balzarini, J., Baba, M., et al. (1988) *J. Virol. Methods* 20, 309–321.

Phillipson, J.D., and Anderson, L.A. (1989) *J. Ethnopharmacol.* 25, 61–72.

Phillipson, J.D., and Wright, C.W. (1991) *J. Ethnopharmacol.* 32, 155–165.

Piot, P., Plummer, F.A., Mhalu, F.S., et al. (1988) *Science* 239, 573–579.

Plotkin, M. (1988) *Pharmacotherapy* 8, 257–262.

Powell, R.G., Weisleder, D., and Smith, C.R. Jr. (1972) *J. Pharm. Sci.* 61, 1227–1230.

Prance, G.T. (1991) *J. Ethnopharmacol.* 32, 209–216.

Principe, P.P. (1989) in *Economic and Medicinal Plant Research*, vol. 3 (Wagner, H., Hikino, H., and Farnsworth, N.R., eds.), pp. 1–17, Academic, London.

Rasoanaivo, P. (1990) *Ambio* 19, 421–424.

Rivier, L. and Bruhn, J.G. (1979) *J. Ethnopharmacol.* 1, 1.

Rowinsky, E.K., Cazenave, L.A., and Donehower, R.C. (1990) *J. Natl. Cancer Inst.* 82, 1247–1259.

Saul, R., Chambers, D.P., Molyneux, R.J., and Elbein, A.D. (1983) *Arch. Biochem. Biophys.* 221, 265–275.

Schultes, R.W. (1972) in *Plants in the Development of Modern Medicine* (Swain, T., ed.), pp. 103–124, Harvard University Press, Cambridge, MA.

Slichenmeyer, W.J., and Von Hoff, D.D. (1990) *J. Clin. Pharmacol.* 30, 770–778.

Sneader, W. (1985) *Drug Discovery: The Evolution of Modern Medicines*, John Wiley & Sons, Chichester, U.K.

Soejarto, D.D., and Farnsworth, N.R. (1989) *Perspect. Biol. Med.* 32, 244–256.

Soudijn, W. (1991) *Pharmaceutisch Weekblad, Scientific Edition* 13, 161–166.

Spjut, R.W., and Perdue, R.E. Jr. (1976) *Cancer Treat. Rep.* 60, 979–985.

Steiner, R.P., Ed. (1986) *Folk Medicine: The Art and the Science*, American Chemical Society, Washington, DC.

Suffness, M., and Douros, J. (1982) *J. Nat. Prod.* 45, 1–14.

Takahashi, I., Nakanishi, S., Kobayashi, E., et al. (1989) *Biochem. Biophys. Res. Commun.* 165, 1207–1212.

Tan, G.T., Pezzuto, J.M., Kinghorn, A.D., and Hughes, S.H. (1991) *J. Nat. Prod.* 54, 143–154.

Trigg, P.I. (1989) in *Economic and Medicinal Plant Research*, vol. 3 (Wagner, H., Hikino, H., and Farnsworth, N.R., eds.), pp. 19–55, Academic, London.

Tyler, V.E. (1986) *Econ. Bot.* 40, 279–288.

Tyler, V.E., Brady, L.R., and Robbers, J.E. (1988) *Pharmacognosy* 9th ed., Lea & Febiger, Philadelphia.

Tyms, A.S., Berrie, E.M., Ryder, T.A., et al. (1987) *Lancet* 2, 1025–1026.

Vanden Berghe, D.A., Vlietinck, A.J., and Van Hoof, L. (1986) *Bull. Inst. Pasteur* 84, 101–147.

Vane, J., and Cuatrecasas, P. (1984) *Nature* 312, 303–305.

Verpoorte, T., Harkes, P.A., and Hoopen, H.J.G. (1987) in *Topics in Pharmaceutical Sciences 1987*, (Breimer, D.D., and Speiser, P., eds.), pp. 263–281, Elsevier Science Publishers, Amsterdam.

Vidensek, N., Lim, P., Campbell, A., and Carlson, C. (1990) *J. Nat. Prod.* 53, 1609–1610.

Vlietinck, A.J., and Vanden Berghe, D.A. (1991) *J. Ethnopharmacol.* 32, 141–153.

Wall, M.E., Wani, M.C., Cook, C.E., et al. (1966) *J. Am. Chem. Soc.* 88, 3888–3890.

Wang, Z.-G., and Liu, G.-Z. (1985) *Trends Pharmacol. Sci.* 6, 432–436.

Wani, M.C., Taylor, H.L., Wall, M.E., Coggon, P., and McPhail, A.T. (1970) *J. Am. Chem. Soc.* 93, 2325–2327.

Weislow, O.S., Kiser, R., Fine, D.L., et al. (1989) *J. Natl. Cancer Inst.* 81, 577–586.

Witherup, K.M., Look, S.A., Stasko, M.W., et al. (1990) *J. Nat. Prod.* 53, 1249–1255.

Woerdenbag, H.J., Lugt, C.B., and Pras, N. (1990) *Pharmaceutisch Weekblad, Scientific Edition* 12, 169–181.

Xiao, P.-G., and Chen, K.-J. (1987) *Phytother. Res.* 1, 53–57.

Xiao, P.-G., and Chen, K.-J. (1988) *Phytother. Res.* 2, 55–62.

Xiao, P.-G., and Fu, S.-L. (1987) in *Herbs, Spices and Medicinal Plants: Recent Advances in Botany, Horticulture, and Pharmacology*, vol. 2 (Craker, L.E. and Simon, J.E., eds.), pp. 1–55, Oryx Press, Phoenix, AZ.

Yoshida, T., Hatano, T., Ahmed, A.F., Okonogi, A., and Okuda, T. (1991) *Tetrahedron* 47, 3575–3584.

The Discovery of Marine Natural Products with Therapeutic Potential

Oliver J. McConnell
Ross E. Longley
Frank E. Koehn

There is a tremendous level of worldwide interest in marine natural products with therapeutic potential in industry, academia, and government research labs, largely because natural products generally continue to be viewed as one of the few de novo sources of drug discovery, yielding unorthodox and often unexpected chemical structures that offer novel points of departure for molecular modification leading to clinically available drugs (de Souza et al. 1982).

We would like to thank Dr. Gabriel Saucy, Prof. Bob Jacobs (University of California, Santa Barbara), and Dr. Shirley Pomponi (HBOI) for reviewing this chapter and making helpful comments and suggestions, Dr. Ed Armstrong (HBOI), Dr. Sarath Gunasekera (HBOI), Prof. Bob Jacobs, and John Reed (HBOI) for supplying useful information, Prof. Paul Scheuer (University of Hawaii) for permission to reprint Figure 5-1, Drs. Neal Burres and Jacob Clement (Abbott Laboratories) for permission to reprint Tables 5-1, 5-3, and 5-4, Prof. Sidney Hecht (University of Virginia and HBOI board member) for catalyzing thoughtful discussions among ourselves regarding our drug discovery program, and Mr. Seward Johnson, Jr. (HBOI, CEO), for providing us the opportunity to conduct research on marine natural products with therapeutic potential. This chapter constitutes HBOI contribution number 908.

In the United States, most major pharmaceutical companies are actively acquiring terrestrial and marine natural products, extracts from terrestrial and marine macroorganisms and microorganisms, and/or cultures of terrestrial and marine microorganisms for the purpose of finding new, small molecule lead structures. To our knowledge, many of the major pharmaceutical companies in Japan and Europe are involved as well. The U.S. government has provided substantial support since the 1970s for academic research on biologically active marine-derived compounds; government agencies include the National Institutes of Health (NIH)–the National Cancer Institute (NCI) and the National Institute of Allergic and Infectious Diseases (NIAID), as well as the National Sea Grant College Program/National Oceanic and Atmospheric Administration (NOAA) (Persinos 1989c, 1990c, 1991c). Increased interest by the U.S. government in biologically active marine and terrestrial natural products has recently led, for example, to the formation by NIH of National Cooperative Natural Product Drug Discovery Groups, which include most of the prominent academic labs engaged in marine and terrestrial natural products research, several of the major pharmaceutical companies, and representatives from NIH for the purpose of discovering new structural leads with activity against cancers, human immunodeficiency virus (HIV), and opportunistic infections associated with acquired immune deficiency syndrome (AIDS) (Persinos 1990b, 1990c). Strong programs in natural products research have historically existed in Europe and Japan. The tremendous level of interest by Japan in the development of marine resources has recently led their government, industry, and universities to undertake major, long term initiatives in "marine biotechnology," which include efforts to produce marine-derived pharmaceuticals (Persinos 1989c, 1990a). In fact, two major facilities were recently built expressly for this purpose, one in the city of Kamaishi, to take advantage of organisms found in the cold currents off Japan's northeast coast, and the other in the city of Shimizu, where the currents are warm.

Current interest in natural products as a source of therapeutic agents is not surprising considering the major role that natural materials and natural products have played in the development of modern therapeutics. Throughout the 5000 years of recorded history of medicines, which includes records from China (Kim 1967), the Middle East, South America, Europe, and Mexico, until the 1800s, the primary source of medicines consisted of terrestrial plants (in the form of herb preparations; "Galenic" medicine) (Burger 1986). Indeed, the development of modern therapeutics is reported to have begun with digitalis, a mixture of steroid (cardiac) glycosides from several plants, including purple foxglove, *Digitalis purpurea*, which was used in 1785 by a British physician to treat "dropsy," a condition caused by congestive heart failure and characterized by an accumulation of fluid (Burger 1986). Since the 1800s, numerous naturally occurring compounds have been purified and used for medicinal and health purposes, including plant metabolites, steroid hormones (androgens, estrogens, progestins), glucocorticoids (cortisone), pituitary hormones (antidiuretic hormone (ADH), adrenocorticotropin (ACTH), vasopressin, oxytocin), and vitamins (vitamin D, cod liver oil) (*Physicians Desk Reference* 1991; Scrip 1991; Guyton 1991).

With the recognition of the microbial basis for infectious diseases in the latter half of the 1800s and the discovery of the "lysis factor" (penicillin) from *Penicillium notatum* in 1928, the stage was eventually set in the 1940s for the use of microorganism-derived natural products as medicines, initially as antibiotics (see discussion in Reuben and Wittcoff 1989). Shortly after the publication of studies on penicillin, an intense search ensued to find new and better microorganism-derived antibiotics (Burger 1986). Near a sewage outfall in the Mediterranean, an antibiotic-producing fungi, *Cephalosporium acremonium*, was isolated (Abraham 1962; Abraham and Newton 1967). The biologically active components were named the cephalosporins, which are structurally related to the penicillins (Reuben and Wittcoff 1989; Caprile 1988), and some would claim are *marine-derived* (Rinehart 1988a; Torres 1988). The cephalosporins turned out to be significantly more active than the penicillins against Gram-negative microorganisms and are resistant to degradation by staphylococcal β-lactamase (Caprile 1988). The cephalosporins are broadly used therapeutically as well as prophylactically (prior to surgery) (Kaiser 1988). Since 1964, which is when the first member of three generations of cephalosporins was introduced, over 20 cephalosporins have been marketed (Caprile 1988). Numerous additional microorganism-derived compounds are used or are being developed not only as antibiotics, but in most major disease areas (Scrip 1991; Hall 1989). Notable examples are the immunosuppressive agents, cyclosporin (Dreyfuss et al. 1976) and FK506 (Kino et al. 1987a, 1987b), numerous antitumor antibiotics including bleomycin, calicheamicin, esperamicin, mitomycin C, CC-1065, and actinomycin D (Williams et al. 1989 and references therein), and the cholesterol-lowering agent, mevinolin (Alberts et al. 1980).

Toward the end of the 1980s, approximately 75% of the top 20 hospital drugs (in total U.S. dollar sales) were derived from natural sources (mostly antibiotics), and approximately 20% of the top 100 most widely prescribed ethical drugs were also derived from natural sources, i.e., antibiotics from fermentation broths and alkaloids from plants (Reuben and Wittcoff 1989). Microorganism-derived antibiotics are also among the largest volume pharmaceuticals consumed, approximately 15% of the total annual tonnage. (Vitamins and aspirin/acetaminophen comprise most of the remaining tonnage.) Fermentation yields all antibiotics except chloramphenicol and it yields biotechnology-related products as well. In contrast, chemical synthesis and semisynthesis yield the majority of heart drugs, central nervous system agents, antihistamines, analgesics, and anti-inflammatory drugs.

The world's oceans represent an enormous resource for the discovery of potential therapeutic agents and will ultimately yield as many pharmaceutical leads as have been obtained from the terrestrial biosphere. This prediction is based on the following information.

1. The world's oceans cover more than 70% of the earth's surface (Barnes 1980) and contain well over 200,000 invertebrate and algal species (George and George 1979; Winston 1988). There exist nearly 150,000 species (Winston 1988) of algae (seaweed), green (Chlorophyta), red (Rhodophyta),

and brown (Phaeophyta), and some groups of marine invertebrates in which new chemical structures or biological activities have been reported: sponges (Porifera), cnidarians or coelenterates (corals, octocorals (including sea fans), hydroids, and sea anemones), nemerteans (worms), bryozoans, ascidians (tunicates including sea squirts), molluscs (sea snails and sea slugs), and echinoderms (brittlestars, sea urchins, starfish, and sea cucumbers). All but two of the 28 major animal phyla are represented in aquatic environments; eight phyla, including the Cnidaria, Porifera, Bryozoa, and Echinodermata, are exclusively aquatic, and primarily marine (Barnes 1980).

2. In terms of known species and habitats, and excluding insects, the world's oceans represent our greatest resource of new natural products (Munro et al. 1987). Indeed, since about 1975, approximately 4000 new marine natural products have been reported from a relatively small number of research labs (Faulkner 1984a, 1984b, 1986, 1987, 1988, 1990, 1991). Unfortunately, only a small percentage of these compounds has been fully evaluated for biological activity relevant to humans and only a few have been evaluated in clinically relevant assays. A clinically relevant assay is loosely defined as one in which a positive (potent) response would elicit very serious interest by a (pharmaceutical) company or organization that has sufficient resources and experience to successfully develop a therapeutic agent.

3. Most nonmoving (sessile) marine invertebrates (e.g., Figure 5–1: Porifera-Demospongia, sponges; Coelenterata, sea fans, soft corals, hard corals), and algae (Figure 5–1: Rhodophyta, red algae; Phaeophyta, brown algae; and Chlorophyta, green algae) contain a primitive immune system, as do plants and microorganisms found in the terrestrial biosphere (vide supra). It has been hypothesized (Hall 1989; Williams et al. 1989) that some terrestrial, and, by analogy, some marine organisms that do not contain sophisticated immune systems use natural products, or allelochemicals (Whittaker and Feeny 1971), in response to a variety of ecological, behavioral, and physiological factors (Bakus et al. 1986; Davis et al. 1989; Geiselman and McConnell 1981; McConnell et al. 1982; Hay and Fenical 1988; Hay et al. 1987; Paul 1988; Paul and Fenical 1987; Paul et al. 1987). In order to survive and proliferate in their environments, these primitive organisms must produce compounds that are used to chemically defend themselves and/or to assist in the prevention of overgrowth or fouling. These metabolites also have therapeutic potential against human diseases because of very specific interaction with receptors and enzymes (Hall 1989; Williams et al. 1989). The rationale for this specificity of biological activity is that specialized vertebrate cells such as those of the humoral, nervous, immune, and vascular systems, appeared relatively recently in evolution, while the transmitter molecules they use, such as hormones, neuropeptides, and biological response modifiers as well as allelochemicals, may have arisen much earlier in unicellular or simple multicellular organisms (Gilbert 1977; Nisbet and Porter 1989; Roth et al. 1986). Indeed, it is believed that algae and bacteria have existed up to possibly 3.5 billion years ago, and relatively

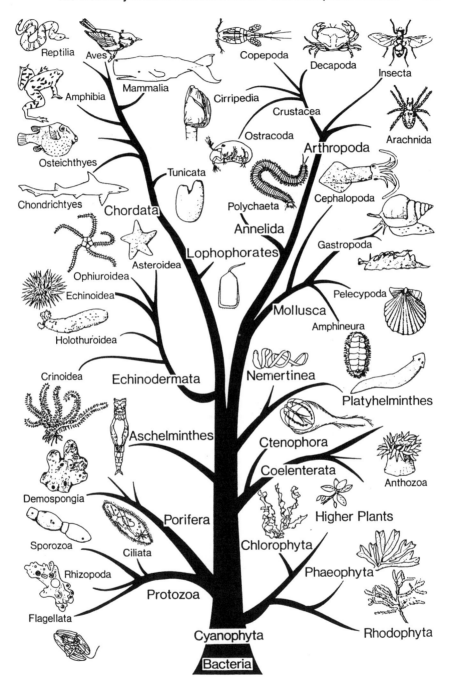

FIGURE 5–1 The phylogenetic tree. Reproduced with permission from Scheuer (1973).

simple invertebrates have existed for approximately 0.8 to 1.0 billion years, whereas the first primitive vertebrates did not develop until 500 million years ago, and early mammals have existed only for 180 million years (Tullar 1972).

Because of the rather short period of time (20 years) that a relatively intense effort has been made isolating and identifying new marine-derived natural products, and of the even shorter period of time, perhaps 5 to 10 years, that the majority of the research efforts in this area have clearly shifted from finding new compounds to finding biologically active compounds in clinically relevant assays, it is not surprising that only a handful of compounds has made it to clinical trials or beyond. Among these compounds are bryostatins, didemnin B, manoalide, the pseudopterosins, and Ara-A and Ara-C, synthetic analogs of marine sponge-derived compounds. Numerous, additional marine-derived compounds, however, have been found to be useful biological probes or biochemical tools, or have been studied because of their unusual structures and/or biological activities, including marine-derived prostaglandins, palytoxin, ciguatoxin, brevetoxin, okadaic acid, tetrodotoxin, saxitoxin, calyculins, and kainic acid. Manoalide is also being sold commercially as a biochemical probe.

In this chapter, we hope to show that marine natural products represent a promising source of potential therapeutic agents, to review some of the strategies and approaches by academic and industrial labs that have led to and are leading to the discovery of marine natural products with therapeutic potential, to highlight the sources of biologically active marine natural products and collection techniques of their respective source organisms, to review the history and status of selected marine-derived compounds that are at various points along the discovery and development path, and to present and discuss some of the unique difficulties in the therapeutic development of marine-derived compounds. We have chosen to focus our discussion on relatively low molecular weight compounds derived primarily from marine macroorganisms, such as invertebrates and algae, and from marine microorganisms; we have chosen not to discuss marine-derived biopolymers, i.e., agglutinins, polysaccharides, and glycoproteins (Shimizu and Kamiya 1983), high molecular weight toxins, i.e., conotoxins, sea anemone polypeptide toxins, jellyfish toxins, and marine-derived neurotoxins (Hall and Strichartz 1990), diagnostic reagents from marine organisms, for example, the horseshoe crab (Cohen 1979), and compounds from marine vertebrates. This chapter covers the literature to approximately the middle of 1992.

5.1 SOURCES OF MARINE NATURAL PRODUCTS WITH THERAPEUTIC POTENTIAL

A successful drug discovery program clearly needs continuous sources of new compounds, and a program planning for continued success cannot afford to exhaust its sources. Samples for screening may consist of in-house compound libraries, compounds from various chemical companies and universities, and/

or collections of natural materials. The acquisition of natural materials from the marine environment was slow to develop for various reasons; however, this situation has changed dramatically over the past 10 to 20 years.

5.1.1 Collection of Macroorganisms

The first step in the discovery of marine-derived compounds with therapeutic potential is the collection of marine organisms or marine material from which microorganisms are isolated (Pomponi 1988; Reed and Pomponi 1989). An intensive and systematic collection of marine organisms for compounds with therapeutic potential was hampered in the past by the general lack of (1) familiarity chemists engaged in this research had with the marine environment, (2) lack of documented ethno-natural history of marine organisms for medicinal purposes (Scheuer 1988), (3) appropriate collection tools, techniques, and protocols, and (4) access to clinically relevant assays by those who study marine-derived compounds. These problems have largely been overcome.

Three developments that greatly aided the systematic collection of marine organisms for compounds with therapeutic potential include the advent of scuba about 40 years ago, the development of deep water collection tools over the past 20 years (for example, at Harbor Branch Oceanographic Institution, HBOI), and, probably most importantly, a much closer collaboration among scientists in a variety of disciplines, i.e., marine biologists, ecologists, chemical ecologists, cell biologists, biochemists, pharmacologists, and natural products chemists.

Prior to scuba, shallow water samples could be obtained by wading or snorkeling. Collection by these techniques is limited to a depth of 20 feet (~6 m) or less and is generally inefficient. Scuba, however, allows routine collections to depths of 120 feet (~36.5 m) for 15 minutes of bottom time with no decompression stops. Using mixtures of nitrogen and oxygen (nitrox) or helium and oxygen instead of air allows scuba collections of extended duration and depth (Miller 1979). For example, helium and oxygen mixtures may allow dives to 800 feet (~244 m) or greater, and nitrox mixtures (32% oxygen/68% nitrogen) may double bottom time at depths of 50 to 120 feet (15 to 36.5 m). Mixed gas diving is not routine, however, and only a few sample collectors, mainly professional divers, use it. In general, most marine organisms that have been studied to date have been collected by scuba diving.

Deep water collections have been made by dredging and trawling. These are both cost-effective collection methods if the substratum does not cause damage to or snag the gear. There are several disadvantages to these approaches: it is difficult to photograph the organisms in their habitat; encrusting organisms or organisms that grow in crevices, under ledges or on steep rock faces cannot be easily collected unless the hard substrate that supports the organism is collected as well; dredging and trawling put all collected samples in close contact with each other, therefore, some organisms may chemically contaminate others because of exudations or secretions of various compounds;

and the environmental impact of dredging or trawling can be detrimental because the sampling is nonselective and habitats can be damaged or destroyed.

Deep water collections can also be made by manned and unmanned submersibles or remotely operated vehicles (ROVs). The cost of this form of collection is obviously much higher than dredging or trawling (approximately double or triple); however, the environmental impact of this form of collection is negligible. ROVs are used extensively by the oil industry for survey and repair of oil platforms, and by marine biologists, geologists, and archeologists for survey of marine biota, geologic formations, or artifacts, respectively; however, the technology has not progressed sufficiently to render the ROVs cost effective in marine sample collection (S. Pomponi, personal communication). In contrast, manned submersibles, as exemplified by those at HBOI, which is located on the southeast coast of Florida, have performed very well in the selective and pin-point collection (and recollection) of deep water marine samples for pharmacological evaluation. The HBOI-Johnson-Sea-Link submersibles, for example, can accommodate up to a total of four passengers, and operate at depths of 3000 feet (915 m); they have an array of collection devices and associated sample storage containers to facilitate collection of organisms displaying a wide range of shapes and sizes, and they have various video and 35 mm cameras for purposes of documentation. Using the manned submersibles, the research group in the Division of Biomedical Marine Research (DBMR) at HBOI, for example, has collected samples at 100 to 3000 feet (30 to 915 m) from the Galapagos Islands, various locations in the Caribbean, Bahamas, Gulf of Mexico, and in the eastern Atlantic at the Canary and Madiera islands.

5.1.2 Collection Strategy

Because the collection program in the DBMR at HBOI to collect marine macroorganisms has been considered a very good one in the marine field, the collection strategy will be reviewed briefly. For the first several years of the research effort, beginning in 1984, the strategy was to collect representatives of as many shallow and deep water benthic organisms as could be found. As the research program evolved, the effort began to focus more on shallow and deep water sponges, due to their high incidence of biological activity in a variety of mechanism-based and whole cell assays and high abundance (especially as compared with such organisms as tunicates, some bryozoans, and hydroids). Because of the specialized tools available for collecting of deep water samples, an important niche of this sample collection effort has been *deep* water sponges. Out of approximately 150 described species from deep water environments in the West Indian biogeographic region (Van Soest and Stentoft 1988), the DBMR/HBOI collection contains nearly all these; further, the collection contains at least another 20 to 40 new, undescribed species of deep water sponges. For comparison, the total number of described shallow and deep water species from the West Indian biogeographic region is approximately 600 (Pulitzer-Finali 1986).

As a consequence of their rather enormous collection effort, the current DBMR/HBOI collection consists of a total of several thousand deep water and well over 10,000 shallow water macroorganism samples in sufficient quantities for chemical fractionation. To our knowledge, this is the largest marine sample collection in the world. The number of discrete samples (with unique taxonomy) is probably less than one-third, and is an estimate based on the taxonomic identification to the genera level on less than one-quarter of the samples. (Taxonomic identification of these samples has been driven by biological activity of the respective extracts; inactive samples have not been identified unless the taxonomic identification has been trivial.) Other marine research groups have successfully focused their collection and drug discovery efforts on cnidaria (see discussions on the pseudopterosins and the fuscosides), blue green algae (Moore and Patterson 1988; Moore et al. 1988), and marine-derived microorganisms (Gustafson et al. 1989; Kobayashi 1989; Okami 1988).

5.1.3 Marine-Derived Microorganisms

Microorganisms in the terrestrial environment are a proven source of antibiotics and other therapeutic agents. It would seem likely that microorganisms derived from the marine environment should also yield therapeutic agents; however, chemical and biological studies have been slow to develop (Faulkner 1990, 1991). In general, marine-derived microorganisms have been isolated from environmental samples, i.e., sediments, water samples, decaying pieces of wood, etc., or from macroorganisms. To date, new structures have been found in marine-derived bacteria (Burkholder et al. 1966; Gustafson et al. 1989; Takahashi et al. 1987; Umezawa et al. 1983; Wratten et al. 1977), including actinomycetes (Nakamura et al. 1977; Okami et al. 1979; Sato et al. 1978), and in marine-derived fungi (Poch and Gloer 1989a, 1989b).

Isolation techniques for marine-derived microorganisms are relatively straightforward. For example, to obtain microorganisms from macroorganisms, macroorganisms are typically rinsed with sterile seawater, blotted, ground in artificial seawater or filtered seawater from the site, diluted, and plated onto several different agar media (Wilkinson 1978). Several general and several selective media are usually used for primary isolation cultures. For example, general media may include Zobell's marine agar (Zobell 1941) or peptone-yeast extract-seawater (Weyland 1969); actinomycete-selective media may include starch-yeast, extract-seawater, chitin-seawater, or glycerol-arginine agar-seawater with trace metals and vitamins added (Nonomura and Ohara 1969) as well as cycloheximide to inhibit fungal growth. Agar cultures can be incubated at different temperatures. Cultures are examined periodically and colonies are transferred to appropriate media for purification, identification, and biological and chemical studies.

A controversial facet of marine-derived microorganisms is their putative role with respect to the origin of bioactive natural products from marine macroorganism-microorganisms associations (Ireland et al. 1988). Symbiotic microorganisms have been repeatedly suggested as being the direct or indirect sources of bioactive metabolites in marine sponges (Frincke and Faulkner 1982; McCaffrey and Endean 1985) and other invertebrates, i.e., tunicates (Wright et al. 1990) and bryozoans (Fenical 1991). Symbiotic microorganisms, in the broadest working definition of the term, may be mutualistic or parasitic with the associated host animal and the symbiosis may be obligatory or facultative (Isenberg and Balows 1981). Actinomycetes are ubiquitous in the marine environment (Goodfellow and Haynes 1984; Weyland 1969), and marine invertebrates contain diverse communities of microorganisms (Bergquist 1978; Colwell 1991), including actinomycetes (E. Armstrong, personal communication) and cyanobacteria (Wilkinson 1980). Only several specific examples exist, however, in which secondary metabolites initially extracted and identified from macroorganisms, i.e., sponges, could be produced by fermentation of specific microorganism strains isolated from the respective macroorganism (Stierle et al. 1988; Elyakov et al. 1991). A noteworthy case in regard to marine vertebrates is that of the fish-derived N-3 fatty acids, which seem to have great value for preventing cardiovascular disease (Leaf 1985); they may indeed be produced by associated marine microorganisms (Kitagawa et al. 1989).

5.1.4 Distribution of Biologically Active Marine Natural Products

As evidenced by the numerous reviews of marine natural products (Faulkner 1984a, 1984b, 1986, 1987, 1988, 1990, 1991) and as will be shown in this chapter, the marine environment houses a vast number of new types of chemical structures; many of these structures represent new classes unknown previously from nature or synthesis. Analysis of these reviews indicates that a substantial amount of research has been conducted by chemists in Europe, especially Italy, and in Australia, New Zealand, Israel, Japan, and North America. The greatest number of marine-derived compounds have been isolated from shallow water organisms and their structures published by North American and Japanese researchers, who are located at academic institutions. To our knowledge, a greater number of academic and industrial research groups have become involved in the past several years in the search for biologically active and novel compounds from marine-derived microorganisms, and we expect that this increased effort will be reflected in both the scientific and patent literature in the next several years. Collection sites for the organisms that contain these compounds span the globe and include the Caribbean, Indo-Pacific, Mediterranean, Red Sea, and the Antarctic. Relatively few deep water marine organisms have been investigated, although they appear to be fertile grounds for biologically active compounds. Most shallow and deep water macroorganisms have been collected in tropical and subtropical latitudes, presumably because of milder conditions for sample collection, and because of the greater biodiversity of available organisms (Ricklefs 1973). These factors certainly facilitate the collection of a greater number of different organisms per collecting effort, though sample collection in higher latitudes has also been productive.

The earlier reviews by Faulkner show a phyletic distribution of all marine natural products reported between 1977 and 1985 (Ireland et al. 1988). During this time period, most marine natural products were derived from algae > sponges > cnidaria > echinoderms > tunicates > bryozoans > microbes. In part, this trend is related to the relative abundance of the respective organisms; i.e., tunicates, bryozoans, and hydroids are prolific producers of structurally unusual and biologically active natural products. However, these animals are not as abundant as algae, sponges, cnidarians, and echinoderms. With respect to their biosynthetic capabilities of secondary metabolites, algae, sponges, and cnidaria produce a greater proportion of terpenoids than any other compound type; in all three groups of organisms, the second most abundant biosynthetic class of compounds is polyketides. From our own analysis of the compounds reported since 1985, the phyletic distribution of marine natural products is approximately the same. However, the total number of compounds reported from sponges is now about the same as those reported in algae, and there have been significant increases in the number of compounds reported from tunicates and marine-derived microorganisms.

Analyses of reviews that have emphasized biologically active marine natural products (Fusetani 1987; Krebs 1986; Munro et al. 1987) show that virtually all sessile organisms are represented, i.e., sponges, cnidarians, tunicates, bryozoans, hydroids, zoanthids, and micro- and macroalgae. Therefore, although our own experience has led us to focus our efforts on sponges, we believe it is premature to conclude that there may be preferred organisms for active marine-derived compounds for any particular disease state; further, deep water macroorganisms, marine-derived microorganisms from any environment, and organisms from stressed environments, i.e., hydrothermal vents, anaerobic environments, etc., have not been sufficiently studied in any case. The search continues.

5.2 APPROACHES TO THE DISCOVERY OF BIOLOGICALLY ACTIVE MARINE NATURAL PRODUCTS

With few exceptions, intense and systematic bioassay-guided studies of extracts and compounds from marine organisms by academic, government, and industrial research groups using clinically relevant assays to discover naturally occurring substances with therapeutic potential have only been underway for less than 10 years. Because of the tremendous advances in the understanding of the biology of certain diseases and the concomitant explosion of new assays, researchers are now in a position to explore fully the potential of marine natural products. These biological advances highly complement the development of new chemical technology, i.e., new separation and structure elucidation techniques.

In general, drug discovery strategies can be trivially separated into three categories:

1. Chemically driven, finding biological activities for purified compounds.
2. Biologically driven, bioassay-guided approach beginning with crude extracts.
3. Combination of chemically and biologically driven approaches (vide infra).

For marine-derived drug discovery, strategies may involve one or more of the following elements:

1. In vivo screens
2. Mechanism-based screens
3. Functional, whole cell, or tissue-based assays
4. On-site assays versus post-collection assays
5. "Dereplication," via biological profiles or chemical profiles, e.g., thin layer chromatography (TLC), nuclear magnetic resonance (NMR), and high pressure liquid chromatography (HPLC)

The chemically driven or "traditional grind-and-find" approach (Persinos 1991b) has been pursued vigorously, primarily by academic research groups. In fact, in the 1950s through the 1970s, the majority of the academic effort in studies of marine natural products was "chemistry driven," i.e., the object of the search was novel compounds from marine sources. TLC (for use of TLC in the field, see Norris and Fenical 1985) and ^{1}H NMR (D.J. Faulkner, personal communication) have been used extensively to "screen" crude extracts and solvent partitions for unusual, and, therefore, interesting patterns. More recently, ^{13}C NMR has been used (Manes et al. 1985; Crews et al. 1986; Adamczeski et al. 1988). The structural elegance and number of the compounds found in various reviews on marine natural products attest to the success of this approach. It should be noted that isolation and structure elucidation of new, structurally complex compounds have provided immense gratification to marine natural products chemists and, indeed, to academic synthetic chemists who thrive on new synthetic challenges.

With respect to discovering potential therapeutic agents, the chemically driven approach is a viable one. Through a marine pharmacology program at the University of California-Santa Barbara (UCSB) and Scripps Institution of Oceanography (SIO), which has been funded for nearly 20 years by Sea Grant/NOAA, over 1200 pure compounds isolated primarily in the laboratories of W. Fenical (SIO) and D.J. Faulkner (SIO) have been evaluated by

R. Jacobs (UCSB). Although many of these compounds were isolated based on the antibacterial, cytotoxic, and/or anti-inflammatory activities of the respective extracts, many were isolated based on purely chemical, chemotaxonomic, or chemical ecological interests and also screened by R. Jacobs (Jacobs et al. 1985). The anti-inflammatory properties of manoalide (see Section 5.5.3), for example, were discovered through an essentially chemically driven process. In contrast, the anti-inflammatory properties of the pseudopterosins (Section 5.5.4) and the fuscosides (Section 5.4.3) were discovered from a more activity-directed effort. The chemically driven approach has also served other valid scientific purposes, i.e., the structural information can be used to help solve taxonomic problems associated with a variety of marine organisms (Bergquist and Wells 1983; Fenical and Norris 1975; Lawson et al. 1984; Liaaen-Jensen et al. 1982; Pomponi et al. 1991).

Beginning in the 1970s through today, the majority of the academic-based research efforts has become essentially "biologically driven," i.e., the object of the search has shifted to discover marine-derived compounds with biological activity. The biological activities include exploring their potential as agrochemicals (Crawley 1988) and pharmaceuticals, as well as their possible chemical ecological roles (Bakus et al. 1986; Hay and Fenical 1988). In addition, a few marine-derived compounds have been isolated and their structures elucidated based on "ethno-natural history" (Scheuer 1988); compounds and organisms that have been implicated as public health hazards (Haddad et al. 1983), i.e., tetrodotoxin (Nakanishi et al. 1975), red tide and shellfish toxins (Hall and Strichartz 1990), and ciguatoxin (Yasumoto and Murata 1990) are prime examples.

The screening programs for antitumor agents at NCI have been biologically driven since their inception in 1958 (Suffness et al. 1989). However, the majority of the (U.S.) academic-based marine natural products chemistry groups were not funded by NCI on a grant or contract basis until the 1970s and 1980s. The current screening program utilizes an extensive human tumor panel that includes a number of cell lines which are representative of solid tumors having different tissue origin and clinical responsiveness, as well as murine and human leukemia cell lines. This approach is predicated on the assumption that agents can be identified that exhibit selective activity toward particular tumor histiotypes, and such selectivity is detectable in long-term, tissue-culture-adapted cell lines (Johnson and Hertzberg 1989). Numerous marine-derived, pure compounds have been screened by this panel, and several compounds have current preclinical interest (D. Lednicer, personal communication).

In general and by comparison, drug discovery in industry has "evolved to the use of specific assays with target receptors and enzymes involved in pathogenesis of disease rather than cellular or tissue assays" (Johnson and Hertzberg 1989) and has benefitted immensely from breakthroughs in receptor technology (Hall 1989; Reuben and Wittcoff 1989). These assays reflect new opportunities due to the recent identification of previously unrecognized biomolecular targets for therapy (Larson and Fischer 1989). More specifically, this approach for most disease areas is characterized in industry by:

1. Essentially exclusive reliance on biological activity of crude extracts in numerous target-specific assays, i.e., enzyme assays and receptor-binding assays, for selection of crude extracts and bioassay-guided fractionation of the crude extract (prioritization criteria emphasize selectivity and potency).
2. High volume, automated screening, i.e., thousands of samples per year for smaller companies and thousands of samples per week for larger companies.
3. The use of "functional" or whole-cell assays to confirm activity in a particular disease state and to further prioritize samples for fractionation.
4. The use of genetically engineered microorganisms, enzymes, and receptors.

For example, the targeted approach in the antitumor area includes assays for multidrug resistance reversing agents, mitogenic growth factors, protein tyrosine kinase antagonists, topoisomerase inhibitors, DNA intercalators, purine/pyrimidine nucleotide biosynthesis inhibitors, tubulin polymerization inhibitors, signal transduction agonists or antagonists (phospholipase a, phosphatidlyinositol kinase, protein kinase c), and assays to find compounds that interfere with the expression or function of oncogene products (Larson and Fischer 1989; Johnson and Hertzburg 1989). One major pharmaceutical company reportedly screens every marine- and terrestrially derived extract against 11 to 12 different assays that have indications in the areas of cancer, AIDS, and cardiovascular and inflammatory diseases (Persinos 1989a). The assays are prioritized, and after one has provided five to six lead structures, the assay is dropped or its priority lowered, and another (perhaps unproven) assay is brought on-line, or the priority of another assay is raised.

In the antitumor area, selectivity, potency, and high volume screening are elements common to both the NCI and industry; however, NCI uses whole cell assays and industry uses primarily mechanism-based assays. Both approaches appear to be valid, or, in the case of NCI, yet to be disproven, and arguments have been put forth to support the use of either approach. Arguments against using mechanism-of-action screens as primary screens are that serendipity and pharmacology are lacking, and useful activities may be missed (Suffness et al. 1989). The arguments for whole-cell systems are that general pharmacokinetic problems are addressed; if a compound can't cross a membrane, it will test negative; also, if metabolic activation is required for activity, then it has a chance of occurring in whole-cell systems. In some areas such as immune modulation, whole cell assays, e.g., mixed lymphocyte reaction (MLR), mitogen response, and lymphokine production assays, rather than receptor or enzyme assays, continue to be used routinely by academic, government, and industrial research groups.

At least in the antitumor area, arguments against the use of whole cell (cytotoxicity) assays are that (1) they are uneconomical and unproductive because of the requirement to discriminate bona fide leads with therapeutic potential from the vast number of known cytotoxic natural products and of the large number of cell lines against which extracts must be tested, and (2) the overwhelming majority of cytotoxic natural products have minimal or no significant activity in certain chemosensitive animal tumor models (Johnson and Hertzberg 1989). In addition, a practical concern often faced by pharmaceutical companies when deciding which compounds to develop is whether or not the mechanism(s)-of-action is (are) known, i.e., if it is not known, vis-à-vis mechanism-based screening efforts, and a substantial effort is required to determine the mode of action, or if it cannot be determined, the compound, in all likelihood, will not be pursued (S. Hecht, personal communication).

Because of the low correlation between cytotoxicity and antitumor activity, a number of programs have utilized in vivo tumor models directly for drug discovery (Johnson and Hertzberg 1989). From 1958 through 1985, NCI used in vivo L1210 and P-388 murine leukemia assays as primary screens (Suffness et al. 1989; Boyd et al. 1988) and was successful primarily in identifying compounds possessing clinical activity against leukemias and lymphomas. Unfortunately, they were not very successful in finding compounds active against slow growing tumors in humans. Further, these in vivo assays were expensive, time consuming, and relatively insensitive (Suffness et al. 1989). Our experience at HBOI in other disease areas has shown that activity in in vitro antiviral assays does not translate well to in vivo activity, e.g., using *Herpes simplex*. In contrast, reasonable correlations exist between in vitro and in vivo antifungal activity, e.g., using *Candida albicans*, and between in vitro and in vivo immune suppressive activity, e.g., using the MLR as the in vitro assay and the graft-versus-host reaction (GVHR) or splenomegaly model as the follow up in vivo assay.

Chemically and biologically driven approaches can be combined. In our experience, this combined approach has meant selecting extracts for chemical fractionation based on the biological activity profile of the crude extract. However, instead of using a bioassay-guided approach to purify the compounds responsible for the activity of the extract, NMR and TLC are used to isolate the chemically most interesting substances. Ideally, the structurally unusual or novel compounds are also responsible for the activity of the extract. This approach works well when the active compounds are present in high concentration and the assay turnaround time is longer than a couple of weeks. If the concentration is low, as was the case for the ecteinascidins (Wright et al. 1990), structurally complex isoquinoline alkaloids found at a level of 1 mg compound/1 kg of the tunicate, *Ecteinascidia turbinata*, the active compounds would have been overlooked completely, had a purely chemical approach been employed. This approach is indeed productive with respect to isolating numerous new compounds, at least some of which usually express some of the activity observed for the crude extract, but is obviously not the best method to identify the most active compounds if they are present in low concentration.

An element of the biologically driven approach that has been used by some research groups is on-site or shipboard assays (Rinehart 1988a, 1988b; Rinehart and Shield 1988). Proponents of on-site bioassays point out that recollections of organisms whose extracts have been found active can be made immediately, organisms whose extracts prove to be inactive during the collecting trip can be subsequently avoided, and chances are improved for observing biological activity in extracts/samples that may undergo degradation by enzymes, air, or heat. The requirements for on-site bioassays are that they do not consume much time; they are simple, rugged, and do not involve pathogenic organisms. Arguments against the use of on-site bioassays include the following:

1. Sample collection is an expensive activity, especially with large research vessels and associated submersibles. Therefore, a more efficient and productive plan is to focus efforts on the collections themselves, and fill ship berths with experts in sample collection and organism identification rather than experts in conducting assays.
2. Because the optimum collection strategy is to collect samples in as many different locales and habitats as possible, the charted course for a collecting trip generally does not involve retracing parts of the route; therefore, it is not practical to count on recollecting larger amounts of a particular organism whose extract has exhibited activity. Further, some assays take several days, and waiting at one collecting site or in the vicinity until assay results are evaluated is a waste of collecting resources.
3. A valid strategy to optimize finding new marine-derived compounds with therapeutic potential is to have the respective extracts evaluated in as many clinically relevant assays as possible and making collection decisions on a few simple (usually unsophisticated), rugged on-site assays imposes an unnecessary constraint.
4. Worrying about sample degradation after taking reasonable precautions, e.g., freezing samples or storing them in solvent before biological evaluation, also imposes unnecessary constraints. In our experience, biological activity in most frozen samples remains intact for up to 9 years; it is simply not worth the effort to concern oneself with a minority of samples that are somewhat labile.

An important element for both chemically and biologically driven approaches is dereplication, which has been defined as the elimination of replicates or repeats of materials already isolated and known in the literature (Suffness et al. 1989; Technical Resources Inc. 1987). In a biologically driven approach, the concept is simply to determine as quickly as possible the chemical nature of the compounds responsible for the biological activity of the crude extract. With this information in hand, informed decisions can be made whether or not to expend precious resources on the (bioassay-guided) isolation and structure elucidation of the active compounds. Numerous databases are available to assist in dereplication (Technical Resources, Inc. 1987), including those assembled by J. Berdy (Institute for Drug Research, Budapest, Hungary), J. French (Warner Lambert pharmaceutical company), D.J. Faulkner (SIO), NAPRALERT (Natural Products Alert, a literature database assembled by N.R. Farnsworth, University of Illinois, Chicago), *Chemical Abstracts*, and a database assembled initially by P. Crews (University of California-Santa Cruz; Crews et al. 1991), which has been further embellished by our group in the DBMR/HBOI and other researchers in the marine natural products field.

In our hands, dereplication of marine-derived compounds from macroorganisms has consisted of (1) taxonomy, (2) chemical analysis, and (3) biological profile. Our approach has been to first use the biological activity profile of the extract in order to decide whether or not to continue effort on a sample; if the decision is affirmative, then solvent partitions of the extract are prepared and biological activity profiles are obtained for organic and aqueous layers of solvent partitions. If the activity persists in one or more solvent partitions, the taxonomy of the source organism is ascertained by an HBOI expert in sponge, gorgonian, or algal taxonomy or by collaborators whose taxonomic expertise cover other phyla. Concurrently, NMR spectra and TLC data are obtained for the crude extract and solvent partitions. Based on these data and the biological activity profiles of the crude extracts and solvent partitions, several databases are consulted to better understand the nature of the compounds that may be responsible for the biological activity in the source organism:

1. Our own, which has 9 years of biological activity and taxonomy data about marine organisms.
2. The database assembled by D.J. Faulkner (SIO), which has information retrievable by organism type, name of researcher, compound name, genus or species name, molecular formula, or molecular weight.
3. The database assembled by P. Crews.
4. *Chemical Abstracts*.

The Crews database can be searched by a combination of structure/substructure, number of carbons and hydrogens in molecule (so-called Attached Proton Test (APT) formula), molecular weight, and taxonomy, if known.

With respect to the various approaches that have been used to isolate marine natural products with therapeutic potential, a somewhat obvious truism can be put forth—one usually finds what one is looking for. If novel compounds are sought (chemically driven approach) (which may subsequently be evaluated for biological activity), NMR, TLC, HPLC, and/or liquid chromatography/mass spectrometry (LC/MS) can be used to "screen" extracts, and taxonomic identification of the respective organism can be used as well to dereplicate. If biologically active compounds are sought (biologically driven approach), mechanism-based, functional whole cell, tissue-based or in vivo assays are used to screen extracts and guide chemical fractionation. In regard to marine natural products, the biologically driven approach is very appropriate because many compounds whose structures have been published have not been thoroughly evaluated for biological activity. Further, as is well appreciated, a new assay redefines an otherwise exhausted compound and extract library into a whole new sample set. The particulars of a biologically driven approach are critical, however, if marine-derived compounds are sought with therapeutic potential. Indeed, discriminating and predictive assays are required to find compounds that will be used therapeutically not within the next year but rather in 5, 10, or 15 years from now!

To optimize a biologically driven approach, interorganizational collaboration between groups that have expertise in chemistry, biology, research, and development is important, in our view; few academic, government, or industrial research groups truly have the resources, expertise, and interest to be completely autonomous in taking a compound from the ocean to the market. Academic and government research groups typically have some expertise or ready access to expertise in marine biology and ecology, sample collection or procurement, sometimes fermentation, isolation, and structure elucidation or synthesis (usually not both), and biological evaluation in a limited number of disease areas. Many industrial concerns usually have expertise in isolation, structure elucidation, synthesis, biological evaluation, process chemistry, fermentation, preclinical and clinical development, patenting, and marketing. Pharmaceutical companies, however, view natural products as one source in their drug discovery programs, and, therefore, their resources are spread between natural products screening and chemical synthesis.

From our experience, the best relationship to strive for between academic and industrial programs is a scientific collaboration, i.e., one in which extensive exchange of technical information takes place. From an academic point of view, the needs are usually modest, i.e., scientific gratification and monetary support for research. From an industrial point of view, the samples obtained and information exchanged through such a collaboration must ultimately have a commercial benefit (in order to stay in business). Further, a related industrial concern is confidentiality, which, unfortunately, is not the best fit with the academic goal of publishing (or perishing!). A distinct advantage of working with a pharmaceutical company is the possibility that if a marine-derived compound expresses a suitable profile from the company's point of view, resources and expertise to conduct structure-activity optimization studies may become available to study the compound, and to supply sufficient quantities of optimized compound through fermentation or synthesis for preclinical and clinical studies. Typically, academic groups are simply not in a position to accomplish these tasks by themselves; if the goal of a marine-derived drug is to become a reality, some sort of appropriate partnership between an academic group and an interested development partner is necessary and inevitable.

5.3 NOTEWORTHY AND COMMERCIALLY AVAILABLE MARINE NATURAL PRODUCTS

In the last 25 years, several marine-derived compounds have generated considerable interest scientifically, commercially, and from a public and health point of view; these include prostaglandins (5.1A, 5.1B, 5.2A, 5.2B, 5.3, 5.4), the potent and structurally complex toxin, palytoxin (5.5), and one of the major causative agents of fish poisoning, ciguatoxin (5.6). Further, because of their unique and potent biological activities, several marine-derived compounds have already found use as biological probes or biochemical tools and are sold commercially, i.e., palytoxin (5.5), brevetoxins (e.g., brevetoxin A, PbTX-1) (5.7), okadaic acid (5.8), tetrodotoxin (5.9), saxitoxin (5.10), calyculin A (5.11), manoalide (5.12), and kainic acid (5.13).

5.3.1 Prostaglandins

Prostaglandins are a group of monocyclic fatty acid metabolites that possess diverse and potent biological activities that include control of blood pressure, renal blood flow, smooth muscle contraction, and gastric acid secretion, and are involved in inflammation (Berkow and Flethcher 1987). In 1969, the Caribbean gorgonian, *Plexaura homomalla* (an animal related to sea fans and sea whips), was reported to contain large concentrations of 15-*epi*-PGA$_2$ (5.1A) and its diester (5.1B) (Weinheimer and Spraggins 1969). Unfortunately, the 15-*epi*-configuration rendered these compounds physiologically

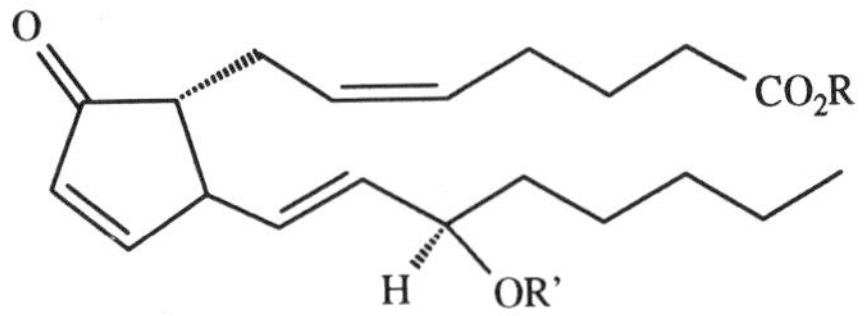

A   R = H
B   R = Ac   (5.1)

inactive. However, because of the limited supplies of mammalian prostaglandins from sheep seminal tissues at the time, there was great interest in *P. homomalla* as a potential commercial-scale source of synthetic precursors (Weinheimer and Spraggins 1969; Corey and Matsuda 1987). Further, it was determined that *P. homomalla* grew well at shallow depths in the Caribbean on artificial substrates (Corey and Matsuda 1987). Subsequently, the industrial process chemistry to synthesize active prostaglandins improved such that interest waned in harvesting *P. homomalla* from the ocean.

From a scientific viewpoint, *P. homomalla* continues to generate interest because the prostaglandins found in this animal and a soft coral, i.e., the clavulones (e.g., clavulone I; (5.2A)) found in *Clavularia viridis* (Kikuchi et

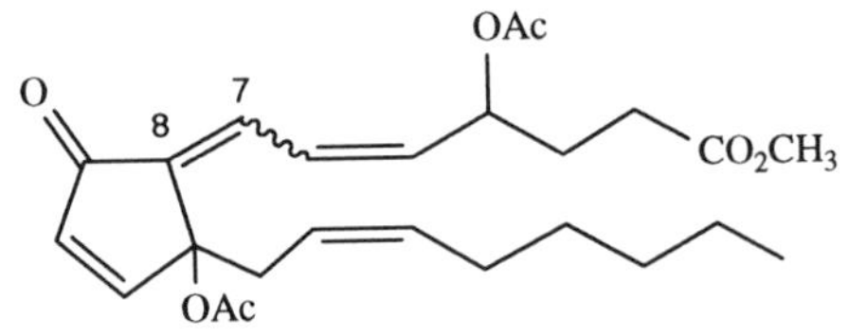

$$\text{A} \quad 7,8 - E$$
$$\text{B} \quad 7,8 - Z$$

(5.2)

al. 1982), are not produced by the mammalian cyclooxygenase/endoperoxide route (Corey et al. 1975); rather, they appear to be produced from arachidonic acid by lipoxygenation at C-8, and subsequent transformation into an allene oxide, followed by oxidopentadienyl cation intermediates. These intermediates readily account for the observed 5,6-*cis*- and 5,6-*trans*-prostanoids as well as a cyclopropyl compound that are produced when arachidonic acid is incubated with an acetone powder of *P. homomalla* (Corey and Matsuda 1987; Baertshi et al. 1989).

In addition to the physiologically inactive $(15R)$-PGA_2, its methyl ester and the clavulones, other marine-derived prostanoids have subsequently been reported from *P. homomalla*, including (physiologically active) $(15S)$-PGA_2, $(15S)$-PGE_2, $(15S)$-5,6-*trans*-PGA_2, $(15S)$-13,14-*cis*-PGA_2, 15-acetate, $(15S)$-13,14-dihydro-PGA_2 acetate methyl ester, and $(15S)$-13,14-dihydro-PGA_2 (Bundy et al. 1972; Schneider et al. 1972, 1977a, 1977b), $PGF_{2\alpha}$ from the gorgonean *Euplexaura erecta* (Komoda et al. 1979), PGE_2 and $PGF_{2\alpha}$ from the red alga *Gracilaria lichenoides* (Gregson et al. 1979), $(15S)$-$PGF_{2\alpha}$ 11-acetate methyl ester and its 18-acetoxy derivative as well the two corresponding free carboxylic acids from a soft coral *Lobophyton depressum* (Carmely et al. 1980), the claviridenones (e.g., claviridenone-a; (5.2B)) from *C. viridis* (Kobayashi et al. 1982), the chlorovulones (e.g., chlorovulone I; (5.3)) from

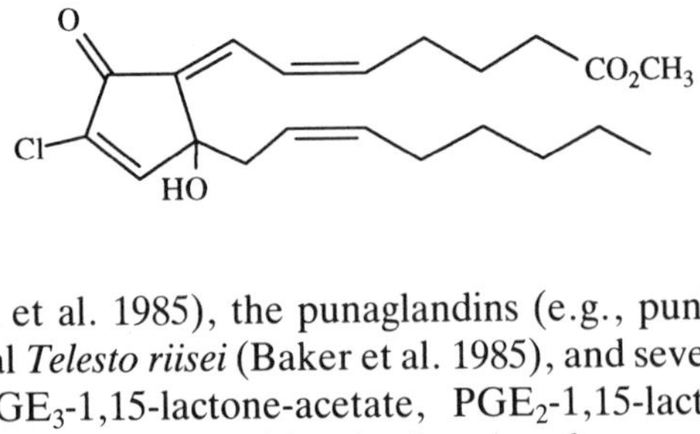

(5.3)

C. viridis (Iguchi et al. 1985), the punaglandins (e.g., punaglandin 1; (5.4)) from the octocoral *Telesto riisei* (Baker et al. 1985), and several prostaglandin-1,15-lactones, PGE_3-1,15-lactone-acetate, PGE_2-1,15-lactone, and PGE_3-1,15-lactone, which were isolated for the first time from a natural source, the nudibranch mollusk *Tethys fimbria* (Cimino et al. 1989).

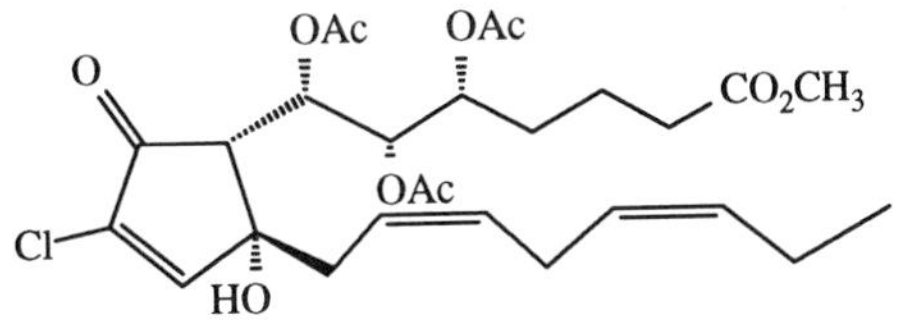

$$(5.4)$$

Punaglandin 3, which is the $\Delta^{7,8}$ analog of (5.4) was reported to express cytotoxicity against L1210 with an IC_{50} of 20 ng/mL (Baker et al. 1985); chlorovolune I (5.3) was reported to be cytotoxic against human promyelocytic leukemia (HL-60) cells with an IC_{50} of 0.01 µg/mL (Iguchi et al. 1985); the clavulones showed anti-inflammatory effects at 30 µg/mL in the fertile egg assay, which uses the chorio-allantoic membrane of the chick embryo as the site of induced inflammation (Kikuchi et al. 1982, 1983).

5.3.2 Palytoxin

The independent structure elucidation of palytoxin (5.5) by R. Moore and Y. Hirata and their respective research groups (Moore and Bartolini 1981;

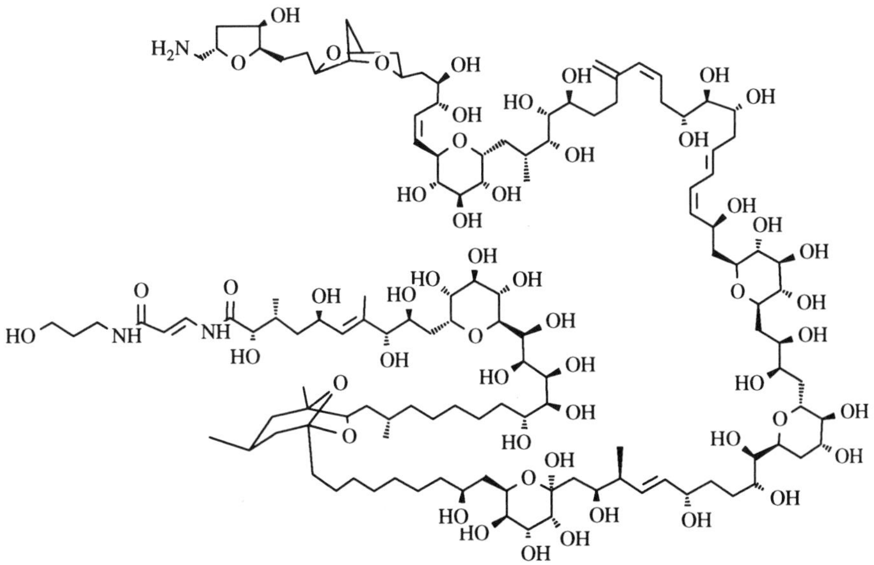

$$(5.5)$$

Uemura et al. 1981), the subsequent completion of the structure determination of the compound's stereochemistry, and the synthesis of palytoxin carboxylic acid by Y. Kishi and his research group (Armstrong et al. 1989a, 1989b and references therein), must be considered quintessential scientific achievements. Palytoxin, a large polyoxygenated metabolite (molecular weight, 2681) obtained from the zoanthid *Palythoa toxica* and other *Palythoa* species, posed an extreme challenge; the structure elucidation was achieved by oxidatively degrading the molecule and elucidating the degradation products, which by themselves are larger than many natural products. From a biological activity viewpoint, it is especially noteworthy because it is the most potent nonproteinaceous toxin known, having a minimum lethal dose in guinea pigs of 0.15 µg/kg (Haddad et al. 1983). (By comparison, the most potent proteinaceous toxin, botulinus toxin A, has a minimum lethal dose of 0.00003 µg/kg.) In the two-stage mouse skin promotion assay, palytoxin is also a potent tumor promoter, although it fails to induce ornithine decarboxylase in mouse skin and does not bind to protein kinase C (Fujicki et al. 1984). Similar to phorbol ester-type tumor promoters, palytoxin has been found to stimulate arachidonic acid metabolism in rat liver cells (Levine and Fujicki 1985). Palytoxin also acts synergistically with phorbol ester-type tumor promoters to promote prostaglandin release (Levine and Fujicki 1985; Levine et al. 1986) and stimulate histamine release from rat peritoneal mast cells (Ouchi et al. 1986). Further, palytoxin has been found to inhibit epidermal growth factor (EGF) binding through a pathway that is not dependent upon protein kinase C (Wattenberg et al. 1987). Palytoxin is commercially available.

5.3.3 Ciguatoxin

Ciguatoxin (5.6) is the primary causative agent of ciguatera, which is a type of food poisoning that occurs intermittently and unpredictably from eating certain coral reef fish (Haddad et al. 1983; Murata et al. 1990; Yasumoto and Murata 1990). The characteristic clinical syndrome includes severe gastrointestinal and neurologic symptoms (Haddad et al. 1983), and affects annually approximately 20,000 people worldwide (Murata et al. 1990). Although the origin of ciguatoxin has been the subject of speculation for centuries (Haddad et al. 1983), the causative organism, the dinoflagellate *Gambierdiscus toxicus*,

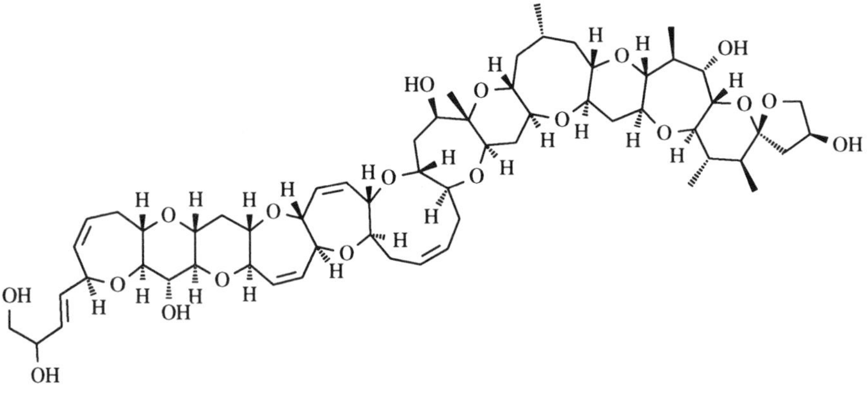

(5.6)

was not identified until 1977 (Yasumoto et al. 1977). Ciguatoxin is transferred through the food chain among coral reef biota and accumulates in carnivorous fish, including barracuda, grouper, amberjack, dolphin-fish, pompono, morey eels, and others (Murata et al. 1990).

Although research on the isolation and structure elucidation has been on-going since the 1960s (Scheuer et al. 1967), the toxin is always present in such trace quantities in affected fish that the structure was finally solved only recently, on less than 1 mg from 125 kg of moray eel viscera (Murata et al. 1990). Using primarily NMR, the structure was determined to be a polyether, structurally similar to some of the other polyether toxins from dinoflagellates, such as brevetoxin A (vide infra).

Numerous marine-derived compounds (in addition to palytoxin) are commercially available. A very brief description of the substance, source, and biological activity or utility for each of these compounds follows.

5.3.4 Brevetoxins A (PbTx-1), (5.7)

These are dinoflagellate toxins (e.g., brevetoxin A, Shimizu et al. 1986), which express toxicity by site-specific binding associated with voltage-sensitive sodium channels, and cause contractile paralysis in animal models (Trainer et al. 1990).

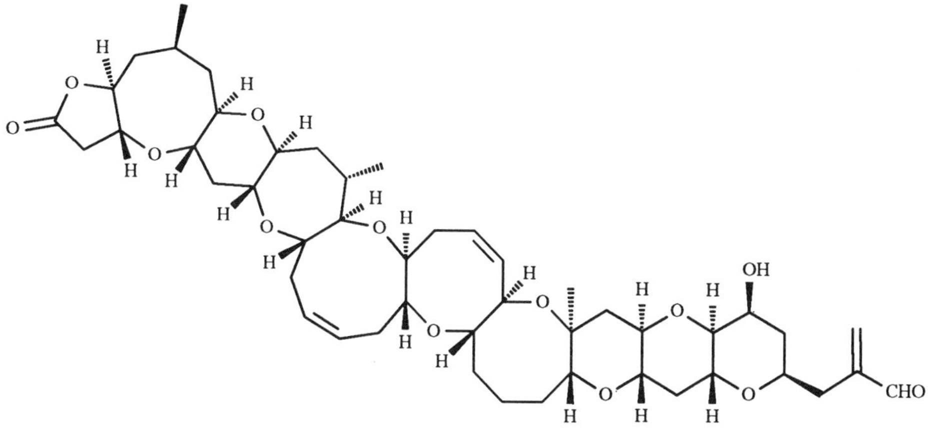

(5.7)

5.3.5 Okadaic Acid (5.8)

This is dinoflagellate toxin (Murata et al. 1982), which was first isolated from a sponge (Tachibana et al. 1981). It has been implicated as a major causative agent of diarrhetic shellfish poisoning. It is a potent tumor promoter, but is *not* an activator of protein kinase C, and is a potent inhibitor of protein phosphatases-1 and -2A. This toxin expresses marked contractile effect on smooth muscles and heart muscles (Haystead et al. 1989).

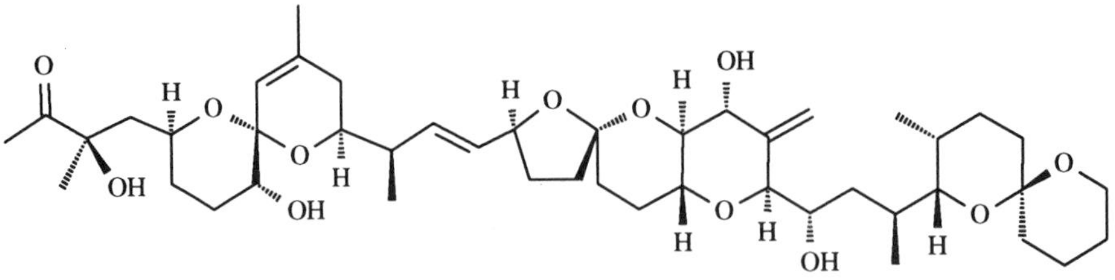

(5.8)

5.3.6 Tetrodotoxin (5.9)

This bacterial metabolite (Tamplin 1990) is also found in fish (pufferfish), starfish, crab, octopus, frog, newt, salamander, goby, gastropod, mollusk, flatworm, annelid, zooplankton, and algae (Nakanishi et al. 1975). It is a potent toxin that reversibly blocks sodium channels but has no effect on potassium channel blockage. It is useful as a marker of sodium channels in excitable tissues and can be used to block sodium channels in multiple conducting systems, such as the study of the calcium channel in cardiac muscle (Hu and Kao 1985).

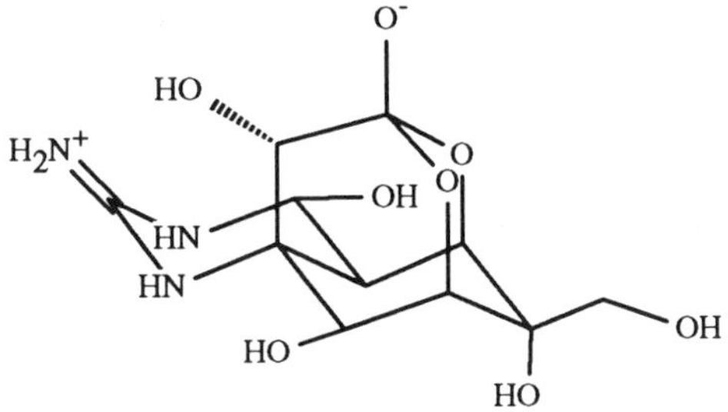

(5.9)

5.3.7 Saxitoxin (5.10)

This bacterial metabolite/dinoflagellate toxin (Hall et al. 1990; Tamplin 1990) is one of the main causative agents of paralytic shellfish poisoning (Hall et al. 1990). It is a potent, reversible channel blocking agent and similar in mechanism-of-action to tetrodotoxin (Kao 1966; Hall et al. 1990).

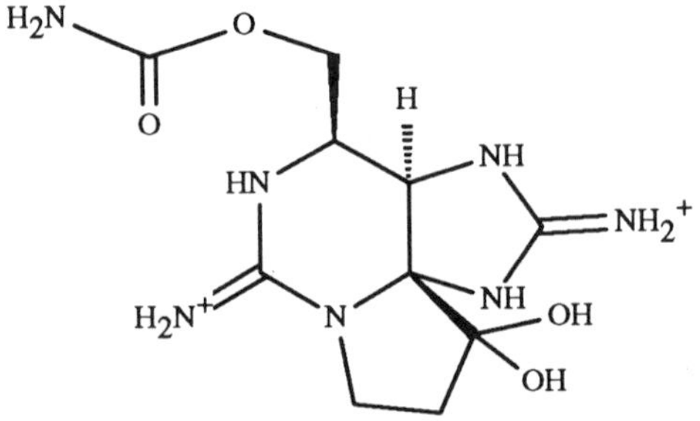

(5.10)

5.3.8 Calyculin A (5.11)

This sponge metabolite (Kato et al. 1986; Matsunaga and Fusetani 1991) inhibits protein phosphatases A and 2A (which also serve as okadaic acid receptors), promotes tumor growth on CD-1 mouse skin, and induces ornithine decarboxylase in mouse skin (Suganuma et al. 1990; Matsunaga et al. 1991). It also exhibits antitumor activity against Ehrlich and P-388 leukemia in mice (Kato et al. 1986, 1988).

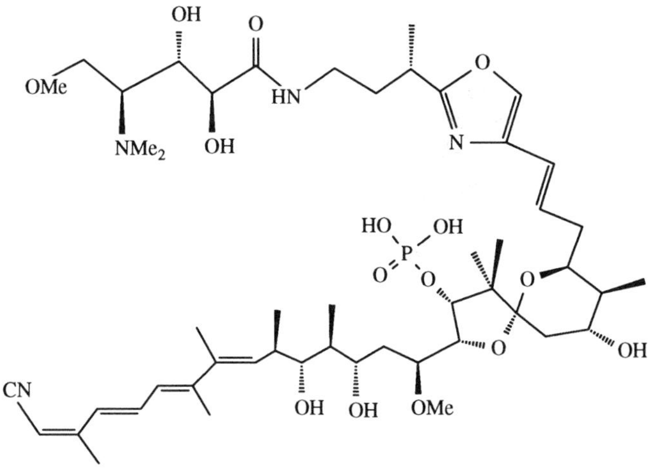

(5.11)

5.3.9 Manoalide (5.12)

This compound is a sponge metabolite that inhibits phospholipase A_2 and C and intervenes in the arachidonic acid cascade (Wheeler et al. 1987). (See Section 5.5.4.)

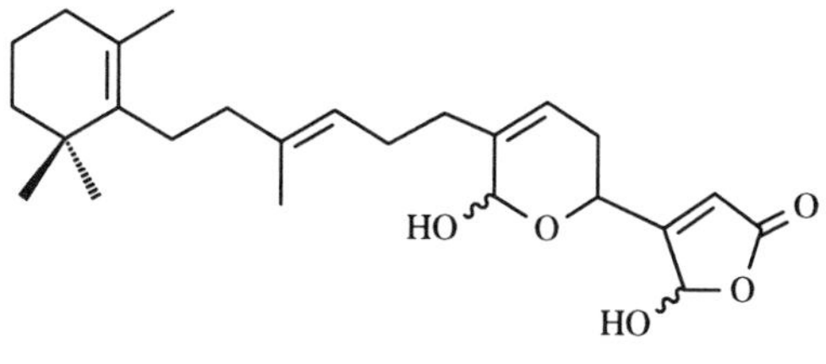

(5.12)

5.3.10 Kainic Acid (5.13)

This red algal metabolite is a well-known anthelmintic (Fattorusso and Piatelli 1980). It is an excitatory amino acid that affects mammalian and amphibian central nervous systems and is used as a neurobiological tool in the study of Huntington's disease and epilepsy (McGeer et al. 1978).

By serving as research tools, these marine-derived compounds have an important spin-off effect of facilitating the discovery of therapeutic agents.

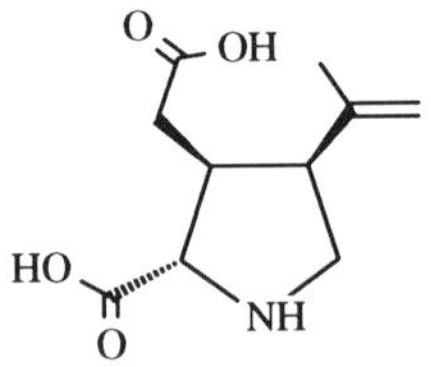

(5.13)

An example of this role can be found in the case of marine neurotoxins such as tetrodotoxin and saxitoxin. These compounds, each of which has a specific and distinct action on the function of the voltage-sensitive sodium channel of excitable tissues, have for years served as probes to investigate the structure and function of the sodium channel protein itself and its role in the overall function of neuromuscular tissue. Such studies lead to the identification of new biochemical targets for chemical intervention and thus facilitate new drug discovery.

5.4 SELECTED COMPOUNDS
WITH THERAPEUTIC POTENTIAL

Few marine-derived compounds have been studied biologically beyond a simple assessment of cytotoxicity and antimicrobial activity; therefore, it has been difficult to fully assess their therapeutic potential. The purpose of this section is to share the history and results of in-depth biological studies of several marine-derived compounds that have therapeutic potential: dercitin (5.14), which expresses antitumor activity; mycalamides A (5.26A) and B (5.26B) and the structurally related compound, onnamide (5.28), which also exhibits antitumor activity as well as antiviral activity; discodermolide (5.19) and microcolins A (5.25A) and B (5.25B), which display promising immunosuppressive activity; and fuscoside A (5.23), which is a promising anti-inflammatory agent. These compounds were derived from a variety of sessile, shallow and deep water marine organisms: sponges, gorgonians, and blue green algae, that were collected throughout the world. All these compounds have been or can be synthesized.

5.4.1 Dercitin
Dercitin (5.14) is one of many acridine heterocycles isolated and identified in the last 5 to 10 years from ascidians collected in the Indo-Pacific and Red Sea and shallow and deep water sponges collected in the Indo-Pacific and Caribbean. Dercitin itself was isolated from a deep water Caribbean sponge, *Dercitus* sp. (Gunawardana et al. 1988). Structurally related compounds include nordercitin (5.15A), dercitamine (5.15B), dercitamide (kuanoniamine C; Carroll and Scheuer 1991) (5.15C), cyclodercitin (5.16), dehydrocyclodercitin (5.17), and stellettamine (5.18) (Gunawardana et al. 1989, 1992).

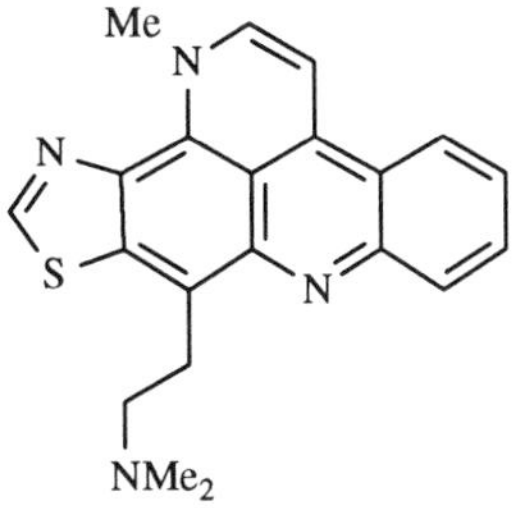

(5.14)

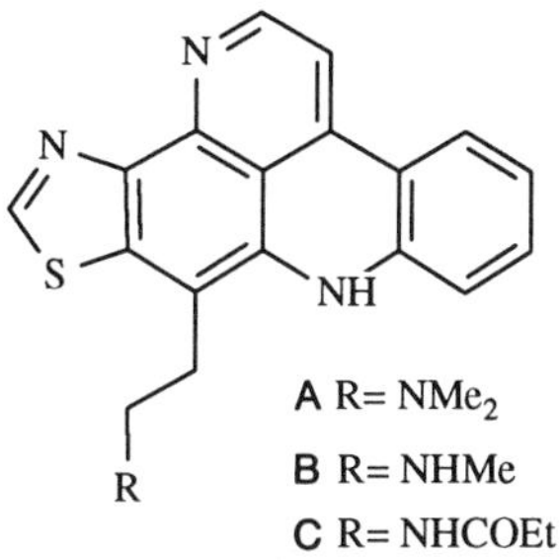

A R= NMe₂
B R= NHMe
C R= NHCOEt

(5.15)

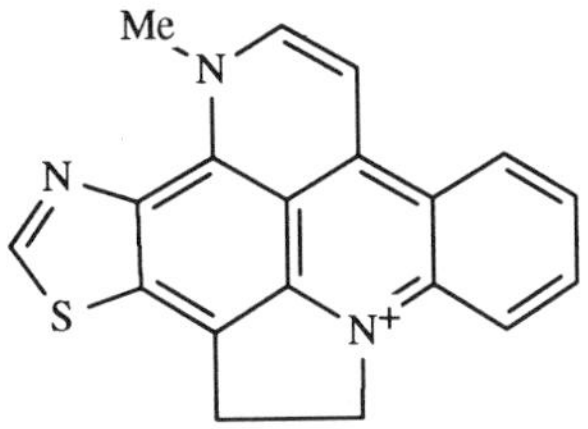

(5.16)

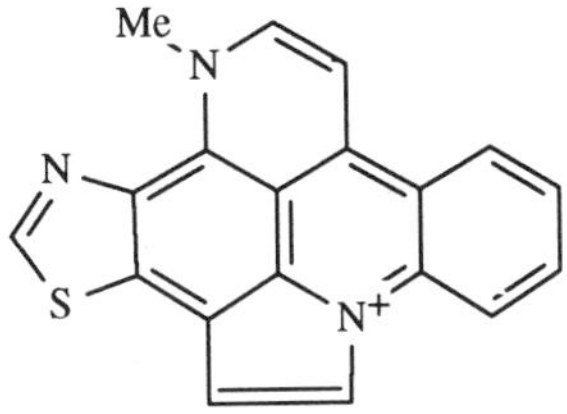

(5.17)

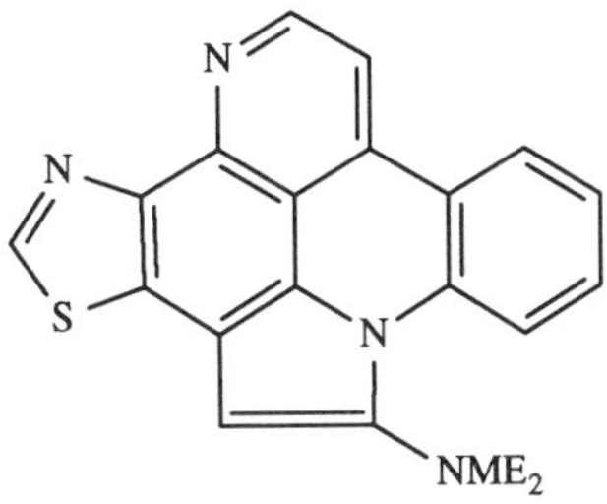

(5.18)

TABLE 5–1 Biological Activity of Dercitin (5.14)[1]

Cytotoxicity[2]	IC_{50} (nM)	Antiviral Activity (μM)
P-388 murine leukemia	81	
A549 human lung carcinoma	75	
HT-29 human colon carcinoma	63	
HL-60 human promyelocytic leukemia	150	
HL-60/AR (adriamycin resistant)	240	
A59 (mouse corona virus)		$EC_{50} = 1.8$
		$TC_{50} = 5.5$

[1] Activity for dercitin was not detected in the following assays: *Herpes simplex* I, PR8-influenza virus, feline leukemia virus, *E. coli*, *P. aeruginosa*, and *A. nidulans*.
[2] Cytotoxicity data adapted with permission from Burres et al. (1989).

Dercitin was isolated using a bioassay-guided approach. The *Dercitus* sp. sponge was selected as an antitumor lead based on the in vitro and in vivo P-388 activity of the crude extract, i.e., 100% inhibition of in vitro cell proliferation at 20 μg/mL, and percent treatment control (%T/C) = 140 at 100 mg/kg (intraperitoneal, i.p., QD1-9). Purification of dercitin was achieved by following in vitro P-388 activity.

The concentration of dercitin in *Dercitus* sp. is high, approximately 1% wet weight, i.e., 1 g of compound can be obtained from 100 g of (frozen) sponge. However, the sponge is difficult to collect (S. Pomponi, personal communication); the sponge is encrusting, which means that it requires substantial effort to scrape off the rocks on which it resides, and it is deep water, which means that it can only be collected by submersible. It is not possible to collect this encrusting sponge by trawling or dredging.

The in vitro biological profile of dercitin is found in Table 5–1 (Gunawardana et al. 1988; Burres et al. 1989; S. Cross and P. McCarthy, unpublished data): dercitin (5.14) is 10 to 100 times more active in vitro against P-388 than the congeners (5.15A–5.18) nordercitin (5.15A) $IC_{50} = 4.8$ μM, dercitamine (5.15B) $IC_{50} = 26.7$ μM, dercitamide (5.15C) $IC_{50} = 12.0$ μM, cyclodercitin (5.16) $IC_{50} = 1.9$ μM, dehydrocyclodercitin (5.17) $IC_{50} = 9.9$ μM, and stelletamine (5.18) $IC_{50} > 60$ μM.

The antitumor properties and mechanism-of-action of dercitin have been studied (Burres et al. 1989); dercitin prolonged the life of mice-bearing ascites P-388 tumors (%T/C = 170, 5 mg/kg, i.p., QD1-9), was active against (i.p.) B16 melanoma (%T/C = 125, 1.25 mg/kg, i.p. QD1-9), and modestly inhibited the growth of (subcutaneous, s.c.) Lewis lung carcinoma (median tumor volume/untreated tumor volume = 0.49, 1.25 mg/kg, i.p. QD1-9).

The fused-ring acridine structure of dercitin suggested the possibility of intercalation (Burres et al. 1989). Exogenous (calf thymus) DNA (<1 mg/mL) was found to protect P-388 cells when exposed to dercitin at concentrations <1 μM. From incorporation studies, dercitin disrupted DNA and RNA synthesis, whereas effects on protein synthesis were less pronounced; this profile is similar to the known intercalators actinomycin D and daunomycin. Further, the addition of either (calf thymus) DNA or (calf liver) RNA to solutions of dercitin altered the visible spectrum of dercitin in a manner consistent with binding, and equilibrium dialysis experiments revealed that dercitin binding to calf thymus DNA was saturable with an affinity of 3.1 μM and maximal binding of 0.2 mol dercitin for each base pair. Dercitin was found to relax covalently closed supercoiled Φx174 DNA with a concentration of 35 nM required for half-maximal relaxation; relaxation was reversible. DNA intercalation by dercitin was not found to be related to affects on topoisomerase II. Dercitin completely inhibited DNA polymerase I/DNase nick translation at 1 μM; however, the effects of dercitin on enzyme activity appeared to be secondary to changes in DNA conformation.

5.4.2 Discodermolide

Discodermolide (5.19) is a new polyhydroxylated lactone from a shallow water Caribbean sponge, *Discodermia dissoluta*, that expresses immune suppressive activity (Gunasekera et al. 1990). The sponges, *Discodermia* species, have been collected in the Caribbean and in waters adjacent to Okinawa, and have been proven to be prolific producers of structurally diverse, biologically active

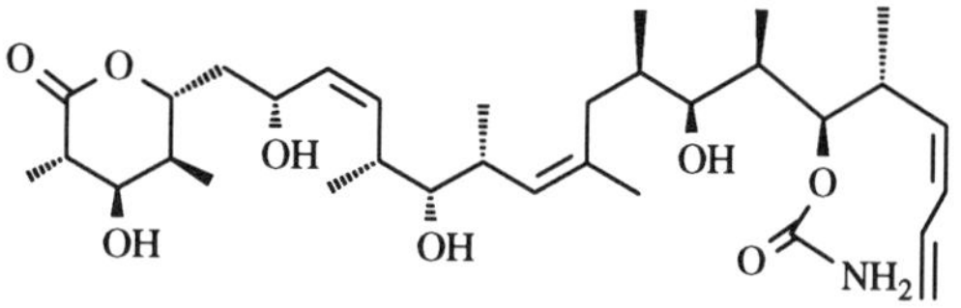

(5.19)

compounds, including the discodermins (Matsunaga et al. 1985a, 1985b), which are cyclic polypeptides containing 13 residues, the calyculins (e.g., calyculin A (5.11) Section 5.3.8), discodermide (5.20) (Gunasekera et al. 1991), polydiscamide (5.21) (Gulavita et al. 1992), and discodermindole (5.22) (Sun and Sakemi 1991). The discodermins are cytotoxic and exhibit antimicrobial activity; the calyculins are potent cytotoxins and were recently found to be protein phosphatase inhibitors; discodermide expresses antifungal ac-

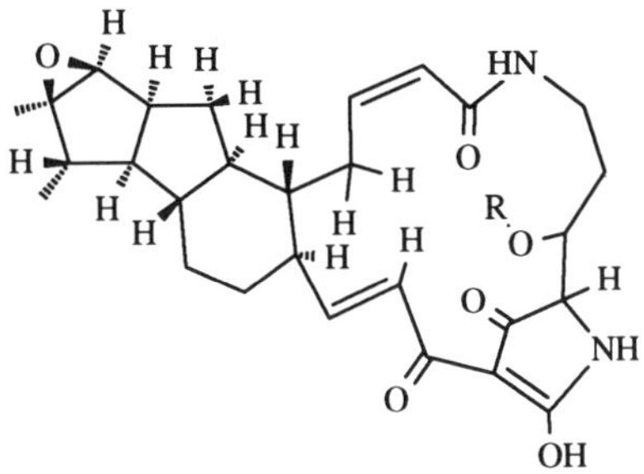

(5.20)

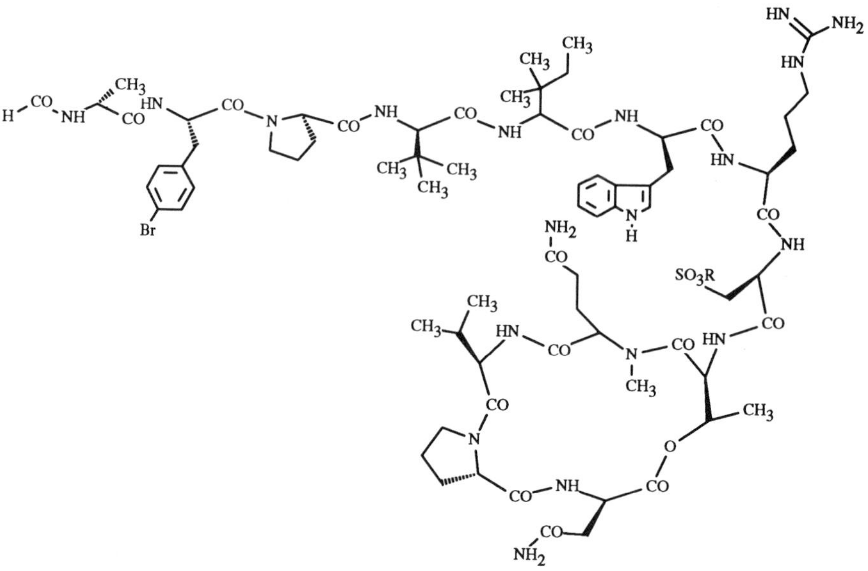

(5.21)

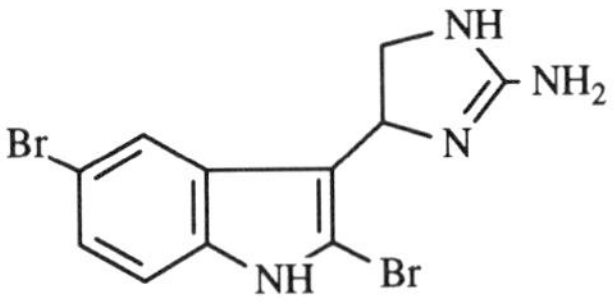

(5.22)

tivity and is related structurally to the antimicrobial agent ikarugamycin, which was isolated from *Streptomyces phaeochromogenes* var. *ikaruganensis* Sakai (Ito and Hirata 1977; Jomon et al. 1972); polydiscamide exhibits in vitro cytotoxicity against P-388 murine leukemia and A549 human lung cancer (Gulavita et al. 1992); discodermindole also expresses cytotoxicity.

The structural types and diversity of the *Discodermia* metabolites argue for some involvement of microorganisms associated with these sponges. Efforts are underway at HBOI to isolate microorganisms that produce these or related compounds (Armstrong et al. 1991).

D. dissoluta was selected as a lead based initially on the antifungal properties of a crude alcohol extract against *C. albicans*. The crude extract also expressed cytotoxicity against P-388 and modest immune suppressive activity; however, the cytotoxic properties relegated the extract as a lower priority immunosuppressive lead. Although in vitro P-388 and *C. albicans* activities were followed, fractions were also tested in the MLR assay. The cytotoxic fractions also exhibited potent immunosuppressive activity, and ultimately discodermolide was isolated from these fractions. Consequently, discodermolide was found to inhibit the in vitro proliferation of cultured murine

P-388 leukemia cells with an IC_{50} of 0.5 μg/mL and to suppress the two-way mixed lymphocyte response of both murine splenocytes and human peripheral blood lymphocytes at 0.5 and 5 μg/mL, respectively. Very importantly, the immunosuppressive activity was accompanied by greater than 85% viability of the splenocyte cells at these concentrations (Gunasekera et al. 1990).

Discodermide was found to be responsible for the antifungal activity of the *D. dissoluta* extract: *C. albicans* minimum inhibitory concentration (MIC) = 12.5 μg/mL. Polydiscamide was isolated from a different, possibly new deep water species of *Discodermia*, and was isolated simply because it is in high concentration in the water partition of water/butanol solvent partition of an ethanol extract. Its biological activities were subsequently discovered: P-388 IC_{50} = 0.9 μg/mL, A549 (human lung cancer) IC_{50} = 0.7 μg/mL. Discodermindole was isolated from *D. polydiscus* and expressed cytotoxicity: P-388 IC_{50} = 1.8 μg/mL; A549 IC_{50} = 4.6 μg/mL; HT29 (human colon cancer) IC_{50} = 12.2 μg/mL.

Further quantitative studies with discodermolide (Longley et al. 1991a) revealed potent suppression of the murine two-way MLR and concanavalin A (Con A) stimulation of splenocyte cultures, with IC_{50} values of 0.24 and 0.19 μM, respectively. These values are approximately 8- to 20-fold less potent than those that are commonly reported for suppression of murine MLR responses by the immunosuppressive standard, cyclosporine A (CsA). There was no evidence of cytotoxicity for murine splenocytes at discodermolide concentrations as high as 1.26 μM. Discodermolide was also active in suppressing the two-way MLR of human peripheral blood lymphocytes (PBL) (IC_{50} = 5.7 μM), of Con A (IC_{50} = 28.0 μM), and phytohemagglutinin (PHA) (IC_{50} = 30.1 μM)-stimulated PBL. Discodermolide was not cytotoxic to human PBL at concentrations as high as 80.6 μM. In a comparison with the standard immunosuppressive agent, CsA, discodermolide was equally effective in inhibiting T-cell receptor (TCR) and accessory cell independent, calcium dependent PMA-ionomycin induced proliferation of purified murine T-lymphocytes, with IC_{50} values of 9.0 and 14.0 nM for discodermolide and CsA. These findings suggest that discodermolide's suppressive activity reaches beyond T-cell activation via an inhibition of TCR mediated activation or interference with cell-to-cell interaction of B-cells or macrophages. Discodermolide did not significantly suppress the production of IL-2 in PMA-ionomycin stimulated murine splenocytes; however, the expression of the p55 associated IL-2 receptor as measured by fluorescence microscopy and flow cytometric analyses, using 7D4 antibody in these same cells was reduced. Similar results were obtained with PHA-stimulated blast cells obtained from human PBL in which the expression of p55 "Tac antigen" (human IL-2 receptor) was evaluated. The p55 receptor in mouse and human normally has low affinity for IL-2 binding (Malek et al. 1986; Waldman 1986). However, this chain associates with an additional cell surface, a 75 kDa (p75) protein to form an IL-2 receptor of high affinity. Before this apparently unique mechanism of action can be ascribed to discodermolide, it would be of great interest to determine if such suppression extends to the high affinity, p55-p75 complex.

Additionally, whereas discodermolide suppressed the PMA-ionomycin stimulation of murine T-cells, when examined microscopically, these cells appeared to have already undergone blast transformation, indicating that activation had already taken place (Longley et al. 1991a). Discodermolide's mechanism of action may resemble that of rapamycin rather than FK506 or CsA (Metcalf and Richards 1990) in that the blockage may occur at G_2 rather than at the G_0/G_1 interface, as for FK506 or CsA. Finally, discodermolide appears not to inhibit protein synthesis per se as indicated by ^{35}S methionine incorporation studies (R.E. Longley, unpublished observations). The precise in vitro mechanism of action of discodermolide still remains to be elucidated.

The in vivo activity of discodermolide has also been studied (Longley et al. 1991b). The compound was evaluated in the GVHR splenomegaly assay, using the BALB/c → CB6F$_1$ (BALB/c × C57BL/6J)F$_1$ model. In initial experiments, discodermolide was effective suppressing the GVHR in allogeneic grafted mice at 5.0, 2.5, and 1.25 mg/kg when administered as daily, i.p. injections for 7 days. While the two higher dosages resulted in the greatest suppression of the response (219% and 150%, respectively), these dosages were associated with a degree of morbidity in each group (two out of five and four out of five survivors, respectively). However, the lower dosage groups (1.25 mg/kg) continued to exhibit substantial suppression of their response (93%), but with no associated morbidity. Mice similarly treated with 150 mg/kg of CsA exhibited 80% suppression of the splenomegaly response. In order to determine an endpoint for the in vivo immunosuppressive action of discodermolide, a second dose-ranging experiment was initiated in which mice were treated with 2.5, 1.25, 0.312, and 0.156 mg/kg. Discodermolide suppressed the GVHR at all dosages tested with some morbidity once again associated with the 2.5 mg/kg group. However, all remaining groups exhibited suppression of the response, with the 0.156 group continuing to express 76% suppression. As a comparison, CsA-treated allogeneic grafted mice demonstrated 90% suppression. An endpoint for this suppression by discodermolide has yet to be determined.

Splenocytes obtained from discodermolide-treated, allogeneic grafted mice were suppressed in their response to Con A and in their ability to lyse YAC-1 tumor cells in natural killer cell assays compared with their vehicle-treated, allogeneic grafted controls (Longley et al. 1991b). Production of Con A-induced-IL-2 was suppressed only in the highest discodermolide dosage group (5.0 mg/kg), in contrast to suppression of IL-2 by CsA. Discodermolide was also evaluated for its in vivo effect on the primary serum antibody response to the T-dependent antigen, sheep red blood cells (SRBC). Mice were treated with daily i.p. injections of 5.0 mg/kg discodermolide and 150 mg/kg of CsA as a comparative standard. The rather surprising result was that discodermolide *did not* suppress the SRBC antibody response, even at the highest dosage. By contrast, CsA-treated mice were suppressed in their ability to mount such a response, a result consistent with the known mechanism of action of CsA. This result indicates that the immunosuppressive effect of discodermolide appears not to be of a general nature and, in fact, seems to be unique.

These results collectively indicate that the in vivo mechanism of action of discodermolide is very different from that of CsA and, in fact, may indicate an entirely novel mechanism of its immunosuppressive effects. An endpoint for suppression of the in vivo GVHR has yet to be reached, however, a conservative estimate of discodermolide's in vivo comparison with CsA (via i.p. routes) indicates that the compound is from 100 to 1000 times more potent in vivo compared with CsA. Further work regarding discodermolide's apparent unique in vivo mechanism of action and evaluation of additional in vivo models of graft rejection are currently underway.

5.4.3 Fuscosides A and B

Fuscosides A (5.23) and B (5.24) are diterpenoid arabinose glycosides, which were isolated from the Caribbean gorgonian *Eunicea fusca* (Shin and Fenical 1991) through a collaboration between Sea Grant/NOAA-funded research groups at SIO and UCSB. Fuscoside A (5.23), one of the major components of the extract, was found to reduce phorbol ester (PMA)-induced mouse ear

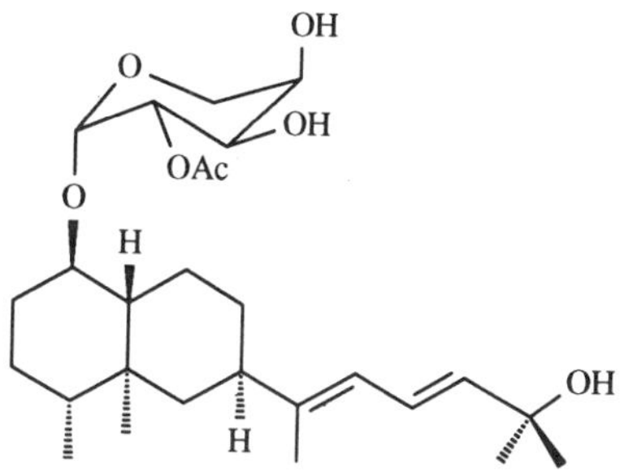

(5.23)

edema comparable to indomethacin, whereas fuscoside B (5.24) is slightly less active (Jacobson and Jacobs 1991). However, fuscoside B appears to selectively inhibit leukotriene synthesis (i.e., blocks the production of LTC_4 in murine macrophages activated by a calcium ionophore), but has no effect on the production of PGE_2. The anti-inflammatory effects of the fuscosides are long lasting and have been found to be active in human neutrophil studies as well (Jacobsen and Jacobs 1991).

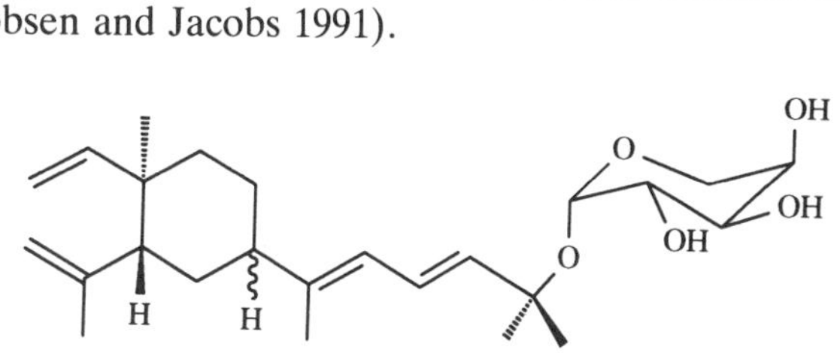

(5.24)

5.4.4 Microcolins A and B

Microcolins A (5.25A) and B (5.25B) are lipopeptides from the filamentous blue green algae *Lyngbya majuscula* (*Microcoleus lyngbyaceous*) that exhibit potent immunosuppressive activity (Koehn et al. 1992). Samples containing the compounds (5.25A and 5.25B) were collected in the southern Caribbean, off of the coast of Venezuela; samples recently collected near Miami also contain the same suite of lipopeptide metabolites. *Lyngbya majuscula* is notable in that it appears to exist in multiple, chemically distinct, varieties. For example, one or more shallow water strains cause "swimmers itch," a contact dermatitis caused by the lyngbyatoxins (Cardellina et al. 1979; Aimi et al. 1990) or aplysiatoxins (Moore et al. 1984). Shallow water samples of *L. majuscula* also contain the epimeric lipopeptides, majusculamides A and B (Marner et al. 1977), whereas the deep water variety contains antifungal cyclic depsipeptides, majusculamide C (Carter et al. 1984) and normajusculamide C (Mynderse et al. 1988). Deep water samples of *L. majuscula* have also furnished two cytotoxic lipopeptides, majusculamide D (5.25C) and deoxy-majusculamide D (5.25D) (Moore and Entzeroth 1988), which share structural similarities to microcolins A and B, since they contain an N,O-dimethyl tyrosine instead of an N-methyl leucine.

A crude ethanol extract of the algae showed in vitro activity against murine P-388 leukemia with an IC_{50} of 0.4 µg/mL and immunosuppressive

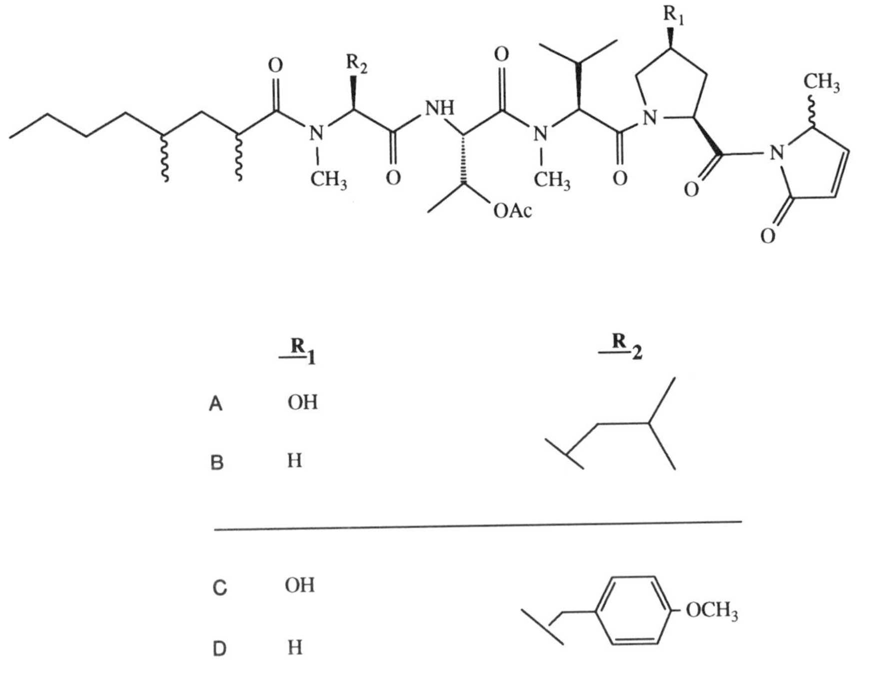

	R₁	R₂
A	OH	
B	H	
C	OH	
D	H	

(5.25)

activity in the MLR; however, the extract expressed lymphotoxicity in the lymphocyte viability assay (LCV) (Koehn et al. 1992). Because of the lymphotoxicity, bioassay-guided purification was undertaken using cytotoxicity as the primary indicator. Later, semipure chromatographic fractions containing microcolins A and B were subsequently evaluated in the MLR and LCV assays at nontoxic doses and their impressive immunosuppressive properties were recognized. The in vitro biological profiles of the compounds are presented in Table 5–2.

In an experiment whereby purified T-lymphocytes were induced to proliferate by the combined action of the calcium ionophyore, ionomycin, and the phorbol ester, 12-myristate 13-acetate (PMA), microcolin A yielded an EC_{50} value of 7.4 nM; cyclosporin yielded an EC_{50} of 37.5 nM in the same experiment. This experimental result suggested that the apparent mechanism of action of the microcolins may not directly involve the T-lymphocyte in relation to its suppressive action in the MLR experiment (Longley and Koehn 1991). Further studies on the effect of microcolin A on IL-2 production and IL-2 receptor expression of Con A-stimulated splenocytes indicate that the mechanism of microcolin A suppressive action does not involve the modulation of IL-2 production or IL-2 receptor expression per se.

From several in vivo experiments (Longley et al. 1993) using Simonsen's GVHR splenomegaly model, microcolin A at 1.0 mg/kg suppressed the splenomegaly response. A higher dosage regimen (5.0 mg/kg) resulted in death of the animals. Doses lower than 1.0 mg/kg were not associated with suppression of the splenic response. However, suppressed responses of splenocytes

TABLE 5–2 Biological Activity of Microcolins A (5.25A) and B (5.25B)[1]

	A	B
Immunomodulation		
MLR (two-way) EC_{50} (nM)	1.5	42.7
TC_{50} (nM)	22.6	191
Con A mitogen assay EC_{50} (nM)	1.7	—
LPS mitogen assay EC_{50} (nM)	1.9	—
PHA mitogen assay EC_{50} (nM)	1.1	—
Cytotoxicity		
P-388 IC_{50} (nM)	1.3	123
Antiviral activity		
HSV-1 ED_{50} (ng/mL)	2.5	146
TD_{50} (ng/mL)	480	1185

[1] Activity for microcolins A and B was not detected against the following pathogens: A549 mouse corona virus, PR8-influenza virus, feline leukemia virus, *E. coli*, *B. subtilis*, *A. nidulans*, and *C. albicans*.

that were obtained from these two lower dosage groups (0.2 and 0.04 mg/kg) were observed for PHA but not for Con A stimulation, indicating a specific cell subset (possibly B-cells or macrophages) that is affected in vivo.

5.4.5 Mycalamides A and B, and Onnamide

Mycalamides A (5.26A) and B (5.26B) are polyoxygenated compounds from the New Zealand shallow water sponge, *Mycale* sp. (Perry et al. 1988, 1990), that exhibit various types of biological activity, including in vivo antitumor activity in a wide variety of tumor models (Burres and Clement 1989) and antiviral activity. The sponge, *Mycale* sp. was collected by the research group at the University of Canterbury during a collaborative effort between this group and the research group at HBOI. The crude sponge extract exhibited potent cytotoxicity and in vitro antiviral activity in assays conducted at the University of Canterbury. Using a bioassay-guided approach, mycalamides A and B were purified and shown to be responsible for the biological activity of the extract (Perry et al. 1988).

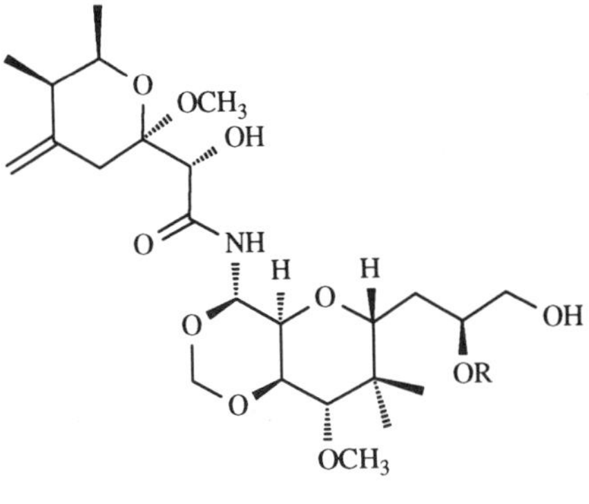

A R = H
B R = Me

(5.26)

Interestingly, the mycalamides resemble pederin (5.27) and related compounds, which were isolated from the terrestrial beetle *Paederus fuscipes* (see references in Perry et al. 1988).

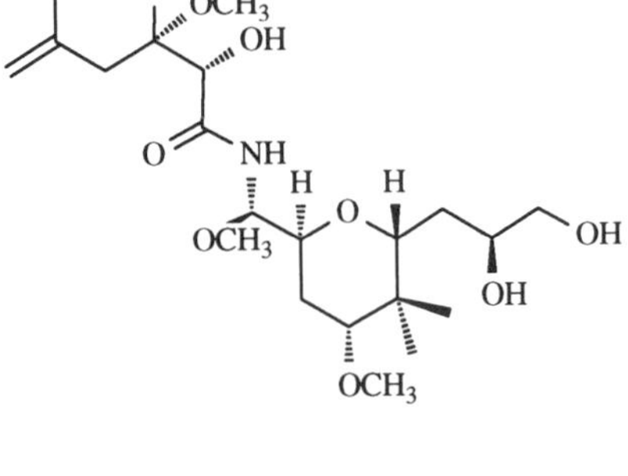

(5.27)

From a concurrent independent investigation of the shallow water Okinawan sponge, *Theonella* sp., at Harbor Branch Oceanographic Institution by the research group from the University of the Ryukus (Sakemi et al. 1988), the structurally related compound, onnamide (5.28) was isolated using a combined chemically and biologically driven approach.

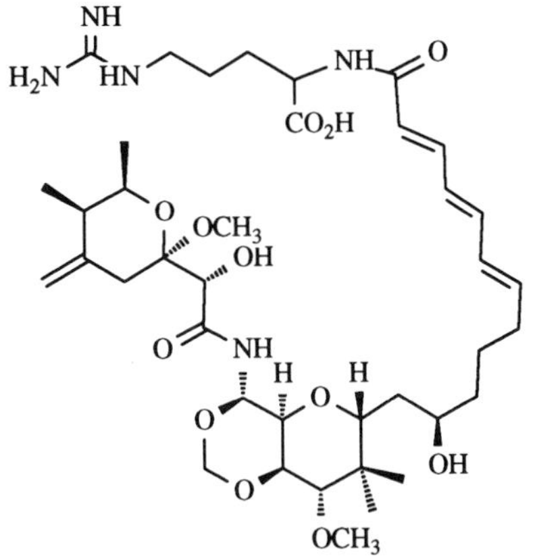

(5.28)

The mycalamides and onnamide express potent cytotoxicity (Burres and Clement 1989), as shown in Table 5–3. All three compounds are highly cytotoxic toward P-388. However, compared with mycalamides A and B, onnamide was found to be less cytotoxic against the HL-60 human promyelocytic leukemia cell line, HT-29 human colon cancer cell line, and the A549 human lung cancer cell line. None of the compounds induced HL-60 cell differentiation (Burres and Clement 1989).

Mycalamides A and B were found to exhibit similar in vivo efficacy in the i.p. P-388 murine leukemia model (Burres and Clement 1989); the best responses for mycalamides A and B were T/C = 140% (10 μg/kg) and 150% (2.5 μg/kg), respectively, whereas onnamide was found to be virtually inactive (best response, T/C = 115% (40 μg/kg)), presumably because of the ion-

TABLE 5–3 **In Vitro Cytotoxicity of Mycalamides A (5.26A) and B (5.26B) and Onnamide (5.28)[1]**

	Mycalamide A	Mycalamide B	Onnamide
P-388 IC_{50} (nM)	5.2	1.3	2.4
HL-60 IC_{50} (nM)	3.0	1.5	25
HT-29 IC_{50} (nM)	2.8	1.5	180
A549 IC_{50} (nM)	3.6	0.6	170

Adapted with permission from Burres and Clement (1989).

[1] P-388, murine leukemia; HL-60, human leukemia; HT-29, human colon cancer; A549, human lung cancer.

izable arginine group in the molecule, which may result in poor uptake/passage across membranes. All three compounds expressed efficacy in the i.p. M5976 reticulum cell sarcoma model (Burres and Clement 1989). The best responses for mycalamides A and B, and for onnamide were T/C = 233% (60 μg/kg, i.p., Q4Dx3), 218% (30 μg/kg, i.p., Q4Dx3) and 183% (20 μg/kg, i.p., QD1-9), respectively. From further in vivo evaluation, mycalamide A (5.26A) has been found to be substantially more active against solid tumors of murine and human origin than against P-388 murine leukemia (Burres and Clement 1989) (Table 5–4).

From experiments aimed toward elucidating the cytotoxic mechanism(s) of action of mycalamides A and B and onnamide, all three compounds were found to markedly inhibit protein and DNA synthesis in P-388 cells, with intermediate effects on RNA synthesis (Burres and Clement 1989). Mycalamide B was found to be the most potent protein synthesis inhibitor and onnamide the least potent. Further, all three compounds were found to inhibit translation in cell-free extracts, using Brome mosaic virus as an RNA substrate in lysed rabbit reticulocytes; mycalamide B was found to be most potent and mycalamide A was found to be least potent.

In experiments to determine whether mycalamide A intercalated into DNA, mycalamide A was found to alter neither the electrophoretic mobility of supercoiled DNA nor lose its antiproliferative effects when combined with thymidine, deoxycytidine, deoxyadenosine, and deoxyguanosine (Burres and Clement 1989).

TABLE 5–4 Antitumor Efficacy of Mycalamide A (5.26A): Best Response

Tumor Type	Site	Schedule	Best Response	Dose (μg/kg)
P-388 leukemia	i.p.	QD1-9	56% ILS[1]	10.0
B16 melanoma	i.p.	Q4Dx9	145% ILS (40)[2]	30.0
M5076 reticulum cell	i.p.	Q4Dx3	133% ILS (40)	60.0
Colon 26 carcinoma	i.p.	Q4Dx3	49% ILS (20)	60.0
B16 melanoma	i.p.	Q4Dx9	43% ILS	7.5
B16 melanoma	s.c.	Q4Dx9	0.77 T/C TWI[3]	3.8
Lewis lung	s.c.	QD1-9	0.15 T/C TWI	20.0
M5076 reticulum cell	s.c.	Q4Dx3 (i.v.)	0.15 T/C TWI	120.0
MX-1 mammary xenograft	s.c.	Q4Dx3	0.21 T/C TWI	80.0
CX-1 colon xenograft	s.c.	Q4Dx3	0.29 T/C TWI	60.0
LX-1 lung xenograft	s.c.	Q4Dx3	0.76 T/C TWI	7.5
Burkitts lymphoma	s.c.	Q4Dx3	0.19 T/C TWI	20.0

Adapted with permission from Burres and Clement (1989).

[1] Efficacy expressed as increased life span (ILS) of treated mice compared with untreated tumor-bearing mice (>25% ILS = significant antitumor effect).

[2] Numbers in parentheses indicate percentage of cure rate for treated group.

[3] Tumor volume calculated as $(L \times W^2/2)$ and expressed as median tumor weight index (TWI), which represents the ratio of tumor volumes of treated/control tumors.

TABLE 5-5   Antiviral Activity of Mycalamide A (5.26A)

	100% Virus Inhibition (ng/ml)
In vitro:	
Herpes simplex Type 1	2
Vesicular stomatitis	2
Polio vaccine Type 1	2
Coronavirus A59	10
Human immunodeficiency virus	6/16[1]
In vivo—Murine Coronavirus A59:	
Mycalamide A[2] (5.26A) QD1-9, i.p., 100 μg/kg	100% Survivors, day 14
Ribavarin QD1-9, i.p., 100 mg/kg	100% Survivors, day 14
Control	0% Survivors, day 14

[1] IC_{50}/IC_{90} (in ng/ml)
[2] Approximately 10% concentration of (5.26A).

Biological evaluation of mycalamide (5.26A) in antiviral and antimicrobial assays has revealed that it expresses antiviral activity but does not exhibit antibacterial activity (Perry et al. 1988; S. Cross and P. McCarthy, unpublished observations). Mycalamide A (5.26A) exhibited in vitro antiviral activity (Table 5-5) against *Herpes simplex* Type 1, vesicular stomatitis, polio vaccine Type 1, coronavirus A59, and HIV; the latter assessment was carried out at NCI/NIH. In vivo activity was observed against murine coronavirus A59.

The mycalamides and onnamide have been synthesized (Hong and Kishi 1990, 1991).

5.5   COMPOUNDS IN CLINICAL TRIAL

The structures and biological activities of several marine-derived compounds that have been or are currently in clinical trials, including bryostatin 1 (5.29A), didemnin B (5.30B), manoalide (5.12), and the pseudopterosins (including A, (5.32A) and E, (5.32B)) are presented in the following sections.

5.5.1   Bryostatins

The bryostatins are a series of 15 structurally related macrocyclic lactones that are derived from the bryozoan, *Bugula neritina*, and have been studied for their antitumor and immunomodulatory properties for over 20 years by G.R. Pettit and his research group at the University of Arizona, scientists at NCI, and collaborators. The history of bryostatin research has been previously reviewed (Suffness et al. 1989). The bryostatins, e.g., bryostatin 1 (5.29A)

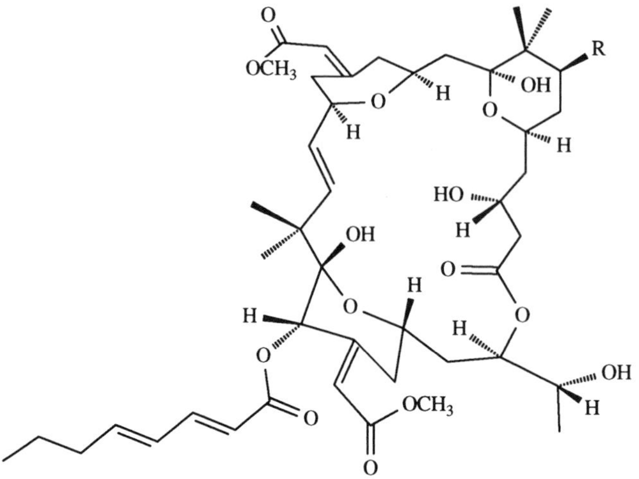

A   R = OAc

B   R = OH   (5.29)

(Pettit et al. 1982), exhibit antitumor activity in several tumor models, selective cytotoxicity against leukemia cell lines (Suffness et al. 1989), and
modulate the immune system in various ways (vide infra).

The antitumor properties of extracts from *B. neritina* were first demonstrated in 1969 using P-388 murine leukemia (Suffness et al. 1989). In one
experiment, a %T/C of 168 (400 mg/kg, QD1-10) was observed. However,
cytotoxicity was not observed in the KB cell line, which was used routinely
at NCI in other projects to monitor fractionation of active extracts. Further,
reproducible activity could not be achieved from the evaluation of extracts
obtained from different collections of *B. neritina*, and research on this organism was put aside for several years.

Ten years passed before samples could be found that were sufficiently
potent in in vitro assays, reproducible in in vivo antitumor assays, and in
sufficient quantities to permit bioassay-guided isolation of the active components (Suffness et al. 1989). The bryostatins were also discovered in three
other organisms in or on which *B. neritina* grew, i.e., the bryozoan *Amathia
convoluta* (Pettit et al. 1985), the sponge *Lissodendoryx isodictyalis* (Pettit et
al. 1986a), and the tunicate *Aplidium convoluta* (Pettit et al. 1986b).

In the early 1980s, bryostatins 1 and 2 (5.29B) were evaluated in a number
of in vivo tumor models. Activity was demonstrated in several experimental
leukemia and sarcoma models (Suffness et al. 1989); however, none of the
animals were cured of their respective tumors. At the same time, similar
activity was not observed against melanoma, mammary, colon, or lung cancers. This profile of biological activity was not encouraging until the results

of some of the first mechanism-of-action studies were reported beginning in 1985 (Berkow and Kraft 1985). It was also then that bryostatins 1 and 2 showed very strong specificity in NCI's disease-specific panel against human cell lines with differences in IC_{50} values between sensitive and resistant lines equal to or greater than 3 log units (Suffness et al. 1989).

The first mechanistic studies (Berkow and Kraft 1985; Kraft et al. 1986) showed that bryostatin 1 (1) binds to the phorbol ester receptor of human polymorphonuclear leukocytes and HL60 cells, which is the equivalent of the calcium, phospholipid-dependent protein kinase, i.e., protein kinase C, (2) blocked the phorbol ester-induced macrophage-like differentiation of HL-60 cells, although bryostatin 1 was unable to induce differentiation of HL-60 cells, and (3) decreased cytoplasmic protein kinase C activity. In later studies (Dale et al. 1989; Dale and Gescher 1989), the bryostatins were found to inhibit the growth of A549 lung cancer cells and to inhibit DNA replication, causing translocation of protein kinase C from the cytosol to membranes, and down regulation of protein kinase C, activity similar to that induced by phorbol esters. However, at relatively high concentrations, the bryostatins blocked their own inhibitory effect and the antireplicative action of phorbol esters. Therefore, the relationship between growth arrest and protein kinase C activities was not viewed as a straightforward one.

In addition to acting as cytotoxic and antitumor agents, and as antitumor promoters, the bryostatins modulate the immune system, presumably by affecting protein kinase C, an important mediator of signal transduction-associated cell regulation and transformation (Gescher and Dale 1989). In one study (May et al. 1987), several of the bryostatins and recombinant human granulocyte macrophage colony-stimulating factor (rHGM-CSF) were found to stimulate the formation of granulocyte-macrophage and pure granulocyte colonies in bone marrow. However, unlike rHGM-CSF, bryostatin 1 (but not bryostatin 13), was found to rapidly activate mature neutrophils, which are involved in nonspecific immune defense (Schindler 1990). This study concluded that the bryostatins may represent clinically attractive agents to complement other compounds used to treat clinical situations resulting in bone marrow failure. Bryostatin may also be useful as an adjuvant in these clinical situations. In another study (Trenn et al. 1988), the bryostatins, like the phorbol esters, were found to:

1. Lower the amount of recombinant IL-2 (rIL-2) required to generate antigen-specific, cytotoxic T-lymphocytes from in vivo primed spleen cells.
2. Cause antigen-*non*specific lysis of EL-4 cells through stimulation of antigen primed, cytotoxic T-cells when used in combination with rIL-2 and in the presence of the nonspecific T-cell activator, Con A.
3. Cause antigen-*non*specific lysis of EL-4 cells with the T-cell clone 2C.
4. Synergize with recombinant B cell stimulatory factor 1/interleukin 4 (BSF-1/IL-4), with and without rIL-2, to differentiate unprimed, resting T-cells into cytotoxic T lymphocytes.

The study concluded that the antineoplastic activities of the bryostatins themselves may be related to their immunoenhancing properties, and that clinically, they may be useful in combination with recombinant IL-2 to promote tumor rejection at IL-2 doses sufficiently low to avoid side effects. In neither study were the bryostatins found to be lymphotoxic at concentrations in which they expressed their immunomodulating effects.

Currently, through a license from G.R. Pettit/Arizona State University and under a Cooperative Research and Development Agreement (CRADA) with NCI, Bristol-Myers Squibb is pursuing large-scale isolation of the bryostatins and development (Persinos 1990d). Responding to Pettit's encouragement in the mid-1980s, the Cancer Research Campaign in Great Britain conducted preclinical toxicology in small animals (Persinos 1989b), and clinical trials began in Manchester and Cambridge, England in February 1991 (Persinos 1991a). Bristol-Myers Squibb plans to initiate studies in dogs (Persinos 1990d), and NCI has planned more in vivo studies in animals (Persinos 1991a).

5.5.2 Didemnins

From several hundred shallow water samples of various marine organisms collected in 1978 throughout the western Caribbean, organic extracts from the tunicate *Trididemnum solidum* Van Name were found to be active in a shipboard assay against *Herpes simplex* I and cytotoxic against the viral host CV-1 cell line (monkey kidney tissue) (Rinehart et al. 1981a). In subsequent land-based assays, the tunicate extracts also exhibited in vitro activity against L1210 murine leukemia and against various RNA and DNA viruses, i.e., HSV-2, vaccinia, influenza (PR8), parainfluenza-3 (HA-1), Coxsackie (A-21), and equine rhinovirus. Three compounds were isolated initially, didemnins A through C (5.30A, 5.30B, and 5.30C), which contained both the antiviral activity and the cytotoxicity of the crude extracts (Rinehart et al. 1981b, 1981c). Structurally related compounds didemnins D and E were subsequently isolated and their structures elucidated (Rinehart et al. 1988). A total of 12 didemnins have been detected (Munro et al. 1987). The didemnins are all cyclic depsipeptides with five or more amino acid units in common. Didemnin D, the largest compound in the series whose structure has been published, contains 12 amino acid units, including standard amino acids (L-Glu, L-Thr, L-Leu, L-Pro), methylated amino acids (D-MeLeu, N,O-dimethyl-L-Tyr), and unusual groups (L-Lac-L-Pro, (3S,4R,5S)-isostatine, and (2S,4S)-hydroxyisovalerylpropionic acid). The isostatine unit was originally thought to be (3S,4R)-statine (Rinehart et al. 1988), and the hydroxyisovalerylpropionic acid was tentatively originally assigned as the 2R,4R-diastereomer and subsequently revised (Ewing et al. 1986).

In addition to antiviral activity and cytotoxicity, the didemnins exhibit antitumor and immunomodulatory activity; the antiviral activity was pursued initially. In the original reports (Rinehart et al. 1981b, 1981c), didemnins A

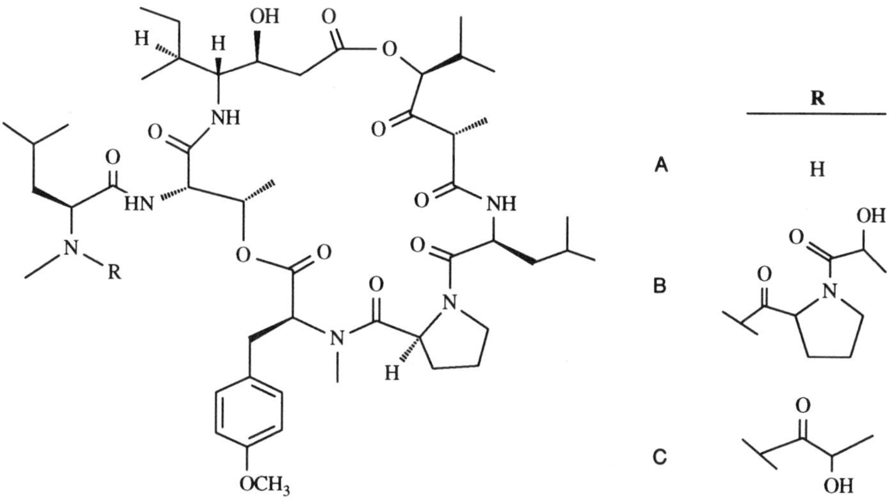

(5.30)

and B exhibited good dose-response relationships against the viruses in which the crude extracts were active (vide supra). Subsequent in vitro studies revealed activity against a broad range of RNA viruses, including Rift Valley fever, Venezuelan equine encephalomyelitis virus, yellow fever, sandfly fever, and a Pichinde virus (Canonico et al. 1982). In vivo activity was demonstrated by didemnin B against lethal challenges of HSV-2 (DNA virus) in the mouse vaginal *Herpes* model (Canonico et al. 1982; Weed and Stringfellow 1983), and against Rift Valley fever (RNA virus) (Rinehart et al. 1988). Unfortunately, didemnin B was only effective in vivo against *Herpes* if topical applications were started prior to infection, and was not effective if applied after infection or given intraperitoneally. In general, the didemnins were found to be too toxic and not sufficiently selective for antiviral therapy (Rinehart et al. 1988; Munro et al. 1987).

The immunomodulatory properties of the didemnins were studied initially by researchers at the University of Arizona (Montgomery and Zukoski 1985; Montgomery et al. 1987a, 1987b; Russell et al. 1987). In the first published study by this group (Montgomery and Zukoski 1985), didemnin B was found to inhibit the alloantigen response in the MLR, inhibit Con A-induced blastogenesis of T lymphocytes, and inhibit bacterial lipopolysaccharide (LPS)-induced blastogenesis of B lymphocytes at nanogram to picogram/milliliter concentrations, which were nontoxic to unstimulated lymphocytes. Didemnin B also inhibited protein synthesis in normal resting splenic mononuclear cells, but at concentrations much greater than those affecting Con A- and LPS-induced mitogenesis, and did not affect RNA synthesis at microgram/milliliter concentrations. Didemnin B also inhibited splenomegaly in the in vivo Simonsen GVHR at 0.05 to 0.3 mg/kg/day; however, concentrations of didemnin B of ≥ 0.2 mg/kg/day caused mortality, primarily through adverse effects on hepatic function.

Follow-up in vivo studies of the inhibition of LPS-induced mitogenesis were conducted (Montgomery et al. 1987a, 1987b). Observed changes in bone marrow function due to didemnin B were thought to be related to increased production and release of lymphocytes and neutrophils from bone marrow. From a study by a different research group to investigate the mechanism of inhibition of T-lymphocyte proliferation (LeGrue et al. 1988), didemnin B was found to inhibit lymphocyte proliferation through a cytostatic affect; its mechanism of inhibition appears to be distinct from that of cyclosporine, which affects lymphokine production.

Other in vivo studies have yielded mixed results. In one study, didemnin B was found to significantly prolong survival of rat heart allografts with doses as low as 0.005 mg/kg; however, long-term survival of rat heart allografts could not be achieved, even at near-lethal dosages, which contrasts with the effects of cyclosporine (Stevens et al. 1989). In another study (Yuh et al. 1989), didemnin B was evaluated in mouse and rat heterotopic cardiac transplantation models and the mouse popliteal lymph node hyperplasia assay and found to possess a low therapeutic index, i.e., doses of didemnin B that prolonged allograft survival or inhibited lymphocyte proliferation were quite toxic, evidenced by significant (reversible) body weight loss. Toxicity due to didemnin B appeared to be cumulative, which was presumed to be related to the lipophilicity and, therefore, accumulation of the compound in certain body tissues. The latter study concluded that the in vitro potency of didemnin B was (unfortunately) not sufficient to predict its efficacy as an immunosuppressant in vivo and that a suitable mechanism of action, i.e., as shown by the clinically used agent, cyclosporine, is highly important.

The most advanced area of biological evaluation of the didemnins is as an antitumor agent. The cytotoxic and antitumor properties and results from mechanism-of-action studies of the didemnins have been previously reviewed (Chun et al. 1986; Munro et al. 1987; Rinehart et al. 1988). Didemnins A and B are potent cytotoxins against L1210 with ED_{50} values of 30 and 2 ng/mL, respectively. In vivo activity was established for didemnin B in (i.p.) P-388 murine leukemia (%T/C = 199, 1 mg/kg, i.p., Q1,5,9, (i.p.) B16 murine melanoma (%T/C = 160, 1 mg/kg, i.p., Q1,5,9; %T/C = 172, 0.3 mg/kg, i.p., QD1-9), and (i.p.) M5076 murine sarcoma (%T/C = 209, Q1,5,9,13). Didemnin B also exhibited in vivo activity in (i.p.) Yoshida ascites tumors in rats (%T/C = 369, 0.06 mg/kg, i.p., Q1, 11 deaths/20 rats total) (Fimiani 1987). In vivo efficacy was not observed in experiments with subcutaneous (s.c.) CD8F murine mammary tumor, (s.c.) colon 38 tumor, (i.p.) L1210 murine leukemia, intravenous (i.v.) Lewis lung carcinoma, CX-1 human colon tumor (xenograft), LX-1 human lung tumor (xenograft), and MX-1 human mammary tumor (xenograft). Didemnin A has not been tested extensively in in vivo tumor models, and in experiments in which it has been evaluated, it has proven not to be as efficacious as didemnin B.

Antitumor activity against B16 murine melanoma was not found for didemnin B with subcutaneous, intravenous, or oral administration; none of the in vivo experiments yielded long-term survivors, and the dosages in the P-388 and B16 experiments that provided the greatest treatment values, i.e., 1 mg/kg, also caused the greatest amount of weight loss (Chun et al. 1986; Rinehart et al. 1988).

Didemnin B has also been evaluated in the human tumor stem cell or human tumor colony forming assays using fresh human tumor surgical or biopsy specimens (results also reviewed in Munro et al. 1987 and Chun et al. 1986). Activity at nanogram/milliliter to microgram/milliliter concentrations were observed against carcinomas of the breast, ovary, and kidney, mesothelioma and sarcoma, ovarian cancer, hairy cell leukemia, breast cancer, and oligodendroglioma.

Didemnin B appears to exert its antiproliferative effects in various cell lines primarily through inhibition of protein synthesis and to a lesser extent through inhibition of DNA synthesis (reviewed in Chun et al. 1986). Inhibition of protein synthesis is not related to amino acid uptake and didemnin B does not bind to DNA. The compound effects exponentially growing B16 cells to a much greater extent than plateau-phase cells. Low doses of didemnin B stop B16 cell progression at the G_1/S boundary while cell progression from S to G_2 and M to G_1 is relatively unaffected, except at high doses when complete inhibition of cell progression or growth occurs.

Didemnin B has completed Phase I clinical trials as an anticancer agent (Dorr et al. 1988). The dose-limiting toxicity was nausea and vomiting, which could be ameliorated somewhat by pretreatment with an aggressive antiemetic regimen; no objective antitumor response was observed.

Broad Phase II clinical trials for didemnin B as an antitumor agent are ongoing (Persinos 1991c).

5.5.3 Manoalide

The sesterterpenoid manoalide (5.12) and structurally related compounds were obtained from the marine sponge, *Luffariella variabilis* (de Silva and Scheuer 1980, 1981); the compound was named in recognition of the location in which the chemistry work was completed (Manoa Valley, Hawaii). Although manoalide and the crude methylene chloride extract from which it was derived demonstrated in vitro antibiotic activity against *Streptomyces pyogenes* and *Staphylococcus aureus* (de Silva and Scheuer 1980), it was the analgesic and anti-inflammatory activity of manoalide (Faulkner 1992; Jacobs et al. 1985) and a related series of compounds (Albizati et al. 1987; Kernan et al. 1987) that elicited substantial scientific and pharmaceutical interest. The first total synthesis of manoalide and the related compound, seco-manoalide, was accomplished soon after the structure was published (Katsumura et al. 1985) and thus ameliorated any anticipated difficulties regarding supply for in-depth biological evaluation and facilitated the possibility of preparing more efficacious analogs.

The analgesic activity of manoalide was demonstrated in the phenylquinone-induced writhing assay in mice, with an $ED_{50} = 0.36$ mg/kg, i.p. (Jacobs et al. 1985). Manoalide was also shown to be active in antagonizing the inflammatory response of mouse epidermis induced by PMA, with potency greater ($ED_{50} = 100$ μg) than that of the standard nonsteroidal compound, indomethacin ($ED_{50} = 250$ μg) but less than that of hydrocortisone ($ED_{50} = 20$ μg, Burley et al. 1982). Manoalide, like hydrocortisone, did not prevent the inflammation response normally induced by arachidonic acid, and it was postulated that the compound might inhibit enzyme reactions that were located upstream from arachidonic acid release and prostaglandin synthesis. This hypothesis was supported, in part, by in vivo time course studies showing that manoalide was most effective when applied within 5 minutes of PMA application, a very early time point considering the normal 3-hour period it required for the inflammatory reaction to manifest itself.

Speculation at that time on the mechanism of action of both indomethacin, which prevents prostaglandin production by inhibition of the cyclooxygenase pathway, and hydrocortisone, which prevents prostaglandin production by inhibiting arachidonic acid release, led to the hypothesis that manoalide might serve to inactivate or inhibit various actions of the enzyme PLA_2. In one study (de Freitas et al. 1984), manoalide (2×10^{-5} M) was shown to prevent the neurotoxic action of β-bungarotoxin on rat phrenic nerve-diaphragm preparations when preincubated with the toxin for as little as 25 minutes. In the same study, manoalide inhibited the hydrolysis of phosphatidylcholine by bee venom-derived PLA_2, when preincubated with the enzyme for 1 hour. Subsequent studies demonstrated the irreversibility of the reaction and its pH dependence (Glaser and Jacobs 1984, 1986). Manoalide has also been shown to inhibit the activity of PLA_2 from other sources, including cobra venom (Lombardo and Dennis 1985), while manoalide blocked the hydrolysis of both phosphatidylcholine and phosphatidylethanolamine by bee venom PLA_2.

Manoalide has also been reported to be 300- to 100-fold less active against PLA_2 obtained from mammalian microsomal preparations (guinea pig lung and uterus rat basophilic leukemia cells and smooth muscle) than against PLA_2 obtained from cobra venom (Bennett et al. 1987). These results suggest that PLA_2 alone is not the sole principal intracellular target for manoalide. Further, manoalide was found to be capable of inhibiting phosphoinositide-specific phospholipase C (PLC), within an IC_{50} value range of from 3 to 6 μM; this inactivation was calcium and pH dependent and irreversible.

While the purported mechanism of action of manoalide's anti-inflammatory activity appeared to involve direct binding of the molecule to PLA_2, manoalide was observed to form a chromophore when mixed with bee venom PLA_2 (Glaser and Jacobs 1986). A similar reaction occurred when manoalide was incubated with monomeric lysine, cysteine, and tryptophan, but not with their respective N-alpha-amino-blocked analogs. Through a series of straightforward but elegant experiments, a correlation between chromophore production and the specific amino acid residues modified upon binding of manoalide to PLA_2 and corresponding enzymatic activity guided the team to this discovery. The results of these and subsequent experiments (Glaser et al. 1988) indicated that of the 11 lysine residues available for modification by manoalide, only three were actually modified upon binding by manoalide to bee venom PLA_2. This contrasted with the modification of four of the six lysines present in cobra venom PLA_2 (Lombardo and Dennis 1985). Subsequent investigations (Glaser et al. 1989) have revealed a number of structure activity relationships associated with the manoalide molecule and related analogs that may prove useful in determining the exact nature of the binding of manoalide to PLA_2. These include the γ-hydroxybutenolide ring system (selectivity for various sources of PLA_2), the C-24 aldehyde (the irreversibility of PLA_2 binding), and the hydrophobic alkyl regions of the molecule (cooperative binding with the ring system and PLA_2 molecule).

Manoalide's mechanism of the inhibition of phospholipid hydrolysis led to a number of investigations into its utility as a probe to study the role and importance of phospholipases and the role of Ca^{2+} in signal transduction-mediated events during cellular growth and replication. A key component of many inflammatory mediators and growth factors is their ability to bind to cellular receptors that, in turn, stimulate phosphoinositide turnover and Ca^{2+} mobilization as part of the signal transduction pathway. The major role of PLC in this process includes the activation of phosphatidylinositol-4,5-biphosphate (PIP_2), which gives rise to inositol 1,4,5-triphosphate (IP_3) and 1,2 diacylglycerol (DAG) (Berridge 1983). The IP_3 serves as a second messenger and binds to a receptor on the rough endoplasmic reticulum, which causes the release of Ca^{2+} from intracellular stores. DAG, on the other hand, is the physiological receptor for protein kinase C, an important second messenger enzyme involved in the regulation of cellular proliferation. While manoalide's primary pharmacological action involves inhibition of PLA_2 activity, as previously noted, manoalide's less potent, but measureable inhibition of PLC activity prompted further investigations into its role as an antagonist of calcium-mediated, signal transduction-mediated control of cellular growth and regulation.

Manoalide was found to be a potent inhibitor of Ca^{2+} mobilization in several cell lines (Wheeler et al. 1987). Epidermal growth factor (EGF) stimulates a rise in intracellular Ca^{2+} in A431 human epidermoid carcinoma cells. Manoalide blocked both the EGF-mediated entry and release of Ca^{2+} from intracellular stores in a time-dependent manner with an IC_{50} of 0.4 μM. However, the production of inositol monophosphate (phosphoinositide metabolism) did not coincide with the EGF-mediated response. Manoalide also blocked the thyrotropin-releasing hormone (TRH)-dependent release of Ca^{2+} from intracellular stores in GH_3 rat pituitary cells (0.5 μM for 20 minutes or 3.0 μM for 5 minutes), without concomitant inhibition of the formation of inositol phosphates from PIP_2. In addition, manoalide also inhibited the K^+

depolarization-activated Ca^{2+} channel in these cells and the activation of the channel by BAY K8644, with an $IC_{50} = 1 \mu M$. Manoalide also inhibited Ca^{2+} uptake of murine splenocyes stimulated with concanavalin A in a time- and temperature-dependent manner with an $IC_{50} = 0.07 \mu M$, thus indicating that at least in the cells examined in the study, manoalide's inhibitory action appeared to involve blockage of the calcium channel.

The role of manoalide in the regulation of various eicosanoids and their precursors and the role they play in the inflammatory process began to be established several years ago (Mayer and Jacobs 1988; Mayer et al. 1988). Manoalide inhibited the production of prostaglandin E_2 (PGE_2) in a dose-dependent fashion by PMA-zymosan stimulated murine peritoneal macrophages. The release of leukotriene C_4 (LTC_4) in murine peritoneal macrophages stimulated with the calcium ionophore A23187 was similarly inhibited. Curiously, LTC_4 production was enhanced by manoalide when the same cells were stimulated with zymosan. Both PGE_2 and LTC_4 production were reduced in resident (unstimulated) mouse peritoneal macrophages, indicating a direct inhibitory effect of manoalide on the cyclooxygenase pathway. Additionally, the release of ^{3}H-arachidonic acid was shown to be partially reduced (37% inhibition) in the presence of $0.05 \mu M$ of manoalide. These results pointed to the possibility that manoalide's mechanism of action appeared to include all of the following:

1. Inhibition of PLC and PLA_2;
2. The resulting inhibition of arachidonic release;
3. Inhibition of PGE_2 production; and
4. Blockage of Ca^{2+} mobilization.

Subsequent studies (Lister et al. 1989) confirmed and extended the findings of Mayer using the murine macrophage cell line $P388D_1$. Lister found that manoalide, unlike other PLA_2 inhibitors, was equally effective in inhibition of both intracellular and extracellular (isolated) PLA_2.

While manoalide continues to be a useful tool in helping to elucidate the complex interplay of eicosanoid regulation and membrane phospholipid-signal transduction mediated events, the in vivo anti-inflammatory activity of the compound and its related analogs suggest the potential clinical usefulness of the compound in disorders that manifest themselves through altered production of inflammatory eicosanoid products (i.e., prostaglandins). In vitro studies (Jacobson et al. 1990) in which manoalide was demonstrated to inhibit PLA_2 from human synovial fluid, the site of pathological damage in rheumatoid arthritis, and the demonstration of manoalide's ability to inhibit the release of elastase from human neutrophils stimulated by fMLP (fMet-Leu-Phe) further underscore the potential of the compound as a potential next generation anti-inflammatory agent. In the past 5 years, a tremendous number of synthetic analogs of manoalide have been prepared (Lee 1990a, 1990b; Maullem et al. 1989), including manoalogue (5.31) (Reynolds et al. 1988).

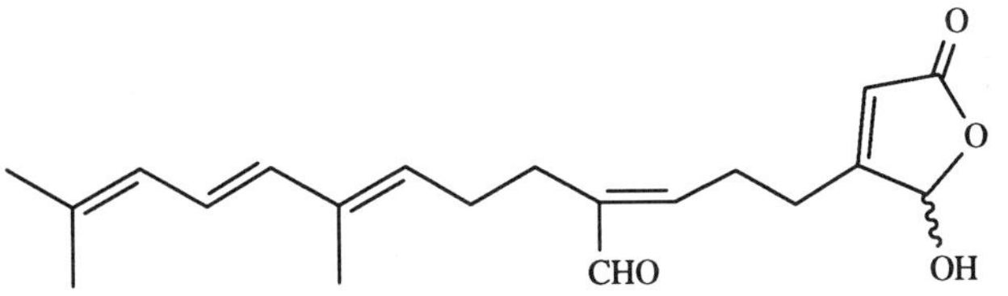

(5.31)

Currently, manoalide and selected analogs are being evaluated in clinical trials as topical anti-inflammatory agents (R. Jacobs, personal communication).

5.5.4 Pseudopterosins

The pseudopterosins are a series of 12 tricyclic diterpene glycosides (A–L) that were isolated from the Caribbean gorgonian, *Pseudopterogorgia elisabethae* (Look et al. 1986a, 1986b, Roussis et al. 1990), based on the biological activity of the crude extract. Pseudopterosin A (5.32A) contains a 3-O-β-D-xylopyranose, which is attached to the diterpene group at C-9; pseudopterosins B through D are structurally identical, except that the xylopyranose is monoacetylated at the various hydroxyl moieties. Pseudopterosins G through J are diastereomers of A through D with a β-CH_3 at C-7 and contain regioisomeric acetates of an α-L-fucose group attached at C-9; pseudopterosins K and L also contain an α-L-fucose group, which is attached at C-9, and in the case of L, is monoacetylated. However, they are diastereomers of A

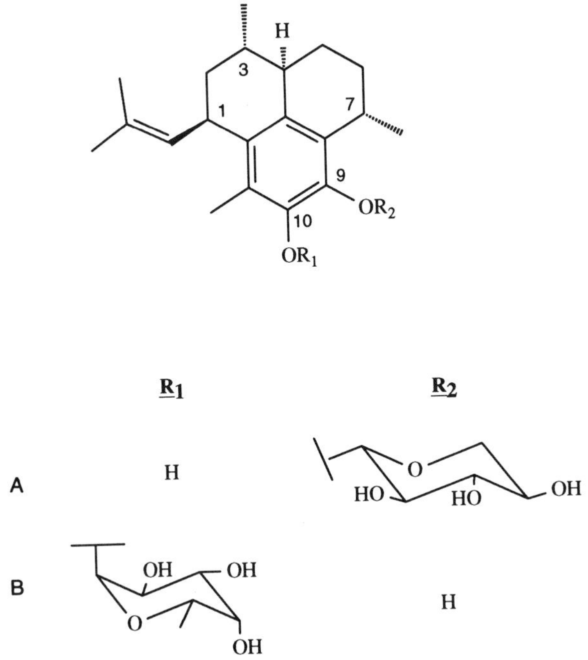

(5.32)

through D with an α-isopropenyl group at C-1, a β-CH$_3$ at C-3, and a β-CH$_3$ at C-7; pseudopterosins E (5.32B) and F contain identical diterpenoid groups as found in A through D; however, E contains an α-D-arabinose, which is attached to C-10 and F contains an α-L-fucose at C-10. The biological activities first reported about the purified compounds described both antimicrobial activity and cytotoxicity directed toward fertilized eggs from the California sea urchin (Ettouati and Jacobs 1987; Jacobs et al. 1981; Jacobs and Wilson 1986). However, potent anti-inflammatory and analgesic activities of pseudopterosin A (5.32A) (Look et al. 1986a) further drove the biological and chemical characterization of these compounds.

Initial studies (Look et al. 1986a) indicated that pseudopterosin A (5.32A) was significantly more potent in blocking the phorbol myristate acetate, topically induced inflammatory response of mouse skin (K = 8.9×10^{-4} M) compared with the standard anti-inflammatory drug indomethacin (40 mM). Concentrations of pseudopterosin A greater than 100 μM did not inactivate the hydrolysis of phosphatidyl choline by phospholipase A$_2$ (PLA$_2$) from bee venom or liver microsomal preparations; however, pancreatic PLA$_2$ activity was shown to be inhibited, with IC$_{50}$ values ranging from 3.0 to >80 μM.

Pseudopterosin A (5.32A) was also found to be more potent than indomethacin as an analgesic for blocking the stretch-flex response of phenylquinone injected mice (ED$_{50}$ obtained at 3.1 mg/kg of pseudopterosin A versus 10 mg/kg for indomethacin). Pseudopterosin A did not mimic morphine in electrically driven preparations, nor antagonized histamine or bradykinin at concentrations up to 30 μM, and did not prevent phorbol dibutyrate inhibition of bradykinin-induced contractions of the guinea pig ileum.

While pseudopterosins A through D expressed some in vivo toxicity (acute toxicity in the range of 50 mg/kg in mice), pseudopterosin E (5.32B) demonstrated very low acute toxicity in mice (LD$_{50}$ > 300 mg/kg) and an anti-inflammatory potency equivalent to that of pseudopterosin A (Roussis et al. 1990). In addition, the preliminary biological activities of pseudopterosin E suggest a unique mechanism of action via the inhibition of leukotriene synthesis. While the precise mechanism of action of the compound is unknown, the molecule seems to serve as a potential antagonist of the lipoxygenases or other enzymes involved in the arachidonic cascade.

The superior in vivo and in vitro anti-inflammatory activities of pseudopterosin E prompted several investigations into the possible derivatization and/or synthesis of the molecule. An attempt was made to obtain the compound by the interconversion of pseudopterosin A to E (Roussis et al. 1990), however, the inefficiency of glycosidation reactions lead to poor yields of the compound. The first total enantiospecific synthesis of pseudopterosin E was reported through the direct attachment of an L-fucose unit to the aglycone via a novel method (Corey and Carpino 1989). A method for the synthesis of pseudopterosin A was similarly described. Other investigations revealed alternate pathways for obtaining both the glycosylated form and the aglycone of pseudopterosin E (Broka et al. 1988; Ganguly et al. 1990). The importance of the role of glycosylation of the molecule is underlined by the fact that the aglycone portion, alone, is not active in vivo or in vitro.

With the synthesis of these compounds well in hand, work continues on the synthesis of various analogs of these compounds and the determination of the precise mechanism of action of their anti-inflammatory activities (R. Jacobs, personal communication). Present evidence indicates that one of the primary mechanisms of the anti-inflammatory effects of the pseudopterosins lies in its ability to selectively block degranulation and leukotriene production in human polymorphonuclear neutrophils (R. Jacobs, personal communication). The model compound, pseudopterosin E, has no effect on eicosanoid biosynthesis in cultured mouse peritoneal macrophages (R. Jacobs, personal communication).

Currently, the pseudopterosins are undergoing clinical evaluation for topical use (R. Jacobs, personal communication).

5.6 CLINICAL AGENTS RESEMBLING MARINE-DERIVED COMPOUNDS

5.6.1 Adenine Arabinoside (Ara-A) and Cytosine Arabinoside (Ara-C)

From studies of Caribbean sponge sterols by W. Bergmann and co-workers in the early 1950s (Bergmann and Burke 1956; Bergmann and Feeney 1950, 1951; Cohen 1966), the nucleosides spongouridine (5.33A) and spongothymidine (5.33B) were discovered serendipitously and found to possess antiviral activity. These discoveries aided in the development of antitumor agents and a generation of nucleoside antivirals. Through sporadic synthetic analog efforts to improve the activities of the compounds, cytosine arabinoside (Ara-C) (5.33C) was introduced nearly 20 years later as a clinically useful antitumor agent (Bodey et al. 1969), and nearly 30 years later, adenine arabinoside (Ara-A) (5.33D) was approved for use as an antiviral agent (Buchanan and Hess 1980). Cytosine arabinoside is produced synthetically, currently distributed as Cytosar-U® and indicated in acute nonlymphocytic leukemia, chronic myelocytic leukemia, and meningeal leukemia (*Physicians Desk Reference* 1991). Adenine arabinoside is obtained from fermentation cultures of *Streptomyces antibioticus* and currently distributed as Vira-A® (*Physicians Desk Reference* 1991). Through intravenous administration, Vira-A is indicated for the treatment of *Herpes simplex* virus encephalitis, *Herpes simplex* infections in newborns, and *Herpes zoster* virus due to reactivated varicela-roster virus infections in immunosuppressed patients. As an ointment, Vira-A® is used for the topical treatment of epithelial keratitis caused by *Herpes simplex*. Subsequent second- and third-generation nucleosides, i.e., acyclovir, AZT and dideoxycytidine, are highly useful antiviral agents and the objects of further medicinal chemistry studies to find even more efficacious and selective compounds.

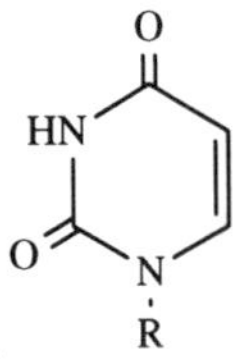

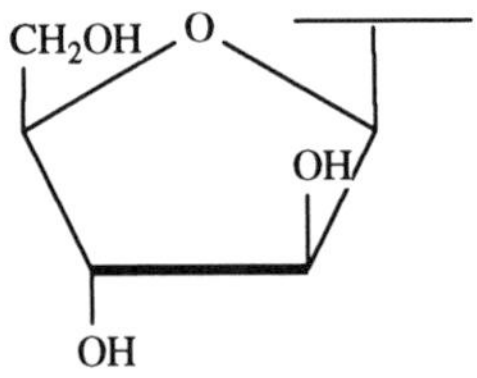

(5.33)

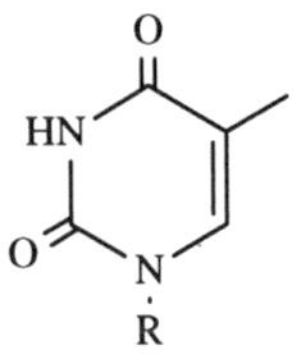

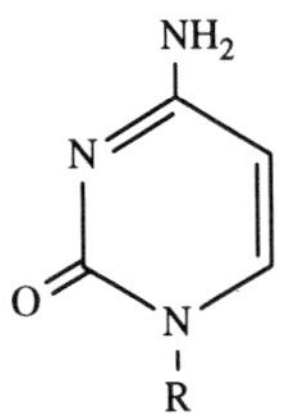

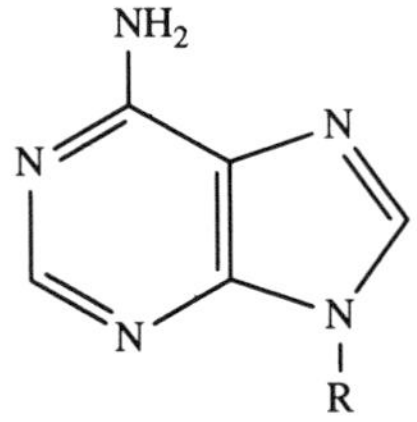

5.7 FUTURE DIRECTIONS

In contrast to biologically active compounds from terrestrial organisms, on which a considerable amount of work has been done and success achieved, the knowledge base of biological information about compounds from marine organisms is very limited. However, for the few compounds that have been studied in detail, the success rate is high and the potential is considerable. Over the past three decades, over 4000 novel compounds have been described from marine organisms (reviews by Faulkner 1984a, 1984b, 1986, 1987, 1988, 1990, 1991), some of which possess striking biological activities, and have been described in this chapter. In addition, recent reports on the results of marine extract screening strongly suggest that the marine environment holds promise for a continued high abundance of biologically active compounds (Munro et al. 1987). Marine natural products, due to their unrivaled degree of structural novelty, have served to inspire many of the great synthetic achievements in organic chemistry. Thus, with the great structural novelty and high degree of biological activity associated with this group of substances, why is it that, in the past 5 to 10 years, there have been no new marine-derived compounds that began as NCEs (new chemical entities) in the United States on the market?

The current role of marine natural products in the therapeutic regime must be considered in light of drug discovery and development as a whole. Given the time and capital investment required to discover, develop, and market new drugs (Niel 1988), it is a natural consequence that therapeutic agents on today's market are, for the most part, the result of research and development efforts of 5, 10, or more years hence, and, unfortunately, much of the *published* biological screening effort during this time period on marine-derived compounds has been mostly classical. Indeed, marine-derived compounds of current clinical interest (see Section 5.5) are the products of marine drug discovery programs smaller in scope and number than those that exist today, and most of the current industrial marine drug discovery efforts were initiated just within the last 5 years. The field is only now beginning to evolve to a state of relative maturity and the total chronology of recent marine drug discovery efforts has yet to run its course.

Consideration of the numerical parameters of drug discovery also reconciles the current minor role played by marine agents in chemotherapy. In 1970, over 703,000 substances were pharmacologically tested by the pharmaceutical industry (Reuben and Wittcoff 1989). The overall composition of this total is somewhat unclear since it is likely comprised of numerous sources such as extracts of fermentation products, natural substances, pure synthetics, etc., and statistical comparisons must be interpreted with caution. It is clear, however, that even today's heightened level of marine-derived compound and extract screening (a few thousand per year perhaps) represents only a small portion of this amount. Consideration of the scope of therapeutic areas screened shows a parallel situation. During the 1970s and early 1980s, marine screening efforts were focused, with some exceptions, in the areas of anti-tumor, cytotoxics, antiviral, and antimicrobials. These areas represent only a portion of the full range of human disease areas. Today, in addition to these, investigations of marine products in immunomodulation, inflammation, cardiovascular, central nervous system, hypertension, to name only a few, are now well underway. These efforts set the stage for marine compounds to play an increasingly important role in disease chemotherapy in the years to come.

Having seen that overall research efforts in marine drug discovery are on the rise, we briefly consider here several possible scenarios in which a marine natural product may lead to a therapeutic agent.

5.7.1 A Marine-Derived Compound that Becomes a Successful Drug

There are diseases such as AIDS, certain forms of cancer, Alzheimer's disease, and others where no truly effective form of chemotherapy exists and where an unoptimized natural product may comprise a suitable treatment. In these cases, the acute need for effective agents would provide the driving force to accelerate clinical investigations of relatively complex, naturally occurring agents.

In the usual commercial scenario, natural products that are biologically active and structurally complex are found in microorganisms and produced in large quantities by fermentation, or they are found in plants and produced in large quantities by cultivation. If the incentives are great enough, total synthesis of a complex natural product is undertaken; for example, the immune suppressant, FK506 (Jones et al. 1990; Nakatsuka et al. 1990). In contrast, pure didemnin B and bryostatin 1 were obtained for clinical trials from massive field collections of the source organisms *Trididemnum solidum* (Rinehart et al. 1988) and *Bugula neritina* (Schaufelberger et al. 1991), respectively. Didemnin B and bryostatin 1 serve well to illustrate a central problem of marine drug development, that of compound supply. While the biological richness of the marine environment is well appreciated, the prospects are usually not good for producing large amounts (kilograms) of marine compounds using field collection of organisms. Only the most abundant marine organisms can be collected in these amounts. An additional complication is the well-established variation in chemistry within a species as a function of geographical location. Thus, only in select cases with highly abundant organisms (i.e., some sponges, bryozoans, gorgonians, or algae) are extremely large-scale collections possible, and even then, environmental consequences can be severe. Therefore, while in some cases it might be possible to produce amounts of marine compound sufficient for preliminary clinical trial (gram quantities) from collection of the source organism, this approach cannot produce quantities needed for extended trials and the market (tens of kilograms).

Several, exciting new technologies hold promise for solving the supply problem, i.e., sponge cell culture and aquaculture, but it is presumed that current technologies of chemical synthesis and microbial fermentation will provide the first solutions. Fermentation is a mainstay in the pharmaceutical industry for compound production, in many cases providing supplies of structurally complex antibiotics, antitumor antibiotics, and immune suppressants, which are economically inaccessible by chemical synthesis. Fermentation of marine microorganisms may provide additional benefits outside drug development. An example of this is the culture of thermophilic bacteria isolated from marine geothermal sites (Borman 1991), whose thermally robust enzymes may have potential utility as industrial catalysts for the production of fuels, pharmaceuticals, or specialty chemicals. Indeed, the origin of several bioactive marine natural products isolated from sponges (Stierle et al. 1988; Elyakov et al. 1991) and other macroorganisms (Tamplin 1990) have been traced to microbial symbionts, and reports of success in the culture of these microorganisms are becoming more frequent.

The central role of chemical synthesis in drug development certainly requires no illustration. Chemical synthesis or modification of natural compounds yields the majority of chemotherapeutic agents in the U.S. market, especially in certain disease areas. Due to the inherent supply problem described previously, even in cases where the original, parent compound might be a suitable drug candidate, chemical synthesis is still likely to become the means by which suitable quantities are produced for advanced clinical investigation and marketing. The natural outgrowth of such a synthetic effort is the investigation of analogs as part of a medicinal chemistry effort. This brings forth another scenario.

5.7.2 A Bioactive Marine Natural Product as a Structural Lead

Many marine natural products have striking molecular complexity to match their biological activity. Certainly, molecules such as palytoxin (Armstrong et al. 1989a, 1989b), mycalamide (Hong and Kishi 1990), onnamide (Hong and Kishi 1991), and okadaic acid (Isobe et al. 1987) represent tremendous challenges to the synthetic organic chemist. The fact that many complex marine-derived molecules have now been synthesized is a testimony to the great advances made in the science and art of synthetic chemistry; however, even if compounds of this type did possess medicinal value competitive with current agents, their structural complexity greatly lowers the likelihood of an economically viable synthesis.

It is more likely that a bioactive, complex marine metabolite will serve as a lead structure for a medicinal chemistry effort. Programs of this type are designed with several goals in mind, not the least of which is the discovery of structural analogs exhibiting improved therapeutic attributes and simpler (i.e., more economical) chemical structures, as well as trying to secure sufficiently broad and appropriate patent coverage. An illustrative case of such a scenario is available in the case of manoalide (see Section 5.5.3), a compound isolated from the marine sponge *Luffariella variabilis* (de Silva and Scheuer 1980). Manoalide was found to be a specific inhibitor of phospholipase A_2 and C and is currently under development as an anti-inflammatory agent (see Section 5.5.3). The synthesis and evaluation of an extensive series of analogs was undertaken by several pharmaceutical companies and some of this work has been reported in the patent literature as well as the open literature (Garst et al. 1986; Katsumura et al. 1985; Lee 1990a, 1990b; Maullem et al. 1989). Consequently, manoalide and selected synthetic analogs have been evaluated as topical anti-inflammatory agents in clinical trials (R. Jacobs, personal communication).

Because of the enormous investment of money (>U.S. $200 million) and time (10 years) involved in the development of a new therapeutic agent that performs well, the desire to develop a better analog becomes very important. Therefore, considerable effort is expended to prepare second- and third-generation compounds that are even more active and less toxic. Similar to the situation with Ara-A and Ara-C, this scenario will certainly also apply to newly discovered marine-derived compounds or analogs that are developed as therapeutic agents.

5.8 CONCLUSIONS

By highlighting the sources, collection methods and strategies, structures, and respective biological activities of numerous marine-derived compounds, we believe we have shown that the marine environment is a rich and viable source of structurally unusual and biologically active compounds with therapeutic potential. Further, we believe that we have touched on many of the important issues related to marine-derived therapeutic agents, including the discovery approaches, supply/synthesis versus collection or fermentation, structure-activity optimization studies, and appropriate collaboration between academic researchers and industry and between research and development groups.

Finally, we believe that an important key to realizing the therapeutical potential of marine natural products is patience; the discovery by an academic, government, or industrial research group of a marine-derived compound with a satisfactory biological profile to become a lead compound may occur tomorrow or may not occur for years, and when it is discovered, the optimized compound may take another 10 to 15 years to reach the market. The wait will be worthwhile, both scientifically and commercially, because of the numerous, inevitable spin-offs: the continued discovery and development of biochemical probes or tools, a better understanding of the biology of particular disease states based on the unique biological activities of marine-derived compounds, a better understanding of the marine organisms that produce these compounds and of their respective environments, and advances in spectroscopy and synthesis as a consequence of conducting research on these often complex and structurally unusual molecules.

REFERENCES

Abraham, E.P. (1962) *Pharmacol. Rev.* 14, 473–500.

Abraham, E.P., and Newton, G.G.F. (1967) in *Antibiotics 2, Biosynthesis* (Gottleib, D., and Shaw, P.D., eds.), pp. 1–16, Springer-Verlag, New York.

Adamczeski, M., Quinoa, E., and Crews, P. (1988) *J. Am. Chem. Soc.* 110, 1598–1602.

Aimi, N., Odaka, H., Sakai, S.-I., et al. (1990) *J. Nat. Prod.* 53, 1593–1596.

Alberts, A.W., Chen, J., Kuron, G., et al. (1980) *Proc. Natl. Acad. Sci. USA* 77, 3957–3961.

Albizati, K.F., Holman, T., Faulkner, D.J., Glaser, K.B., and Jacobs, R.S. (1987) *Experientia* (*Basel*) 43, 949–950.

Armstrong, R.W., Beau, J.-M., Cheon, S.H., et al. (1989a) *J. Am. Chem. Soc.* 111, 7525–7530.

Armstrong, R.W., Beau, J.-M., Cheon, S.H., et al. (1989b) *J. Am. Chem. Soc.* 111, 7530–7533.

Armstrong, J.E., Janda, K.E., Willoughby, R., et al. (1991) in *Second International Marine Biotechnology Conference*, Baltimore, MD, Abstract P10, Society for Industrial Microbiology, Annandale, VA. p. 80.

Baertschi, S.W., Brash, A.R., and Harris, T.M. (1989) *J. Am. Chem. Soc.* 111, 5003–5005.

Baker, B.J., Okuda, R.K., Yu, P.T., and Scheuer, P.J. (1985) *J. Am. Chem. Soc.* 107, 2976–2977.

Bakus, G.J., Targett, N.M., and Schulte, B. (1986) *J. Chem. Ecol.* 12, 951–987.

Barnes, R.D. (1980) *Invertebrate Zoology*, 4th ed., Saunders College/Holt, Rinehart and Winston, Philadelphia.

Bennet, C.F., Mong, S., Clark, M.A., Kruse, L.I., and Crooke, S.T. (1987) *Biochem. Pharmacol.* 36, 733–740.

Bergmann, W., and Burke, D.C. (1956) *J. Org. Chem.* 21, 226–228.

Bergmann, W., and Feeney, R.J. (1950) *J. Am. Chem. Soc.* 72, 2809–2810.

Bergmann, W., and Feeney, R.J. (1951) *J. Org. Chem.* 16, 981–987.

Berkow, R., and Fletcher, A.J. (1987) *Merck Manual*, 15th ed., Merck & Co., Rahway, NJ.

Berkow, R.L., and Kraft, A.S. (1985) *Biochem. Biophys. Res. Commun.* 131, 1109–1116.

Bergquist, P.G. (1978) *Sponges*, pp. 197–198, University of California Press, Los Angeles.

Bergquist, P.R., and Wells, R.J. (1983) in *Marine Natural Products*, vol. V (Scheuer, P.J., ed.), pp. 1–50, Academic Press, New York.

Berridge, M. (1983) *Biochem. J.* 212, 849–858.

Bodey, G.P., Freirich, E.J., Monto, R.W., and Hewlett, J.S. (1969) *Cancer Chemother.* 53, 59–66.

Borman, S. (1991) *Chem. Eng. News* 69, 31–34.

Boyd, M.R., Schoemaker, R.H., Cragg, G.M., and Suffness, M. (1988) in *Pharmaceuticals and the Sea* (Jefford, C.W., Rinehart, K.L., and Shield, L.S., eds.), pp. 27–44, Technomic Publishing Company, Lancaster, PA.

Broka, C.A., Chan, S., and Peterson (1988) *J. Org. Chem.* 53, 1584–1586.

Buchanan, R.A., and Hess, F. (1980) *Science* 127, 143–171.

Bundy, G.L., Daniels, E.G., Lincoln, F.H., and Pike, J.E. (1972) *J. Am. Chem. Soc.* 94, 2124.

Burger, A. (1986) *Drugs and People*, University Press of Virginia, Charlottesville.

Burley, E.S., Smith, B., Cutter, G., Ahlen, J.K., and Jacobs, R.S. (1982) *Pharmacologist* 24, 117.

Burkholder, P.R., Pfister, R.M., and Leitz, F.H. (1966) *Appl. Microbiol.* 14, 649–653.

Burres, N.S., and Clement, J.J. (1989) *Cancer Res.* 49, 2935–2940.

Burres, N.S., Sazesh, S., Gunawardana, G.P., and Clement, J.J. (1989) *Cancer Res.* 49, 5267–5274.

Canonico, P.G., Pannier, W.L., Huggins, J.W., and Rinehart, K.L. (1982) *Antimicrob. Agents Chemother.* 22, 696–697.

Caprile, K.A. (1988) *J. Vet. Pharmacol. Therap.* 11, 1–32.

Cardellina, J.H., Marner, F.-J., and Moore, R.E. (1979) *Science* 204, 193–195.

Carmely, S., Kashman, Y., Loya, Y., and Benayahu, Y. (1980) *Tetrahedron Lett.* 21, 875–878.

Carroll, A.R., and Scheuer, P.J. (1991) *J. Org. Chem.* 55, 4426–4431.

Carter, D.C., Moore, R.E., Mynderse, J.S., Miemczura, W.P., and Todd, J.S. (1984) *J. Org. Chem.* 49, 236–241.

Chun, H.G., Daview, B., Hoth, D., et al. (1986) *Investigational New Drugs* 4, 279–284.

Cimino, G., Spinella, A., and Sodano, G. (1989) *Tet. Lett.* 30, 3589–3592.

Cohen, E., ed. (1979) *Biomedical Applications of the Horseshoe Crab (Limulidae)*, A.R. Liss, Inc., New York.

Cohen, S.S. (1966) in *Progress in Nucleic Acid Research and Molecular Biology*, vol. 5 (Davidson, J.N., and Cohn, W.E., eds.), pp. 1–88, Academic Press, New York.

Colwell, R.R. (1991) *Large-Scale Production of Natural Products*, National Cancer Institute Workshop, Rockville, MD, 4 Mar. 1991, National Cancer Institute, Rockville, MD.

Corey, E.J., and Carpino, P. (1989) *J. Am. Chem. Soc.* 111, 5472–5474.

Corey, E.J., Ensley, H.E., Hamberg, M., and Samuelsson, B. (1975) *J. Chem. Soc. Chem. Commun.* 277–278.

Corey, E.J., and Matsuda, P.T. (1987) *Tetrahedron Lett.* 28, 4247–4250.

Crawley, L.S. (1988) in *Pharmaceuticals and the Sea* (Jefford, C.W., Rinehart, K.L., Shield, L.S., eds.), pp. 101–107, Technomic Publishing Company, Lancaster, PA.

Crews, P., Jimenez, C., and O'Neil-Johnson, M. (1991) *Tetrahedron* 47, 3585–3600.

Crews, P., Manes, L.V., and Boehler, M. (1986) *Tetrahedron Lett.* 27, 2797–2800.

Dale, I.L., Bradshaw, T.D., Gescher, A., and Pettit, G.R. (1989) *Cancer Res.* 49, 3242–3245.

Dale, I.L., and Gescher, A. (1989) *Int. J. Cancer* 43, 158–163.

Davis, A.R., Targett, N.M., McConnell, O.J., and Young, C.M. (1989) in *Bioorganic Marine Chemistry*, vol. 3 (Scheuer, P.J.), pp. 85–114, Springer-Verlag, New York.

de Freitas, J.C., Blankemeier, L.A., and Jacobs, R.S. (1984) *Experientia* 40, 864–865.

de Silva, E.D., and Scheuer, P.J. (1980) *Tetrahedron Lett.* 21, 1611–1614.

de Silva, E.D., and Scheuer, P.J. (1981) *Tetrahedron Lett.* 22, 3147–3150.

de Souza, N.J., Ganguli, B.N., and Reden, J. (1982) in *Annual Reports in Medicinal Chemistry*, vol. 17, pp. 301–310, Academic Press, San Diego, CA.

Dorr, F.A., Kuhn, J.G., Phillips, J., and Von Hoff, D.D. (1988) *Eur. J. Cancer Clin. Oncol.* 24, 1699–1706.

Dreyfuss, M., Harri, E., Hofmann, H., et al. (1976) *Eur. J. Appl. Microbiol.* 3, 125–133.

Elyakov, G.B., Kuznetsova, T., Mikhailov, V.V., et al. (1991) *Experientia* 47, 632–633.

Ettouati, W.S., and Jacobs, R.S. (1987) *Mol. Pharmacol.* 31, 500–505.

Ewing, W.R., Bhat, K.L., and Joullie, M.M. (1986) *Tetrahedron Lett.* 42, 5863–5868.

Fattorusso, E., and Piattelli, M. (1980) in *Marine Natural Products*, vol. III (Scheuer, P.J., ed.), pp. 95–140, Academic Press, Inc., New York.

Faulkner, D.J. (1984a) *Nat. Prod. Rep.* 1, 251–280.

Faulkner, D.J. (1984b) *Nat. Prod. Rep.* 1, 551–598.

Faulkner, D.J. (1986) *Nat. Prod. Rep.* 3, 1–33.

Faulkner, D.J. (1987) *Nat. Prod. Rep.* 4, 539–576.

Faulkner, D.J. (1988) *Nat. Prod. Rep.* 5, 613–663.

Faulkner, D.J. (1990) *Nat. Prod. Rep.* 7, 269–309.

Faulkner, D.J. (1991) *Nat. Prod. Rep.* 8, 97–147.

Faulkner, D.J. (1992) *Oceanus* 35(1), 29–35.

Fenical, W. (1991) *Large-Scale Production of Natural Products*, National Cancer Institute Workshop, Rockville, MD, 4 Mar. 1991, National Cancer Institute, Rockville, MD.

Fenical, W., and Norris, J.N. (1975) *J. Phycol.* 11, 104–108.

Fimiani, V. (1987) *Oncology* 44, 42–46.

Frincke, J.M., and Faulkner, D.J. (1982) *J. Org. Chem.* 104, 265–269.

Fujicki, H., Sugunuma, M., Tahira, T., et al. (1984) in *Cellular Interactions by Environmental Tumor Promoters* (Fujicki, H., Hecker, E., Moore, R.E., Sugimura, T., and Weinstein, I.B., eds.), pp. 37–45, Japan Scientific Society's Press, Tokyo; VNU Science Press, Utrecht, The Netherlands.

Fusetani, N. (1987) in *Bioorganic Marine Chemistry*, vol. 1 (Scheuer, P.J., ed.), pp. 61–92, Springer-Verlag, New York.

Ganguly, A.K., McCombie, S.W., Cox, B., Lin, S., and McPhail, A.T. (1990) *Pure Appl. Chem.* 62, 1289–1291.

Garst, M.E., Tallman, E.A., Bonfiglio, J.N., et al. (1986) *Tetrahedron Lett.* 27, 4533–4536.

George, J.D., and George, J.J. (1979) *Marine Life*, Wiley-Interscience, New York.

Gescher, A., and Dale, I.L. (1989) *Anti-Cancer Drug Design* 48, 93–105.

Gieselman, J.A., and McConnell, O.J. (1981) *J. Chem. Ecol.* 7, 1115–1133.

Gilbert, B. (1977) *Pontif. Acad. Sci. Scr. Varia* 41, 225–252.

Glaser, K.B., De Cavalho, M.S., Jacobs, R.S., Kernan, M.R., and Faulkner, D.J. (1989) *Mol. Pharmacol.* 36, 782–788.

Glaser, K.B., and Jacobs, R.J. (1984) *Fed. Proc.* 43, 954.

Glaser, K.B., and Jacobs, R.J. (1986) *Biochem. Pharmacol.* 35, 449–453.

Glaser, K.B., Vedvick, T.S., and Jacobs, R.S. (1988) *Biochem. Pharmacol.* 37, 3639–3646.

Goodfellow, M., and Haynes, J.A. (1984) in *Biological, Biochemical and Biomedical Aspects of Actinomycetes* (Ortiz-Ortiz, L., Nojalil, L.F., and Yakoleff, V., eds.), pp. 453–472, Academic Press, Orlando, FL.

Gregson, R.P., Marwood, J.F., and Quinn, R.J. (1979) *Tetrahedron Lett.* 4505–4506.

Gulavita, N.K., Gunasekera, S.P., Pomponi, S.A., and Robinson, E.V. (1992) *J. Org. Chem.* 1767–1772.

Gunasekera, S.P., Gunasekera, M., Longley, R.E., and Schulte, G. (1990) *J. Org. Chem.* 55, 4912–4915.

Gunasekera, S.P., Gunasekera, M., and McCarthy, P. (1991) *J. Org. Chem.* 56, 4830–4833.

Gunawardana, G.P., Koehn, F.E., Lee, A.Y., et al. (1992) *J. Org. Chem.* 57, 1523–1526.

Gunawardana, G.P., Kohmoto, S., and Burres, N.S. (1989) *Tetrahedron Lett.* 30, 4359–4362.

Gunawardana, G.P., Kohmoto, S., Gunasekera, S.P., McConnell, O.J., and Koehn, F.E. (1988) *J. Am. Chem. Soc.* 110, 4856–4858.

Guyton, A.C. (1991) in *Medical Physiology*, pp. 810–938, W.B. Saunders Company, Philadelphia, PA.

Gustafson, K., Roman, M., and Fenical, W. (1989) *J. Am. Chem. Soc.* 111, 7519–7524.

Haddad, L.M., Lee, R.F., McConnell, O.J., and Targett, N.M. (1983) in *Poisoning and Drug Overdose* (Haddad, L.M., and Winchester, J.F., eds.), pp. 303–317, W.B. Saunders Company, Philadelphia, PA.

Hall, M.J. (1989) *Biotechnology* 7(5), 427–430.

Hall, S., and Strichartz, G., eds. (1990) *Marine Toxins: Origin, Structure and Molecular Pharmacology*, American Chemical Society, Washington, DC.

Hall, S., Strichartz, G., Moczydlowski, E., Ravindran, A., and Reichardt, P.B. (1990) in *Marine Toxins* (Hall, S., and Strichartz, G., eds.), pp. 29–65, American Chemical Society, Washington, DC.

Hay, M.E., Duffy, J.E., Pfister, C.A., and Fenical, W. (1987) *Ecology* 68, 1567–1580.

Hay, M.E., and Fenical, W. (1988) *Annu. Rev. Ecol. Syst.* 19, 111–145.

Haystead, T.A.J., Sim, A.T.R., Carling, D., et al. (1989) *Nature* 337, 78–81.

Hong, C.Y., and Kishi, Y. (1990) *J. Org. Chem.* 55, 4242–4245.

Hong, C.Y., and Kishi, Y. (1991) *J. Am. Chem. Soc.* 113, 9693–9694.

Hu, S.L., and Kao, C.Y. (1985) *Toxicon* 23, 723–724.

Iguchi, K., Kaneta, S., Mori, K., et al. (1985) *Tetrahedron Lett.* 26, 5787–5790.

Ireland, C.M., Roll, D.M., Molinski, T.F., et al. (1988) in *Biomedical Importance of Marine Organisms* (Fautin, D.G., ed.), pp. 41–58, California Academy of Sciences, San Francisco.

Isenberg, H.D., and Balows, A. (1981) in *The Prokaryotes*, vol. 1 (Starr, M.P., Stolp, H., Truper, H.G., Balows, A., and Schlegel, H.G., eds.), pp. 83–122, Springer-Verlag, New York.

Isobe, M., Ichikawa, Y., Bai, D., Masaki, H., and Goto, T. (1987) *Tetrahedron* 43, 4767–4776.

Ito, S., and Hirata, Y. (1977) *Bull. Chem. Soc. Jpn.* 50, 1813–1820.

Jacobs, R.S., Culver, P., Langdon, R., O'Brien, T., and White, S. (1985) *Tetrahedron* 41, 981–984.

Jacobs, R.S., White, S., and Wilson, L. (1981) *Fed. Proc.* 40, 26–29.

Jacobs, R.S., and Wilson, L. (1986) in *Modern Analysis of Antibiotics* (Aszalos, A., ed.), pp. 481–493, Marcel Dekker, Inc., New York.

Jacobson, P.B., and Jacobs, R.S. (1991) *Fed. Proc.* 805, A510.

Jacobson, P.B., Marshall, L.A., Sung, A., and Jacobs, R.S. (1990) *Biochem. Pharmacol.* 39, 1557–1564.

Johnson, R.K., and Hertzberg, R.P. (1989) in *Annual Reports in Medicinal Chemistry*, vol. 25 (Bristol, J.A., ed.), pp. 129–140, Academic Press, San Diego, CA.

Jomon, K., Kuroda, Y., Ajisaka, M., and Sakai, H. (1972) *J. Antibiot.* 25, 271–280.

Jones, T.K., Reamer, R.A., Desmond, R., and Mills, S.G. (1990) *J. Am. Chem. Soc.* 112, 2998–3017.

Kaiser, A.B. (1988) *Am. J. Surgery* 155, 52–55.

Kao, C.Y. (1966) *Pharmacol. Rev.* 18, 997–1049.

Kato, Y., Fusetani, N., Matsunaga, S., and Hashimoto, K. (1986) *J. Am. Chem. Soc.* 108, 2780–2781.

Kato, Y., Fusetani, N., Matsunaga, S., Hashimoto, K., and Koseki, K. (1988) *J. Org. Chem.* 53, 3930–3932.

Katsumura, S., Fujiwara, S., and Isaoe, S. (1985) *Tetrahedron Lett.* 26, 5827–5830.

Kernan, M.R., Faulkner, D.J., and Jacobs, R.S. (1987) *J. Org. Chem.* 52, 3081–3083.

Kikuchi, H., Tsukitani, Y., Iguchi, K., and Yamada, Y. (1982) *Tetrahedron Lett.* 23, 5171–5174.

Kikuchi, H., Tsukitani, Y., Iguchi, K., and Yamada, Y. (1983) *Tetrahedron Lett.* 24, 1549–1552.

Kim, S. (1967) *New Med. J.* (*Korea*) 10, 56–58.

Kino, T., Hatanaka, H., Hashimoto, M., et al. (1987a) *J. Antibiot.* 40, 1249–1255.

Kino, T., Hatanaka, H., Miyata, S., et al. (1987b) *J. Antibiot.* 40, 1256–1265.

Kitagawa, I., Hayashi, K., and Kobayashi, M. (1989) *Chem. Pharm. Bull.* 37, 849–851.

Kobayashi, J. (1989) *J. Nat. Prods.* 52(2), 225–238.

Kobayashi, M., Yasuzawa, T., Yoshihara, M., et al. (1982) *Tetrahedron Lett.* 23, 5331–5334.

Koehn, F.E., Longley, R.E., and Reed, J.K. (1992) *J. Nat. Prods.* 55, 613–619.

Komoda, Y., Kanayasu, T., and Ishikawa, M. (1979) *Chem. Pharm. Bull.* 27, 2491–2494.

Kraft, A.S., Smith, J.B., and Berkow, R.L. (1986) *Proc. Natl. Acad. Sci. USA* 83, 1334–1338.

Krebs, H.C. (1986) in *Progress in the Chemistry of Organic Natural Products*, vol. 49 (Herz, W., Grisebach, H., Kirby, G.W., and Tamm, C., eds.), pp. 152–400, Springer-Verlag, New York.

Larson, E.R., and Fischer, P.H. (1989) in *Annual Reports in Medicinal Chemistry*, vol. 24 (Allen, R., ed.), pp. 121–128, Academic Press, San Diego, CA.

Lawson, M.P., Berquist, P.R., and Cambie, R.C. (1984) *Biochem. Syst. Ecol.* 12, 375–393.

Leaf, A. (1988) in *Pharmaceuticals and the Sea* (Jefford, C.W., Rinehart, K.L., and Shield, L.S., eds.), pp. 17–22, Technomic Publishing Company, Lancaster, PA.

Lee, G.C.M. (1990a) European patent application no. 372940 A2.

Lee, G.C.M. (1990b) European patent application no. 372941 A2.

LeGrue, S.J., Sheu, T., Carson, D.D., Laidlaw, J.L., and Sanduja, S.K. (1988) *Lymphokine Res.* 7, 21–29.

Levine, L., and Fujiki, H. (1985) *Carcinogenesis (London)* 6, 1631–1634.

Levine, L., Xiao, D.-M., and Fujiki, H. (1986) *Carcinogenesis (London)* 7, 99–103.

Liaaen-Jensen, S., Renstrom, B., Ramdahl, T., Hallenstvet, M. (1982) *Biochem. Syst. Ecol.* 10, 167–174.

Lister, M.D., Glaser, K.B., Ulevitch, R.J., and Dennis, E.A. (1989) *J. Biol. Chem.* 264, 8520–8528.

Lombardo, D., and Dennis, E.A. (1985) *J. Biol. Chem.* 260, 7234–7240.

Longley, R.E., Caddigan, D., Harmody, D., Gunasekera, M., and Gunasekera, S.P. (1991a) *Transplantation* 52, 650–656.

Longley, R.E., Caddigan, D., Harmody, D., Gunasekera, M., and Gunasekera, S.P. (1991b) *Transplantation* 52, 656–661.

Longley, R.E., and Koehn, F.E. (1991) *Fed. Proc.* 5, 1206.

Longley, R.E., Koehn, F., Reilly, B., and Ward, J. (1993) *J. Immunol.* 150(2), 1594.

Look, S.A., Fenical, W., Jacobs, R.S., and Clardy, J. (1986a) *Proc. Natl. Acad. Sci. USA* 83, 6238–6240.

Look, S.A., Fenical, W., Matsumoto, G.K., and Clardy, J. (1986b) *J. Org. Chem.* 51, 5140–5145.

Malek, T.R., Ashwell, J.D., Germain, R.N., Shevach, E.M., and Miller, J. (1986) *Immunol. Rev.* 92, 81–87.

Manes, L.V., Naylor, S., and Crews, P. (1985) *J. Org. Chem.* 50, 284–286.

Marner, F.J., Moore, R.E., Hirotsu, K., and Clardy, J. (1977) *J. Org. Chem.* 42, 2815–2819.

Matsunaga, S., and Fusetani, N. (1991) *Tetrahedron Lett.* 32, 5605–5606.

Matsunaga, S., Fusetani, N., and Konosu, S. (1985a) *Tetrahedron Lett.* 26, 855–856.

Matsunaga, S., Fusetani, N., and Konosu, S. (1985b) *J. Nat. Prods.* 48, 236–241.

Matsunaga, S., Fujicki, H., and Sakata, D. (1991) *Tetrahedron* 47, 2999–3006.

Maullem, S., Sachs, G., and Wheeler, L. (1989) U.S. patent no. 483985.

May, W.S., Sharkis, S.J., Esa, A.H., et al. (1987) *Proc. Natl. Acad. Sci. USA* 84, 8483–8487.

Mayer, A.M., Glaser, K.B., and Jacobs, R.S. (1988) *Biochem. Pharmacol.* 37, 871–878.

Mayer, A.M., and Jacobs, R.S. (1988) in *Biomedical Importance of Marine Organisms* (Fautin, D.G., ed.), pp. 133–142, California Academy of Sciences, San Francisco.

McCaffrey, E.J., and Endean, R. (1985) *Mar. Biol.* 89, 1–8.

McConnell, O.J., Hughes, P.A., Targett, N.M., and Daley, J. (1982) *J. Chem. Ecol.* 8, 1437–1453.

McGeer, E.G., Olney, J.W., and McGeer, P.L., eds. (1978) *Kainic Acid as a Tool in Neurobiology*, Raven Press, New York.

Metcalf, S.M., and Richards, F.M. (1990) *Transplantation* 49, 798–802.

Miller, J.W., ed. (1979) *NOAA Diving Manual*, 2nd ed., pp. 11-1–11-13, Appendices D and E, U.S. Government Printing Office, Washington, DC.

Montgomery, D.W., Celniker, A., and Zukoski, C.F. (1987a) *Transplantation Proc.* 19, 1295–1296.

Montgomery, D.W., Celniker, A., and Zukoski, C.F. (1987b) *Transplantation* 43, 133–139.

Montgomery, D.W., and Zukoski, C.F. (1985) *Transplantation* 40, 49–56.

Moore, R.E., and Bartolini, G. (1981) *J. Am. Chem. Soc.* 103, 2491–2494.

Moore, R.E., Blackman, A.J., Cheuk, C.E., et al. (1984) *J. Org. Chem.* 49, 2484–2489.

Moore, R.E., and Entzeroth, M. (1988) *Phytochemistry* 27, 3101–3103.

Moore, R.E., and Patterson, G.M.L. (1988) in *Pharmaceuticals and the Sea* (Jefford, C.W., Rinehart, K.L., and Shield, L.S., eds.), pp. 95–100, Technomic Publishing Company, Inc., Lancaster, PA.

Moore, R.E., Patterson, G.M., Carmichael, W.W. (1988) in *Biomedical Importance of Marine Organisms* (Fautin, D.G., ed.), pp. 143–150, California Academy of Sciences, San Francisco.

Munro, M.H.G., Luibrand, R.T., and Blunt, J.W. (1987) in *Bioorganic Marine Chemistry*, vol. 1, pp. 93–176, Springer-Verlag, New York.

Murata, M., Leglrand, A.M., Ishibashi, Y., Fukui, M., and Yasumoto, T. (1990) *J. Am. Chem. Soc.* 112, 4380–4386.

Murata, M., Shimatani, M., Sugitani, H., Oshima, Y., and Yasumoto, T. (1982) *Bull. Jpn. Soc. Sci. Fish.* 48, 549–552.

Mynderse, J.S., Hunt, A.H., and Moore, R.E. (1988) *J. Nat. Prod.* 51, 1299–1301.

Nakamura, H., Iitaka, H., Kitahara, T., Okazaki, T., and Okami, Y. (1977) *J. Antibiot.* 30, 714–717.

Nakanishi, K., Goto, T., Ito, S., Naton, S., and Nozoe, S., eds. (1975) *Natural Products Chemistry*, vol. 2, pp. 457–463, Academic Press, New York.

Nakatsuka, M., Ragan, J.A., Sammakia, T., et al. (1990) *J. Am. Chem. Soc.* 112, 5583–5601.

Niel, G.L. (1988) in *Pharmaceuticals and the Sea* (Jefford, C.W., Rinehart, K.L., and Shield, L.S., eds.), pp. 45–48, Technomic Publishing Company, Inc., Lancaster, PA.

Nisbet, L.J., and Porter, N. (1989) *Symp. Soc. General Microbiol.* 44, 309–342.

Nonomura, H., and Ohara, Y. (1969) *J. Ferment. Technol.* 47, 463–469.

Norris, J.N., and Fenical, W. (1985) in *Handbook of Phycological Methods* (Littler, M.M., and Littler, D.S., eds.), pp. 121–144, Cambridge University Press, Cambridge, U.K.

Okami, Y. (1988) in *Horizon of Antibiotic Research* (Davis, B.D., Ichikawa, T., Maeda, K., and Midscher, L.S., eds.), pp. 213–227, Japan Antibiotic Research Association, Act Research Corporation, Tokyo, Japan.

Okami, Y., Hotta, K., Yoshida, M., et al. (1979) *J. Antibiot.* 32, 964–966.

Ouchi, K., Hirasawa, N., Takahashi, C., et al. (1986) *Biochim. Biophys. Acta* 887, 94–99.

Paul, V.J. (1988) in *Biomedical Importance of Marine Organisms* (Fautin, D.G., ed.), pp. 23–27, California Academy of Sciences, San Francisco.

Paul, V.J., and Fenical, W. (1987) in *Bioorganic Marine Chemistry*, vol. 1 (Scheuer, P.J., ed.), pp. 1–29, Springer-Verlag, New York.

Paul, V.J., Hay, M.E., Duffy, J.E., Fenical, W., and Gustafson, K. (1987) *J. Exp. Marine Biol. Ecol.* 114, 249–260.

Perry, N.B., Blunt, J.W., Munro, M.H.G., and Pannell, L.K. (1988) *J. Am. Chem. Soc.* 110, 4850–4851.

Perry, N.B., Blunt, J.W., Munro, M.H.G., and Thompson, A.M.J. (1990) *J. Org. Chem.* 55, 4242–4245.

Persinos, G.J. (1989a) *Washington Insight*, March 15, North Bethesda, MD.

Persinos, G.J. (1989b) *Washington Insight*, September 15, North Bethesda, MD.

Persinos, G.J. (1989c) *Washington Insight*, December 15, North Bethesda, MD.

Persinos, G.J. (1990a) *Washington Insight*, March 15, North Bethesda, MD.

Persinos, G.J. (1990b) *Washington Insight*, June 15, North Bethesda, MD.

Persinos, G.J. (1990c) *Washington Insight*, September 15, North Bethesda, MD.

Persinos, G.J. (1990d) *Washington Insight*, December 15, North Bethesda, MD.

Persinos, G.J. (1991a) *Washington Insight*, March 15, North Bethesda, MD.

Persinos, G.J. (1991b) *Washington Insight*, September 15, North Bethesda, MD.

Persinos, G.J. (1991c) *Washington Insight*, December 15, North Bethesda, MD.

Pettit, G.R., Herald, C.L., Doubek, D.L., et al. (1982) *J. Am. Chem. Soc.* 104, 6846–6848.

Pettit, G.R., Kamano, Y., Aoyagi, R., et al. (1985) *Tetrahedron* 41, 985–994.

Pettit, G.R., Kamano, Y., Herald, C.L., Schmidt, J.M., and Zubrod, C.G. (1986a) *Pure Appl. Chem.* 58, 415–421.

Pettit, G.R., Leet, J.E., Herald, C.L., Kamano, Y., Doubek, D.L. (1986b) *J. Nat. Prods.* 49, 231–235.

Physicians Desk Reference (1991) Medical Economics Company, Inc., Oradell, NJ.

Poch, G.K., and Gloer, J.B. (1989a) *Tetrahedron Lett.* 30, 3483–3486.

Poch, G.K., and Gloer, J.B. (1989b) *J. Nat. Prod.* 52, 257–260.

Pomponi, S.A. (1988) in *Biomedical Importance of Marine Organisms* (Fautin, D.G., ed.), pp. 7–11, California Academy of Sciences, San Francisco.

Pomponi, S.A., Wright, A.E., Diaz, M.C., and Van Soest, R.W.M. (1991) in *Fossil and Recent Sponges* (Reitner, J., and Keupp, H., eds.), pp. 150–158, Springer-Verlag, New York.

Pulitzer-Finali, G. (1986) in *Annali del Museo Civico di Storia Naturale di Genova*, V. LXXXVI, Monotipia Erredi, Genova, Italy.

Reed, J.K., and Pomponi, S.A. (1989) in *Proceedings of the American Academy of Underwater Sciences* (Lang, M.A., and Jaap, W.C., eds.), pp. 273–287, American Academy of Underwater Sciences, Costa Mesa, CA.

Reuben, B.G., and Wittcoff, H.A. (1989) *Pharmaceutical Chemicals in Perspective*, John Wiley & Sons, New York.

Reynolds, L.J., Morgan, B.P., Hite, G.A., Mihelich, E.D., and Dennis, E.A. (1988) *J. Am. Chem. Soc.* 110, 5172–5177.

Ricklefs, R.E. (1973) *Ecology*, Chiron Press, Inc., Newton, MA.

Rinehart, K.L. (1988a) in *Pharmaceuticals and the Sea* (Jefford, C.W., Rinehart, K.L., and Shield, L.S., eds.), pp. 3–15, Technomic Publishing Company, Lancaster, PA.

Rinehart, K.L. (1988b) in *Biomedical Importance of Marine Organisms* (Fautin, D.G., ed.), pp. 13–22, California Academy of Sciences, San Francisco.

Rinehart, Jr., K.L., Gloer, J.B., Hughes, Jr., R.G., et al. (1981b) *Science* 212, 933–935.

Rinehart, Jr., K.L., Gloer, J.B., and Cook, Jr., C. (1981c) *J. Am. Chem. Soc.* 103, 1857–1859.

Rinehart, K.L., Kishmore, V., Bible, K.C., et al. (1988) *J. Nat. Prod.* 512, 1–21.

Rinehart, Jr., K.L., Shaw, P.D., Shield, L.S., et al. (1981a) *Pure Appl. Chem.* 309–385.

Rinehart, K., and Shield, L.S. (1988) in *Horizons on Antibiotic Research* (Davis, B.D., Ichikawa, T., Maeda, K., and Midscher, L.A., eds.), pp. 194–212, Japan Antibiotics Research Association, Act Japan Corporation, Tokyo, Japan.

Roth, J., LeRoith, D., Collier, E.S., Watkinson, A., and Lesniak, M.A. (1986) *Ann. N.Y. Acad. Sci.* 463, 1–11.

Roussis, V., Wu, Z., Fenical, W., et al. (1990) *J. Org. Chem.* 55, 4916–4922.

Russell, D.H., Buckley, A.R., Montgomery, D.W., et al. (1987) *J. Immunol.* 138, 276–284.

Sakemi, S., Ichiba, T., Kohmoto, S., and Saucy, G. (1988) *J. Am. Chem. Soc.* 110, 4851–4853.

Sato, K., Okazaki, T., Maeda, K., and Okami, Y. (1978) *J. Antibiot.* 31, 632–635.

Schaufelberger, D.E., Koleck, M.P., Beutler, J.A., et al. (1991) *J. Nat. Prods.* 54, 1265–1270.

Schneider, W.P., Bundy, G.L., Lincoln, F.H., Daniels, E.G., and Pike, J.E. (1977a) *J. Am. Chem. Soc.* 99, 1222–1232.

Schneider, W.P., Hamilton, R.D., and Rhuland, L.E. (1972) *J. Am. Chem. Soc.* 94, 2122–2123.

Schneider, W.P., Morge, R.A., and Henson, B.E. (1977b) *J. Am. Chem. Soc.* 99, 6062–6066.

Scheuer, P.J. (1973) *Chemistry of Marine Natural Products*, Academic Press, New York.

Scheuer, P.J. (1988) in *Biomedical Importance of Marine Organisms* (Fautin, D.G., ed.), pp. 37–40, California Academy of Sciences, San Francisco.

Scheuer, P.J., Takashi, W., Tsutsumi, J., and Yoshida, T. (1967) *Science* 155, 1267.

Schindler, L.W. (1990) *Understanding the Immune System*, NIH Publication no. 90-529, National Institutes of Health, U.S. Department of Health and Human Services, Bethesda, MD.

Scrip (1991) *Cancer Chemotherapy Report*, Pharmabooks, Ltd., New York.

Shin, J., and Fenical, W. (1991) *J. Org. Chem.* 56, 3153–3158.

Shimizu, Y., Bando, H., Chou, H.-N., van Duyne, G., and Clardy, J.C. (1986) *J. Chem. Soc., Chem. Commun.* 1656–1658.

Shimizu, Y., and Kamiya, H. (1983) *Marine Natural Products*, vol. V (Scheuer, P.J.), pp. 391–427, Academic Press, New York.

Stevens, D.W., Jensen, R.M., and Stevens, L.E. (1989) *Transplantation Proc.* 21, 1139–1140.

Stierle, A.C., Cardellina, J.G., and Singleton, F.L. (1988) *Experientia* 44, 1021.

Suffness, M., Newman, D.J., and Snader, K. (1989) in *Bioorganic Marine Chemistry*, vol. 3 (Scheuer, P.J., ed.), pp. 131–168, Springer-Verlag, New York.

Suganuma, M., Fujiki, H., Furuya-Suguri, H., et al. (1990) *Cancer Res.* 50, 3521–3525.

Sun, H.H., and Sakemi, S. (1991) *J. Org. Chem.* 56, 4307–4308.

Tachibana, K., Scheuer, P.J., Tsukitani, Y., et al. (1981) *J. Am. Chem. Soc.* 103, 2469–2471.

Takahashi, A., Nakamura, H., Kameyama, T., et al. (1987) *J. Antibiot.* 40, 1671–1676.

Tamplin, M.L. (1990) in *Marine Toxins* (Hall, S., and Strichartz, G., eds.), pp. 78–86, American Chemical Society, Washington, DC.

Technical Resources, Inc. (1987) in *Workshop on Dereplication of Bioactive Natural Products*, May 28–29, 1987, Natural Products Branch, Developmental Therapeutics Program, Division of Cancer Treatment, National Cancer Institute, Bethesda, MD.

Torres, J.C. (1988) in *Pharmaceuticals and the Sea* (Jefford, C.W., Rinehart, K.L., and Shields, L.S., eds.), pp. 109–112, Technomic Publishing Company, Lancaster, PA.

Trainer, V.L., Edwards, R.A., Szment, A.M., et al. (1990) in *Marine Toxins* (Hall, W., and Strichartz, G., eds.), pp. 166–175, American Chemical Society, Washington, DC.

Trenn, G., Pettit, G.R., Takayama, H., Hu-Li, J., and Sitkovsky, M.V. (1988). *J. Immunology* 140, 433–439.

Tullar, R.M. (1972) *Life: Conquest of Energy*, Holt, Rinehart and Winston, Inc., New York.

Uemura, D., Ueda, K., Hirata, Y., Naoki, H., and Iwashita, T. (1981) *Tetrahedron Lett.* 22, 2781–2784.

Umezawa, H., Okami, Y., Kurasawa, Y., et al. (1983) *J. Antibiot.* 36, 471–477.

Van Soest, R.W.M., and Stentoft, N. (1988) *Studies Fauna Curacao Other Caribbean Islands* 70, 1–175.

Waldman, T.A. (1986) *Science* 232, 727–729.

Wattenberg, E.V., Fujiki, H., and Rosner, M.R. (1987) *Cancer Res.* 47, 4618–4622.

Weed, S.D., and Stringfellow, D.A. (1983) *Antiviral Res.* 3, 269–274.

Weinheimer, A.J., and Spraggins, R.L. (1969) *Tetrahedron Lett.* 5185–5188.

Weyland, H. (1969) *Nature* 223, 858.

Wheeler, L.A., Sachs, G., De Vries, G., et al. (1987) *J. Biol. Chem.* 262, 6531–6538.

Whittaker, R.H., and Feeny, P.P. (1971) *Science* 171, 757–770.

Wilkinson, S.R. (1978) *Marine Biol.* 49, 161–167.

Wilkinson, S.R. (1980) in *Endocytobiology, Endosymbiosis and Cell Biology*, vol. 1 (Schwemmler, W., and Schenk, H.E.A., eds.), pp. 553–563, Walter de Gruyter, New York.

Williams, D.H., Stone, M.J., Hauck, P.R., and Rahman, S.K. (1989) *J. Nat. Prod.* 52(6), 1189–1208.

Winston, J.E. (1988) in *Biomedical Importance of Marine Organisms* (Fautin, D.G., ed.), pp. 1–6, California Academy of Sciences, San Francisco.

Wratten, S.J., Wolfe, M.S., Andersen, R.J., and Faulkner, D.J. (1977) *Antimicrob. Agents Chemother.* 11, 411–414.

Wright, A.E., Forleo, D.A., Gunawardana, G.P., et al. (1990) *J. Org. Chem.* 55, 4512–4515.

Yasumoto, T., and Murata, M. (1990) in *Marine Toxins* (Hall, S., and Strichartz, G., eds.), pp. 120–132, American Chemical Society, Washington, DC.

Yasumoto, T., Nakamjima, I., Bagnis, R., and Adachi, R. (1977) *Bull. Jpn. Soc. Sci.* 43, 1021–1026.

Yuh, D.D., Zurcher, R.P., Carmichael, P.G., and Morris, R.E. (1989) *Transplantation Proc.* 21, 1141–1143.

Zobell, C.E. (1941) *J. Marine Res.* 4, 42–75.

PART
II

New Targets for Screening Assays

Mechanism-Based Screening for the Discovery of Novel Antifungals

Donald R. Kirsch
Beth J. DiDomenico

In recent years, fungi have played an increasingly expanded role in the clinical setting as disease-causing agents. This has been due primarily to an increase in the population of patients with deficient immune systems. Many fungi, some of which are not normally pathogenic, are capable in the immune-suppressed patient of causing severe, life-threatening diseases such as Aspergillosis, Blastomycosis, Histoplasmosis, and Candidosis. Unlike the wide range of antibiotics available for the treatment of bacterial infections, very few antifungal antibiotics are available for the treatment of systemic fungal infections. As a result, there is a great need for novel compounds having both

We would like to thank our colleagues in the field of antifungal screen design for providing screen design "tidbits" in their publications because this information was crucial for the production of this review. We gratefully acknowledge Sandy Silverman, Mark Pausch, Jonathan Greene, Claude Nash, and Karen Shaw for their critical review of this manuscript and for helpful discussions. We also thank Carol Bollinger for high quality secretarial services, David Gang for graphic design, and Robert Schenkel of American Cyanamid, and Roberta Hare and George Miller of Schering-Plough Research Institute for supporting our writing this review.

enhanced efficacy and selectivity against the most clinically important fungal pathogens.

This chapter will review methods that have been employed for the discovery of new antifungal agents, with an emphasis on mechanism-based screening. First, general techniques for pharmaceutical discovery will be compared and contrasted. Second, general methods for mechanism-of-action-based screening are described with an overview of both in vitro and in vivo methods. Finally, we will list a variety of molecular targets for both clinically useful or potentially useful antifungal agents. In vitro and in vivo screening strategies are described for these targets based upon (1) published screening protocols, (2) publications describing genetic screens or biochemical assay methods from which chemical screening methods can be easily inferred, and (3) methods revealed by colleagues in the area of antifungal screen design. This review is not intended to highlight the mechanism of action of available antifungal drugs since a number of excellent reviews on this topic are available (e.g., Vanden Bossche et al. 1987; Koller 1992; Sutcliffe and Georgopapadakou 1992). The focus of this review is to highlight both traditional screening methods and the most current methodologies that will lead to the discovery of novel antifungal agents in the future.

6.1 GENERAL STRATEGIES FOR THE DISCOVERY OF ANTIFUNGAL AGENTS

6.1.1 Traditional Screening

Classically, antifungal agents were identified by the ability to inhibit fungal (yeast) growth in agar solidified media. There are a number of variations on this basic assay design, including the "Walksman streak" assay, agar overlay method, and disk or agar plug diffusion assay (Walksman 1945; Betina 1983). All these methods resemble one another in that the identification of the antibiotic substance depends upon its ability to inhibit the growth of the detection organism while diffusing through an agar solidified medium. The earliest researchers in this field rapidly appreciated that they could routinely detect antifungal activities from soil microorganisms. For example, in an early study by Burkholder (1946) approximately 514 of 7369 actinomycetes tested (~7%) produced antibiotic activity against *Candida albicans*. Repetitive discovery of antibiotics was clearly recognized early and rediscovery of known antibiotics rapidly became a problem as the field matured. Advanced chemical and biological methods for "dereplication" and the rapid identification of known antibiotics have proven to be the major strategy for dealing with this problem. Random screening of chemical libraries and natural products has led to the discovery of most of the currently available antifungal agents and still remains the primary method for discovering agents with activity against unanticipated or unknown targets (Silver and Bostian 1990; Koller 1992).

6.1.2 Mechanism-Based Screening

Another approach to discover novel antibiotics is the use of mechanism-based screening techniques. Such methods identify antibiotics based upon their action against selected enzymatic targets or molecular processes. As a result, many known antibiotics will not be detected and rediscovered in such assays. This approach is particularly valuable in antifungal screening since fungi, like their hosts, are eukaryotes, and most agents that inhibit fungal growth are also active against mammalian cells. Mechanism-of-action-based screening thus reduces the detection of toxins and the subsequent potentially laborious characterization of a large number of substances that will be therapeutically useless. Additionally, a high level of sensitivity can be achieved as a result of this type of screen design. This can result in the discovery of activities that previously escaped detection either because they are present at levels too low to inhibit fungal growth or because they have intrinsically low activity against the target. Compounds of this latter type could prove to be useful as lead structures for the preparation of synthetic or semisynthetic compounds with enhanced therapeutic properties (e.g., Sykes et al. 1981).

6.1.3 Medicinal Chemistry

The use of chemical synthesis has proven to be extremely valuable in the discovery and development of a wide variety of pharmaceutical agents. One of the strengths of this approach is that a synthetic program can proceed either without knowledge of the target from a random lead or from theoretical targets based upon existing naturally occurring endogenous ligands (Burger 1983; Wolff 1982). Examples of the success of this approach are abundant in the areas of cardiovascular, renal, and central nervous system pharmacology (Wolff 1982; deStevens 1986). This strategy has proven to be valuable in the anti-infective area as well. Flucytosine is a synthetic compound with clinical use against *Candida* infections and is especially useful when employed in combination with amphotericin B (Scholer 1980). The imidazoles and triazole antifungals also are synthetic compounds and inhibit ergosterol biosynthesis. These agents comprise one of the most important groups of clinically used antifungals (Janssen and Vanden Bossche 1987; Saag and Dismukes 1988).

6.1.4 Rational Drug Design

Rational drug design is a relatively new method by which the development of inhibitory compounds proceeds from a comprehensive understanding of the structure of the proposed drug target. The appeal of this approach is that it derives from fundamental principles of biochemistry and protein structural analysis, and in theory does not depend upon serendipity or stochastic methodologies. While methods in protein structure and enzyme function analyses have seen great progress in recent years (Deisenhofer and Michel 1989; Benkovic et al. 1988), there remains no marketable pharmaceutical that has been initially identified through this approach. Rather, rational drug design is a tool which currently aids in identifying structure-activity relationships and in designing enhanced target-drug interactions. True de novo discoveries from rational drug design are still some years in the near future, and are eagerly anticipated (Hardy et al. 1987).

6.2 OVERVIEW OF DESIGN STRATEGIES FOR MECHANISM-OF-ACTION-BASED SCREENS

6.2.1 In Vitro Screens

The use of in vitro enzyme assays to identify low molecular weight inhibitors from culture broths was initiated by Hamao Umezawa in the mid-1960s (Umezawa 1982). In his initial studies, compounds were identified that showed no antimicrobial activity but were active as protease inhibitors. This observation suggested that there might be a large class of previously undetected molecules present in fermentation broths and set the stage for the development of protocols to screen fermentation broths for a wide range of enzyme inhibitory activities. Subsequently, inhibitors of a large number of enzyme activities have been identified from fermentation broths, including inhibitors of 3-hydroxy-3-methylglutaryl coenzyme A reductase (Endo 1981) and glycosidase (Umezawa 1982). More recently, in vitro screening has also been carried out to identify ligands for mammalian receptors (Creese 1978; Goetz et al. 1985; Kase et al. 1986; Ihara et al. 1991).

This methodology, athough initially directed toward the discovery of pharmacological agents, is also useful for antimicrobial screening. Although some targets for antifungals are complex processes that are not conveniently assayed in vitro (e.g., RNA splicing), many targets are enzymes for which there are convenient assays. This approach combines the benefits of biochemical target selectivity, potentially high throughput for screening, and the ability to identify novel lead structures that may be inactive against the whole cell because of penetration or metabolic stability problems.

The detection of enzyme activity (and thus inhibitor action) depends either on measuring the change in the concentration of a substrate or product as the reaction proceeds. This change can be determined spectrophotometrically using the crude reaction mixture when the spectral properties of the substrate(s) and product(s) are distinct or, alternatively, after the substrate(s) and product(s) are physically separated. If the substrate(s) and product(s) cannot be readily detected by spectral methods, or if their extinction coefficients do not allow accurate measurements to be made, radiochemical methods can be employed. In radiochemical detection systems, a radioactively labeled substrate is employed, the enzymatic reaction is run, substrate(s) and products are physically separated, and the radioactivity present in each fraction is determined. Limitations to the application of this approach for fermentation screening are due to the presence of compounds in fermentation broths having spectral properties that resemble the enzyme substrate(s)/product(s) and the presence of proteases and denaturing activities that nonspecifically inhibit enzyme action. This second type of artifact, however, can be eliminated through the parallel use of multiple enzyme target screening systems, or pretreating fermentation broths (e.g., heat denaturation treatment) to eliminate certain inhibitory activities.

6.2.2 In Vivo (Molecular Genetic) Screens

It is also possible to employ intact cells in the design of mechanism-based screens for enzyme inhibitors or receptor ligands. The advantages of employing target-directed screens based upon intact living mammalian or microbial cells have recently been discussed by Tomoda and Omura (1990). These include (1) maintaining the in vivo microenvironment of the enzyme, (2) decreasing the detection of false positives that can inhibit the isolated enzyme but cannot penetrate the intact cell, and (3) increasing the likelihood of detecting a prodrug that might be a functional inhibitor only after being activated by some cellular process. Initial efforts to design such screening systems were in the areas of antibacterial and antitumor agent discovery (White 1982; Sykes and Wells 1985; Kitano et al. 1975). These assay systems provided the starting point from which antifungal screens have been developed.

One of the first issues to be considered in the design of molecular genetic screens is the decision of whether to employ a "model system." Fungi comprise a morphologically and physiologically diverse group of lower eukaryotic organisms. In nature, they are major agents of disease and decomposition. They are also economically important as producers of secondary metabolites. However, fewer than 200 species of fungi are recognized as human pathogens, and only a small number of species are frequently recovered from infection sites (Odds 1988). Traditionally, a number of test isolates were chosen as representative of this diverse group of organisms and screened for antifungal activity. Minimal inhibitory concentration (MICs) determinations demonstrate that fungi exhibit a wide range of susceptibilities for certain classes of drugs (Rippon 1980; D. Loebenberg, personal communication). Therefore, depending upon the particular isolates chosen for screening, some compounds would invariably remain undetected. In this regard, classic screening techniques can be viewed as model systems even though they utilize pathogenic species.

With recent advances in recombinant DNA technology, it has been possible to "create" novel test organisms for screening purposes. Without a doubt, *Saccharomyces cerevisiae* provides the most well-characterized fungal system for this purpose, although *Aspergillus nidulans, Neurospora crassa, Candida albicans,* and a number of other fungi, some of which are known pathogens, can now be manipulated in many of the same ways, expanding the possibilities for the design of specific test systems. Whether *Saccharomyces* or some other fungal organism is used for designing new screens is a strategic decision based on the ease or difficulty of the specific screen design and the potential benefits derived from screening with a pathogen versus a model species. For example, DNA sequence analysis indicates close evolutionary relationships between certain fungal genera with sequence similarities for a number of target proteins being greater than 70% between *Saccharomyces* and *Candida*. The single most difficult challenge that remains is overcoming permeability differences among various fungi. It has been widely demonstrated that some drugs are substantially less effective against some fungal iolates than they are against others, limiting their usefulness in the clinical setting. These differences have been assumed to be due to differences in cell wall architecture, although this assertion remains largely unproven (DeNobel and Barnett 1991).

In the earliest approaches to generate in vivo genetic systems for the identification of antifungals, test fungi were screened in order to isolate strains with enhanced sensitivity to a particular class of compound. These strains were then used in differential screen strategies. Frequently, such supersensitive strains were capable of detecting metabolites at very low concentrations. In screening for supersensitive strains, however, the nature of the genetic lesion leading to the supersensitive phenotype was often not determined. This approach also relied upon the existence of isolated, characterized compounds of a desired class for use in the isolation of the supersensitive screening strain. More recently, the combination of classical genetic techniques and recombinant DNA technology have been used for screen design. Specific genetic lesions can now be introduced into select fungal species, creating novel test organisms. Researchers have recently described strategies using new technologies to create strains with phenotypes they desire. The benefits of these types of screens are their sensitivity and specificity. In this section, a few general design strategies (some of which are purely hypothetical at this time) will be described. The following sections show where these approaches have been utilized in screening for inhibitors of specific targets.

In *S. cerevisiae*, the combination of classical and molecular genetics has led to the design of elegant systems for the isolation of mutations that block specific cellular functions. In academic research programs, this budding yeast has proven to be an excellent model system for dissecting a variety of complex cellular functions, many of which have counterparts in metazoan species. With its ease of long-term culture, well-developed systems for genetics and molecular biology, and its stable haploid and sexual lifestyles, *S. cerevisiae* has been the model of choice for the analysis of numerous cellular processes, including signal transduction, regulation of gene expression, protein sorting, mechanisms of recombination, and synthesis and assembly of cell wall components. In all these areas, the combination of classical genetics and recombinant DNA technology has led to increases in the identification of key components.

Molecular genetic strategies are equally adaptable to the search for novel antifungal compounds. Genetic screens used to isolate mutations that disrupt specific cellular processes may be employed to identify compounds that disrupt those same targets. A *mutation* is defined as an alteration in an organism's genetic makeup that can result in a heritable and identifiable change or phenotype. Treatments that mimic the effect of a mutation without changing the organism's genome are known as *phenocopies*. A compound applied to a yeast strain that causes an alteration in the physiology of that strain mimics or phenocopies a mutant allele. The parallel between the effects of mutations and the potential for antifungal compounds to phenocopy these effects makes this approach extremely attractive. The well-understood, extensive genetic system available in *S. cerevisiae* makes this approach practical. However, the rapid advances being made in genetics and transformation systems of other fungi will extend the application of this approach to pathogenic species.

We describe several assay designs and strategies that have been (or can be) adapted from preexisting genetic screens targeted to specific fungal processes. Like the biochemical assays, genetics-based assays are specific for a chosen cellular target and are simple to run, allowing for high throughput. Unlike the biochemical assays, they may be designed to target a broad-based process rather than a single enzyme in order to identify an increased number of inhibitor types. The rationale for broader-based mechanism-of-action screening stems from the paucity of truly unique fungal targets. Fungi are eukaryotic organisms, having many similarities to their more complex hosts. While it may seem more reasonable to limit the search to inhibitors of truly unique targets, it should be pointed out that compounds currently in clinical use are, in fact, inhibitors of enzymes that are present in mammalian systems (Koller 1992). A useful therapeutic index for these compounds is achieved due to quantitative differences in target binding.

Last, most, if not all, the assays described below may be run in an agar plate format. This method has as its singular, most crucial advantage, the fact that a single compound applied to the agar surface diffuses outward, creating a continuous concentration gradient over which inhibition is tested. No other method of large-scale screening can come close to testing the large number of different inhibitor concentrations that can be tested with such high efficiency and economy in agar diffusion assays.

6.2.3 Pairwise Screens (Supersensitive versus Resistant)

Pairwise screens can be designed in a number of related formats in which the relative sensitivity to a compound is compared using two genetically related strains. If the strains differ at a single locus, then a compound specific for that target can be identified by comparing each strain's sensitivity to the inhibitor. For example, inhibitors specific to the target will be more active against a supersensitive test strain when compared with an otherwise isogenic sister strain. In an agar diffusion format, this is determined by measuring the size of the zone of inhibition surrounding the disc or well carrying the compound. General antifungals, or antifungals with mechanisms of action other than the desired one are generally observed as having similar activities against the two strains.

This approach has some limitations that must be considered when deciding which strategy to employ for a given target. Firstly (and ideally) one will already have a lead compound acting on the target of interest. Secondly, screening for supersensitive mutants can be labor intensive if there is no independent means for constructing a sensitive mutant. Thirdly, since the strain being used is selected for sensitivity to a particular compound or class of compounds, it may be possible that one will only find compounds of the same class, albeit, with increased potency, or with more desirable pharmacokinetic properties. For example, large, target proteins may have multiple domains, and the sensitivity elicited by one allele may be limited to compounds acting at the same site as the original compound. Inhibitors acting with novel mechanisms at other domains of the same target molecule may be lost if differential sensitivity is not conferred by a compound acting at another reactive center on the protein (e.g., substrate mimics versus compounds interfering with protein/protein interactions).

One solution to this type of problem relies upon another type of molecular genetic screen, involving pairs of strains where a cloned gene product is overexpressed in one strain compared with a control strain. The rationale behind this type of assay is that the strain containing an elevated quantity of the target protein should be more resistant to inhibitors specific to the cloned gene product than an isogenic strain containing normal amounts of the target protein. In fact, a number of genes encoding target proteins have been cloned by virtue of their producing enhanced resistance to antifungal compounds when genes for these targets were present on high copy plasmids (reviewed by Rine 1991). In an agar diffusion assay, the zone size surrounding a specific compound is expected to be smaller in the strain overexpressing the target protein as compared with an otherwise isogenic strain. Compounds not affecting the target protein itself, but other proteins in the cell are observed to have similar zone sizes in both strains. A potential artifact results from inhibitors of eukaryotic promoters or regulatory components of the translational apparatus that can mimic inhibitors of the protein. Another type of artifact that may be encountered results from the fact that overexpressing some proteins may cause subtle, otherwise undetectable changes in membrane permeability that indirectly produce differential zone sizes in this type of screen. Appropriate secondary screens may be employed to identify artifacts of this type. A strategy related to the above approach is to selectively decrease the expression of the target protein (e.g., through promoter mutation) resulting in a selective, supersensitive screening strain. This approach combines the advantages of the supersensitive and overexpression methods.

There are many advantages to utilizing pairwise screens. First, because of the very nature of the supersensitive versus resistant screen, the test strain has an enhanced sensitivity and is capable of detecting compounds at much reduced levels in culture broths. Second, this approach is applicable to use in species that are less genetically well defined than *S. cerevisiae*. Screening for super-sensitive strains versus wild type in any fungal species is possible, subject to the limitation that one may end up with strains that differ, not only in the target locus desired, but in permeability or in other metabolic pathways that can lead to the detection of nonspecific inhibitors.

6.2.4 Rescue Screens

The use of recombinant DNA technology has made it possible to regulate the expression of cloned genes in *S. cerevisiae* by placing them under the control of an inducible promoter such as *GAL1-10* (St. John and Davis 1981; Guarente et al. 1982), *ADC1* (Denis et al. 1983), or *PHO5* (Miyanohara et al. 1983). Transcription from each of these promoters is regulated by specific factors; galactose up-regulates transcription from the *GAL1-10* promoter; glycerol/ethanol up-regulates *ADC1*, and low levels of inorganic phosphate increase expression from *PHO5*. It has been shown in a number of instances that overexpression of a gene product in *S. cerevisiae* can be lethal. Examples of genes/proteins that exhibit overexpression lethality include *ras*, *KAR1*, and β-tubulins. Rescue screens are designed to detect compounds that inhibit the activity of such targets to the extent that some growth is restored.

Lethality due to overexpression of a single protein can arise by several mechanisms: in some cases, the increased level of the proteins may result in increased activity, which in turn, disrupts normal cellular functions. This is believed to be the mechanism that kills yeast cells harboring activated alleles of *ras*, and *v-src* oncogenes (Kataoka et al. 1985; Brugge et al. 1987). For proteins that are an integral part of a multiprotein complex, it is thought that overexpression of any one of the components disrupts the stoichiometry or saturates minor components, disrupting function. Examples include *KAR1* (Rose and Fink 1987), histone gene pairs (Meeks-Wagner and Hartwell 1986; Meeks-Wagner et al. 1986), and β-tubulin (Huffaker et al. 1987).

In genetic selections, rescue screens have been used successfully to define interacting gene products in a variety of yeast cellular systems, including signal transduction, the cytoskeleton, and transcriptional regulation (reviewed by Huffaker et al. 1987; Rine 1991). In the same way that a reversion event or a secondary mutation in another protein in the pathway can alleviate a lethal phenotype caused by mutation, compounds can be identified that inhibit the overexpressed gene product directly or produce a downstream change in the pathway blocking the effect of overexpression. It is possible to distinguish between direct inhibition and downstream effects using biochemical or genetic methods appropriate for the specific rescue system.

In rescue screens, the yeast screening strain is grown overnight under noninducing conditions and seeded in an agar plate under inducing conditions, and compounds are applied. Compounds specific for the target system of interest are identified by the production of a ring of growth surrounding the disk or well. Rescue assays are not limited to phenotypes such as rescue of overexpression lethality. Mutant screens have been designed to uncover mutations that cause a "sick" or debilitated strain to resume vigorous growth. Such screens can often be designed to cover a broader pathway or spectrum of targets. For example, in the field of intracellular protein traffic, screens have been devised that essentially utilize the signal sequence of one gene product to misdirect a second gene product to an incorrect compartment, causing an easily identifiable phenotype: limited growth on a particular substrate. Mutations in any number of genes regulating specific import of proteins into the organelle cause the engineered strains to grow due to the mislocalization of the fusion protein. Examples include *VPS* genes (Banta et al. 1988), *MAS1* and *MAS2* (Yang et al. 1988; Hawlitschek et al. 1988), and *NPL* genes (Silver and Hall 1988). In the same way, these systems are adaptable for identifying compounds that block steps in intracellular trafficking. Care must be taken to not overinterpret initial data of this type, since a large number of mechanisms are known to produce protein mislocation (for example, mitochondrial energy uncouplers and membrane disrupters acting on the vacuole).

6.2.5 Gene Induction Screens

Compounds that cause fungal proteins to be synthesized at the incorrect time or cause the synthesis of essential proteins to be repressed may be utilized as antifungals. The general strategy for these types of screens is to construct a hybrid gene in vitro containing the fungal promoter target of interest fused to a reporter gene such as β-galactosidase, luciferase, catechol dioxygenase, etc. (Control promoters fused to the same reporter gene are used to monitor the specificity of the compounds' effects.) Screening strains are treated with test compounds, and promoter repression is determined by monitoring the level of reporter gene activity compared to control promoter constructs. Compounds causing specific induction in promoter activity can also be monitored in similar assays.

Gene induction screens can also be used to monitor specific inhibitor action not directly related to transcription/translation. In such schemes, the inhibitor produces an alteration in cellular physiology that is monitored by the activation or repression of a reporter gene. For example, in signal transduction pathways, a cascade occurs as the result of a ligand-receptor interaction that ultimately leads to up or down regulation of certain gene products. If the receptor-ligand interaction is the target, a downstream promoter β-galactosidase fusion may be used to monitor receptor-ligand interactions (King et al. 1990). The advantages to this type of screen, as compared with biochemical receptor/ligand binding screens, are threefold.

1. Any untoward, toxic cellular response produced by a test compound can be monitored through the production of secondary toxic effects. General inhibitors that block protein-protein interactions would be screened out as general antifungal agents in this set-up.

2. Proteases, which are commonly found in fermentation broths, are active in receptor/ligand binding assays but would be inactive in this type of screen.
3. These assays can be run using an agar diffusion format exploiting the multiple advantages of agar diffusion assays.

6.2.6 Physiological Response Screens

Assays of this type are designed to take advantage of some aspect of the physiology of the test organism. A special set of growth conditions is chosen that allows the normal growth of the test organism or supports the growth of a morphologically modified form of the test organism in the presence of a desired antifungal. Active compounds of interest are identified by their inhibitory activity under "normal" growth conditions, which is reversed under "special" physiological circumstances. Perhaps the first examples of this type of assay were the screens employed almost 20 years ago to detect antimetabolites (Scannell and Pruess 1974). In these screens, test compounds were applied to microorganisms grown in minimal and rich media. Antimetabolite actives inhibit growth on minimal media but not on rich media, where the presence of nutrients reverses the action of compounds that inhibit the biosynthesis of those nutrients. This design has been the basis for screens for inhibitors of ergosterol biosynthesis.

Another example that illustrates the detection of antifungals via their alteration of cellular morphology is the *os-1* screen for cell wall synthesis inhibitors (Selitrennikoff 1983; Kirsch and Lai 1986). In this assay, fungal cells are converted to protoplasts by the action of a cell wall synthesis inhibitor. The protoplasts are able to survive due to growth medium formulation and the genetic background of the test organism so that active compounds of interest are detected in a single plate by protoplast formation. It is possible to design other assays in which the alteration of cellular morphology or metabolism are exploited as the basis for the detection of activities of interest.

6.3 SCREENING APPROACHES FOR ANTIFUNGALS ACTING ON CELL WALL BIOSYNTHESIS

A major goal of antifungal discovery efforts has been to identify a "fungal penicillin," an effective antifungal agent that is nontoxic to mammalian cells due to its inhibitory action on the synthesis of cell wall components that are unique to fungi. The cell walls of fungi such as *Candida albicans* are composed of many components, including mannoproteins, chitin, and β-glucans (Cabib et al. 1982; Ruiz-Herrera 1992) (Figure 6–1). Enzymes that produce chitin, a $\beta(1\text{-}4)$ polymer of *N*-acetyl-glucosamine, and those that synthesize the β-glucans, $\beta(1\text{-}3)$ and $\beta(1\text{-}6)$ polymers of glucose, are all potential targets for the discovery of selective antifungals. Prior efforts in these areas have led to

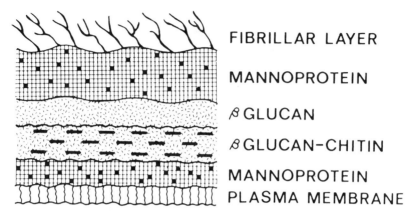

FIBRILLAR LAYER

MANNOPROTEIN

β GLUCAN

β GLUCAN-CHITIN

MANNOPROTEIN

PLASMA MEMBRANE

FIGURE 6–1 Yeast cell wall structure and composition. Reprinted with permission from M.G. Shepherd (1987) *Crit. Rev. Microbiol.* 15, 7–25. Copyright CRC Press, Inc., Boca Raton, FL.

the identification of a number of compounds that inhibit the synthesis of these polymers in vitro. The polyoxins and nikkomycins are nucleoside antibiotics that are highly effective competitive inhibitors of yeast chitin synthase in vitro (Cabib 1991). However, these compounds are not effective antifungals against zoopathogenic fungi, presumably because they are unable to achieve sufficiently high intracellular concentrations (Becker et al. 1983).

Greater success has been achieved in the identification of medically useful antifungals acting via the inhibition of β-glucan synthesis. A group of antibiotics composed of two structural classes, lipoglycosides and lipopeptides, have been identified that inhibit the synthesis of β-glucan. Two lipoglycoside inhibitors of β-glucan biosynthesis, papulacandin and chaeticandin, have been identified. However, neither of these compounds nor a number of derivatives of papulacandin have thus far been found to be clinically useful (Baguley et al. 1979; Komori et al. 1985; Traxler et al. 1987). A number of lipopeptide inhibitors of β-glucan biosynthesis have been identified. The first reported antibiotics in this class, aculeacin A and echinocandin B, are not employed clinically for the treatment of fungal infections (Mizoguchi et al. 1977; Sawistowska-Schroder et al. 1984). However, more recently, two compounds in this structural class were identified as having potential as clinical antifungal agents. Cilofungin (LY121019) is a semisynthetic derivative of echinocandin B that has potent activity versus *Candida* species and lowered mammalian toxicity, resulting in an improved therapeutic index (Gordee et al. 1984). L-671,329 is a recently described glycopeptide inhibitor of β-glucan biosynthesis with demonstrated activity against *Pneumocystis carinii* as well as against *C. albicans* (Schmatz et al. 1990).

6.3.1 Biochemical Screening Approaches for Antifungals Acting on Chitin Biosynthesis

Chitin is found in the cell walls of the great majority of zoopathogenic and phytopathogenic fungi, as well as in the skeletal structures of arthropods. Chitin is notably absent from the cells of plants and vertebrates (Gooday 1977), making chitin biosynthesis an attractive target for the development of highly specific, nontoxic antifungals. Chitin synthase is a membrane-bound enzyme and, in the absence of a purification method for this protein, all in vitro assays depend upon the use of crude membrane preparations as a source of enzyme.

The basic biochemical screen design for identifying chitin synthase inhibitors is to directly assay enzyme activity in vitro. Cell free extracts are incubated in the presence of radiolabeled substrate, UDP-*N*-acetyl-glucosamine, and the amount of radioactivity incorporated into insoluble chitin polymers is determined (Duran et al. 1975; Cabib et al. 1987). This simple assay was used successfully to characterize enzyme preparations from many fungi and invertebrates (reviewed by Gooday 1983).

The nucleoside peptides polyoxin D and nikkomycin Z are known competitive inhibitors of chitin synthases from many sources (Endo and Misato 1969; Endo et al. 1970; Muller et al. 1981; Furter and Rast 1985). It should be pointed out that compounds with other primary cellular targets are known to inhibit chitin synthase preparations. For example, amphotericin B and nystatin (known membrane-acting antifungal agents) are noncompetitive inhibitors of chitin synthase from *C. albicans* and *M. rouxii* (Rast and Bartnicki-Garcia 1981; Hanseler et al. 1983; Nozawa et al. 1985) and tunicamycin is an inhibitor of glycoprotein-lipid biosynthesis that inhibits chitin synthase from Neurospora (Selitrennikoff 1979). These findings suggest the presence of a complex system for chitin biosynthesis. Since none of the known targets, other than the enzyme itself, are likely to be unique to fungi, it is not surprising that both the polyenes and tunicamycin are toxic to mammalian cells in vitro and in vivo.

This assay has also been used to characterize the differences in the inhibition of polyoxin D and nikkomycin Z on the different isoforms of chitin synthase in *S. cerevisiae*. Under experimental conditions where each of the enzyme preparations was measured under conditions for optimal activity, using Mg^{2+} as cofactor for Chs1 and Co^{2+} as cofactor for Chs2 (Sburlati and Cabib 1986), the Chs1 activity was found to be more sensitive than Chs2 activity to both polyoxin D and nikkomycin Z (Cabib 1991). Furthermore, Chs2 inhibition by nikkomycin Z, but not polyoxin D, was affected by which ion was present in the assay buffer, although Chs1 activity was always significantly more sensitive to each inhibitor (Cabib 1991). Recently a third chitin synthase activity has been identified in *S. cerevisiae* and a gene controlling the production of the activity has been isolated (Valdivieso et al. 1991). It is not clear whether these are the only chitin synthase isoforms present in fungi.

However, recent reports (Bowen et al. 1990) indicate that almost all fungal species tested thus far have more than one gene.

6.3.2 Biochemical Screening Approaches for Antifungals Acting on Glucan Biosynthesis

The most abundant cell wall polysaccharides in many fungal species are the β(1-3) and β(1-6)-linked-D-glucans. Unlike chitin, which in many species is present predominantly at bud scars and at septa, glucan is found throughout the cell wall matrix (Molano et al. 1980; Fleet and Phaff 1981; Fleet 1985; Shepherd 1987). These polymers are attractive targets for antifungal drug discovery due to the absence of glucan in mammalian cells.

The standard biochemical assay for glucan is similar to the chitin synthase assay: radiolabeled substrate (in this case, UDP-glucose) is incubated in the presence of partially purified yeast extracts, and ATP or GTP. The amount of UDP-glucose converted into acid-insoluble glucan is determined by filter binding (Lopez-Romero and Ruiz-Herrera 1978; Shematek et al. 1980; Cabib 1987; Cabib and Kang 1987). This general assay format has been used to characterize glucan synthase activities from a variety of fungi and plants (Shematek et al. 1980; Orlean 1982; Jabri et al. 1989; Kottutz and Rapp 1990), and is suitable for high throughput natural products screening. The method of preparation for the enzyme fraction varies among labs and according to the organism used. More importantly, it has been shown that the method of enzyme preparation can alter the requirements for maximal glucan synthase activity. For example, lysis of fungal protoplasts in the presence of 1 mM EDTA results in enzyme preparations that require the addition of exogenous factors and are stimulated by certain nucleotides (Shematek and Cabib 1980; Cabib and Kang 1987; Orlean 1982; Larriba et al. 1981; Kottutz and Rapp 1990; Kang and Cabib 1986; Quigley and Selitrennikoff 1984). The use of high concentrations (10 mM) of ethylenediaminetetraacetic acid (EDTA) in the lysis buffer resulted in glucan synthase preparations that were almost fully active in the absence of added nucleotides (Shematek et al. 1980). The addition of one molar sucrose to the lysis buffer has also been shown to yield high levels of glucan synthase activities in the absence of exogenous factors (Larriba et al. 1981). It is not clear yet whether the EDTA simply binds excess Mg^{2+} ions, or whether it inhibits a negative regulator of glucan synthase that is loosely associated with the complex and is not isolated with the membrane fraction under some procedures. A more complete understanding of the enzyme requirements will require dissociation of the complex, analysis of its components, and finally, reconstitution.

The lipoglycoside antifungal antibiotics papulacandins A through D and the lipopeptide antifungal antibiotics aculeacin A, echinocandin B, and cilofungin, inhibit the synthesis of β-glucans in a variety of fungal species (Traxler et al. 1977, 1987; Mizoguchi et al. 1977; Mizuno et al. 1977; Yamaguchi et al. 1987; Elorza et al. 1987; deMora et al. 1991; Taft et al. 1988; Taft and

Selitrennikoff 1988). Cilofungin (LY121019), the only member of the echin-ocandins to have been in clinical trials, inhibits glucan synthase activity from *C. albicans* in vitro with a Ki-app (apparent inhibition constant) of 2.5 μM (Taft et al. 1988), and from *N. crassa* with a Ki-app of 16 μM (Taft and Selitrennikoff 1988). All the glucan inhibitors produce dramatic effects on the morphology of fungal cells and alter the composition of cell walls from regenerating protoplasts (deMora et al. 1991; Elorza et al. 1987; Yamaguchi et al. 1987; Miyata et al. 1980; Mizoguchi et al. 1977). The compounds are not active in vitro against chitin synthetase under conditions where polyoxin B shows activity.

Basic research in this field has been hampered by the inability to purify β(1-3) glucan synthase by traditional biochemical methods or to clone the gene(s) encoding β(1-3) glucan synthase activity (Kang and Cabib 1986). However, recent progress has been made in the cloning and characterization of a number of genes involved in the synthesis of β(1-6) glucan by Bussey and co-workers (Boone et al. 1990; Meaden et al. 1990; Roemer and Busser 1991; Hill et al. 1992). Recently, Becker and co-workers reported the isolation of a *Saccharomyces* DNA fragment that, in high copy, confers a resistant phenotype to aculeacin A and exhibits elevated levels of glucan synthase activity in vitro (Mason et al. 1989). A similar approach was taken with papulacandin B. However, restriction mapping data suggest that the papu-lacandin B resistant and aculeacin A resistant genes were different from one another (F. delRey, personal communication). Four additional loci resistant to aculeacin A were described by deMora et al. (1991): three of the four mutants were also resistant to papulacandin B, although none of the mutants displayed altered glucan synthetase activity.

The possibility that multiple genes may encode glucan synthase activity is also supported by biochemical data. Kang and Cabib (1986) identified two proteinaceous fractions from *Hansenula* and *Neurospora,* one soluble and the other membrane associated, that are both required for reconstitution of β(1-3) glucan synthase activity in vitro. In plant cells, two separable glucan synthase activities have been reported that differ in their sensitivity to Mg^{2+} ion concentration and to added UDP-glucose (Robinson et al. 1982). Taken together, the data suggest that glucan synthase exists as a complex in fungal cells with other, as yet, unknown and uncharacterized regulatory proteins.

6.3.3 Molecular Genetic Screens for Identifying Inhibitors of Cell Wall Biosynthesis

Since the discovery of penicillin in the 1930s, there has been a continual search for novel antibacterial antibiotics, especially for antibiotics that act via the inhibition of cell wall biosynthesis because antibiotics of this type would be expected to be highly effective antimicrobial agents and have low toxicity to the host. The effort to find agents of this type has involved the use of screening techniques with increasing specificity, sensitivity, and sophistication. These techniques have served as models for the development of screening methods for detecting antifungal antibiotics directed at cell wall biosynthesis.

The development of in vivo screening systems has been stimulated by the observation that there is little correlation between the cell-free chitin synthase inhibition by nikkomycin and polyoxin and the susceptibility of the corresponding species in culture (Bartnicki-Garcia and Lippman 1972; Endo et al. 1970). This effect is produced partly by the poor rate of uptake of polyoxins by fungal peptide permease systems (Becker et al. 1983; McCarthy et al. 1985; Yadan et al. 1984), partly by the presence of competing compounds present in fungal growth media (Hori et al. 1977; Becker et al. 1983), and partly by differences in activity exhibited by the various isoforms of chitin synthase (Cabib 1991). Design strategies employed in molecular genetic screens bypass some (but not all) of these problems since, in order to be active, test compounds must be stable to media components and must be taken up by the test organism.

6.3.4 Screens Based on Protoplast Formation

One type of screen for the identification of antibacterials acting on cell wall biosynthesis is to screen for agents that are capable of protoplasting bacterial cells (Imanaka 1985; Lederberg 1956). Two screen designs of this type for antifungal cell wall inhibitors have been described in the literature. These assays are based upon the exploitation of a temperature sensitive os-1 mutant of N. crassa (Selitrennikoff 1983; Kirsch and Lai 1986). Mutations at the os-1 locus in N. crassa produce a very complex phenotype.

1. Cells carrying the os-1 mutation are sensitive to high salt concentrations (osmotic sensitive).
2. These cells show high-level resistance to dicarboximide fungicides such as vinclozolin (Grindle 1983).
3. The os-1 cells are able to divide after the removal of the cell wall.

In N. crassa, the cell wall may be eliminated either by treatment with digestive enzymes or with antibiotics that inhibit cell wall biosynthesis (e.g., polyoxins and aculeacin). Slime strains of N. crassa carry three mutations, including os-1, which in combination produce a stable wall-free phenotype after "training" a wall-containing precursor (Emerson 1963).

This last phenotype of os-1 mutants, the ability to divide in the absence of a cell wall, forms the basis of the screens designed initially by Selitrennikoff (1983) and later modified by Kirsch and Lai (1986). The basic screen designed by Selitrennikoff is performed by preparing protoplasts from a temperature sensitive os-1 strain of N. crassa by incubating the strain at a nonpermissive temperature in the presence of sorbose (to inhibit glucan biosynthesis) and polyoxin B (to inhibit chitin biosynthesis). The resulting protoplasts can be maintained as wall-less cells at the restrictive temperature and will regenerate walls when shifted to a permissive temperature. The regeneration of the wall, however, is blocked by the presence of agents that inhibit wall polymer biosynthesis such as nikkomycins and polyoxins. (Unexpectedly, papulacandin, an inhibitor of β(1-3) glucan biosynthesis, does not block regeneration in this assay and instead produces a cytotoxic response.) In this screen, protoplasts are plated in agar solidified media at the permissive temperature in order to induce regeneration, and test samples are placed on the surface of the agar. The inhibition of wall generation is readily scored in this assay via the production of turbid zones of inhibition that, upon microscopic examination, are seen to contain protoplasts and not hyphal cells.

The modification of this assay by Kirsch and Lai (1986) more closely resembles traditional bacterial protoplasting screens. *N. crassa* can be protoplasted by the combined action of a glucan synthesis inhibitor (e.g., sorbose) and a chitin synthesis inhibitor (e.g., polyoxin). In this screen, spores from an *os-1* strain of *N. crassa* are plated at the restrictive temperature in agar solidified media containing sorbitol as an osmotic support and sorbose as an inhibitor of glucan biosynthesis. Under these conditions, growth occurs as walled, hyphal cells. Test samples are applied to the agar surface. Most antifungal agents produce a clear zone of inhibition, which is scored as inactive. Chitin synthesis inhibitors such as polyoxin or nikkomycin, however, produce a turbid zone that, upon microscopic examination, is seen to be composed of protoplasts. The production of protoplast zones is also accompanied by the production of an orange-colored amorphous material that can be readily observed by gross visual examination. This simplifies the scoring of the assay so that microscopic examination is only required for confirmation. This assay is sensitive to polyoxins and appears to be specific for chitin synthesis inhibitors. In a pilot screening program, polyoxins were detected in actinomycete fermentations at a high rate (~0.7% of fermentations tested) and no other antibiotics were detected (Kirsch and Lai 1986).

6.3.5 Screens Based on Antifungal Resistance by Wall-less Cells

Screens for antibacterial agents that inhibit cell wall biosynthesis have been based on the detection of agents that inhibit bacterial growth but do not inhibit the growth of wall-less mycoplasma cells (Omura et al. 1979). In a similar fashion, the wall-less slime mutant of *N. crassa* can be used to screen for antifungal antibiotics that inhibit cell wall biosynthesis. Slime is a strain of *N. crassa* carrying three mutations that in conjunction, produce a phenotype that allows the strain to grow as protoplasts (Emerson 1963). A pair-wise screen was devised based upon comparing the action of samples on a slime strain to a control, wall-containing wild type *N. crassa* strain (A.M. Gillum and D.R. Kirsch, unpublished observations). Actives are identified as samples that are inhibitory to the growth of the wild type control strain, but inactive versus the slime strain. A panel of antifungal agents with varied mechanisms of action was tested in this assay and only polyoxin and nikkomycin were scored as an active, suggesting that the assay was highly specific for inhibitors of cell wall biosynthesis.

This assay was used for extensive fermentation screening using the assay described by Kirsch and Lai (1986) to eliminate polyoxins. Actives were recovered at a very low rate (~1/10,000). Surprisingly, the active compounds (mycophenolic acid, pyrrolnitrin, phenazine carboxylic acid, clavam TVM18, phosphazomycin) were all previously reported antifungals, some having clearly defined mechanisms of action that are unrelated to cell wall biosynthesis. For example, mycophenolic acid is a well-characterized inhibitor of inosine monophosphate (IMP) dehydrogenase (Franklin and Cook 1969). It is difficult to explain the lack of specificity for cell wall synthesis inhibitors seen in this screen unless the active compounds have multiple sites of action, including cell wall biosynthesis. However, a toxic response against the slime strain would have been expected due to action against processes unrelated to cell wall biosynthesis. Another possible explanation is that mutations carried in the slime strain produce resistance to certain antibiotics in addition to contributing to the protoplast growth phenotype seen in slime. For example, the os-1 mutation has been reported to produce resistance to dicarboximide antifungals (Grindle 1983) and this resistance may extend to other unrelated antifungals. It is also possible that the cell wall is required for the efficient transport of certain types of molecules so that the removal of the wall produces resistance to toxic molecules in certain structural classes.

A related assay was devised by M. Kurtz based on the cell wall regeneration properties of S. cerevisiae (personal communication). In S. cerevisiae, protoplasts will regenerate a cell wall if plated in osmotically supported media containing high agar concentrations but will not regenerate a cell wall in liquid media or in media containing low concentrations of agar (Necas 1971). One can therefore conveniently compare the action of compounds on growing protoplasts (which will grow to some extent as syncytial cells) and protoplasts, which are regenerating a cell wall. Protoplasts are prepared by the enzymatic digestion of the cell wall and plated in osmotically supported low agar media (0.5%) and osmotically supported high agar media (2.5%). Duplicate test samples are placed on the two media, and the activity of samples is compared on the protoplasts and regenerated cells after a period of growth. Actives are identified as compounds that inhibit the growth of the regenerating protoplasts in the high concentration agar medium but do not inhibit protoplast growth in the low concentration agar medium. Compounds that inhibit cell wall synthesis, such as aculeacin, cilofungin, and nikkomycin, are active in this assay. In natural products screening, the antifungal monorden was identified as an active. This was unanticipated because monorden had not been described as an inhibitor of cell wall biosynthesis. It is possible that additional screening with this assay may lead to the identification of other antifungal agents.

Workers from Upjohn (Zaworski and Gill 1990) recently described a screen for cell wall active fungicides based upon the release of an intracellular enzyme from cells following a brief osmotic shock. In this screen, S. cerevisiae cells expressing the lacZ gene driven off the SUC2 promoter are treated with test compounds, osmotically shocked, and assayed for the release of β-galactosidase activity. The assay is based on the observation that antibiotics that act on the cell wall weaken the wall and make the cell more susceptible to lysis following osmotic shock. Lysis can then be easily followed by measuring extracellular β-galactosidase. A possible artifact in this assay might be agents that act on the plasma membrane and sensitize the cell to lysis following osmotic shock. However, Zaworski and Gill argue that this may not be a limitation in practice because nystatin, an antibiotic having a direct action on the plasma membrane, is inactive in this screen.

6.3.6 Screens Based on Genetically Engineered Strains of Saccharomyces cerevisiae

Saccharomyces cerevisiae provides the most well-characterized model system in which to set up a molecular genetic screen for the identification of chitin synthase inhibitors. Three genes, producing different forms of chitin synthase, have been identified and characterized (Bulawa et al. 1986; Silverman et al. 1988; Valdivieso et al. 1991). Chs1 is a "repair enzyme" and is not required for vegetative growth whereas Chs2 is the chitin synthase that is active during cell growth and is responsible for primary septum formation. The Chs3 activity appears to lay down chitin early during bud emergence. While both Chs1 and Chs2 chitin synthase activities are sensitive to polyoxin D in vitro (Sburlati and Cabib 1986), strains containing a null allele of chs1 are known to be more sensitive to polyoxin D than the isogenic parental strain (Bulawa et al. 1986). Thus far, there have been no reports of the sensitivity of CHS2 null strains to polyoxins, leaving the use of this construction for screening an open question.

The supersensitive phenotype of the *chs1* mutant suggests a pairwise screen format for chitin synthase inhibitors using whole cells. One potential strategy is to use a Δ*chs1* mutant as a supersensitive strain and a strain carrying a multicopy construct of *CHS2* as a resistant strain (Shaw and Cramer 1991). Specific activities would be expected to produce a larger zone of inhibition against the sensitive strain as compared with the resistant strain. Because of the multiplicity of chitin synthase isoforms, the choice of screen format, and even assay conditions, is somewhat problematic. It may even be possible to assay directly in a pathogenic organism such as *Candida albicans,* since multiple isoforms have been identified and two chitin synthase genes have been cloned (Bowen et al. 1990; Au-Young and Robbins 1990; Chen-Wu et al. 1992). However, the functional relationships between the identified *Candida* and *Saccharomyces* genes have not yet been described in detail.

6.3.7 Screens Based on Cell Wall Binding of Killer Toxin

Mutants defective in glucan biosynthesis have been isolated and provide a framework whereby molecular genetic screens can be used in high throughput assay formats. In particular, type 1 killer toxin of *Saccharomyces cerevisiae* has been shown to bind to β(1-6) glucans and to kill sensitive cells by collapsing the chemical proton gradient (Hutchins and Bussey 1983; Kagan 1983a, 1983b). Killer strains produce both the toxin and immunity factor from a precursor polypeptide that transits the yeast secretory pathway. Bussey and colleagues (Boone et al. 1990; Meaden et al. 1990) demonstrated that mutants resistant to K1 type killer toxin are deficient in β(1-6) glucans. The original screen used to identify these mutants is applicable to high throughput screening. In theory, a compound that alters (or decreases) the content of β(1-6) glucan should also promote "resistance" to killer toxin. In this type of screen the inhibitor will "rescue" the lethal phenotype caused by another agent (in this case, killer toxin).

The assay may be set up in any number of ways using isolated toxin, in microtiter format or in an agar diffusion scheme. Known inhibitors of β-glucan biosynthesis such as papulacandin and aculeacin are active in this screen, while inhibitors of chitin synthase (polyoxin B) are not. Inhibitors of protein maturation and the secretory pathway are expected to score positive in this screen format (if in vivo synthesized toxin is employed) since the toxin is synthesized through this pathway. However, compounds of this type may be identified by testing for prepro-alpha factor synthesis inhibition, a peptide that has many synthetic steps in common with toxin maturation. Douglas and co-workers (Douglas et al. 1989) described a high throughput quantitative assay using crude toxin applied to sensitive cells in a microtiter format. This strategy overcomes the limitation caused by the biology of killer toxin and should limit the percentage of false positives considerably.

6.3.8 Cell Wall Biosynthesis Antibiotic Synergy Screens

Synergy assays have proven useful in the discovery of novel antibacterial agents. Similar formats have been applied to the discovery of antifungals. Combinations of nikkomycin (a chitin synthesis inhibitor) and papulacandin B (a glucan synthesis inhibitor) have been shown to exert a synergistic effect against a number of *Candida albicans* isolates (Hector and Braun 1986). At sublethal concentrations of nikkomycin, glucan synthesis inhibitors, but not other general inhibitors, result in an enhanced sensitivity as compared with fungi grown in the absence of nikkomycin. Caution should be observed in interpreting results from these types of schemes due to the complex nature of biological systems. In a screen designed to find chitin synthetase inhibitors using a synergy format with aculeacin, Sitrin and co-workers (Sitrin et al. 1988) reported the discovery of several novel cerebrosides that potentiated the activity of aculeacin on plates, but had no effect whatsoever in chitin synthase in vitro assays.

6.4 SCREENING APPROACHES FOR ANTIFUNGALS ACTING ON ERGOSTEROL BIOSYNTHESIS

The major membrane sterol in fungi is ergosterol (Table 6–1) whereas cholesterol is present in the membranes of animal cells (Henry 1982; Mercer

TABLE 6-1   Ergosterol Biosynthetic Pathway[1]

Enzyme[2]	Gene	Substrate[2]	Inhibitor[3]
Thiolase	ERG10	Acetate (CoA)	
HMGCoA synthase	ERG11	Acetoacetate (CoA)	
HMGCoA reductase	HMG1,HMG2	3-hydroxy-3-methylglutaryl CoA	Mevinolin
Mevalonate kinase	ERG12	Mevalonate	
Phosphomevalonate kinase	ERG8	Phosphomevalonate	
		Mevalonate pyrophosphate	
		Isopentenyl pyrophosphate	
		Geranyl pyrophosphate	
Squalene synthase	ERG9	Farnasyl pyrophosphate	
Squalene monooxygenase	ERG1	Squalene	Allylamines
Squalene cyclase	ERG7	2,3 Oxidosqualene	
Lanosterol demethylase	ERG16	Lanosterol	Imidazoles
Δ14 Reductase			Morpholines
C4 Demethylase			
C24 Methyl transferase	ERG6		
Δ8–Δ7 isomerase	ERG2		Morpholines
C5 Desaturase	ERG3		
C22 Desaturase	ERG5		
Δ24 Reductase	ERG4		

[1] Modified from Henry (1982), and Karst and LaCroute (1977).
[2] Only enzyme or substrate common names are shown.
[3] See text for inhibitor references.

1984). This difference has proven to be the basis for the selective action of most of the fungicides currently used in clinical practice. The polyene antifungals (e.g., nystatin and amphotericin B) inhibit fungal growth by binding to ergosterol in the plasma membranes of fungi and forming channels that lead to the loss of cell integrity (Kerridge 1986). The polyenes used clinically have a higher binding affinity for ergosterol than cholesterol and therefore show specificity as antifungal agents. The imidazole and triazole antifungals (e.g., clotrimazole, ketoconazole, and fluconazole) are inhibitors of the enzyme lanosterol 14α-demethylase, an enzyme in the ergosterol biosynthesis pathway (Janssen and Vanden Bossche 1987; Saag and Dismukes 1988). These inhibitors are selective for fungal lanosterol 14α-demethylase and as a result do not inhibit sterol biosynthesis in mammalian cells. The allylamine and thiocarbamate antifungals (e.g., naftifine and tolnaftate) are inhibitors of the enzyme squalene monooxygenase, an enzyme acting in the sterol biosynthesis pathway prior to lanosterol 14α-demethylase (Ryder 1985; Ryder et al. 1986). Analogous to the imidazoles and triazoles, these inhibitors show selectivity for squalene monooxygenase from fungi. Based upon the clinical success of inhibitors targeted to this pathway, there has been interest in screening for structurally novel inhibitors of a number of enzymes involved in sterol biosynthesis.

6.4.1   Biochemical Screens for Identifying Ergosterol Biosynthesis Inhibitors

A number of strategies can be taken to design in vitro screening assays for identifying ergosterol biosynthesis inhibitors. These assays exploit the ability to measure enzyme activity or label intermediates in the sterol biosynthesis pathway. Some of the enzymes in this pathway can be assayed directly and the assays used for screening. 3-Hydroxy-3-methylglutaryl coenzyme A (HMGCoA) reductase activity can be conveniently assayed using radioactively labeled substrate. (This method was used to identify the HMGCoA reductase inhibitor mevinolin (Alberts et al. 1980).) Lanosterol 14α-demethylase is a cytochrome P450 enzyme that can be assayed by standard reduced-difference spectral methods (Janssen and Vanden Bossche 1987). Imidazole and triazole antifungals, which inhibit this enzyme, are identified by the production of Type II difference spectra (Janssen and Vanden Bossche 1987). Cell free extracts can be incubated with radioactively labeled pathway intermediates such as mevalonic acid or isopentenyl pyrophosphate, which are converted to lanosterol. Inhibitors of any of the enzymes in this portion of the pathway should cause a decrease in lanosterol synthesis (Ryder 1985; Kelly et al. 1990). It is also possible to label ergosterol and its metabolic precursors by incubating intact yeast cells with radioactive acetate. Solvent extraction of nonsaponifiable lipids separates radioactively labeled ergosterol from its hydrophilic metabolic precursors, which can then be analyzed by a number of methods, including thin layer chromatography. Inhibitors of ergosterol biosynthesis should decrease the radioactive labeling of ergosterol and metabolites downstream from the site of inhibition (Ryder 1985; Kelly et al. 1990).

6.4.2   Molecular Genetic Screens for Identifying Ergosterol Biosynthesis Inhibitors

S. cerevisiae provides an excellent system for the development of molecular genetic screens for the identification of sterol biosynthesis inhibitors. Strains carrying mutations in sterol biosynthesis have been known for many years, because such mutations produce resistance to polyene antifungals. Mutations that reduce the level of ergosterol in membranes result in polyene resistance (reviewed by Henry 1982). More recently, work on the ergosterol biosynthesis pathway has identified a number of genes encoding pathway enzymes; many have been cloned and characterized (Basson et al. 1988; Oulmouden and Karst 1991; Tsay and Robinson 1991; Anderson et al. 1989a, 1989b; J. Tkacz, personal communication; Lorenz and Parks 1990; Kalb et al. 1987; Jennings et al. 1991). These mutants and clones have provided basic materials for designing a number of screening protocols. In addition, genes from *Candida albicans* that encode homologous enzymes in this pathway have been cloned, resulting in opportunities for developing screens that directly employ *C. albicans* (J. Tkacz, personal communication; Kirsch et al. 1988; Kelly et al. 1990).

6.4.3   3-Hydroxy-3-Methylglutaryl Coenzyme A (HMGCoA) Reductase

Although HMGCoA reductase is not a target for commercial antifungals, inhibitors of the enzyme have antifungal activity against some yeast species and are relatively nontoxic towards mammalian cells, which can partially nutritionally bypass the need for HMGCoA reductase (Ikeura et al. 1988). HMGCoA reductase has been characterized in *S. cerevisiae* by Rine and co-workers (Basson et al. 1986), who determined that: (1) there are two structural genes for HMGCoA reductase in *S. cerevisiae*, *HMG1* and *HMG2*; (2) either of these genes is adequate for viability, although the deletion of both genes is lethal; (3) *HMG1* encodes the major portion (~83%) of the HMGCoA reductase activity in *S. cerevisiae*; and (4) overexpression of HMGCoA reductase via high copy cloning results in increased resistance to inhibitors, whereas under expression via disruption of the *HMG1* locus leads to increased sensitivity to inhibitors. These last observations directly suggest a screening assay for HMGCoA reductase inhibitors in which the activity of samples on an HMGCoA reductase underexpressing strain is compared with the activity on an overexpressing strain. Compounds that are less active against the high level expressing strain as compared with the underexpressing strain are presumptive inhibitors of HMGCoA reductase.

In a separate study, Rine and co-workers (Basson et al. 1988) expressed the human HMGCoA reductase gene in a yeast strain mutant for both *HMG1* and *HMG2*. In this strain, growth depends upon human HMGCoA reductase activity. Basson et al. (1988) suggest that this strain could be used as the basis for a microbiological screen to identify inhibitors of the human enzyme. Since mevalonate is produced through the action of HMGCoA reductase, and strains lacking HMGCoA reductase activity can be grown in media supplemented with high levels of mevalonolactone, inhibitors of human HMGCoA reductase can be identified in a screen that employs a physiological response strategy. Compounds that suppress the growth of the recombinant strain on normal media but not on mevalonolactone-containing media are identified as actives specific for enzyme inhibition.

6.4.4   Squalene Monooxygenase

Karst and Lacroute (1977) originally identified the gene for squalene monooxygenase in genetic studies that investigated the early portion of the ergosterol biosynthetic pathway. The squalene monooxygenase mutant isolated in this study (*erg1*) was exploited in further genetic and molecular characterization studies and for the design of screening protocols by J. Tkacz and co-workers (personal communication). It was determined that the mutation recovered by Karst and Lacroute resulted in only a partial loss of function of squalene monooxygenase activity, so that viability was retained even though ergosterol biosynthesis was depressed. Based upon this, Tkacz and co-workers tested the *erg1* mutant for sensitivity to allylamine antifungals, expecting strains with decreased enzyme activity to show increased sensitivity to inhibitors of that enzyme. As expected, the *erg1* mutant was found to be supersensitive to naftifine and terbinafine, providing the basis for cloning the squalene monooxygenase gene as well as developing a screening assay for the discovery of other squalene monooxygenase inhibitors.

Disruption of the gene blocks the sterol biosynthesis pathway beyond squalene and produces a strict auxotrophic requirement for ergosterol (J. Tkacz, personal communication; Lorenz and Parks 1990). This sequence also proved to be extremely useful for screen design that is similar to the pairwise screen format described for chitin synthesis inhibitors (see above). Overexpression of the *ERG1* gene leads to an increased the level of squalene monooxygenase activity and resistance to allylamines. It should be possible to identify squalene monooxygenase inhibitors by their differential action on *erg1* strain and a control strain that carries a wild type *ERG1* gene at high copy. No novel antibiotics were identified in initial screening experiments with this assay. In subsequent work these clones were used as hybridization probes to isolate the squalene monooxygenase gene from *C. albicans* (J. Tkacz, personal communication). Lastly, the recently reported DNA sequence of the *S. cerevisiae ERG1* gene may be useful for rational drug design approaches (Jandrositz et al. 1991).

6.4.5 Squalene Cyclase

Squalene cyclase catalyses a reaction in the ergosterol biosynthesis pathway that occurs just after the action of squalene monooxygenase and just prior to lanosterol 14α-demethylase. Although this enzyme is not currently the target of any commercial antifungal agent, it remains a potential target considering that the previous and subsequent reactions in the pathway are both the targets of commercial antifungals (Cattel et al. 1986). Mutations in the gene for squalene cyclase (*ERG7*) were recovered independently by two groups in the mid-1970s (Karst and Lacroute 1977; Gollub et al. 1977). The mutation isolated by Gollub and co-workers is a nonconditional allele and results in ergosterol auxotrophy. This mutation was recovered together with a heme pathway mutation (*hem3*), which allowed the uptake of ergosterol to satisfy the sterol requirement of the strain. The mutation isolated by Karst and Lacroute (*erg7-1*) is a temperature sensitive lethal. R. Kelly and co-workers (personal communication) observed that the *erg7-1* mutation produces super-sensitivity to inhibitors of squalene cyclase (U18666A and AMO1618) presumably because this mutant is deficient in squalene cyclase activity even at the permissive temperature. This observation was used by Kelly's group to devise a pairwise plate assay for inhibitors of squalene cyclase comparing the sensitivity of the *erg7-1* mutant and its wild type parent to various compounds and fermentation broths (R. Kelly, personal communication). Unfortunately, the assay was never critically evaluated in mass screening so that its value as a screen is unknown.

Subsequent work by Kelly and co-workers focused on the isolation and characterization of the corresponding gene from *C. albicans* (Kelly et al. 1990). *C. albicans* sequences were isolated based upon their ability to complement the sterol auxotrophic phenotype of an *erg7, hem3* strain. The cloned sequence was identified as the *C. albicans ERG7* gene based upon the complementation of two additional *erg7* mutations in *S. cerevisiae* and the restoration of squalene cyclase activity in the transformed strain, demonstrated by radioactive tracer studies performed both in vitro and in vivo. The cloning of this gene raises the prospect of developing *C. albicans*-based assays for identifying squalene cyclase inhibitors.

6.4.6 Lanosterol 14α-Demethylase

Mutations in the gene for lanosterol 14α-demethylase were originally isolated in *S. cerevisiae* through studies of mutations that lowered membrane ergosterol content and resulted in resistance to polyene antibiotics (reviewed by Henry 1982). The nature of the structural gene for lanosterol 14α-demethylase was further defined through the work of Loper's group, who cloned and characterized the *S. cerevisiae* lanosterol 14α-demethylase gene (Kalb et al. 1986, 1987). This gene was isolated as an insert from a high copy vector that produced resistance to ketoconazole, an inhibitor of lanosterol 14α-demethylase. DNA sequencing studies indicated that, as anticipated, the gene encoded a cytochrome P450 enzyme. Gene disruption of this locus results in cells that are obligate anaerobes due to an absolute nutritional requirement for ergosterol. (*S. cerevisiae* wild type cells do not take up ergosterol when grown aerobically, but will take up ergosterol under anaerobic conditions.)

The cloned *S. cerevisiae* lanosterol 14α-demethylase gene was subsequently used as a hybridization probe to isolate the *C. albicans* lanosterol 14α-demethylase gene (Kirsch et al. 1988; Lai and Kirsch 1989). The cloned *C. albicans* sequence was able to complement a lanosterol 14α-demethylase disruption in *S. cerevisiae* although expression of the *C. albicans* sequence in *S. cerevisiae* was inefficient. *S. cerevisiae* strains disrupted for their own lanosterol 14α-demethylase gene and carrying a single copy of the *C. albicans* sequence are viable, but produce significantly less lanosterol 14α-demethylase activity than wild type *S. cerevisiae* strains. Not surprisingly, these strains are supersensitive to imidazole antifungals. Depending upon the specific imidazole tested, inhibition of this strain occurs at 10 to 100 times lower concen-

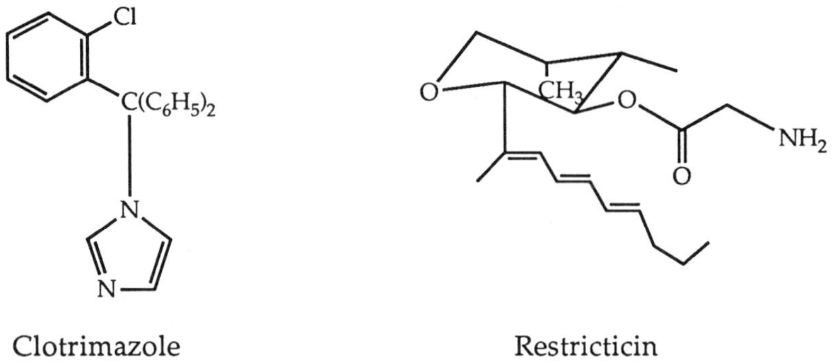

Clotrimazole   Restricticin

FIGURE 6–2 Comparison of the structures of clotrimazole and restricticin. Both compounds, although structurally unrelated, are inhibitors of the fungal enzyme lanosterol 14α-demethylase.

trations of the compound than wild type strains. In addition, when the *C. albicans* gene is present in *S. cerevisiae* at multiple copy, both lanosterol 14α-demethylase activity and imidazole resistance are increased relative to wild type *S. cerevisiae* strains. These recombinant strains can be used to screen for lanosterol 14α-demethylase inhibitors by comparing the action of test samples on single copy and multicopy transformants. Lanosterol 14 α-demethylase inhibitors strongly inhibit the single copy transformant and are relatively inactive towards the multicopy transformant. Screening of natural product fermentations with this assay led to the identification of antifungal compounds present in fermentations from *Scopulariopsis*, *Penicillium*, and *Pycnidiophora* species. These fermentations produce a family of structurally related antibiotics, including restricticin and lanomycin (Schwartz et al. 1991; Matsukuma et al. 1992; Aoki et al. 1992; O'Sullivan et al. 1992). Restricticin and lanomycin produce antibiotic spectra similar to ketoconazole, which may not be surprising considering that these compounds share a common mechanism of action. Although restricticin and related antibiotics are structurally unrelated to the imidazole antifungals (Figure 6–2) both classes are inhibitory to lanosterol 14α-demethylase in vivo and in vitro (O'Sullivan et al. 1992; Aoki et al. 1992).

6.4.7 Ergosterol Reversal Assays

Screens for biosynthetic pathways inhibitors were first described many years ago (reviewed by Scannell and Pruess 1974). All these screens are based on detecting compounds that inhibit growth only in the absence of one or more specific nutrient(s), since inhibitors of a biosynthetic pathway should not be active if the end product of the pathway is provided as a nutrient. In these assays, the test organism is grown in minimal chemically defined media and samples are tested for growth inhibition. As a control, the assay is repeated using media containing end products from pathway(s) of interest (e.g., glutamine). Active compounds are defined as samples that inhibit growth in minimal media but not in media to which nutrient(s) have been added.

In principle, this strategy should be applicable to the design of screens for any biosynthetic pathway. However, this technique is not directly applicable to ergosterol biosynthesis because yeast cells do not take up ergosterol under normal conditions (Henry 1982; Lewis et al. 1985). There are three conditions under which *S. cerevisiae* will take up exogenous ergosterol: (1) under conditions of anaerobic growth, (2) under aerobic growth conditions if the strain carries a heme mutation, and (3) under aerobic growth conditions if the strain carries a *upc* mutation. Therefore, either Hem⁻ or *upc* strains of *S. cerevisiae* can be used as the basis for a conventional screen that would identify ergosterol synthesis inhibitors based on growth inhibition reversal by ergosterol. In practice, ergosterol biosynthesis inhibitors do reverse the growth inhibition of both Hem⁻ and *upc* mutants in the presence of ergosterol, suggesting that this might be a practical screening method. Unfortunately, in screening experiments, the sensitivity of these strains to a number of antibiotics was altered by the presence or absence of ergosterol (J. Tkacz, unpublished observations). Therefore, although it is possible to identify ergosterol inhibitors by using this type of assay, secondary screening methods would be required to eliminate actives with other mechanisms of action. Finally, the analysis of a *C. albicans* Hem⁻ mutant indicated no ergosterol uptake, suggesting that this procedure could not be modified for use with *Candida* (Kurtz and Marrinan 1989).

6.4.8 Heme Auxotrophy Screen for Ergosterol Biosynthesis Inhibitors

Kurtz and Marrinan (1989) constructed and characterized a *C. albicans* strain that was homozygous for mutations at the *hem3* locus (uroporphyrinogen I synthase) and thus auxotrophic for heme. Unexpectedly, the phenotype of this strain differed from the analogous *S. cerevisiae* mutant. As mentioned earlier, the strain did not take up exogenous ergosterol, making the strain unsuitable for use in a sterol reversal assay. However, the *Candida* heme auxotrophs were observed to be slow-growing, suggesting that the utilization of exogenously provided heme may be inefficient in these strains. Since heme is a cofactor for multiple enzymes in the ergosterol biosynthesis pathway (reviewed by Mercer 1984), it is possible that ergosterol biosynthesis might proceed inefficiently in *Candida* heme auxotrophs. Consistent with this, Kurtz and co-workers observed that these strains showed a pronounced sensitivity to a number of ergosterol biosynthesis inhibitors and particularly to a number of triazole antifungals (M.B. Kurtz and J. Marrinan, personal communication), leading to the development of a screen in *Candida*. However, actual screening studies were not carried out to determine the utility of this screen.

6.5 SCREENING APPROACHES FOR ANTIFUNGALS ACTING ON THE CYTOSKELETON

Microtubules play a key role in a variety of processes, including cell division, motility, cytoplasmic transport, secretion, and morphogenesis (for reviews, see Huffaker et al. 1987; Thomas et al. 1986). Microtubules are long cylindrical 250 Å diameter structures primarily composed of tubulin, a 110,000 dalton heterodimeric protein that contains two similar size subunits designated α and β. In most eukaryotes, tubulin proteins are encoded by a small family of related genes. Although eukaryotic tubulins are highly homologous, for example, with 70% identity between yeast and chicken β-tubulin (Neff et al. 1983; Thomas et al. 1986), the differences among species are sufficient for the development of agents showing species selectivity. (Although one can not be certain a priori that an observed difference in amino acid sequence will be sufficient for the development of a selective agent, the discovery of selective antimicrotubule agents by random screening proves that microtubule proteins show levels of divergence that are adequate for antifungal development.) Drugs known to bind tubulins in vitro and cause lethality in cell culture affect tubulins from diverse sources differently. For example, colchicine binds with lower affinity to fungal tubulins as compared with mammalian tubulins (Davidse 1986). Low affinity binding may be partly responsible for the failure of colchicine to inhibit fungal growth, although poor uptake of colchicine into fungal cells may also explain the low antifungal activity of this compound. Therefore, although toxicity would appear to be a general concern for the development of antifungal compounds acting on microtubule assembly, the cidal nature of effective compounds and the existence of well-characterized mammalian systems for comparison studies make tubulins an attractive antifungal target.

6.5.1 Biochemical Screens for Antimicrotubule Agents

There are a large number of compounds that inhibit the assembly of disassembly of microtubules. The plant alkaloids colchicine, vinblastine, and vincristine, and the benzimidazole carbamates nocodazole, methyl-bezimidazole-carbamate (MBC), and benomyl are the most well-characterized inhibitors of fungal, plant, and mammalian microtubules. These compounds have been shown to bind to tubulin dimers in vitro and destabilize microtubules in vitro and in vivo (Davidse and Flach 1977, 1978; DeBrabander et al. 1976; Hammerschlag and Sisler 1973; Schiff and Horwitz 1980). Griseofulvin, a clinically useful antimicrotubule agent, also binds to tubulin dimers and inhibits fungal mitosis (Gull and Trinci 1973). Griseofulvin is particularly potent against the dermatophytes and is used clinically against these organisms in superficial infections. Taxol is another compound gaining acceptance in clinical use, in this case as an antitumor agent. It has been shown to bind tubulin, resulting in profound antimitotic properties, but it inhibits microtubule function differently: mammalian cells treated with taxol accumulate hyperstable microtubule arrays (Schiff and Horwitz 1980). Inhibition of fungal growth has not been reported for taxol.

Biochemical assays for antimicrotubule agents fall primarily into two groups: methods to analyze the kinetics of microtubule assembly and disassembly in vitro and in vivo and methods that directly measure the binding of compounds to microtubules. Polymerization/depolymerization can be measured using a double isotope labeling procedure to differentially label the ends of microtubules, which are then incubated in the presence of compounds and analyzed for the net rates of tubulin addition and loss at opposite ends of the microtubules (Wilson and Farrell 1986). Using this method, colchicine and the vinca alkaloids such as vinblastine and vincristine were shown to inhibit tubulin assembly (Wilson and Farrell 1986; Luduena 1979), whereas taxol enhanced both the rate and yield of microtubule assembly (Schiff et al. 1979). In another type of experiment, the assembly of radiolabeled yeast and brain tubulin was used to determine the relative inhibition of various drugs in vitro, including carbendazim (MBC), nocodazole, and colchicine in vitro. Assembly of brain tubulin is essentially insensitive to MBC and very sensitive to colchicine (50% inhibition occurs at $> 1.3 \times 10^{-3}$ M versus 3×10^{-6} M) whereas assembly of yeast tubulin is sensitive to MBC and insensitive to colchicine (50% inhibition in 3×10^{-5} M versus $> 1 \times 10^{-6}$ M; Kilmartin 1981). A third method uses indirect immunofluorescence techniques and monoclonal antibodies to analyze the rates of addition or loss of tubulin in vivo (Schiff and Horwitz 1980; Mitchison and Kirschner 1985a, 1985b).

Measurements of tubulin/drug interactions most often involve the incubation of labeled compounds with isolated extracts or purified tubulins. In general, the extent of binding is determined by differential gel filtration or filter disk methods. Compounds that destabilize microtubules often bind tubulin dimers in solution (see Dustin 1984 for a review). Competition experiments have been used to determine the number and specificity of the binding sites on tubulin (Davidse and Flach 1978). Another useful procedure takes advantage of the changes in the intrinsic fluorescence of tubulin upon drug binding (Andreu and Timasheff 1986; Engelborghs and Fitzgerald 1986). All the methods cited above have been used to analyze drug interactions using tubulins from a great number of different sources (Hammerschlag and Sisler 1973; Quinlan et al. 1981; Hasek et al. 1986; Kilmartin and Adams 1984). Although these methods are suitable for natural products screening, some of these assays may be too labor intensive for mass screening efforts.

6.5.2 Molecular Genetic Screens

Fungal mutations that produce alterations in the level of benomyl resistance generally do not produce alterations in permeability to benomyl but rather produce changes in the genes encoding the α- and β-tubulin subunits. Mutations in these genes, in addition to producing benomyl resistance, cause pleiotropic changes in cell morphology, chromosome and nuclear movement, and growth temperature defects (Kanbe et al. 1990; Schatz et al. 1988; Stearns and Botstein 1988; Huffaker et al. 1988; Thomas et al. 1985; Morris et al. 1979; Oakley et al. 1987; Neff et al. 1983). Furthermore, extensive genetic analysis has shown that mutations that confer resistance and supersensitivity to benomyl can occur in the same β-tubulin gene (Stearns et al. 1990; Schatz et al. 1988; Stearns and Botstein 1988; Thomas et al. 1985; Oakley et al. 1987; Oakley and Morris 1981). These data can be exploited for designing molecular genetic screens to identify antimicrotubule agents.

Two groups, working with *Aspergillus* and *Saccharomyces*, have isolated numerous benomyl resistant mutants that were shown to be altered in the structural genes encoding β-tubulin (among them, *benA33* in *Aspergillus*, Oakley and Morris 1981; and *tub2-216* in *Saccharomyces*, Thomas et al. 1985). These mutations lead to the inhibition of nuclear movement and nuclear division at the restrictive temperature, resulting in high temperature lethality. The effects of certain mutations produce hyperstable microtubles at elevated temperatures, which leads to resistance to a number of antimicrotubule agents including benomyl, nocodazole, and thiabendazole, as well as griseofulvin and *p*-fluorophenylalanine. (The mutants are insensitive to colchicine, as are wild type strains.) Interestingly, the temperature sensitive phenotype caused by these β-tubulin mutations has been shown to be rescued by the addition of benomyl at concentrations that otherwise cause cell death of wild type strains (Thomas et al. 1985). Because lethality is the direct result of microtubule hyperassembly, agents that decrease microtubule assembly suppress the lethal effect of the mutation. This phenotype suggests a format for natural products screening. In an agar plate or microtiter format, compounds that destabilize microtubules should allow fungal cells to grow at elevated temperatures. In practice, nocodazole, thiabendazole, and griseofulvin partially reverse the temperature sensitive phenotype of the mutant strains (Oakley and Morris 1981).

The β-tubulin gene has been cloned from a number of other fungal species including *C. albicans, N. crassa*, and *Schizosaccharomyces pombe* (Smith et al. 1988; Orbach et al. 1986; Hiraoka et al. 1984). It may now be possible to generate mutants possessing similar properties to those described for *Aspergillus* and *Saccharomyces* in these other fungal species and to devise analogous screens for antimicrotubule agents in these organisms.

6.5.3 Pairwise Screens for Antimicrotubule Agents

Many conditional lethal mutations in genes that interact with *Saccharomyces* β-tubulin and several suppressors of *Aspergillus benA33* produce supersensitivity to the benzimidazole agents when present in a wild type genetic background (Oakley et al. 1987; Schatz et al. 1988; Stearns and Botstein 1988). These mutations are found primarily in the genes encoding α-tubulin. A pairwise screen with these mutants and an otherwise isogenic strain carrying a β-tubulin allele that confers resistance to benomyl provides a simple plate assay that can be utilized to discover novel compounds of this type. The advantage of this approach lies in the enhanced sensitivity of the supersensitive strain to compounds that may be present at very low concentration in fermentations.

Other pairwise screens are suggested by the variety of chromosome segregation mutants and the suppressors that are generated in mutant hunts for genes affecting mitosis. For example, mutations in chromosome instability genes, *CIN1, CIN2*, and *CIN4* produce supersensitivity to benomyl (Stearns et al. 1990; Hoyt et al. 1990) although their primary defect does not lie in any of the known tubulin genes. It may be possible that microtubule associated proteins will offer a wider net for the discovery of novel compounds; these proteins may also provide a source of unique fungal targets and thus lower the chances of toxicity associated with known microtubule inhibitors. The clinically used antifungal griseofulvin is believed to act through an as yet uncharacterized microtubule associated protein.

6.5.4 Other Components of the Cytoskeleton

The previous sections discussed microtubules as antifungal targets and described screens for the detection of antimicrotubule agents. However, microtubules comprise only part of the structure of the cytoskeleton, a dynamic and essential organelle that also contains microfilaments, intermediate filaments, and associated proteins. It is difficult to define specific targets in this area since many of the components of the fungal cytoskeleton have significant homology with their mammalian counterparts (e.g., *SPA1, ACT1, CDC31*). Until recently, the 10 nm filaments encoded by the *CDC3, CDC10, CDC11*, and *CDC12* genes were believed to comprise a unique type of microfilament limited to fungi, but present in a number of distantly related fungal species. These data made these gene products excellent candidates as differential targets for antifungal agents (Byers and Goetsch 1976; Haarer and Pringle 1987; Kim et al. 1991; J.R. Pringle, personal communication; B. DiDomenico and Y. Koltin, unpublished observations). However, the protein sequences predicted by this family of genes have recently been shown to have homology with proteins from *Drosophila* and mouse, limiting their theoretical usefulness as antifungal targets (J.R. Pringle, personal communication). It might be possible, however, to use the sequences encoding the mammalian analogues to design screens for identifying agents with selective antifungal action.

Since different components of the cytoskeleton affect the same processes, some of the molecular genetic screens described for inhibitors of microtubule function may be expected to uncover inhibitors of other components of the cytoskeleton. Genetic screens designed to detect mutations in non-tubulin proteins of the cytoskeleton have included suppressor analysis of temperature-sensitive conditional-lethal alleles in α- and β-tubulin genes (Weil et al. 1986; Thomas et al. 1985) and screens designed to detect increased chromosome loss or deficiencies in karyogamy (nuclear fusion) (Huffaker et al. 1987). With minor modifications, each of these screens may be used in a natural products screening program. Genetic analyses have shown that these screens will detect mutations that produce pleiotropic phenotypes, which is expected of cytoskeletal components.

6.6 SCREENING APPROACHES FOR ANTIFUNGALS ACTING ON OTHER TARGETS

6.6.1 Dihydrofolate Reductase

Dihydrofolate reductase (DHFR) is a target for antibacterial, antiprotozoal, and antitumor agents (Gilman et al. 1985). Dihydrofolate reductase from *C. albicans* differs from mammalian DHFR in molecular weight, pH optimum, amino acid sequence, turnover number, and inhibitor binding (Baccanari et al. 1989), suggesting that DHFR might also be a target for the development of antifungals. The *C. albicans* DHFR gene was isolated independently by three groups (Kurtz et al. 1987; Baccanari et al. 1989; Franceschi et al. 1991). Kurtz and co-workers overexpressed the *Candida* DHFR gene in *C. albicans* by introducing the gene into *C. albicans* on a high copy vector (Kurtz et al. 1987). Two groups have exploited DHFR as a selectable marker for transformation in *S. cerevisiae* using methotrexate in the presence of sulfanilamide as a selective agent and high copy DHFR as a resistance determinant (Miyajima et al. 1984; Zhu et al. 1985). Similar to *S. cerevisiae*, *Candida* strains are sensitive to methotrexate in the presence of high levels of sulfanilamide and strains that overexpress DHFR show increased resistance to DHFR inhibitors. This observation was exploited in order to devise a screen to identify novel inhibitors of DHFR. Although *C. albicans* is fairly resistant to DHFR inhibitors such as methotrexate, the level of resistance drops dramatically when sulfanilamide is added to the medium at high, but subinhibitory levels (5 mg/ml). At this level of sulfanilamide, as little as 1 ng of methotrexate can be detected in a disk diffusion assay using a wild type *C. albicans* strain. To specifically identify DHFR inhibitors, test compounds can be assayed against a second control strain that carries multiple copies of the *C. albicans* DHFR gene. These strains overexpress DHFR and show greatly increased resistance to DHFR inhibitors even in the presence of sulfanilamide (Kurtz et al. 1987). This assay was used to screen a large number of fermentations and no actives were obtained (M. Kurtz, unpublished observations). It is likely that the tested fermentations did not contain in vivo inhibitors of *C. albicans* DHFR because of the high sensitivity of the assay to methotrexate. It might be possible to use the strain that overexpresses DHFR as a source of enzyme for an in vitro screen. Alternatively, it might be useful to screen plant extracts, "unusual Actinomycetes," or chemical bank compounds with this assay in order to identify novel inhibitors of *C. albicans* DHFR.

6.6.2 Thymidylate Synthase

Thymidylate synthase has been pursued as a target for the development of antifungals because the enzyme product (dTMP) can only be synthesized de novo in yeasts that lack thymidine kinase and efficient uptake mechanisms for thymine, thymidine, and dTMP (Singer et al. 1989). Singer and co-workers (1989) isolated the *C. albicans* thymidylate synthase gene by complementing a *tmp1* mutation in *S. cerevisiae*. In *S. cerevisiae, tmp1* mutants (in a Tup⁻ genetic background that allows dTMP uptake) have a nutritional requirement for dTMP. The *S. cerevisiae tmp1* mutant carrying the *C. albicans* gene could be used as the basis for an antimetabolite screen in which active compounds are identified by antifungal activity on unsupplemented media, which is reversed by media containing dTMP. Singer et al. (1989) also demonstrated that the *C. albicans TMP1* gene would express in *Escherichia coli* and complement a thyA (thymidylate synthase) mutation without the addition of an *E. coli* promoter. High levels of *Candida* thymidylate synthase were produced when the *Candida* coding region was expressed from a T7 promoter in an *E. coli* strain expressing T7 RNA polymerase. This result raises the possibility of screening for *C. albicans* thymidylate synthase inhibitors in *E. coli* either by a dTMP reversal method or by comparing the action of inhibitors on *E. coli* strains with low level and high level thymidylate synthase.

6.6.3 Plasma Membrane ATPase

The yeast plasma membrane H^+-ATPase is an essential electrogenic ion pump that structurally resembles animal membrane ATPases such as Ca^+-ATPase, K^+-ATPase, and Na^+, K^+-ATPase. This enzyme, however, differs from its animal analogs in both subunit composition and ion specificity (Serrano et al. 1986). The yeast plasma membrane ATPase, therefore, is a potential target for the action of antifungal agents since inhibitors are expected to be fungicidal and could potentially be selective for the yeast enzyme. The development of molecular genetic screens for inhibitors of the yeast plasma membrane H^+-ATPase is frustrated by the fact that the level of this enzyme is under extremely tight control, making it difficult to generate genetic variants that vary in enzyme level for screening (Eraso et al. 1987). Another approach to screen for inhibitors of the plasma membrane H^+-ATPase would be to employ the conventional enzyme assay, which has been used to study enzyme inhibitors (Dufour et al. 1980). Lastly, although biochemical characterization data are available for the *C. albicans* plasma membrane H^+-ATPase, genetic and molecular biological studies have been limited largely to *S. cerevisiae* and *S. pombe*. Recently, however, Perlin and co-workers have isolated the structural gene for the *C. albicans* plasma membrane H^+-ATPase by using the *S. cerevisiae PMA1* gene as a hybridization probe (Monk et al. 1991). The *C. albicans* sequence is very similar to the *S. cerevisiae* sequence at the amino acid level (83%) although the *Candida* sequence surprisingly does not complement *pmal* mutations in *S. cerevisiae*. With the availability of the *C. albicans* gene it may now become possible to design molecular genetic screens for inhibitors of the *Candida* enzyme and to use sequence information for rational design approaches.

6.6.4 Virulence Factors

A great deal of effort has gone into identifying factors that enhance the virulence of fungal infections. Many pathogenic fungi grow in either yeast-like or filamentous form, depending on environmental conditions. Among these factors are a growing list of agents that cause dimorphic fungi to shift to a predominantly mycelial phase of growth. It has long been postulated that the virulent stage of an organism's life cycle is the mycelial phase (reviewed by Odds 1988; Rippon 1974, 1980). However, this hypothesis is largely supported by indirect, anecdotal evidence. These reviews argue that either yeast or mycelial cells can produce a disease state and all cite numerous reports of both yeast and hyphal forms found at the sites of infection. The hyphal form may play an important role in pathogenicity with regard to tissue invasion since hyphal cells have been shown to adhere better to human epithelia than do yeast cells (Segal et al. 1984). Since genetically characterized, non-reverting yeast-only or mycelial-only forms of *C. albicans* are not yet available, reliable data on this topic remain controversial and subject to much argument. However, with the recent discovery of genetic loci that enable *S. cerevisiae* to undergo a yeast-to-mycelial conversion, this limitation may at last be amenable to experimental manipulation (Gimeno et al. 1992).

High throughput screens have been used to study both the yeast-to-mycelial transition and adherence. In the former, *Candida* cells are grown under conditions favoring the yeast form and are harvested and resuspended in medium that induces hyphal formation. A pairwise screen format is used to detect compounds that inhibit the transition but not the growth of yeast cells per se. However, it is widely known that hyphal cells are considerably more sensitive to many classes of drugs than are yeast cells, resulting in a significant number of false positives.

Assays for adherence include measuring the percentage of radiolabeled whole cells that bind to fibrin-platelet matrices (Calderone and Scheld 1987) or to concanavalin-A coated surfaces (Sandin et al. 1982). The data from these types of studies suggest that cell wall mannoproteins are the main mediators of tissue adherence. While a good correlation exists between adherence competence and pathogenicity in in vivo models, antifungals acting on these targets are not likely to be cidal and will require extensive and continuous treatment in order to manage the infection. To date, no drug is in clinical use (or in trials) that directly decreases fungal adherence in vitro or in vivo.

Another fungal virulence factor that has received attention is the secreted acid proteases. All fungi are acidophiles and as they grow they acidify their environment. In response to acidic growth conditions, *C. albicans* synthesizes and secretes an acid-active protease that has been shown to enhance the infectivity of the organism in vivo (Kwon-Chung et al. 1985; Macdonald and Odds 1980; Kondoh et al. 1987; Shimizu et al. 1987). Assays for general proteinases rely on traditional biochemical methods that use a radiolabeled peptide substrate incubated with enzyme extract (Billich et al. 1990). The cleaved peptide products are separated electrophoretically and analyzed by autoradiography. More recent advances have utilized synthetic peptides that may be analyzed by high pressure liquid chromatography (HPLC) methods, increasing throughput. The acid protease described in *C. albicans* can be easily scored in vivo by using agar plates supplemented with hemoglobin: a zone of clearing surrounding the yeast colony indicates the level of activity. However, the absence of a zone may indicate defects associated with post-translational modification or secretion, as well as inhibition of the protease activity itself. *C. albicans* proteinase genes have been cloned and these sequences may prove to be useful for screen design (Lott et al. 1989; Hube et al. 1991).

6.6.5   RNA Splicing

In eukaryotes, many genes contain non-coding information or introns that are removed during maturation of the RNA. Examples of intron-containing RNAs that represent all three classes, mRNA, rRNA, and tRNA, have been found. This universal process involves a number of well-characterized biochemical reactions collectively called splicing (for review, see Abelson et al. 1986). However, several of the reactions differ between yeast and mammalian cells. Analogs of the mammalian U1 and U2 small nuclear RNAs that direct splicing in higher eukaryotes are not known to be present in yeast. Also, yeast appear to be more restrictive in their sequence requirements for splice junctions: the yeast 5′ splice junction is almost always GTATGT whereas mammalian junctions are more variable (Teem et al. 1984). Yeast also require a TACTAAC sequence approximately 30 base pairs upstream from the 3′ splice junction, and single point mutations in this sequence abolish all splicing in vitro (Langford and Gallwitz 1983). For these reasons, splicing provides an attractive target for the development of antifungal inhibitors (Silver and Bostian 1990).

In vitro biochemical assays using both yeast and HeLa (a known cancerous human cell line) extracts have been developed and described in detail (see Dahlberg and Abelson 1989). These assays have been used to analyze individual reaction steps and to identify components that are essential for the splicing reaction and are suitable for high throughput screening. A genetic assay was devised to study the effect of a temperature sensitive mutation, *rna2*, on splicing in yeast (Teem and Rosbash 1983). The yeast rp51 intron was fused *within* the coding region of β-galactosidase, interrupting the open reading frame. Yeast transformants harboring this fusion express β-galactosidase at levels comparable to strains bearing wild type *E. coli* lacZ constructs without introns. Yeast *rna2* mutants fail to express β-galactosidase when grown at a restrictive temperature and β-galactosidase expression is unaffected at the permissive temperature. Inhibitors of the splicing reaction are also expected to inhibit expression of β-galactosidase under the conditions described above, although general transcription/translation inhibitors would also be expected to be active in this assay. General transcription/translation inhibitors may quickly be eliminated by using a pairwise screening strategy comparing β-galactosidase expression in transformants containing an intronless lacZ construct with those containing the rp51 intron.

6.6.6 Transcriptional Termination

The proper maturation of mRNA includes capping, splicing, termination, and polyadenylation. Compared to the splicing field, relatively little is known concerning the biochemistry of termination and poly A addition. However, comparisons between yeast and mammalian termination suggest that the two processes may be rather dissimilar. For example, the AAUAAA consensus sequence required for proper mRNA maturation in mammalian cells is notably absent from the ends of many yeast genes (Zaret and Sherman 1982). Also, the site of termination lies several hundred nucleotides downstream from the poly A site in mammalian cells whereas the two sites are less than 100 nucleotides apart in yeast cells (Osborne and Guarente 1989; Russo and Sherman 1989). Finally, mutations in the AAUAAA consensus sequence known to reduce proper 3'-end maturation in mammalian cells concomitantly cause drastic reductions in gene expression (Higgs et al. 1983), suggesting inhibition of termination may be a lethal event in eukaryotic cells and may provide targets of opportunity for drug discovery in the antifungal field.

In vitro assays for termination use S1 nuclease protection or run-on transcription analysis to determine changes in the position of the 3' ends of transcripts (Osborne and Guarente 1989). These assays are labor intensive and would be difficult to run on the massive scale required for a natural products screening program. In vivo assays have also been developed to study transcription termination in yeast. These assays are based on changes in plasmid stability caused by "run-through" transcription complexes that obscure the function of sequences placed downstream of the terminators (Russo and Sherman 1989; Snyder et al. 1988). Compounds inhibiting fungal termination may be expected to permit run-through transcription to occur, reducing the efficiency of sequences placed near terminators and causing changes in plasmid stability in the absence of selective pressure.

Although the assays as described in these papers are too cumbersome for natural products screening, minor changes in plasmid construction (e.g., insertion of a reporter gene) can be made to more easily monitor plasmid stability. It is expected that this format will also identify compounds that induce chromosome loss in fungi. However, compounds that promote plasmid instability through some other mechanism might be eliminated through the use of pairwise screens containing otherwise identical plasmids (containing or missing the fungal terminator).

6.6.7 Elongation Factor 3 (EF-3)

A number of well-known compounds in clinical use inhibit protein synthesis in prokaryotes. However, of all the protein synthesis inhibitors known to be effective against eukaryotic ribosomes, none are specific to fungal ribosomes. A unique translation factor called elongation factor 3 (or EF-3) has been identified in diverse species of fungi that is absent from mammalian cells (for review, see Chakraburtty and Kamath 1988). Antibodies to *Saccharomyces* EF-3 react with a single, high-molecular weight peptide from different fungi, including *Aspergillus, Neurospora, Cryptococcus*, etc. (Chakraburtty and Kamath 1988; Uritani and Miyazaki 1988). EF-3 has been isolated and purified and shown to be essential for elongation of both synthetic and natural mRNAs (Kamath and Chakraburtty 1989; Herrera et al. 1984; Dasmahapatra and Chakraburtty 1981) and disruption of the gene is lethal in *Saccharomyces* (Qin et al. 1987; Sandbaken et al. 1990a). Inhibitors of this protein are predicted to be cidal and nontoxic.

Several properties of this protein suggest assay formats for natural products screening. First, EF-3 exhibits ribosome-dependent ATP hydrolysis that can be assayed in vitro by measuring the amount of inorganic radioactive phosphate released in the presence and absence of added yeast ribosomes (Kamath and Chakraburtty 1989; Sandbaken et al. 1990b). Second, EF-3 has also been shown to stimulate EF1α-dependent binding of aminoacyl-tRNAs to the ribosome during the elongation cycle, a reaction that is typically assayed by filter binding (Kamath and Chakraburtty 1989). Third, yeast transformants containing elevated levels of EF-3 exhibit an enhanced sensitivity to the aminoglycoside antibiotics hygromycin and paromomycin (Sandbaken et al. 1990b). Inhibitors of EF-3 would be expected to "rescue" this phenotype, as would compounds causing decreased plasmid stability or a reduction in the level of EF-3 expression. A pairwise screen format may be used to distinguish between EF-3 inhibition and plasmid curing. The recent isolation and characterization of the gene encoding EF-3 from *C. albicans* may provide useful information for the rational design of EF-3 inhibitors (Myers et al. 1992; DiDomenico et al. 1992).

6.7 CLOSING COMMENTS

While we do not believe that this review will reveal anything of significant novelty to professionals in the area of antifungal screen design, we hope that it might serve, for those working in other disciplines, as a useful introduction to this area. We anticipate that future reviews on this topic will present screens of increased sophistication directed at a larger number of novel targets. It is also interesting that in preparing this review, we have come to the perhaps not surprising realization that the technology pursued by competing pharmaceutical companies is highly comparable. At the very least, we hope that this chapter will make the task of reviewing this field easier for those who write subsequent reviews on this topic.

REFERENCES

Abelson, J.N., Brody, E.N., Cheng, S.-C., et al. (1986) *Chemica Scripta* 26, 127–137.

Alberts, A.W., Chen, J., Kuron, G., et al. (1980) *Proc. Natl. Acad. Sci. USA* 77, 3957–3961.

Anderson, M.S., Muehlbacher, M., Street, I.P., Proffitt, J., and Poulter, C.D. (1989a) *J. Biol. Chem.* 264, 19169–19175.

Anderson, M.S., Yarger, J.G., Burck, C.K., and Poulter, C.D. (1989b) *J. Biol. Chem.* 264, 19176–19184.

Andreau, J.M., and Timasheff, S.N. (1986) in *Dynamic Aspects of Microtubule Biology* (D. Soifer, ed.), The New York Academy of Sciences, New York.

Aoki, Y., Yamazaki, T., Kondoh, M., et al. (1992) *J. Antibiot.* 45, 160–170.

Au-Young, J., and Robbins, P.W. (1990) *Mol. Microbiol.* 4, 197–207.

Baccanari, D.P., Tansik, R.L., Joyner, S.S., et al. (1989) *J. Biol. Chem.* 264, 1100–1107.

Baguley, B.C., Rommele, G., Gruner, J., and Wehrli, W. (1979) *Eur. J. Biochem.* 97, 345–351.

Banta, L.M., Robinson, J.S., Klionsky, D.J., and Emr, S.D. (1988). *J. Cell Biol.* 107, 1369–1383.

Bartnicki-Garcia, S., and Lippmann, E. (1972) *J. Gen. Microbiol.* 71, 301–309.

Basson, M.E., Thorsness, M., Finer-Moore, J., Stroud, R.M., and Rine, J. (1988) *Mol. Cell. Biol.* 8, 3797–3808.

Basson, M.E., Thorsness, M., and Rine, J. (1986) *Proc. Natl. Acad. Sci. USA* 83, 5563–5567.

Becker, J.M., Covert, M.L., Shenbagamurthi, P., Steinfeld, A.S., and Naider, F. (1983) *Antimicrob. Agents Chemother.* 23, 926–929.

Benkovic, S.J., Fierke, A.M., and Naylor, A.M. (1988) *Science* 239, 1105–1110.

Betina, V. (1983) in *The Chemistry and Biology of Antibiotics*, pp. 59–114, Elsevier Scientific Publishing Co., New York.

Billich, A., Hammerschmid, F., and Winkler, G. (1990) *Biol. Chem. Hoppe Seyler* 371, 265–272.

Boone, C., Sommer, S.S., Hensel, A., and Bussey, H. (1990) *J. Cell. Biol.* 110, 1833–1843.

Bowen, A.R., Cheng, R., Cohen, S., et al. (1990) *15th Int. Conf. on Yeast Genetics and Molecular Biology,* The Hague, John Wiley and Sons Ltd. (Abstract 04-39B).

Brugge, J.S., Jarosik, G., Andersen, J., et al. (1987) *Mol. Cell. Biol.* 7, 2180–2187.

Bulawa, C.E., Slater, M.C., Au-Young, J., Sburlati, A., Adair, W.L. Jr., and Robbins, P.W. (1986) *Cell* 46, 213–225.

Burger, A. (1983) *A Guide to the Chemical Basis of Drug Design*, John Wiley and Sons, Inc., New York.

Burkholder, P.R. (1946) *J. Bacteriol.* 52, 503–504.

Byers, B., and Goetsch, L. (1976) *J. Cell. Biol.* 69, 717–721.

Cabib, E. (1987) *Adv. Enzymol.* 59, 59–101.

Cabib, E. (1991) *Antimicrob. Agents Chemother.* 35, 170–173.

Cabib, E., and Kang, M.S. (1987) *Methods Enzymol.* 138, 637–642.

Cabib, E., Kang, M.S., and Au-Young, J. (1987) *Methods Enzymol.* 138, 643–649.

Cabib, E., Roberts, R., and Bowers, B. (1982) *Annu. Rev. Biochem.* 51, 763–793.

Calderone, R.A., and Scheld, W.M. (1987) *Rev. Infect. Dis.* 9, S400–S403.

Cattel, L., Ceruti, M., Biola, F., et al. (1986) *Lipids* 21, 31–38.

Chakraburtty, K., and Kamath, A. (1988) *Int. J. Biochem.* 20, 581–590.

Chen-Wu, J.L., Zwicker, J., Bowen, A.R., and Robbins, P.W. (1992) *Mol. Microbiol.* 6, 497–502.

Creese, I. (1978) in *Neurotransmitter Receptor Binding* (Yamamura, H.I. ed.), pp. 142–169, Raven Press, New York.

Dahlberg, J.E., and Abelson, J.N. (eds.) (1989) *Methods in Enzymology*, vol. 180, Academic Press, San Diego.

Dasmahapatra, B., and Chakraburtty, K. (1981) *J. Biol. Chem.* 256, 9999–10004.

Davidse, L.C. (1986) *Annu. Rev. Phytopathol.* 24, 43–65.

Davidse, L.C., and Flach, W. (1977) *J. Cell Biol.* 72, 174–193.

Davidse, L.C., and Flach, W. (1978) *Biochim. Biophys. Acta* 543, 82–90.

DeBrabander, M.J., Van de Veire, R.M.L., Aerts, F.E.M., Borgers, M., and Janssen, P.A.J. (1976) *Cancer Res.* 36, 1011–1018.

Deisenhofer, J., and Michel, H. (1989) *Science* 245, 1463–1473.

deMora, F., Gil, R., Sentandreau, R., and Herrero, E. (1991) *Antimicrob. Agents Chemother.* 35, 2596–2601.

deMora, J., Valentiin, E., Herrero, E., and Sentandreu, R. (1990) *J. Gen. Microbiol.* 136, 2251–2259.

Denis, C.L., Ferguson, J., and Young, E.T. (1983) *J. Biol. Chem.* 258, 1165–1171.

DeNobel, J.G., and Barnett, J.A. (1991) *Yeast* 7, 313–323.

deStevens, G. (1986) *Prog. Drug Res.* 30, 189–203.

DiDomenico, B.J., Lupisella, J., Sandbaken, M., and Chakraburtty, K. (1992) *Yeast* 8, 337–352.

Douglas, C.M., Frommer, B.R., and Bostian, K.A. (1989) *Yeast Cell Biology*, Cold Spring Harbor Press, Cold Spring Harbor, New York.

Dufour, J., Boutry, M., and Goffeau, A. (1980) *J. Biol. Chem.* 255, 5735–5741.

Duran, A., Bowers, B., and Cabib, E. (1975) *Proc. Natl. Acad. Sci. USA* 72, 3952–3955.

Dustin, P. (1984) *Microtubules* (Porter, K.R., ed.), Springer-Verlag, New York.

Elorza, M., Murgui, A., Rico, H., Miragall, F., and Sentandreu, R. (1987) *J. Gen. Microbiol.* 133, 2315–2325.

Emerson, S. (1963) *Genetica* 34, 162–182.

Endo, A. (1981) *Trends Biochem. Sci.* 16, 10–12.

Endo, A., Kakiki, K., and Misapt, T. (1970) *J. Bacteriol.* 104, 189–196.

Endo, A., and Misato, T. (1969) *Biochem. Biophys. Res. Commun.* 37, 718–722.

Engelborghs, Y., and Fitzgerald, T.J. (1986) in *Dynamic Aspects of Microtubule Biology* (Soifer, D., ed.), The New York Academy of Sciences, New York.

Eraso, P., Cid, A., and Serrano, R. (1987) *FEBS Lett.* 224, 193–197.

Fleet, G.H. (1985) *Curr. Top. Med. Mycol.* 1, 24–56.

Fleet, G.H., and Phaff, H.J. (1981) in *Fungal Glucans-Structure and Metabolism*, pp. 416–440, Springer-Verlag, Berlin.

Franceschi, M., Denaro, M., Irdani, T., et al. (1991) *FEMS Microbiol. Lett.* 80, 179–182.

Franklin, T.J., and Cook, J.M. (1969) *Biochem. J.* 113, 515–524.

Furter, R., and Rast, D.M. (1985) *FEMS Micro. Lett.* 28, 205–211.

Gilman, A.G., Goodman, L.S., Rall, T.W., and Myrad, F. (1985) *Goodman and Gilman's The Pharmacological Basis of Therapeutics*, Macmillan Publishing Co., New York.

Gimeno, C.J., Ljungdahl, P.O., Styles, C.A., and Fink, G.R. (1992) *Cell* 68, 1077–1090.

Goetz, M.A., Lopez, R.L., Chang, R.S.L., Lotti, V.J., and Chen, T.B. (1985) *J. Antibiotics* 38, 1633–1637.

Gollub, E.G., Liu, K., Dayan, M.J., Adlersberg, M., and Sprinson, D.B. (1977) *J. Biol. Chem.* 252, 2846–2854.

Gooday, G.W. (1977) *J. Gen. Microbiol.* 99, 1–11.

Gooday, G.W. (1983) in *Microbial Polysaccharides. Progress in Industrial Microbiology* (M.E. Bushell, ed.), pp. 85–127, Elsevier, Amsterdam.

Gordee, R.S., Zeckner, D.J., Ellis, L.F., Thakkar, A.L., and Howard, L.C. (1984) *J. Antibiotics* 37, 1054–1065.

Gordee, R.S., Zeckner, D.J., Howard, L.C., Alborn, W.E.J., and Debono, M. (1988) *Ann. N.Y. Acad. Sci.* 554, 294–309.

Grindle, M. (1983) *Pest. Sci.* 14, 481–491.

Guarente L., Yocum, R.R., and Gifford, P. (1982) *Proc. Natl. Acad. Sci. USA* 79, 7410–7414.

Gull, K., and Trinci, A.P.J. (1973) *Nature* 244, 292–293.

Haarer, B.K., and Pringle, J.R. (1987) *Mol. Cell. Biol.* 7, 3678–3687.

Hammerschlag, R.S., and Sisler, H.D. (1973) *Pestic. Biochem. Physiol.* 3, 42–54.

Hanseler, E., Nyhlen, L.E., and Rast, D.M. (1983) *Exp. Mycol.* 7, 17–30.

Hardy, L.W., Finer-Moore, J.S., Monfort, W.R., et al. (1987) *Science* 235, 448–455.

Hasek, J., Svobodova, J., and Streiblova, E. (1986) *Eur. J. Cell Biol.* 41, 150–156.

Hawlitschek, G., Schneider, H., Schmidt, B., et al. (1988) *Cell* 53, 795–806.

Hector, R.F., and Braun, P.C. (1986) *J. Clin. Microbiol.* 24, 620–624.

Henry, S. (1982) in *The Molecular Biology of the Yeast Saccharomyces: Metabolism and Gene Expression* (Strathern, J.N., Jones, E.W., and Broach, J.R., ed.), pp. 101–158, Cold Spring Harbor Laboratory Press, Cold Spring Harbor, New York.

Herrera, F., Martinez, J.A., Moreno, N., et al. (1984) *J. Biol. Chem.* 259, 14347–14349.

Higgs, D.R., Goodbourn, S.E.Y., Lamb, J., et al. (1983) *Nature* 306, 398–400.

Hill, K., Boone, C., Goebl, M., et al. (1992) *Genetics* 130, 273–283.

Hiraoka, Y., Toda, T., and Yanagida, M. (1984) *Cell* 39, 349–358.

Hori, M., Kakiki, K., and Misato, T. (1977) *J. Pest. Sci.* 2, 139–150.

Hoyt, A., Stearns, T., and D., B. (1990) *Mol. Cell. Biol.* 10, 223–234.

Hube, B., Turver, C.J., Odds, F.C., et al. (1991) *J. Med. Vet. Mycol.* 29, 129–132.

Huffaker, T.C., Hoyt, M.A., and Botstein, D. (1987) *Annu. Rev. Genet.* 21, 259–284.

Huffaker, T.C., Thomas, J.H., and Botstein, D. (1988) *J. Cell. Biol.* 106, 1997–2010.

Hutchins, K., and Bussey, H. (1983) *J. Bacteriol.* 154, 161–169.

Ihara, M., Fukuroda, T., Saeki, T., et al. (1991) *Biochem. Biophys. Res. Comm.* 178, 132–137.

Ikeura, R., Murakawa, S., and Endo, A. (1988) *J. Antibiot. (Tokyo)* 41, 1148–1150.

Imanaka, H. (1985) *Recent Adv. Chemother. Proc. Int. Congr. Chemother.* 14th issue, pp. 15–17, University of Tokyo Press, Tokyo.

Jabri, E., Quigley, D.R., Alders, M., et al. (1989) *Curr. Microbiol.* 19, 153–161.

Jandrositz, A., Turnowsky, F., and Hogenauer, G. (1991) *Gene* 107, 155–160.

Janssen, P.A.J., and Vanden Bossche, H. (1987) *Arch. Pharmacol. Chem.* 15, 23–40.

Jennings, S.M., Tsay, Y.H., Fisch, T.M., and Robinson, G.W. (1991) *Proc. Natl. Acad. Sci. USA* 88, 6038–6042.

Kagan, B.L. (1983a) *Nature* 302, 709–710.

Kagan, B.L. (1983b) *Nature* 302, 710–711.

Kalb, V.F., Loper, J.C., Dey, C.R., Woods, C.W., and Sutter, T.R. (1986) *Gene* 45, 237–245.

Kalb, V.F., Woods, C.W., Turi, T.G., et al. (1987) *DNA* 6, 529–537.

Kamath, A., and Chakraburtty, K. (1989) *J. Biol. Chem.* 264, 15423–15428.

Kanbe, T., Hiraoka, Y., Tanaka, K., and Yanagida, M. (1990) *J. Cell Sci.* 96, 275–282.

Kang, M.S., and Cabib, E. (1986) *Proc. Natl. Acad. Sci. USA* 83, 5808–5812.

Karst, F., and Lacroute, F. (1977) *Mol. Gen. Genet.* 154, 269–277.

Kase, H., Fujita, H., Nakamura, J., et al. (1986) *J. Antibiotics* 39, 354–363.

Kataoka, T., Powers, S., Cameron, S., et al. (1985) *Cell* 40, 19–26.

Kelly, R., Miller, S.M., Lai, M.H., and Kirsch, D.R. (1990) *Gene* 87, 177–183.

Kerridge, D. (1986) *Adv. Microb. Physiol.* 27, 1–72.

Kilmartin, J.V. (1981) *Biochemistry* 20, 3629–3633.

Kilmartin, J., and Adams, A. (1984) *J. Cell Biol.* 98, 922–933.

Kim, H.B., Haarer, B.K., and Pringle, J.R. (1991) *J. Cell. Biol.* 112, 535–544.

King, K., Dohlman, H.G., Thorner, J., Caron, M.G., and Lefkowitz, R.J. (1990) *Science* 250, 121–123.

Kirsch, D.R., and Lai, M.H. (1986) *J. Antibiotics* 39, 1620–1622.

Kirsch, D.R., Lai, M.H., and O'Sullivan, J. (1988) *Gene* 68, 229–237.

Kitano, K., Kintaka, S., Suzuki, K., Katamoto, K.M., and Nakoa, Y. (1975) *J. Ferment. Technol.* 53, 327–338.

Koller, W. (1992) *Target Sites of Fungicide Action*, CRC Press, Boca Raton, FL.

Komori, T., Yamshita, M., Tsurumi, T., and Kohsaka, M. (1985) *J. Antibiotics (Tokyo)* 38, 455–459.

Kondoh, Y., Shimizu, K., and Tanaka, K. (1987) *Microbiol. Immunol.* 31, 1061–1069.

Kopecka, M. (1984) *Folia Microbiol.* 29, 115–119.

Kottutz, E., and Rapp, P. (1990) *J. Gen. Microbiol.* 136, 1517–1523.

Kurtz, M.B., Cortelyou, M.W., Miller, S.M., Lai, M., and Kirsch, D.R. (1987) *Mol. Cell. Biol.* 7, 209–217.

Kurtz, M.B., and Marrinan, J. (1989) *Mol. Gen. Genet.* 217, 47–52.

Kwon-Chung, K.J., Lehman, D., Good, C., and Magee, P.T. (1985) *Infect. Immunol.* 49, 571–575.

Lai, M.H., and Kirsch, D.R. (1989) *Nucleic Acids Res.* 17, 804.

Langford, C.J., and Gallwitz, D. (1983) *Cell* 33, 519–527.

Larriba, G., Morales, M., and Ruiz-Herrera, J. (1981) *J. Gen. Microbiol.* 124, 375–383.

Lederberg, J. (1956) *Proc. Natl. Acad. Sci. USA* 42, 574–577.

Lewis, T.A., Taylor, F.R., and Parks, L.W. (1985) *J. Bacteriol.* 163, 199–207.

Lorenz, R.T., and Parks, L.W. (1990) *Antimicrob. Agents Chemother.* 34, 1660–1665.

Lopez-Romero, E., and Ruiz-Herrera, J. (1978) *Biochim. Biophys. Acta* 500, 372–384.

Lott, T.J., Page, L.S., Boiron, P., Benson, J., and Reiss, E. (1989) *Nucleic Acids Res.* 17, 1779.

Luduena, R.F. (1979) in *Microtubules* (Roberts, K., and Hyams, J.S., eds.), pp. 65–116, Academic Press, London.

Macdonald, F., and Odds, F.C. (1980) *J. Med. Microbiol.* 13, 423–435.

Mason, M.M., Becker, J.M., and Naider, F. (1989) *Yeast Genetics and Molecular Biology*, Atlanta, GA, Genetics Society of America (Abstract 64a).

Matsukuma, S., Ohtsuka, T., Kotaki, H., et al. (1992) *J. Antibiotics* 45, 151–159.

McCarthy, P., Troke, P.F., and Gull, K. (1985) *J. Gen. Microbiol.* 131, 775–780.

Meaden, P., Hill, K., Wagner, J., et al. (1990) *Mol. Cell. Biol.* 10, 3013–3019.

Meeks-Wagner, D.W., and Hartwell, L. (1986) *Cell* 44, 43–52.

Meeks-Wagner, D.W., Wood, J.S., Garvik, B., and Hartwell, L. (1986) *Cell* 44, 53–63.

Mercer, E.I. (1984) *Pestic. Sci.* 15, 133–155.

Mitchison, T.J., and Kirschner, M.W. (1985a) *J. Cell Biol.* 101, 755–765.

Mitchison, T.J., and Kirschner, M.W. (1985b) *J. Cell Biol.* 101, 766–777.

Miyajima, A., Miyajima, I., Arai, H., and Arai, N. (1984) *Mol. Cell. Biol.* 4, 407–414.

Miyanohara, A., Toh-e, A., Nozaki, C., et al. (1983) *Proc. Natl. Acad. Sci. USA* 80, 1–5.

Miyata, M., Kitamura, J., and Miyata, H. (1980) *Arch. Microbiol.* 127, 11–16.

Mizoguchi, J., Saito, T., Mizuno, K., and Hayano, K. (1977) *J. Antibiotics* 30, 308–313.

Mizuno, K., Yagi, A., Satoi, S., Takada, M., and Hayashi, M. (1977) *J. Antibiotics* 30, 297–302.

Molano, J., Bowers, B., and Cabib, E. (1980) *J. Cell. Biol.* 85, 199–212.

Monk, B.C., Kurtz, M.B., Marrinan, J.A., and Perlin, D.S. (1991) *J. Bacteriol.* 173, 6826–6836.

Morris, N.R., Lai, M.H., and Oakley, C.E. (1979) *Cell* 16, 437–442.

Muller, H., Furter, R., and Zahner, H.R., D.M. (1981) *Arch. Microbiol.* 130, 195–197.

Myers, K.K., Fonzi, W.A., and Sypherd, P.S. (1992) *Nucleic Acids Res.* 20, 1705–1710.

Necas, O. (1971) *Bacteriol. Rev.* 35, 149–170.

Neff, N.F., Thomas, J.H., Grisafi, P., and Botstein, D. (1983) *Cell* 33, 211–219.

Nozawa, Y., and Yaginuma, H.O.K. (1985) in *Filamentous Microorganisms. Biomedical Aspects*, (Arai, T., ed.), pp. 345–354, Japan Science Society Press, Tokyo.

Oakley, B.R., and Morris, N.R. (1981) *Cell* 24, 837–845.

Oakley, B.R., Oakley, C.E., and Rinehart, J.E. (1987) *Mol. Gen. Genet.* 208, 135–144.

Odds, F.C. (1988) *Candida and Candidosis: A Review and Bibliography*, Bailliere Tindall, London.

Omura, S., Tanaka, H., Oiwa, R., et al. (1979) *J. Antibiotics (Tokyo)* 32, 978–984.

Orbach, M.J., Porro, E.B., and Yanofsky, C. (1986) *Mol. Cell. Biol.* 6, 2452–2461.

Orlean, P.A.B. (1982) *Eur. J. Biochem.* 127, 397–403.

Osborne, B.I., and Guarente, L. (1989) *Proc. Natl. Acad. Sci. USA* 86, 4097–4101.

O'Sullivan, J., Phillipson, D.W., Kirsch, D.R., et al. (1992) *J. Antibiotics* 45, 306–312.

Oulmouden, A., and Karst, F. (1991) *Curr. Genet.* 19, 9–14.

Qin, S., Moldave, K., and McLaughlin, C.S. (1987) *J. Biol. Chem.* 262, 7802–7807.

Quigley, D.R., and Selitrennikoff, C.P. (1984) *Exp. Mycol.* 8, 320–333.

Quinlan, R.A., Roobol, A., Pogson, C.I., and Gull, K. (1981) *J. Gen. Microbiol.* 122, 1–6.

Rast, D.M., and Bartnicki-Garcia, S. (1981) *Proc. Natl. Acad. Sci. USA* 78, 1233–1236.

Rine, J. (1991) in *Guide to Yeast Genetics and Molecular Biology: Methods in Enzymology,* (Guthrie, C., and Fink, G.R., eds.), pp. 239–251, Academic Press, Inc., New York.

Rippon, J.W. (1974) *Medical Mycology: The Pathogenic Fungi and The Pathogenic Actinomycetes,* pp. 531–542, W.B. Saunders Company, Philadelphia.

Rippon, J.W. (1980) *Crit. Rev. Microbiol.* 8, 49–97.

Robinson, D.G., Eberle, M., Hafemann, C., Wienecke, K., and Graebe, J.E. (1982) *A. Pflanzenphysiol. Bd.* 105, 323–330.

Roemer, T., and Bussey, H. (1991) *Proc. Natl. Acad. Sci. USA* 11295–11299.

Rose, M., and Fink, G.R. (1987) *Cell* 48, 1047–1060.

Russo, P., and Sherman, F. (1989) *Proc. Natl. Acad. Sci. USA* 86, 8348–8352.

Ruiz-Herra, J. (1992) *Fungal Cell Wall: Structure, Synthesis and Assembly,* CRC Press, Boca Raton, FL.

Ryder, N.S. (1985) *Antimicrob. Agents Chemother.* 27, 252–256.

Ryder, N.S., Frank, I., and Dupont, M. (1986) *Antimicrob. Agents Chemother.* 29, 858–860.

Saag, M.S., and Dismukes, W.E. (1988) *Antimicrob. Agents Chemother.* 32, 1–8.

St. John, T.P., and Davis, R.W. (1981) *J. Mol. Biol.* 152, 285.

Sandbaken, M., Lupisella, J.A., DiDomenico, B., and Chakraburtty, K. (1990a) *Biochim. Biophys. Acta* 1050, 230–234.

Sandbaken, M., Lupisella, J.A., DiDomenico, B.J., and Chakraburtty, K. (1990b) *J. Biol. Chem.* 265, 15838–15844.

Sandin, R., Rogers, A.L., Patterson, R.J., and Beneke, E.S. (1982) *Infect. Immunol.* 35, 79–85.

Sawistowska-Schroder, E.T., Kerridge, D., and Perry, H. (1984) *FEBS Lett.* 173, 134–138.

Sburlati, A., and Cabib, E. (1986) *J. Biol. Chem.* 261, 15147–15152.

Scannell, J.P., and Pruess, D.L. (1974) in *Chemistry of Amino Acids, Peptides and Proteins,* vol. 3 (Weinstein, B., ed.), pp. 189–244, Marcel Dekker, Inc., New York.

Schatz, P.J., Solomon, F., and Botstein, D. (1988) *Genetics* 120, 681–695.

Schiff, P.B., Fant, J., and Horwitz, S.B. (1979) *Nature (London)* 277, 665–667.

Schiff, P.B., and Horwitz, S.B. (1980) *Proc. Natl. Acad. Sci. USA* 77, 1561–1565.

Schmatz, D.M., Romancheck, M.A., Pittarelli, L.A., et al. (1990) *Proc. Natl. Acad. Sci. USA* 87, 5950–5954.

Scholer, H.J. (1980) in *Antifungal Chemotherapy* (Speller, D., ed.), pp. 35–106, John Wiley and Sons, Ltd., New York.

Schwartz, R.E., Dufresne, C., Flor, J.E., et al. (1991) *J. Antibiotics* 44, 463–471.

Segal, E., Soroka, A., and Schechter, A. (1984) *J. Med. Vet. Mycol.* 22, 191–200.

Selitrennikoff, C.P. (1979) *Arch. Biochem. Biophys.* 195, 243–244.

Selitrennikoff, C.P. (1983) *Antimicrob. Agents Chemother.* 23, 757–765.

Serrano, R., Kielland-Brandt, M.C., and Fink, G.R. (1986) *Nature* 319, 689–693.

Shaw, K.J., and Cramer, C. (1991) *The Fungal Cell Wall: Potential Targets for Discovery of Antifungal Agents,* Society of Industrial Microbiology, Philadelphia.

Shematek, E.M., Braatz, J.A., and Cabib, E. (1980) *J. Biol. Chem.* 255, 888–894.

Shematek, E.M., and Cabib, E. (1980) *J. Biol. Chem.* 255, 895–902.

Shepherd, M.G. (1987) *Crit. Rev. Microbiol.* 15, 7–25.

Shimizu, K., Kondoh, Y., and Tanaka, K. (1987) *Microbiol. Immunol.* 31, 1045–1060.

Silver, L., and Bostian, K. (1990) *Eur. J. Clin. Microbiol. Infect. Dis.* 9, 455.

Silver, P.A., and Hall, M.N. (1988) in *Protein Transfer and Organelle Biosynthesis* (Das, A., and Robbin, P.W., eds.), p. 749, Academic Press, New York.

Silverman, S., Sburlati, A., Slater, M.L., and Cabib, E. (1988) *Proc. Nat. Acad. Sci. USA* 85, 4735–4739.

Singer, S.C., Richards, C.A., Ferone, R., Benedict, D., and Ray, P. (1989) *J. Bacteriol.* 171, 1372–1378.

Sitrin, R.D., Chan, G., Dingerdissen, J., et al. (1988) *J. Antibiotics* 41, 469–480.

Smith, H.A., Allaudeen, H.S., Whitman, M.H., Koltin, Y., and Gorman, J.A. (1988) *Gene* 63, 53–63.

Snyder, M., Sapolsky, R.J., and Davis, R.W. (1988) *Mol. Cell. Biol.* 8, 2184–2194.

Stearns, T., and Botstein, D. (1988) *Genetics* 119, 249–260.

Stearns, T., Hoyt, A., and Botstein, D. (1990) *Genetics* 124, 251–262.

Sutcliffe, J.A., and Georgopapadakou, N.H. (1992) *Emerging Targets in Antibacterial and Antifungal Chemotherapy,* Chapman and Hall, New York.

Sykes, R.B., Bonner, D.P., Bush, K., Georgopapadakou, N.H., and Wells, J.S. (1981) *J. Antimicrob. Chemother.* 8(Suppl. E), 1–16.

Sykes, R.B., and Wells, J.S. (1985) *J. Antibiot.* 38, 119–121.

Taft, C.S., and Selitrennikoff, C.P. (1988) *J. Antibiotics* 41, 697–701.

Taft, C.S., Stark, T., and Selitrennikoff, C.P. (1988) *Antimicrob. Agents Chemother.* 32, 1901–1903.

Teem, J.L., Abovich, N., Kaufer, N.F., et al. (1984) *Nucleic Acids Res.* 12, 8295–8312.

Teem, J.L., and Roshbash, M. (1983) *Proc. Natl. Acad. Sci. USA* 80, 4403–4407.

Thomas, J.H., Neff, N.F., and Botstein, D. (1985) *Genetics* 112, 715–734.

Thomas, J.H., Novick, P., and Botstein, D. (1986) *Yeast Cell Biology,* Alan R. Liss, Inc., New York.

Tomoda, H., and Omura, S. (1990) *J. Antibiotics (Tokyo)* 43, 1207–1222.

Traxler, P., Gruner, J., and Auden, J. (1977) *J. Antibiotics* 30, 289–296.

Traxler, P., Tosch, W., and Zak, O. (1987) *J. Antibiotics (Tokyo)* 40, 1146–1164.

Tsay, Y.H., and Robinson, G.W. (1991) *Mol. Cell. Biol.* 11, 620–631.

Umezawa, H. (1982) *Annu. Rev. Microbiol.* 36, 75–99.

Uritani, M., and Miyazaki, M. (1988) *J. Biochem. (Tokyo)* 103, 522–530.

Valdivieso, M.H., Mol, P.C., Shaw, J.A., Cabib, E., and Duran, A. (1991) *J. Cell. Biol.* 114, 101–109.

Vanden Bossche, H., Willemsens, G., and Marichal, P. (1987) *CRC Crit. Rev. Microbiol.* 15, 57–72.

Walksman, S.A. (1945) *Microbial Antagonists and Antibiotic Substances,* The Commonwealth Fund, New York.

Weil, C.F., Oakley, C.E., and Oakley, B.R. (1986) *Mol. Cell. Biol.* 6, 2963–2968.

White, R.J. (1982) *Annu. Rev. Microbiol.* 36, 415–433.

Wilson, L., and Farrell, K.W. (1986) In *Dynamic Aspects of Microtubule Biology* (Soifer, D., ed.), pp. 690–708, New York Acad. of Science, New York.

Wolff, M.E. (1982) *Burger's Medicinal Chemistry,* John Wiley and Sons, Inc., New York.

Yadan, J.C., Gonneau, M., Sarthou, P., and Le Goffic, F. (1984) *J. Bacteriol.* 160, 884–888.

Yamaguchi, H., Hiratani, T., Baba, M., and Osumi, M. (1987) *Microbiol. Immunol.* 31, 625–638.

Yang, M., Jensen, R.E., Yaffe, M.P., Oppliger, W., and Schatz, G. (1988) *EMBO J.* 7, 3857–3862.

Zaret, K.S., and Sherman, F. (1982) *Cell* 28, 563–573.

Zaworski, P.G., and Gill, G.S. (1990) *Antimicrob. Agents Chemother.* 34, 660–662.

Zhu, J., Conteras, R., Gheysen, D., Ernest, J., and Fiers, W. (1985) *Bio/Technology* 3, 451–456.

CHAPTER
7

Strategies for Discovering Antiviral Agents from Natural Products

E. Rozhon
R. Albin
J. Schwartz

7.1 ANTIVIRAL THERAPY: A HISTORICAL PERSPECTIVE

Although a large number of antibiotics are in clinical use for treating bacterial infections, relatively few antiviral agents are approved for treating viral infections in humans (Table 7–1). Antiviral drugs have been slower to emerge than antibacterials because of the complex, interactive, and obligatory relationship that viruses have with their hosts. A consequence of this association

We thank Drs. George Miller and Claude Nash, Schering-Plough Research, for their continued encouragement to explore ways in which novel experimental approaches can be combined with classical techniques to discover antiviral drugs. We thank Drs. Philip Rather and Michael Murray, Antiviral Chemotherapy, for critically reading the manuscript and making suggestions for improvement, and Ms. Ann-Marie Mascellino for tracking down references in the library, and Mr. Stuart Cox for final editing of the manuscript. We thank Ms. Nancy Butkiewicz for assisting in the preparation of Figure 7–5. We are also indebted to various other individuals in Antiviral Chemotherapy and Microbial Products at Schering-Plough for their helpful suggestions and discussions about discovering antiviral drugs from natural sources.

TABLE 7-1   FDA-Approved Drugs for Treatment of Viral Diseases in the United States, 1991

Category	Drug	Trade Name	Disease Targeted
Chemotherapeutic	Trifluridine	Viroptic®	Herpesvirus (HSV) keratoconjunctivitis
	Ribavirin	Virazole®	Severe lower respiratory tract respiratory syncytial virus (RSV) disease
	Vidarabine (adenine arabinoside)	Vira-A®	HSV encephalitis; disseminated HSV infection of neonates; HSV keratoconjunctivitis
	Amantadine	Symmetrel®	Prophylaxis of Influenza A (Flu A) and uncomplicated Flu A respiratory tract infection
	Acyclovir (ACV)	Zovirax®	HSV-2 acute and recurrent genital disease; mucocutaneous HSV infections in immunocompromised patients; Varicella zoster virus shingles
	Azidothymidine (AZT, zidovudine); dideoxyinosine (DDI)	Retrovir® Videx®	Acquired immunodeficiency syndrome (AIDS) caused by immunodeficiency virus (HIV-1) infection and advanced AIDS related complex (ARC)

	Ganciclovir	Cytovene®	Cytomegalovirus (CMV)-induced chorioretinitis
	Foscarnet	Foscavir®	
Plant extract	Podophyllotoxin[1]	Condylox®	Condylomata acuminata (genital warts) caused by human papilloma virus (HPV)
Lymphokine (biologic)	Interferon alpha-2b[2]	Intron® A	Condylomata acuminata (genital warts) caused by HPV and hepatitis C virus infection of the liver
Pooled, purified[3] human immune globulin (biologic)	Rabies virus immune globulin	Hyperab®	Post-exposure to rabies virus; rabies encephalitis
	Hepatitis B virus immune globulin	Hyperhep®	Post-exposure to hepatitis B virus; hepatitis B virus-caused hepatitis
	Cytomegalovirus immune globulin		Primary CMV infection in transplant recipients receiving kidney from CMV antibody-positive donor

[1] Podophyllotoxin is not directly antiviral; acts antimitotically against HPV-transformed warty tissues (Rader 1991b). Podophyllotoxin is derived from plant extracts rather than being chemically synthesized.

[2] Interferon alpha-2b, which is prepared using recombinant DNA techniques, is purified from *E. coli* that contains a genetically engineered plasmid containing the interferon alpha-2b gene of human leucocytes. Interferon alpha-2b acts indirectly as an antiviral.

[3] In some countries, hyperimmune sera or pooled normal immunoglobulin (e.g., Gamastan® or Gamimune® in the United States) are licensed for treating the following viral infections: measles, hepatitis A, polio, rubella, mumps, and varicella viruses.

is that agents intended to inhibit specific viral functions often also inhibit related cellular processes, or in some cases, exert their effect only on cellular processes. Such agents, which are toxic in cell culture, are usually precluded from further development as antivirals. Thus, unlike testing an antibacterial agent, the ability of a putative antiviral agent to inhibit viral replication must always be assessed in the presence of the host cell. An additional factor explaining the slow emergence of antiviral drugs is that, despite significant advances in understanding viral replication and pathogenesis, much is still unknown about these processes.

Historically, three avenues have been taken to affect the course of viral disease: avoidance of exposure, prophylaxis (vaccination prior to infection), and post-exposure therapy. The earliest records of vaccination, which date back nearly 1000 years, describe using dried scabs from victims of smallpox to inoculate individuals not yet showing signs of the disease. However, rather than prevent the disease from developing, this practice usually led to new outbreaks of smallpox. Jenner more successfully used immunization 800 years later when he inoculated humans with homogenates of bovine cowpox lesions (Fenner 1990). Since the mid 1950s, when Salk introduced the poliovirus vaccine, safe and effective vaccines have become available to prevent a variety of viral infections (e.g., measles, mumps, yellow fever, and hepatitis B virus). Except for measures taken to prevent exposure, vaccination remains the most effective method for controlling viral diseases. Once infection is established, however, vaccination rarely prevents the disease from developing. The rabies virus is an example of a vaccine that is administered post-exposure and is effective for preventing the development of disease.

Post-exposure therapy and therapy after a viral infection has manifested itself comprises three possible approaches:

1. Administration of pooled human immune globulin, which contains neutralizing antibodies to virus. A drawback to this approach is the possibility of anaphylactic or adverse systemic reactions following repeated treatment. In addition, the possibility of contamination by adventitious agents has made the use of pooled human immune globulin less desirable.
2. Exploiting the natural defense mechanisms of the host: This therapeutic avenue includes naturally occurring cytokines and lymphokines of the host, which exert their antiviral effects by enabling uninfected cells to become resistant to infection or by stimulating the immune response of the host (Notka et al. 1990). Advances in molecular biology and genetic engineering during the 1980s have advanced the identification and cloning of various lymphokines and cytokines and promoted their large-scale production. An example of a lymphokine being used as a therapeutic agent for a viral infection is interferon alpha-2b, which is approved for treating papillomavirus-induced genital warts (Reichman et al. 1988) and hepatitis C virus infection (Davis et al. 1989; Di Besceglie et al. 1989; Rader 1991a; also see Table 7–1).

3. The administration of synthetic or naturally derived chemotherapeutics: Examples include azidothymidine (AZT) for treating human immunodeficiency virus (HIV) infection, acyclovir (ACV) for herpes simplex virus (HSV) infections, and amantadine for influenza virus A prophylaxis (see Table 7–1). Among the post-exposure approaches discussed here, the latter approach probably represents the predominant current effort to discover and develop antiviral agents.

There remain, however, many viral diseases that have considerable impact in regard to human suffering, hospitalization, and time lost from work for which neither vaccines nor effective antiviral treatments are available. For example, there are no drugs approved to treat the common cold (caused by several kinds of viruses; Monto et al. 1987; Couch 1990b), infectious mononucleosis (Epstein-Barr virus), hepatitis A virus, and the many kinds of arbovirus- and enterovirus-induced diseases. Moreover, there is still much room for improvement in regard to potency and pharmacokinetic properties as well as reducing associated toxicity in viral diseases for which chemotherapeutic agents have been approved (see Table 7–1). For example, AZT, the most common drug prescribed for individuals who have acquired immunodeficiency syndrome (AIDS) (HIV-1 infection), is toxic in many patients (Fischl et al. 1987). Furthermore, while ACV, one of the most widely used antiviral drugs, is effective in treating acute infection with HSV, ACV is not able to clear latent virus that often reactivates to produce recurrent disease (Mertz 1990). In addition, viral resistance to both AZT (Larder et al. 1989; Richman 1990) and ACV (Bevilacqua et al. 1991; McLaren et al. 1983; Marrero et al. 1991) has been reported. Finally, while amantadine is effective prophylactically for preventing infection and illness in individuals exposed to influenza A (Flu A) virus, the drug exhibits limited efficacy once Flu A virus infection is established. In addition, because of adverse reactions, amantadine is generally recommended only for individuals in high-risk medical categories (Couch 1990a). Furthermore, amantadine has no effect on Flu B immunotypes (Couch 1990a). Thus, the effort continues to identify new antiviral chemotherapeutics for viral diseases for which no treatment is available as well as to improve existing strategies for treating viral disease.

7.2 ANTIVIRAL RESEARCH TODAY: A COMBINATION OF CLASSICAL VIROLOGY AND MOLECULAR BIOLOGY

The effort to discover antiviral drugs accelerated in the two decades between 1930 and 1950 with advances in the propagation and quantitation of viruses in cultured cells (Sabin and Olitsky 1936; Enders et al. 1949). These in vitro techniques were rapid and reasonably accurate, and effectively replaced animal LD_{50} studies that were expensive, time consuming, and often not reproducible. While some early antiviral screens exploited the ability of a virus to produce cytopathology in infected cell cultures [the cytopathic effect (CPE) assay], the more quantitative plaque assay became the standard for assessing the potential antiviral activity of a test agent. Although the plaque assay seems primitive in light of current capabilities in virology, it continues to be widely used to identify and confirm antiviral activity.

The plaque assay is not, however, a foolproof method for identifying antiviral agents; inhibition of plaque formation may be due to an antiviral activity or it may reflect the activity of a cytotoxic component. Cytotoxic agents can masquerade as antivirals by virtue of their ability to adversely affect cellular processes to the extent that cells are unable to support viral replication. In fact, random screening of synthetic as well as naturally derived samples for antiviral activity often results in the identification of cytotoxic agents. Only further analyses distinguish true antiviral agents from those that are cytotoxic. Therefore, random screening for antiviral agents using the plaque or CPE assays necessitates analyzing large numbers of samples since the probability of finding a bona fide antiviral activity is poor.

During the 1980s an intense effort was directed toward discovering antiviral agents. This was fueled by the increasing incidence of AIDS as well as the development of novel technologies in the fields of molecular biology, molecular genetics, and structural biology, which have facilitated our understanding of viral biology. In fact, major technological advances in antiviral research have also been aided by important discoveries in virology and the nature of infectious disease. Figure 7–1 illustrates the chronological relationship between major advances made in antiviral research, virology, and molecular biology. Although few antiviral agents discovered by these new methodologies are approved for treating viral diseases, many are in clinical trials and appear to be promising.

Compared with earlier methods for discovering antiviral agents, state-of-the-art techniques offer several advantages. For example, recombinant DNA techniques have made it possible to isolate biochemically essential and specific viral functions from the intact virus that can be used in assays to screen for inhibitory agents. For a lethal viral infection such as HIV-1, an obvious advantage is that this approach is safer than working with the infectious agent itself. In addition, unlike inhibitors identified by the plaque assay, which do not reveal what viral process is affected, the newer approaches are designed to detect agents that interfere with specific viral events. Although many of these newer approaches to screening do not preclude the possibility of identifying cytotoxic activities, they theoretically minimize the probability of finding agents that are solely cytotoxic. Inhibitory agents discovered by these technologically advanced test systems must be further evaluated to confirm antiviral activity. Since many new avenues for discovering antiviral drugs rely less on serendipity than earlier methods did, these new approaches have been referred to as the rational design of drugs (McKinlay and Rossmann 1989).

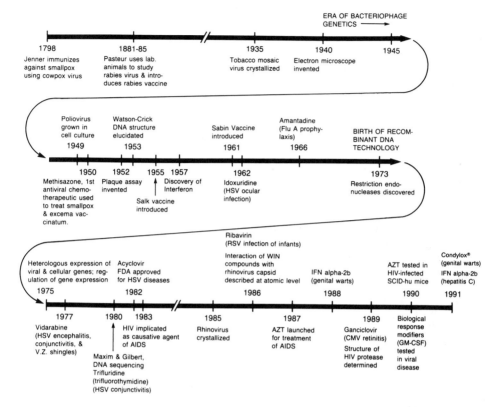

FIGURE 7–1 Milestones in the history of antiviral development. The chronological relationship between selected major advances in molecular and structural biology, infectious disease, virology, and antiviral chemotherapy is shown. Note that dideoxyinosine (DDI) and foscarnet (trisodium phosphonoformate), which do not appear in this figure, were approved for treatment of HIV infection in AIDS patients not responding to AZT and for cytomegalovirus retinitis in AIDS patients late in 1991. Abbreviations used are: Flu A, influenza virus A; HSV, herpesvirus; V.Z., varicella zoster virus; FDA, Food and Drug Administration; HIV, human immunodeficiency virus; AIDS, acquired immunodeficiency syndrome; RSV, respiratory syncytial virus; AZT, azidothymidine; IFN, interferon; CMV, cytomegalovirus; SCID-hu, severe combined immunodeficient-human; GM-CSF, granulocyte macrophage-colony stimulating factor. Adapted with permission from Rozhon et al. (1991).

7.3   TESTING NATURALLY DERIVED SAMPLES FOR ANTIVIRAL ACTIVITY: ADVANTAGES AND DISADVANTAGES

A potential source of antiviral agents may be found in naturally derived samples. Extracts of various herbs, spices, roots, tree barks, leaves, etc. have a rich anecdotal as well as proven history in treating human ailments. In fact, a variety of drugs currently in clinical use originated from natural samples. Today, significant effort in both academia and industry is devoted to testing extracts from plants and algae, marine organisms, fungi, and bacteria for desired pharmacologic activities. Table 7–2 lists examples of antiviral agents that were derived originally from natural sources. Only two of these antiviral agents have made it to the marketplace; the remainder are in various stages of preclinical evaluation.

The identification of antiviral activity in the absence of a demonstrable cytotoxic effect on the host cell remains one of the fundamental difficulties of discovering antiviral agents. This is true whether the source of test sample is a chemically purified, synthetic molecule or an extract from a naturally derived material. However, the magnitude of this difficulty is usually greater with naturally derived samples, since the broth, homogenate, or extract submitted for antiviral testing contains a complex array of molecules. If antiviral activity exists in such a mixture, it may be masked by molecules that exhibit cytotoxic activities. Another possible difficulty is that an antiviral agent may be present at a low concentration, thereby complicating its detection. When applied to naturally derived samples, state-of-the-art approaches to discover antiviral activities are intended to overcome some of these problems. These approaches comprise highly sensitive assays designed to identify inhibitors of specific viral processes. Moreover, the newer assays are intended to rapidly discern cytotoxic agents masquerading as antivirals. While these approaches would seem to offer advantages over the CPE or plaque assays for screening natural samples for antiviral activity, the newer technologies and approaches will only be validated when molecules discovered by these assays are used in the clinic.

The remainder of this chapter describes some of the new screening approaches being used to discover antiviral molecules in samples originating from natural sources. The objective of these approaches is to identify molecules that act by specific and desirable antiviral mechanisms of action. The chapter concludes by providing examples of how other new technologies, such as computer-assisted drug design, may be used to advance antiviral molecules toward clinically useful pharmaceuticals.

7.4 STRATEGIES FOR IDENTIFYING SPECIFIC ANTIVIRAL ACTIVITIES IN NATURAL PRODUCTS

The goal of a screening assay is to discover an agent that acts on a specific viral process, which results in the inhibition of production of infectious particles. This approach is possible for some viruses since sufficient information is known about viral-mediated events occurring after infection of the host cell that allow specific processes to be selected as targets for the action of antiviral agents. The identification of a naturally occurring agent that affects a specific viral process depends on the ability of the assay to detect and quantify a

TABLE 7-2 Examples of Agents Derived from Natural Sources that Have Antiviral Activity or Are Used to Treat Viral Disease

Antiviral Molecule	Activity	Source	Comment	Reference
Amantadine	Influenza A virus	Petroleum	FDA approved for Flu A respiratory infection	Couch 1990a Merck Index 1989
Podophyllotoxin	Antimitotic, genital warts	May apple; mandrake root	FDA approved for HPV-induced genital warts	Rader 1991b
Scopadulcic acid B	HSV-1	Paraguayan medicinal plant	[1]Preclinical studies	Hayashi et al. 1988
[2]Castanospermine	Retroviruses (e.g., HIV-1)	Australian chestnut tree	Preclinical studies	Hudson 1990a
[3]Dichloroflavan	Picornaviruses (e.g., rhinoviruses and enteroviruses)	Xanthorrhea plants	Preclinical studies	Superti et al. 1989 Conti et al. 1988 Hudson 1990b Vrijsen et al. 1987
[3]Methylquercetin	Poliovirus	E. grantii (plant)	Preclinical studies	Castrillo et al. 1986
[3]Chalcones	Rhinoviruses	Chinese medicinal herb (A. rugosa Kuntze)	Preclinical studies	Ishitsuka et al. 1982
SF-2140	Influenza A virus	Actinomadura sp.	Preclinical studies	Ito et al. 1984

[1] Preclinical studies indicates nonhuman studies have been performed and may indicate activity in cell culture systems and small animal models. See referenced literature for details.
[2] Belongs to alkaloid chemical class.
[3] Belongs to flavonoid chemical class.

specific inhibitory activity present in the heterogeneous mixture of the test sample.

7.4.1 Viral-Mediated Transactivation of Gene Expression

When developing assays to detect agents that specifically affect a given viral process, it is important to keep in mind the intimate relationship that exists between the host cell and pathogen. The viral genome is small and has limited coding capability compared with that of the mammalian cell; as a result, the virus depends in large part on the metabolic components and machinery of the host cell to complete its replicative cycle. Some viruses have evolved complex and unique strategies in their exploitation of cellular processes to the extent that cellular gene expression and metabolism are biased in the infected cell to favor viral replication. This is accomplished by viral encoded factors that interact with and alter specific macromolecules required for host biosynthetic processes. The activity of these factors may be exploited in cell-based or cell-free systems to develop specific screens for antiviral agents.

For some double-strand DNA viruses, expression of viral genes occurs in a temporal fashion after the viral genome enters the host nucleus. Regulation of expression is carried out at the transcriptional level by viral proteins referred to as *transactivators* or *trans-acting factors*. These proteins are thought to stimulate specific viral gene expression by interacting with cellular transcription factors and discrete regulatory elements present in the viral promoter DNA. The adenovirus E1A protein(s) and the herpesvirus VMW_{65} and ICP4 factors are examples of viral-encoded trans-acting proteins (Horwitz 1990; Roizman and Sears 1990). Transactivating factors also have been described for retroviruses: for example, the *tat* gene product of HIV (Coffin 1990; Wong-Staal 1990). The tat protein is crucial for stimulating expression of the HIV gene products required for establishing productive infection. Thus, since viral transactivating factors play critical roles in the viral replicative process, agents that inhibit their synthesis or activity may prevent productive viral infection.

Since viral transactivators usually exert their effect on a number of viral, and at times cellular, promoters, it is necessary to devise a strategy for a screening assay whereby a single transactivation event can be isolated and assayed. One possible approach, illustrated in Figure 7–2, relies on two general techniques from molecular biology, cloning and transfection. Using this approach, a recombinant plasmid is constructed in which a gene that encodes a readily assayed product, such as chloramphenicol acetyltransferase (CAT), is placed under the transcriptional control of a viral promoter whose transactivation is targeted for inhibition. Since promoter activity and, hence, synthesis of the CAT reporter gene depend on a transactivation event, a second plasmid must also be constructed that allows for the constitutive synthesis of the appropriate viral transactivator from a heterologous promoter. Both chimeric plasmids are then introduced into tissue culture cells by transfection, and the quantification of CAT activity in cell extracts represents an indirect

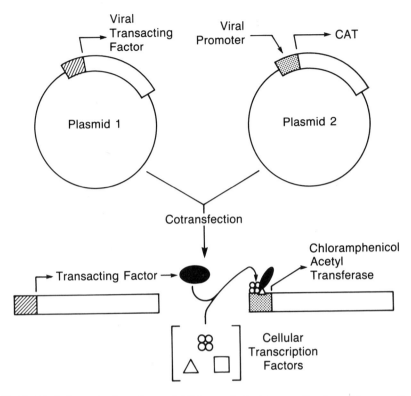

FIGURE 7–2 A method to detect inhibitors of viral transactivation. Tissue culture cells are transfected with two recombinant plasmids. Plasmid 1 encodes a viral trans-activator of gene expression under the control of heterologous promoter sequences (▨). Plasmid 2 contains the viral target promoter (▨) upstream of the CAT reporter gene. Synthesis of CAT, determined by an assay for CAT activity, depends on the viral transacting factor, supplied by plasmid 1, interacting with the viral promoter of plasmid 2 to initiate gene expression.

measure of viral promoter activity. The ability of a natural product to inhibit CAT activity in this type of assay can be used as a preliminary assessment of the potential antiviral activity of the agent.

Transactivation of transcription by viral factors is a complex and still ill-defined process. A complicating factor and a drawback for evaluating the ability of a natural product to interfere with viral transactivation in assay exemplified in Figure 7–2 is the requirement of the host cell, which supplies the necessary metabolic machinery (e.g., cellular transcription factors) required for gene expression. It may be possible that a putative antiviral agent detected in this system affects expression of the reporter gene by affecting a cellular component involved in transactivation rather than directly inhibiting the viral transactivator. Such agents would likely have an adverse effect on cellular gene expression and, therefore, must be eliminated by appropriate cytotoxicity testing.

7.4.2 Viral-Encoded Enzymes

Viruses encode enzymes unique to infected cells that catalyze comparatively simple and well-understood chemical reactions. Exploiting the activities of these enzymes allows for the development of cell-free assay systems that are comprised of the purified viral enzyme of interest and a defined substrate. Advances in molecular biology have made it relatively easy to clone and express large quantities of many viral proteins in bacterial systems, thus providing a source of purified enzyme. Examples of viral enzymes amenable to cell-free assay systems are the HIV reverse transcriptase (RT) and the picornavirus 3C protease.

The RT of HIV-1 is an essential and unique enzyme in the viral replication cycle. The RT, which is carried into the cells as a component of the infecting virus, catalyzes the synthesis of viral DNA from the viral genomic RNA template (Coffin 1990). Ultimately, the viral DNA integrates into the host genome where it is used for transcription of viral gene products. Therefore, an agent that inhibits RT would affect the virus at an early stage in its life cycle to prevent the production of infectious progeny. A simple, cell-free system to measure RT activity can be designed that contains purified RT and a synthetic oligomeric template. The enzyme can be obtained from lysates of a bacterial strain engineered to overproduce RT when transformed with a plasmid containing the cloned RT gene. In a reaction mixture containing template and primer, enzyme activity can be measured by following the incorporation of radioactive nucleoside triphosphate precursors into newly synthesized acid-insoluble material. When the reaction is carried out in the presence of a test sample, a decrease of radioactive incorporation may indicate a potential RT inhibitor.

Another example of how a cell-free assay system can be used to discover antiviral agents from natural sources is exemplified by the 3C protease of picornaviruses. The genome of picornaviruses comprises a single-stranded RNA molecule that, following entry of a virus into the cell and uncoating, is translated directly into a large polyprotein (Rueckert 1985). This polyprotein is subsequently cleaved into essential viral capsid proteins and enzymes. One protein that catalyzes some of these cleavages is the virally encoded 3C protease, which is itself part of the polyprotein. The 3C protease has an auto-catalytic activity that allows its cleavage out of the polyprotein to act at other processing sites within the polyprotein (Hanecak et al. 1984; Palmenberg et al. 1979). Since cellular enzymes are not capable of processing the polyprotein, an inhibitor of 3C protease activity would effectively and specifically prevent processing and thus inhibit viral replication. Like the RT enzyme from HIV, the 3C protease from picornaviruses can be expressed in bacteria and purified (Hanecak et al. 1984; Nicklin et al. 1988). Parts of the polyprotein that contain

3C processing sites also have been cloned and expressed for a substrate (Nicklin et al. 1987; Dasmahapatra et al. 1991; Ypma-Wong et al. 1988). Using these two reagents in a cleavage reaction in vitro, inhibition of 3C protease activity can be detected by sodium dodecyl sulfate-polyacrylamide gel electrophoresis: lack of the expected peptide profile resulting from the cleavage of the polyprotein indicates a potential 3C inhibitor. More efficient assays might utilize an ELISA (enzyme-linked immunosorbent assay) format in which the substrate is affixed to a plastic surface. Intact and cleaved substrate could be distinguished using monoclonal antibody (mAb) to an epitope, which is present only in the uncleaved substrate. In the processed substrate, the epitope is soluble. A second antibody conjugated, for example, to peroxidase and having specificity for the mAb could be used as a colorimetric indicator for uncleaved substrate. With either approach, additional studies must be done to discriminate agents that interact with the 3C protease from those that interact with substrate.

While cell-free assays provide several distinct advantages over cell-based assays, such as the capacity to handle a large number of test samples as well as utilize viral targets with a high degree of specificity, they are not without disadvantages. For example, cell-free systems fail to detect agents that require metabolic activation, such as phosphorylation or glycosylation, for activity. Since cell-free test systems employ isolated viral functions, they are removed from the influences imposed by other viral and cellular processes. Moreover, these systems depend on reaction conditions that, at best, only approximate those within the infected cell. Thus, inherent to the cell-free assay is the possibility that the newly discovered inhibitory activity may fail to inhibit replicating virus because necessary co-factors provided by the cell or other viral processes are absent. Consequently, inhibitory activity must be confirmed using live virus assays. Finally, while the catalytic function of the viral enzymes discussed in this section appears to be unique, there is the possibility that cellular counterparts exist with similar or overlapping activities. Thus, antiviral agents discovered in cell-free systems must be assessed for their effects on the normal metabolic functions of the host cell.

7.4.3 Antiviral Agents that Interact with the Virus Capsid

7.4.3.1 A Brief History. It has been known for many years that some antibodies capable of neutralizing viral infectivity bind to antigenic sites located on the intact viral capsid (Murphy and Chanock 1990). These antibodies neutralize infectivity by preventing viral attachment to the susceptible cell. In addition, as exemplified by the picornaviruses, some neutralizing antibodies may inhibit an early viral process, such as penetration through the cytoplasmic membrane or uncoating (disassembly of the capsid and release of the viral genome; Mandel 1976). By inference from the known effects of neutralizing antibody on virus, it seemed possible that small molecules might exist, or could be synthesized, that interact with the viral capsid to inhibit these essential, early viral processes. This proposition was translated into fact with molecules discovered in the late 1970s and early 1980s at the Sterling-Winthrop Research Institute. Studies on a molecule referred to as WIN 51711, or disoxaril, and its predecessors referred to as arildones, indicated that these molecules interacted with the capsid of some picornaviruses to inhibit early steps in the infectious process (McSharry et al. 1979; Otto et al. 1985; Fox et al. 1986). In 1986, a collaborative effort between Sterling-Winthrop and a group at Purdue University confirmed not only that the disoxaril molecule bound to the capsid of human rhinovirus-14 (HRV14), but also provided details on the precise interaction between disoxaril and the viral capsid at the atomic level (Smith et al. 1986). This was accomplished by infusing crystals of HRV14 with WIN molecules and subjecting the complex to x-ray diffraction analysis. Results showed that the WIN molecules bound within a hydrophobic pocket formed by a beta-barrel antiparallel fold of VP1, one of the four capsid proteins of picornaviruses (Smith et al. 1986; Badger et al. 1989) (Figure 7–3). Since it was already known that disoxaril was able to stabilize some picornaviruses against thermal inactivation (Fox et al. 1986), it was speculated that WIN molecules, occupying the hydrophobic pocket, prevent conformational changes within the capsid, which lead to thermal inactivation (Smith et al. 1986). Similarly, it was proposed that under normal physiological conditions, WIN molecules prevent conformational changes that the picornavirus capsid must undergo during the uncoating process (Smith et al. 1986).

In addition to the arildone and the WIN molecules, several other molecules have been shown to or are suspected to interact with the picornavirus capsid to inhibit viral attachment or uncoating (Alarcon et al. 1986; Chapman et al. 1991; Conti et al. 1988; Eggers 1977; Hudson 1990b; Ishitsuka et al. 1982; Ninomiya et al. 1985, 1990; Superti et al. 1989). At least a few of these molecules are known to originate from natural sources (e.g., dichloroflavan and the chalcones; Ninomiya et al. 1985; see Table 7–2). None of these antipicornavirus agents, originating from either synthetic or natural sources, have been approved for clincial use; however at least one, Janssen R61837, has been tested in humans experimentally infected with human rhinovirus and has shown promise (Al-Nakib et al. 1989; Barrow et al. 1990). Although much remains to be resolved in regard to the pharmacokinetic and toxicologic properties of these kinds of molecules, targeting chemotherapeutic agents to have an effect on the picornavirus capsid appears to be a feasible approach for discovering drugs to inhibit this family of viruses.

7.4.3.2 Examples of Strategies.   One approach to discovering capsid-binding molecules with antipicornavirus activity in naturally derived samples takes advantage of the capsid-stabilizing property of WIN-like molecules. An example of an assay that exploits the capability of capsid-binding molecules to

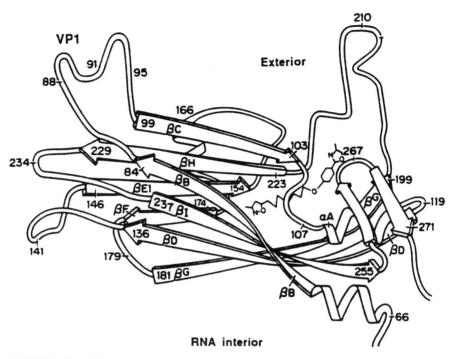

FIGURE 7–3 Diagrammatic representation of anti-picornavirus molecule (WIN 52084) within the VP1 capsid protein of human rhinovirus 14. Ribbon-like structures with terminal arrows (pointing toward carboxyl terminus) indicate antiparallel beta strands that form the hydrophobic binding pocket of the VP1 protein. Numerals shown at various locations indicate the number of the amino acid comprising the VP1 chain. Adapted with permission from T.J. Smith, M.J. Kremer, M. Luo, et al. (1986) *Science* 233, 1286–1293. Copyright 1986 by the AAAS.

prevent loss of viral infectivity when incubated at high temperature is shown in Figure 7–4. This assay is not foolproof since certain combinations of capsid-binding molecules and picornaviruses result in virus stabilization, but not in inhibition of virus (Andries et al. 1989). In any event, the published literature indicates that the majority of capsid-binding compounds that stabilize virus to the effects of heat or acid result in inhibition of early viral processes (Andries et al. 1989; Fox et al. 1986; Tisdale and Selway 1984; Ninomiya et al. 1985; Rombaut et al. 1985, 1990).

Another strategy to screen for agents that interact with the intact virion to interfere with viral attachment to the host cell is exemplified by the binding of the gp120 surface glycoprotein of HIV-1 to the CD4 receptor found on T-cells. Since infection of T-cells by HIV-1 is mediated by the interaction of HIV-1 gp120 and the cellular receptor CD4 (Lasky et al. 1986; Lifson et al. 1986; Sattentau and Weiss 1988), it is probable that an agent interfering with

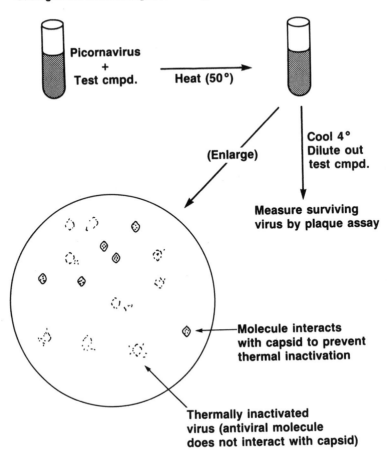

FIGURE 7–4 Example of methodology used for detecting potential anti-picornavirus molecules based on their capsid-stabilizing ability. Molecules that bind to the picornavirus capsid within the VP1 pocket may prevent thermal inactivation of capsid, thereby preventing loss of ability to infect. Detection of such molecules in the plaque assay depends on the ability of the molecule to interact reversibly with the capsid. The enlargement is a pictorial representation of the effect of 50°C heat treatment on a picornavirus to which antiviral molecules have and have not bound. The capsid of viruses that have bound molecules (⊘) remains intact during the 50°C incubation, whereas viruses that have not interacted with the molecules are rendered noninfectious (⦂⟨⟩⦂).

this interaction would be a potent inhibitor of HIV-1 infection of T-lymphoid cells. To identify potential inhibitors of this interaction, a screening assay based on ELISA technology can be readily established using the soluble, purified ligand, gp120, and purified CD4 receptor molecules. One possible design of such an assay is shown in Figure 7–5. In this system, soluble CD4 molecules are immobilized on an inert surface and incubated with solubilized

Immobilize CD4 on plastic surface

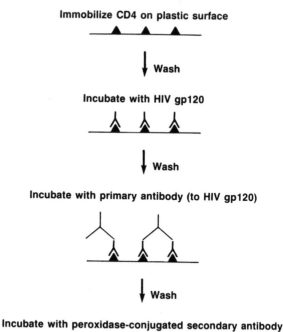

↓ Wash

Incubate with HIV gp120

↓ Wash

Incubate with primary antibody (to HIV gp120)

↓ Wash

Incubate with peroxidase-conjugated secondary antibody

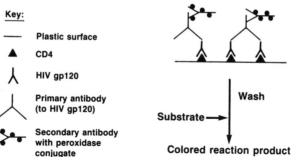

Key:

— Plastic surface

▲ CD4

人 HIV gp120

Primary antibody (to HIV gp120)

Secondary antibody with peroxidase conjugate

Wash

Substrate →

Colored reaction product

FIGURE 7–5 Example of an ELISA-based assay to detect inhibitors of attachment of HIV to the CD4 cellular receptor. As described in detail in Section 7.4.3.2, a rapid, efficient assay having a colorimetric end-point is desirable for screening large numbers of natural products for activities preventing HIV attachment.

gp120 to allow binding. Binding of gp120 to CD4 is detected by initially incubating with antibody specific to gp120 followed by adding a secondary anti-IgG antibody conjugated, for example, to peroxidase. In the presence of an appropriate substrate, the peroxidase generates a colored product that is readily quantified spectrophotometrically. Association of an agent with either gp120 or CD4, which prevents the normal interaction of these proteins, would be detected by a decrease in absorption. Additional studies would have to be performed to determine specificity, since an agent interacting with CD4 might exhibit cytotoxicity and hence be less desirable than an agent interacting with the viral gp120.

7.5 EXAMPLES OF NEW TECHNOLOGIES FOR OPTIMIZING ANTIVIRAL ACTIVITY AND PREDICTING THE EFFECT OF ANTIVIRAL AGENTS PRIOR TO CLINICAL STUDIES

7.5.1 Rational Drug Design Using the Techniques of Structural Biology

Once the structure of an antiviral molecule is identified and a sufficient quantity is available, the molecule may be subjected to chemical alteration to optimize its activities to increase potency or to obtain a broader spectrum activity with less cytotoxicity. This process is traditionally used for many pharmacologic agents isolated from natural sources and provides information necessary to identify structural analogues of the parent molecule having the most desirable biological activities. Unfortunately, this process also consumes valuable time since it depends on trial and error. An alternative approach, which theoretically could hasten the process of optimizing activity, was first used during the 1980s. This approach, which combines methodologies in the fields of structural biology and computer science, is exemplified by studies on the interaction of WIN molecules and the capsid of HRV14.

An important question resulting from x-ray diffraction analysis of HRV14 complexed with WIN molecules (Smith et al. 1986; see Figure 7–3) was whether information about the interaction between a small molecule and its protein target could be used to develop new molecules with improved antiviral activity (as described in the previous section). The answer to this question is not yet known, but it may be answered by investigators who are combining structural information obtained from viral x-ray diffraction analysis with the molecular modeling capabilities of super computers. This technology is complex and highly sophisticated. Prerequisite information includes a knowledge of the structural conformation of the protein target as well as the prototypic antiviral molecule bound within the protein. With this information, it is possible to calculate the interaction between the molecule and the protein based on drug-binding forces and molecular dynamics. Using these data with computer programs for designing chemical structures, new molecules can be designed and modeled within the binding site in an attempt to enhance the binding affinity of the molecule. Although data have not been reported indicating that computer-designed WIN molecules have improved antiviral activity, evidence exists with other WIN molecules that demonstrates improved capsid-binding affinity correlates with improved antiviral potency (Fox et al. 1991).

The atomic structures of other important viruses pathogenic for humans have yet to be elucidated by x-ray diffraction analysis since many of these viruses do not form crystals, which are amenable to this technique. However, x-ray diffraction analysis of essential components, such as the hemagglutinin protein of influenza virus (Weis et al. 1988; Wilson et al. 1981) and the protease of HIV-1 (Navia et al. 1989; Wlodawer et al. 1989), has lead to the determination of the three-dimensional structures of these molecules. Thus, the structures of molecules that bind in the catalytic site of the HIV protease and the viral site for the cellular receptor of influenza can now be predicted by computer. It is not yet certain whether this approach for the rational design of antiviral drugs will be successful. However, it seems highly probable that an improved understanding of the three-dimensional structure of the antiviral target combined with the ability to design potentially inhibitory molecules that fit within the target site will aid in the development of new antiviral drugs. Until the Sterling-Winthrop group or others doing similar studies report improved activity of antiviral molecules designed by computer, computer-assisted design of antiviral drugs must still be considered a science in its infancy.

7.5.2 Recent In Vivo Advances for Evaluating Antiviral Agents: SCID-hu Mice

Although testing antiviral agents in cell culture and other in vitro systems can provide important information about potency and the antiviral mechanism of a drug, data from these tests generally cannot be used to predict activities of drugs in humans. Thus, evaluation of the antiviral agent in animals is necessary to provide essential information regarding toxicity, metabolism, tissue distribution, pharmacokinetic properties, and efficacy. The toxic and pharmacologic properties of a prospective drug are usually tested in a variety of animals. However, the potential efficacy of an antiviral drug must be tested in an animal exhibiting relevant disease. Optimally, an appropriate model is one in which the animal is susceptible to infection and exhibits measurable symptoms or a disease process caused by the viral infection. If the virus produces a disease in an animal that resembles the disease in humans, the chance of predicting the effect of the drug in humans is likely to be more accurate than that from a prediction based on animals in which the disease differs from that occurring in humans. For many viral diseases in humans, relevant animal models are available, but for some viruses, an appropriate model does not exist. For example, although HIV-1 infects certain nonhuman primates, such as chimpanzees and gibbons, infection does not result in overt disease related to immune suppression (Gardner and Luciw 1989; Hirsch and Curran 1990). However, in chimpanzees, the antiviral activity of a prospective anti-HIV drug may be evaluated by the effect the drug has on HIV titers in peripheral blood mononuclear cells or plasma (Gardner and Luciw 1989). The use of nonhuman primates, and chimpanzees in particular, for these kinds of studies may be limited by the difficulties and expenses associated with obtaining these animals as well as considerations regarding animal welfare.

The recent development of the SCID-hu (severe combined immunodeficient-human) mouse has surmounted some of the problems associated with testing prospective HIV drugs in nonhuman primates. The SCID-hu mouse is a SCID mouse that has been engrafted with human hematolymphoid tissues (McCune et al. 1990). Such mice possess an essentially reconstituted and functional human immune system. McCune and associates (McCune et al. 1990; Shih et al. 1991), demonstrated that AZT-treated SCID mice, previously infected with HIV, had a reduced number of HIV genomes as compared with control animals (Figure 7–6). Thus, HIV-infected SCID-hu mice may prove to be useful as a primary in vivo model for evaluating anti-HIV agents prior to studies in nonhuman primates or humans. However, it is too early to know whether these mice will have a major impact in testing anti-HIV agents; in

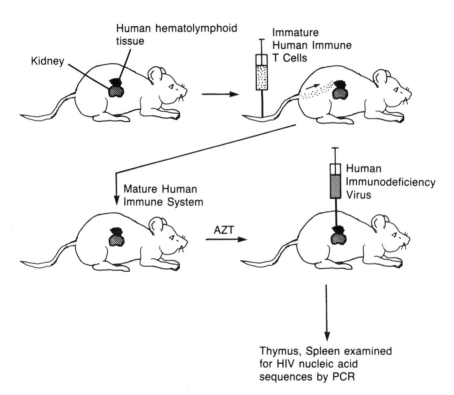

FIGURE 7–6 HIV-infected SCID-hu mice are used to assess efficacy of anti-HIV molecules. An example of one of several procedures to produce SCID-hu mice capable of replicating HIV is shown. SCID mice are engrafted with human hematolymphoid tissue (e.g., thymus and lymph glands) and injected with immature human T lymphocytes to give an animal a reconstituted, functional human immune system. Prophylactic treatment of mice with the antiviral drug AZT, followed by infection with HIV, reveals that viral replication is reduced (McCune et al. 1990). Reproduced with permission from Rozhon et al. (1991).

any event, it seems likely that SCID-hu mice will be used in increasing numbers since they are considerably less expensive and more readily available than chimpanzees. Moreover, it appears that SCID-hu mice will be useful in investigating the mechanisms of HIV pathogenesis. Finally, SCID-hu mice also may be suitable for evaluating antiviral agents directed at other human viruses for which no animal host exists (e.g., Epstein-Barr virus and human cytomegalovirus).

7.6 CONCLUSION

Naturally derived materials provide a rich source of agents that have potential therapeutic benefits and thus play a significant role in the discovery of molecules with antiviral activity. This role has been aided by technological advances in the fields of molecular and structural biology, recombinant DNA, and biochemistry. The approaches used to identify agents that inhibit essential viral functions have become more sophisticated with the ability to manipulate and isolate discrete viral processes. The current state-of-the-art methodologies have led to the development of a variety of assay systems that can detect molecules capable of inhibiting the expression or activity of specific viral gene products. In addition, the use of x-ray crystallography in combination with high speed supercomputer analyses affords a new direction toward the design of drugs tailored for interaction with a particular viral protein structure. Advances in antiviral drug development are not limited to the new in vitro methods. Genetically engineered animals, such as transgenic and SCID mice have provided alternative models for assessing the efficacy of potential antiviral agents directed against human pathogens in vivo. While the success of these novel technologies in drug discovery and development has yet to be fully realized, these approaches provide great promise for the treatment of viral infections previously intractable to chemotherapy.

REFERENCES

Alarcon, B., Zerial, A., Dupiol, C., and Carrasco, L. (1986) *Antimicrob. Agents Chemother.* 30, 31–34.

Al-Nakib, W., Higgins, P.G., Barrow, G.I., et al. (1989) *Antimicrob. Agents Chemother.* 33, 522–525.

Andries, K., Dewindt, B., Snoeks, J., and Willebrords, R. (1989) *Arch. Virol.* 106, 51–61.

Badger, J., Minor, I., Oliveira, M.A., Smith, T.J., and Rossmann, M.G. (1989) *Proteins: Structure, Function, Genet.* 6, 1–19.

Barrow, G.I., Higgins, P.G., Tyrrell, D.A.J., and Andries, K. (1990) *Antiviral Chem. Chemother.* 1, 279–283.

Bevilacqua, F., Marcello, A., Toni, M., et al. (1991) *J. Acquired Immune Deficiency Syndromes* 4, 967–969.

Castrillo, J.L., Berghe, D.V., and Carrasco, L. (1986) *Virology* 152, 219–227.

Chapman, M.S., Minor, I., and Rossmann, M.G. (1991) *J. Mol. Biol.* 217, 455–463.

Coffin, J.M. (1990) in *Virology* (Fields, B.N., and Knipe, D.M., eds.), pp. 1437–1500, Raven Press, New York.

Conti, C., Orsi, N., and Stein, M.L. (1988) *Antiviral Res.* 10, 117–127.

Couch, R.B. (1990a) in *Antiviral Agents and Viral Diseases of Man* (Galasso, G.J., Whitley, R.J., and Merigan, T.C., eds.), pp. 327–372, Raven Press, New York.

Couch, R.B. (1990b) in *Virology* (Fields, B.N., and Knipe, D.M., eds.), pp. 607–629, Raven Press, New York.

Dasmahapatra, B., Rozhon, E.J., Hart, A.M., et al. (1991) *Virus Res.* 20, 237–249.

Davis, G.L., Balart, L.A., Schiff, E.R., et al. (1989) *N. Engl. J. Med.* 321, 1501–1506.

Di Bisceglie, A.M., Martin, P., Kassianides, C., et al. (1989) *N. Engl. J. Med.* 321, 1506–1510.

Eggers, H.J. (1977) *Virology* 78, 241–252.

Enders, J.F., Weller, T.H., and Robbins, F.C. (1949) *Science* 109, 85–87.

Fenner, F. (1990) in *Virology* (Fields, B.N., and Knipe, D.M., eds.), pp. 2113–2133, Raven Press, New York.

Fischl, M.A., Richman, D.D., Greico, M.H. (1987) *N. Engl. J. Med.* 317, 185–191.

Fox, M.P., McKinlay, M.A., Diana, G.D., and Dutko, F.J. (1991) *Antimicrob. Agents Chemother.* 35, 1040–1047.

Fox, M.P., Otto, M.J., and McKinlay, M.A. (1986) *Antimicrob. Agents Chemother.* 30, 110–116.

Gardner, M.B., and Luciw, P.A. (1989) *FASEB J.* 3, 2593–2606.

Hanecak, R., Semler, B., Ariga, H., Anderson, C.W., and Wimmer, E. (1984) *Cell* 37, 1063–1073.

Hayashi, K., Niwayama, S., Hayashi, T., et al. (1988) *Antiviral Res.* 9, 345–354.

Hirsch, M.S., and Curran, J. (1990) in *Virology* (Fields, B.N., and Knipe, D.M., eds.), pp. 1545–1570, Raven Press, New York.

Horwitz, M.S. (1990) in *Virology* (Fields, B.N., and Knipe, D.M., eds.), pp. 1679–1720, Raven Press, New York.

Hudson, J.B. (1990a) *Antiviral Compounds from Plants,* pp. 83–99, CRC Press, Boca Raton, FL.

Hudson, J.B. (1990b) *Antiviral Compounds from Plants,* pp. 119–132, CRC Press, Boca Raton, FL.

Ishitsuka, H., Ninomiya, Y.T., Ohsawa, C., Fujiu, M., and Suhara, Y. (1982) *Antimicrob. Agents Chemother.* 22, 617–621.

Ito, T., Ohba, K., Koyama, M., Sezaki, M., et al. (1984) *J. Antibiotics* 27, 931–934.

Larder, B.A., Darby, G., and Richman, D.D. (1989) *Science* 243, 1731–1734.

Lasky, L.A., Groopman, T.E., Fennie, C.W., et al. (1986) *Science* 233, 209–213.

Lifson, J.D., Feinberg, M.B., Reyes, G.R., et al. (1986) *Nature* 323, 725–728.

Mandel, B. (1976) *Virology* 69, 500–510.

Marrero, M., Alvarez, M., Millan, J.C., et al. (1991) *Acta Virol.* 35, 86–89.

McCune, J.M., Namikawa, R., Shih, C.-C., Rabin, L., and Kaneshima, H. (1990) *Science* 244, 564–566.

McKinlay, M.A., and Rossmann, M.G. (1989) *Annu. Rev. Pharmacol. Toxicol.* 29, 111–122.

McLaren, C., Corey, L., Dekket, C., and Barry, D.W. (1983) *J. Infect. Dis.* 148, 868–880.

McSharry, J.J., Caliguiri, L.A., and Eggers, H.J. (1979) *Virology* 97, 307–315.

Merck Index, 11th ed. (1989) (Budavari, S., ed.), pp. 60–61, Merck & Co., Rahway, NJ.

Mertz, G.J. (1990) in *Antiviral Agents and Viral Diseases of Man* (Galasso, G.J., Whitley, R.J., and Merigan, T.C., eds.), pp. 265–300, Raven Press, New York.

Monto, A.S., Bryan, E.R., and Ohmit, S. (1987) *J. Infect. Dis.* 156, 43–49.

Murphy, B.A., and Chanock, R.M. (1990) *Virology* (Fields, B.N., and Knipe, D.M., eds.), pp. 469–502, Raven Press, New York.

Navia, M.A., Fitzgerald, P.M.D., McKeever, B.M., et al. (1989) *Nature* 337, 615–620.

Nicklin, M.J.H., Harris, K.S., Pallai, P.V., and Wimmer, E. (1988) *J. Virol.* 62, 4586–4593.

Nicklin, M.J.H., Krausslich, H.G., Toyoda, H., Dunn, J.J., and Wimmer, E. (1987) *Proc. Natl. Acad. Sci. USA* 84, 4001–4006.

Ninomiya, Y., Aoyama, M., Umeda, I., Suhara, Y., and Ishitsuka, H. (1985) *Antimicrob. Agents Chemother.* 27, 595–599.

Ninomiya, Y., Shimma, N., and Ishitsuka, H. (1990) *Antiviral Res.* 13, 61–74.

Notka, M.A., Reichman, R.C., and Pollard, R.B. (1990) *Antiviral Agents and Viral Diseases of Man* (Galasso, G.J., Whitley, R.J., and Merigan, T.C., eds.), pp. 49–85, Raven Press, New York.

Otto, M.J., Fox, M.P., Fancher, M.J., et al. (1985) *Antimicrob. Agents Chemother.* 27, 883–886.

Palmenberg, A.C., Pallansch, M.A., and Rueckert, R.R. (1979) *J. Virol.* 32, 770–778.

Rader, R.A. (1991a) *Antiviral Agents Bull.* 4(1), 2–3.

Rader, R.A. (1991b) *Antiviral Agents Bull.* 4(2), 37–38.

Reichman, R.C., Oakes, D., Bonnez, W., et al. (1988) *Ann. Intern. Med.* 108, 675–679.

Richman, D.D. (1990) *Am. J. Med.* 88(Suppl. 5B), 8S–10S.

Roizman, B., and Sears, A.E. (1990) in *Virology* (Fields, B.N., and Knipe, D.M., eds.), pp. 1795–1841, Raven Press, New York.

Rombaut, B., Brioen, P., and Boeye, A. (1990) *J. Gen. Virol.* 71, 1081–1086.

Rombaut, B., Vrijsen, R., and Boeye, A. (1985) *Antiviral Res.* (Suppl. 1), pp. 67–73.

Rozhon, E.J., Albin, R., and Schwartz, J. (1991) *Soc. Ind. Microbiol. News* 41(2), 55–62.

Rueckert, R.R. (1985) in *Virology* (Fields, B.N., and Knipe, D.M., eds.), pp. 705–738, Raven Press, New York.

Sabin, A.B., and Olitsky, P.K. (1936) *Proc. Soc. Exp. Biol. Med.* 34, 357–359.

Sattentau, Q.J., and Weiss, R.A. (1988) *Cell* 52, 631–633.

Shih, C.-C., Kaneshima, H., Rabin, L., et al. (1991) *J. Infect. Dis.* 163, 625–627.

Smith, T.J., Kremer, M.J., Luo, M., et al. (1986) *Science* 233, 1286–1293.

Superti, F., Seganti, L., Orsi, N., et al. (1989) *Antiviral Res.* 11, 247–254.

Tisdale, M., and Selway, J.W.T. (1984) *Antimicrob. Agents Chemother.* 14, 97–105.

Vrijsen, R., Everaert, L., Van Hoof, L.M., et al. (1987) *Antiviral Res.* 7, 35–42.

Weis, W., Brown, J.H., Cusack, S., et al. (1988) *Nature* 333, 426–431.

Wilson, I.A., Skehel, J.J., and Wiley, D.C. (1981) *Nature* 289, 366–373.

Wlodawer, A., Miller, M., Jaskolski, M., et al. (1989) *Science* 245, 616–621.

Wong-Staal, F. (1990) in *Virology* (Fields, B.N., and Knipe, D.M., eds.), pp. 1529–1543, Raven Press, New York.

Ypma-Wong, M.F., Dewalt, P.G., Johnson, V.H., Lamb, J.G., and Semler, B. (1988) *Virology* 166, 265–270.

Mechanism-Based Screens in the Discovery of Chemotherapeutic Antibacterials

Sunil Kadam

Modern antibiotic development began with the serendipitous discovery of penicillin in the late 1920s. The observation that a contaminating *Penicillium* produced a substance that was lytic to staphylococci eventually progressed to the development of manufacturing methods for this compound in the early 1940s. The clinical efficacy of penicillin immediately triggered a search for new antibacterials from natural sources, and many important compounds that even today contribute to our arsenal against the constant threat of bacterial pathogens were discovered by the late 1950s (Table 8–1). In parallel, the significant chemical modification effort initiated in the 1960s has resulted in tens of thousands of chemically stable compounds with improved microbiologic, pharmacokinetic, and therapeutic properties. In excess of 30,000 novel, low-molecular weight microbial metabolites with antibacterial properties have been reported since Fleming's historic discovery. Yet the rate of discovery of chemically novel compounds has declined in recent times. Since 1981, only 30 to 50 new antimicrobials have appeared each year in the *Journal of An-*

TABLE 8–1 Antibiotics Discovered by the Late 1950s

Antimicrobial Agent	Year	Reference
Streptomycin	1944	Schats and Waksman 1944
Chloramphenicol	1947	Ehrlich et al. 1947
Chlortetracycline	1948	Duggar 1948
Neomycin	1949	Waksman and Lechevalier 1949
Oxytetracycline	1950	Finlay et al. 1950
Erythromycin	1952	McGuire et al. 1953
Leucomycin	1953	Hata et al. 1953
Oleandomycin	1955	Sobin 1955
Cycloserine	1955	Harned et al. 1955
Cephalosporin	1955	Newton and Abraham 1955
Kanamycin	1957	Umezawa et al. 1957
Rifamycin	1959	Sensi et al. 1959

tibiotics (Figure 8–1). Whereas the majority of antibiotics were initially discovered because of their ability to inhibit the growth of Gram-positive bacteria, the development of supersensitive strains and selective screening organisms provided stepping stones to the application of molecular biology techniques to target-directed antibacterial screening. In addition to the preeminent human use, antibiotics are now sought for agricultural, veterinary, and marine (aqua-culture) applications. The current annual antibiotic world market is estimated to be several billion dollars (U.S.), and it is not surprising that the majority of screening activity is concentrated in the pharmaceutical

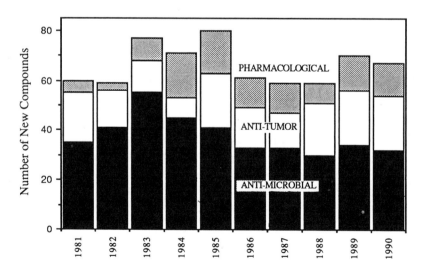

FIGURE 8–1 New microbial metabolites reported in the *Journal of Antibiotics* from 1981–1990.

industry. In this chapter we describe some of the newer technologies available for the development of target specific "mechanistic assays" in the search for new antibacterials.

8.1 HISTORICAL PERSPECTIVE

8.1.1 The Discovery of Antibacterial Agents

In the early days, screening of secondary metabolites from various microorganisms was based primarily on antimicrobial assays such as agar diffusion and broth dilution. The rate of discovery of new antimicrobial metabolites did not change for the two decades after the mid 1940s. Subsequently, high throughput microbiological screens initiated in the pharmaceutical industry using supersensitive cultures and the introduction of automation in the analysis of results produced a dramatic increase in the number of new compounds discovered from 1970–1978 (Berdy 1974). Perlman (1977) reported a rate of 180 to 340 new antibiotics per year during this period. However, the success of this traditional approach was often limited by the chemical source and the reliance on a single pair of isogenic-sensitive and -resistant strains. It is now difficult to find novel compounds using this classical approach unless specific antimicrobial activity is sought, such as *Pseudomonas* or *Clostridium* actives, or where the source of secondary metabolite itself is novel. The decline in the discovery of useful compounds has resulted in a change in strategy. In the 1970s, the primary emphasis was to diversify culture collections, introduce new techniques to isolate and grow potential antibiotic-producing microorganisms, and to examine genera that had received little attention (Conover 1971). In contrast, screening in the 1990s is more target directed, using robotic sample manipulation, computerized data management and analysis using "expert systems," and dereplication of actives using computerized searches of structural data bases.

8.1.2 Screening Technology Today

A successful broad-based screening strategy typically involves an interdisciplinary, multistep process combining at least five major areas; microbiology, fermentation, chemistry, molecular biology, and data management. As a consequence, a screening program has to be collaborative, automated, and technologically flexible in order to accept the continuous influx of new assays and methodology (Nolan and Cross 1988). However, the degree of success of a screening program is not solely dependent on the number of screens or size of its infrastructure, but also on the pool of expertise dedicated to the design of specific screens, the source of the material to be analyzed, and the experience of the isolation chemist. Screening for antibacterials has developed over the last three decades from whole-cell, broad spectrum antibiotic screens to the innovative, target-directed or mechanism-based screens increasingly used today. Among the major contributors to this area of screen development is our understanding of bacterial physiology, advances in molecular biology, and increased sophistication in miniaturization and detection of subtle changes in the bacterial environment. According to Imada and Okazaki (1987), the critical S's in the success of a screening program are simplicity, sensitivity, specificity, software, and sensibility. Many excellent articles have appeared on the development of a screening strategy, methodology, and logistics (Omura 1986; Nolan and Cross 1988; Okami and Hotta 1988; Berdy 1974; Nisbet and Porter 1989). This chapter highlights emerging antibacterial targets and screens that have appeared from about 1985, primarily mechanism-based assays and their role in exploiting potential targets for the development of novel antibacterials.

8.2   ANTIBACTERIALS AND EMERGING TARGETS

8.2.1   Cell Wall Targeting Antibacterials

The bacterial cell wall provides a unique and selective target that is essential for bacterial growth and absent in mammalian cells. It is not surprising that the majority of antibacterials target this structure. β-lactams, glycopeptides, and lipopeptides are among the class of cell wall synthesis inhibitors. Several excellent reviews are available on the discovery and mechanism of action of these compounds (Fukui and Beppu 1986; Imanaka 1985; Yao and Mahoney 1989; Imada and Okazaki 1987; Brown 1987). β-lactams inhibit the D-ala-D-ala carboxy and transpeptidases that are responsible for the cross-linking peptidoglycan strands during the synthesis of bacterial cell walls. Crystallographic, genetic, and kinetic studies show that the penicillin-sensitive transpeptidase in bacteria belong to a family of enzymes that have essentially conserved catalytic domains with subtle changes in amino acids, which can determine its function as a transpeptidase, DD-carboxypeptidase, or β-lactamase. At the molecular level, the β-lactam ring is thought to be a transition state analog in the enzymatic cleavage of acyl-D-ala-D-ala (Kelly et al. 1989). This structural-analog hypothesis predicted the existence of penicillin-binding proteins (PBPs) in penicillin-sensitive targets. When PBPs were verified experimentally, they became the object of intense research as a tool to study the β-lactam target. In the early 1970s, another phase in β-lactam discovery began when the widespread use of these compounds brought about the emergence of resistant bacteria that could cleave the β-lactam ring by secreting β-lactamase. A systematic search for β-lactamase inhibitors resulted in the discovery of new β-lactams such as the carbapenems and clavams. Later the discovery of bacterially derived monobactams, contrary to existing dogma that β-lactams were of fungal and actinomycete origin (Imada et al. 1981), regenerated interest in β-lactam screening.

Unlike the penicillins, glycopeptide antibiotics, such as vancomycin, inhibit cell growth by specifically binding to the D-ala-D-ala portion of the lipid-phosphodisaccharide-pentapeptide and prevent cross-linking of adjacent nascent peptidoglycan chains (Harris et al. 1985; Williams and Waltho 1988). Since this step occurs prior to the reaction inhibited by penicillin there is no cross-resistance between the two types of antibiotics. Lipopeptides differ from glycopeptides in that they consist of a single peptide core coupled to a variety of fatty acid moieties. Daptomycin (LY146032), derived from the core in A 21978C is a bactericidal lipopeptide with a decanoyl fatty acid residue. It is slightly more potent than vancomycin and may also have a different mode of action; namely, the inhibition of lipoteichoic acid synthesis (Canepari et al. 1990; Lakey et al. 1989). Despite our understanding of the bacterial cell wall, events that cause bacteriolysis and cell death after initial inhibition of cell-wall assembly are not clearly understood.

The minimal β-lactam structure containing the bicyclic ring system as it does in penicillin and cephalosporin, Figures 8–2A and 8–2B consists of a 2-azetidinone ring fused with thiazolidine or dihydrothiazine, respectively, and a nonplaner β-lactam nitrogen with high chemical activity thought to be essential for antibiotic action (Johnson et al. 1949). The discovery of nocardicins (Figure 8–2C) and other monobactams, such as sulfazecin (Figure 8–2D), however, shifted the focus from the bicyclic system to the simple 2-azetidinone ring and an ionizable acidic function close to the amide nitrogen. In addition, most β-lactams require a side-chain containing a hydrogen-bonding group. To initiate the inhibitory process, a β-lactam must achieve acylation of the target active-site amino-acid residues, essentially acting as a suicide substrate and rendering the leaving group unable to dissociate.

Screens used for the discovery of new β-lactams represent a unique example of the development of mechanism-based screening as information concerning the mode of action and resistance to these compounds became available. Initial screens relied on the selective activity of β-lactams on hypersensitive bacteria and the induction of morphological changes in the test organism. These screens were simple, high throughput, and tedious yet very successful. Aoki et al. (1976) discovered the first monocyclic β-lactam, nocardicin, whereas Kahan et al. (1979) discovered the carbapenem thienamycin, using β-lactam hypersensitive mutants. A similar approach led to the discovery of sulfazecin and isosulfazecin (Imada et al. 1981). Later, the discovery of β-lactamases led to screens based on sensitivity of the test substance to a β-lactamase producer and inhibition or induction of β-lactamase. Sykes et al. (1981) reported the discovery of *N*-sulfonated monocyclic β-lactams using the induction of β-lactamase in *Bacillus licheniformis*. The inhibition of *Klebsiella aerogenes* β-lactamase in the "KAG assay" (Butterworth et al. 1979) led to the discovery of clavulanic acid (Figure 8–2F), a new type of β-lactam with a clavam structure (Okami and Hotta 1988; Reading and Cole 1977). Clavulanic acid is used clinically as a β-lactamase inhibitor in Augmentin® and Timentin®. The KAG assay was able to detect nanogram levels of the enzyme inhibitor when the β-lactamase-producing culture was seeded on ampicillin-containing plates, causing a zone of inhibition by sparing the ampicillin in that area.

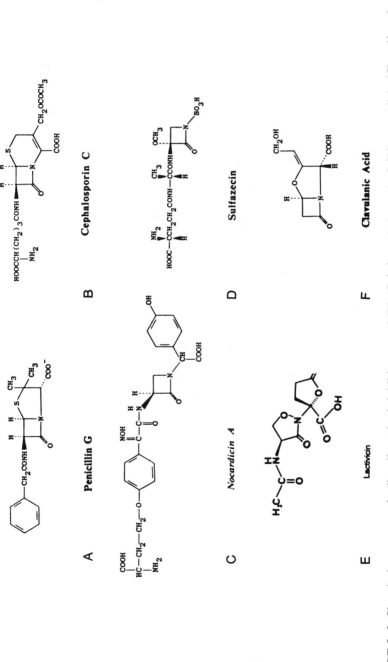

FIGURE 8–2 Chemical structures of cell-wall synthesis inhibitors, penicillin G (A), cephalosporin C (B), nocardicin (C), sulfazecin (D), lactivicin (E), and clavulanic acid (F).

The autolysis assay developed by Patel (1985) is another example of a simple, rapid, high throughput assay designed to detect cell-membrane-active antibiotics. A zone of lysis is triggered by compounds that act on the bacterial cell wall. Ulitzur (1986) improved on the principle by cloning the luminescence system from the β-lactam-resistant *Photobacterium leiognathi* into the sensitive *Bacillus subtilis*. Lysis in bacillus was accompanied by a decrease in luminescence that could be detected with a luminometer upon induction with dodecyl aldehyde. While many of the screening assays described for β-lactams are not directly based on the mechanism of action of these compounds, binding to PBPs and inhibition of DD-carboxypeptidase (DDCase) are examples of such a direct approach. Frere et al. (1980) used the soluble DDCase from *Actinomadura* R39 and monitored UDP-tetra and tri-peptide reaction products by high pressure liquid chromatography (Imada et al. 1981). The use of a fluorogenic substrate by Schindler et al. (1986) further improved the sensitivity and the screening capacity of this assay. Perhaps the discovery of lactivicin (Figure 8–2E), a PBP binding compound discovered at Takeda Laboratories in a screen using hypersensitive mutants, is most notable because lactivicin lacks a β-lactam nucleus (Harada et al. 1986; Okazaki 1987). Although it is weakly active, lactivicin derivatives are synthesized intensely.

The appearance of methicillin-resistant *Staphylococcus aureus,* which often express concomitant aminoglycoside and macrolide resistance has renewed interest in glycopeptide antibiotics related to vancomycin (Figure 8–3A). An elegant screening assay was described by Omura et al. (1979) and Reading and Cole (1977) based on the differential sensitivity of *B. subtilis* and cell wall lacking *Mycoplasma* to microbial broths. Culture broths active against *B. subtilis* but lacking any activity against *Mycoplasma* were tested for their effect on the synthesis of peptidoglycan and protein. Cell-wall inhibitors only affected the synthesis of peptidoglycan. The differential assay resulted in the discovery of the glycopeptides azureomycins and izupeptin, as well as previously known compounds such as D-cycloserine and ristocetin (Figure 8–3B). Rake et al. (1986) developed a targeted screen for glycopeptides based on specific antagonism of glycopeptide activity by the tripeptide diacetyl-L-lysyl-D-alanyl-D-alanine. This functionally directed assay yielded aridicins, synmonicins, and parvocidins (Sitrin et al. 1985; Christensen et al. 1987). Others (Cassani 1989; Goldstein et al. 1987) have adapted the receptor-based assay to solid phase affinity chromatography with bovine serum albumin (BSA) or agarose-ξ-aminocaproyl-D-ala-D-ala. At Abbott Laboratories, an Affi-gel10 bound D-ala-D-ala was used to recover several glycopeptides, including teichomycin-A2, actaplanins, and ristocetins (J. Hochlowski, personal communication). Later, the BSA-tripeptide was conjugated to alkaline phosphatase and inhibition of vancomycin binding by other glycopeptides was monitored by measuring the decrease in alkaline phosphatase activity (Yao and Mahoney 1984). O'Sullivan et al. (1988) established a dual agar plate assay with *B. subtilis* in which one plate was supplemented with enriched *Staphylococcus* peptidoglycan. They followed compounds that had smaller

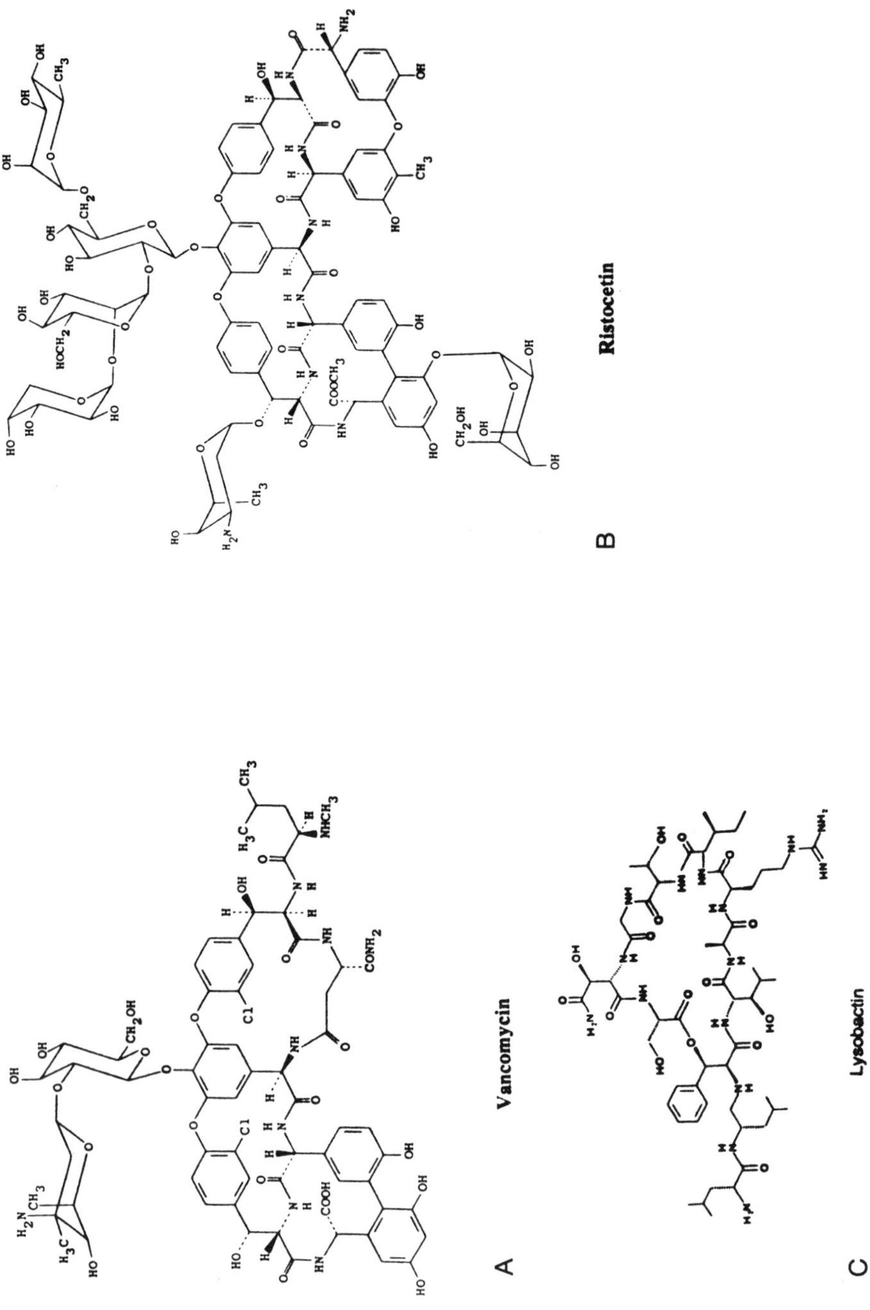

FIGURE 8–3 Chemical structures of the glycopeptides vancomycin (A), ristocetin (B), and lysobactin (C).

zones in the peptidoglycan-containing plates in an attempt to isolate peptides that antagonized binding sites other than L-lysyl-D-ala-D-ala. Two novel peptide-lactone antibiotics, lysobactin (Figure 8–3C) from the gliding bacteria *Lysobacter* (Bonner et al. 1988) and janthinosins from *Janthinobacterium lividum* (O'Sullivan et al. 1990) were isolated using this assay. These compounds are structurally similar to katanosins and are thought to inhibit a step before the formation of UDP-*N*-acetylglucosamine. Other screens that detected cell-wall inhibitors include a differential test using methicillin-resistant *Staphylococcus aureus* (MRSA) and its sensitive L-form, which was used to isolate ramoplanin (Cavalleri et al. 1984; Parenti et al. 1990). Vicario et al. (1987) used a simple assay to detect effectors of alanine racemase and/or D-ala-D-ala ligase.

Other targets in the bacterial cell-wall for which few or no inhibitors are known include transglycolase, an important cell-wall synthesis enzyme and the target of moenomycin (Franklin and Snow 1989) and bulgecin, a bacterial compound coproduced with sulfazecin that causes morphological defects such as bulge formation with β-lactams but has no antibacterial activity of its own (Imada et al. 1982). Studies with mutants blocked either in lateral wall or septum formation (*rodA, pbpA, fts*) and the differential binding of β-lactams to PBPs also suggest additional targets. Several genes involved in the initiation of septum formation have already been isolated, the product of *FtsZ*, for example, is essential for cross-wall formation and is known to be ubiquitous in bacteria (Sitrin et al. 1985). Additionally, cell division, septum formation, and DNA replication appear to be intimately connected with the SOS system in bacteria. Agents that interfere with cross-wall formation are therefore likely to be bactericidal and have a broad spectrum of activity. Consider the product of the *SulA* gene, which is a septum inhibitor and a substrate of the Lon protease, induced during SOS. Studies have shown that upon SOS induction in *lon⁻ sulA⁺* cells, death occurs due to extensive filamentation (Huisman et al. 1984). Compounds capable of *SulA* induction in the absence of SOS should therefore commit a cell to death.

8.2.2 Cell-Membrane Targeting Antibacterials

The outer-membrane is a unique structure in Gram-negative bacteria that is required for growth and virulence. The spread of antibiotic resistance in this important group of pathogens has made the biosynthesis of membrane lipopolysaccharide (LPS) an attractive target for drug discovery. LPS in the outer-membrane is composed of Lipid A, core, and an O-antigen side-chain. Goldman et al. (1987) and others (Hammond et al. 1987) have reported specific inhibitors of CTP:CMP-3-deoxy-*manno*-octulosonate (CMP-KDO) synthetase, an enzyme that activates KDO for incorporation into LPS. Additionally, these inhibitors may also inhibit KDO-containing capsular polysaccharide, which has been implicated in virulence (Jann 1983). The discovery that the least abundant β-pyranose form of KDO was the preferred substrate

Erythromycin Derivatives (Macrolides)

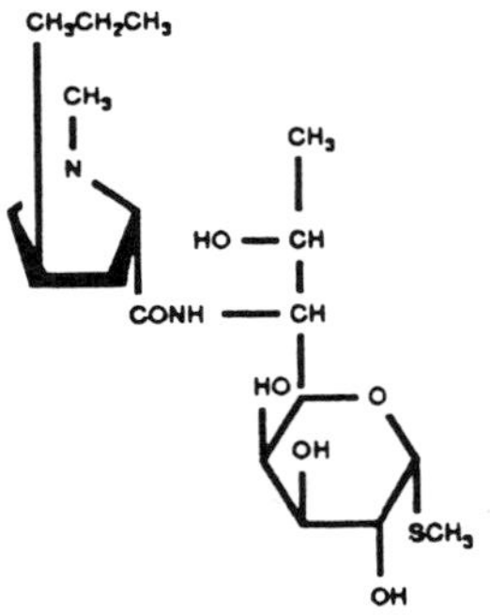

D Lincomycin (Lincosamides)

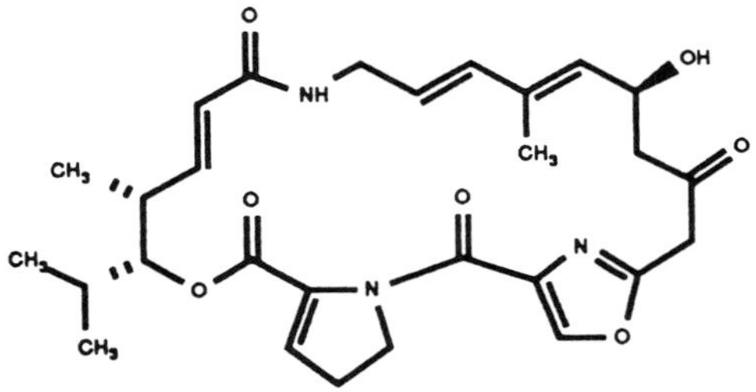

E Virginiamycin M_1 (Strepotogramins)

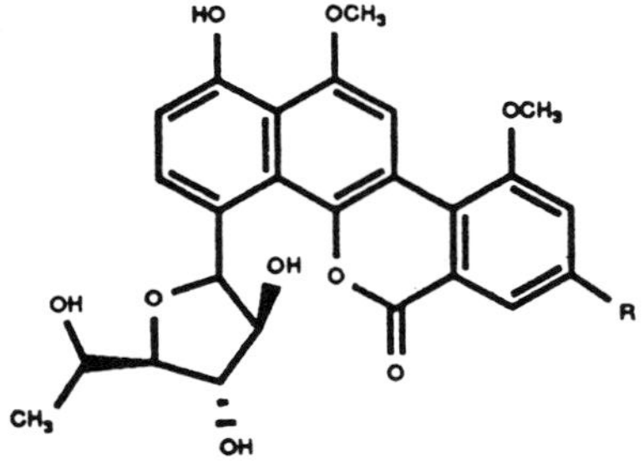

F

Gilvocarcin	R
V	$CH{=}CH_2$
M	CH_3

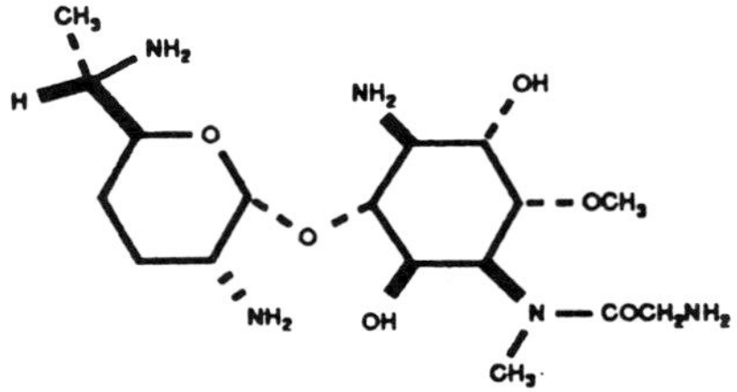

G Fortimicin A

FIGURE 8–4 Chemical structures of 2-deoxy-KDO derivatives (A), erythromycin derivatives (B,C), lincomycin (D), virginiamycin (E), and gilvocarcins (F), and fortimicin A (G).

for incorporation into KDO led to the development of a 2-deoxy derivative of KDO (Figure 8–4A shown on pp. 256–257) locked in the β-pyranose configuration (Ki = 12 μM), where Ki = inhibition constant. An ala-ala dipeptide (see Figure 8–4A) attached to the carboxy terminus of this compound produced an inhibitor that could be tested in whole cells. The addition of an inhibitor to Gram-negative bacteria immediately stopped LPS synthesis, as it does in temperature sensitive mutants of CMP-KDO, which caused the accumulation of Lipid A precursors and bacteriostatis (Goldman et al. 1987; Kadam et al. 1989). Translocation of the precursor to the outer membrane caused increased sensitivity to host defenses and to antibiotics that are normally unable to penetrate the cell (Kadam et al. 1989; Goldman et al. 1987). In addition to KDO, enzymes involved in the synthesis of the LPS core may also present a potential target. Studies using the genes cloned from *Salmonella typhimurium* LT2 (Kadam et al. 1985) suggest that deletions in certain *rfa* genes can result in abnormal cell morphology such as lack of flagella and fimbriae (Parker et al. 1992). The biosynthesis of Lipid A as well as the late stages of addition of the O-antigen to core also appear to be potential targets, as suggested by Sutcliff (1988). Cells lacking a complete O-antigen are known to be lysed by certain phages and may provide a useful basis for a high throughput screen (Hudson et al. 1978).

8.2.3 Protein Synthesis Inhibitors

Protein synthesis inhibitors represent another major group of clinically useful antibacterials, such as erythromycin, tetracycline, chloramphenicol, and aminoglycosides. They selectively interact with the 70S bacterial ribosome and spare the 80S eukaryotic ribosome particle. Macrolide, lincosamide (Figure 8–4D), and streptogramins (Figure 8–4E) (MLS) antibiotics represent three classes of structurally diverse protein biosynthesis inhibitors used clinically against Gram-positive bacteria, chlamydia, mycoplasmas, anaerobes, *Hemophilus, Legionella,* and *Neisseria,* while other macrolides such as tylosin and virginimycin are in veterinary use. Despite continuous interest, such as their effectiveness in the treatment of *Mycobacterium avium* in immune compromised patients and the relatively low level of general clinical resistance, there have been few reports of screening for macrolide protein synthesis inhibitors.

A general assay adopted for protein synthesis inhibition is the incorporation of a radiolabeled amino acid into trichloroacetic acid (TCA) precipitable protein. This assay has been developed using natural mRNA or poly(A) as a template. A major drawback of this assay is the simultaneous detection of nucleic acid inhibitors due to their effect on protein synthesis; hence, the need for a specific secondary assay. Naveh et al. (1984) and Ulitzur (1986) described a sensitive assay for DNA interacting agents as well as protein synthesis inhibitors based on the induction of the luciferase system by acridine dyes in dark variants of *Photobacterium leiognathi*. Yao and Mahoney (1989) developed a direct competitive enzyme-linked immunosorbent assay (ELISA) using a rabbit polyclonal antibody to 23-amino-*O*-mycaminosyl-tolonolide conjugated to alkaline phosphatase. They detected several known compounds in fermentation broths but no new compounds resulted from this assay. However, the antibody approach could be readily applied to detect other antibiotics.

The induction of simultaneous resistance to MLS was exploited to develop a plate assay by Ganguli (1989) in which discs of sublethal concentration of erythromycin were placed adjacent to fermentation broths to detect distorted zones from an MLS compound on an overlay of inducible *Staphylococcus aureus*. Grividomycins and swalpamycin; a neutral 16-membered macrolide with a novel aglycone was discovered by using this assay (Franco et al. 1987; Chatterjee et al. 1987). The plate assay was further modified when induction was described to be caused by translational attenuation in the leader region of the *Erm* gene (Dubnau 1984). The product of the *Erm* gene (methylase) methylates adenine 2058 in the 23S rRNA, conferring simultaneous resistance to MLS antibiotics (Weisblum 1983). A *LacZ* fusion in the *ErmC* gene produced a construct capable of β-galactosidase induction with sublethal concentrations of erythromycin (Gryczan et al. 1984). The modified assay was used to characterize a new series of erythromycin derivatives (Figures 8–4B and 8–4C) that were active against both inducible and constitutively resistant *S. aureus* (Goldman and Kadam 1987; Fernandes et al. 1989). These compounds were also shown to bind to methylated ribosomes (Goldman and Kadam 1987). A similar test was developed to use the thiostrepton resistance determinant of *Streptomyces azureus*, which also confers resistance by methylating an adenine residue in the 23S RNA (Thompson et al. 1982). Neoberninomycin produced by *Micrococcus luteus* was detected in this assay (Biskupiak et al. 1988). The nature of the leader sequence in *ErmC* allowed a way to confirm translational attenuation by a different mechanism; namely, the depletion of charged isoleucine residues caused by pseudomonic acid, a potent inhibitor of isoleucyl-tRNA synthetase (Kadam 1989). As shown in Figure 8–5, the presence of pseudomonic acid adjacent to erythromycin allows growth of erythromycin-sensitive bacteria between the two antibiotic discs. This was demonstrated to be due to the ability of pseudomonic acid, a non-macrolide antibiotic used to induce *ErmC* methylase (Figure 8–6), which in turn generates methylated ribosomes, unable to bind erythromycin and inhibit protein synthesis and cell growth. The leader peptide in *ErmC* has three isoleucines that possibly cause ribosome stalling in several places, while the leader in *ErmD* contains no isoleucine residues. In principle, the assay could be used to detect other tRNA synthetase inhibitors by simply adjusting the amino acid residues in the leader region. Another novel class of protein synthesis inhibitors, the oxazolidinones, was described by Barry (1988). These compounds are thought to bind to ribosomes in a way that permits elongation and termination, but prevents subsequent initiation from natural mRNA.

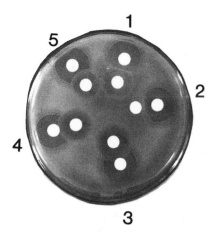

FIGURE 8–5 A plate assay for methylase induction. Filter discs with 10 and 2 μg of erythromycin (1 and 2, respectively) and 10, 5, and 2 μg of pseudomonic acid (3, 4, and 5, respectively) were placed in the inner circle. All outer discs had niddamycin (10 μg/ml). The Luria agar plate contains *Bacillus subtilis* BD170/pE194 grown in Luria broth.

The aminoglycoside antibiotics include a group of structurally related polycationic compounds composed of amino sugars connected by glycosidic linkages. They act on the 30S ribosomal subunit causing inhibition of protein synthesis and subsequent cell lysis. Although their mode of action is not clear, translational misreading appears to be the prevailing consensus opinion. Over the last 30 years, very few new techniques have appeared for the discovery of aminoglycosides. Numata et al. (1986) has described the use of a hyper-

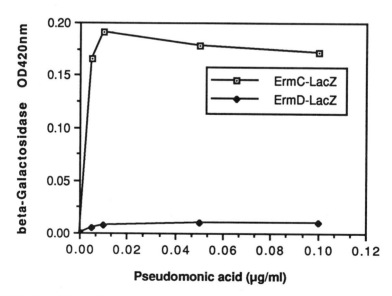

FIGURE 8–6 Effect of pseudomonic acid on β-galactosidase induction in *Bacillus subtilis* strains BD170/pBD246 [*ermC-lacZ*] and BD170/pBD247 [*ermD-lacZ*], grown in amino acid-supplemented Spizizen minimal medium.

sensitive klebsiella to test 20,000 soil isolates. Seven of the ten positives were known aminoglycosides and one produced a novel amino sugar antibiotic (Tsuno et al. 1986). Yao and Mahoney (1984) again used a gentamicin-derived antibody to develop an ELISA to screen fermentation broths for the presence of compounds that could compete with gentamicin. A novel screening procedure was described by Saitoh et al. (1988) in which compounds were selected for the ability to induce the growth of a streptomycin-dependent strain of *Escherichia coli*. They discovered a new nontoxic, broad-spectrum antibiotic, boholmycin, using this assay. Hotta et al. (1983) devised an interesting method in which actinomyces isolated on an aminoglycoside-containing medium were characterized by their pattern of resistance to several aminoglycosides because aminoglycoside producers were previously found to possess individual resistance patterns. This step led to an enrichment for aminoglycoside producers. Researchers at Kyowa and Abbott Laboratories used a specific panel of organisms with a known mechanism of aminoglycoside inactivation to dereplicate for known compounds (Jackson 1988). This procedure led to the isolation of three new compounds; sagamicin, seldomycin, and fortimicin (see Figure 8–4G) (Nara et al. 1977). A similar screen also produced lysinomycin (Kurath et al. 1984).

8.2.4 DNA-Targeting Antibacterials

A significant contribution to our antimicrobial arsenal is the development of quinolone antibacterials that target the DNA gyrase A subunit. Microbial products such as novobiocin and coumermycin also inhibit DNA gyrase by specifically interacting with the gyrase B subunit (Drlica 1989). Based on the mode of action, DNA synthesis inhibitors can be divided into the following categories: (1) inhibitors of DNA topoisomerases (relaxation and supercoiling); (2) DNA binding and/or intercalating agents; (3) inhibitors of DNA synthesis; and (4) inhibitors of RNA synthesis. Numerous assays have been developed to measure the activity of topoisomerases (reviewed in Barrett et al. 1990) and they all can be divided into two groups, catalytic and noncatalytic. Many of the in vitro catalytic assays, such as inhibition of DNA relaxation, supercoiling, unknotting, and the noncatalytic DNA cleavage assays, have been adapted to natural product screening and discussed elsewhere (Barrett et al. 1990). Of the mechanism-based assays used in this area, many utilize the induction of an SOS-related gene to detect DNA damage or inhibition of DNA synthesis. The SOS chromotest in *E. coli* (Janz et al. 1988) based on the induction of an *SfiA-lacZ* fusion and a similar test in *B. subtilis* based on induction of a *Din-LacZ* fusion (Gryczan et al. 1984) are examples of sensitive, high throughput screening assays used to detect DNA damage/ synthesis inhibitors in microbial extracts. The latter assay was used to detect and monitor the isolation of gilvocarcins V and M (see Figure 8–4F). These compounds were found to cause DNA damage by specifically interacting with the gyrase A subunit. A *Bacillus megaterium, gyrA2* mutant that is resistant

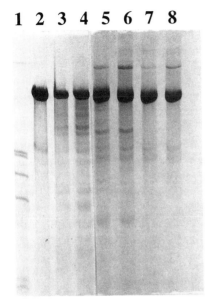

FIGURE 8–7 Autoradiogram of an agarose gel showing cleavage of linearized, [^{35}S] end-labeled pBR322 DNA by DNA gyrase in bacterial extracts of *Bacillus megaterium* (lanes 2–6) and *B. megaterium gyrA2* (lanes 7 and 8). Lane 1, molecular weight markers; lane 2, no compound; lanes 3 and 4, 2 and 5 μg of ciprofloxacin, respectively; lanes 6 and 8 and 5 and 7 contain reactions with 2 and 5 μg of gilvocarcin V, respectively.

to fluoroquinolones was also resistant to gilvocarcin V. Additionally, whole cell extract isolated from the mutant *B. megaterium* showed no DNA cleavage in the presence of gilvocarcin V (Figure 8–7). Gilvocarcins are therefore the first microbial products known to inhibit DNA synthesis by interacting with the gyrase A subunit. The success of synthetic fluoroquinolones is expected to create further interest in screening new DNA targeting compounds.

Omura et al. (1985) developed a method to isolate compounds that inhibit folate metabolism from fermentation broths using *Enterococcus faecium* that is deficient in enzymes required for folate metabolism. Samples active against the organism in limiting pteroate but inactive in thymidine-supplemented media were selected. Diazaquinomycin A and B were discovered in this assay.

8.3 SUMMARY

Numerous assays have been developed over the last 40 years for the detection of novel antibacterial metabolites. I have discussed many of the successful strategies and suggested some potential targets. Although the trend toward mechanism-based assays is relatively recent, it is clear that they have had a profound impact on screening in drug discovery. Often a mechanism-based assay requires construction of specific strains and verification of the antibacterial role of the selected target. Since the conception and development of a mechanism-based screen depends upon knowledge of the specific target and perhaps a compound that affects that target, it is implicit that mode of action studies on compounds discovered through random screening may subsequently lead to new mechanistic assays. While serendipity continues to play a crucial role in any screen, target-directed assays appear to be a worthwhile approach in antibacterial screening.

REFERENCES

Aoki, H., Sakai, H., Kohsaka, M., et al. (1976) *J. Antibiotics* 29, 492–500.

Barrett, J.F., Sutcliffe, J.A., and Gootz, T.D. (1990) *Antimicrob. Agents Chemother.* 34, 1–5.

Barry, A.L. (1988) *Antimicrob. Agents Chemother.* 32, 150–157.

Berdy, J. (1974) *Adv. Appl. Microbiol.* 18, 309–402.

Biskupiak, J.E., Meyers, E., Gillum, A.M., et al. (1988) *J. Antibiotics* 41, 684–692.

Bonner, D.P., O'Sullivan, J., Tanaka, S.K., et al. (1988) *J. Antibiotics* 41, 1745–1751.

Brown, A.G. (1987) *Pure Appl. Chem.* 59, 475–485.

Butterworth, D., Cole, M., Hanscomb, G., et al. (1979) *J. Antibiotics* 32, 287–293.

Canepari, P., Boaretti, M., Lleo, M.M., and Satta, G. (1990) *Antimicrob. Agents Chemother.* 34, 1220–1226.

Cassani, G. (1989) in *Bioactive Metabolites from Microorganisms* (Bushell, M.E., and Grafe, U., eds.), p. 221, Elsevier, Amsterdam.

Cavalleri, B., Pagani, H., Volpe, G., et al. (1984) *J. Antibiotics* 37, 309–318.

Chatterjee, S., Reddy, G.C.S., Franco, C.M.M., et al. (1987) *J. Antibiotics* 40, 1368–1373.

Christensen, S.B., Allaudeen, H.S., Burke, M.R., et al. (1987) *J. Antibiotics* 40, 970–978.

Conover, L.H. (1971) in *Drug Discovery* (Gould, R.F., ed.), p. 33, *Advances in Chemistry*, Series 108, American Chemical Society, Washington, D.C.

Drlica, K., and Coughlin, S. (1989) *Pharmacol. Therapy* 44, 107–112.

Dubnau, D. (1984) *Crit. Rev. Biochem.* 16, 103–119.

Duggar, B.M. (1948) *Ann. N.Y. Acad. Sci.* 51, 177.

Ehrlich, J., Bartz, G.R., Smith, R.M., and Joslyn, S.A. (1947) *Science* 106, 417–425.

Fernandes, P.B., Baker, W.R., Freiberg, L.A., Hardy, D.J., and McDonald, E.J. (1989) *Antimicrob. Agents Chemother.* 33, 78–86.

Finlay, A.C., Hobby, G.L., Pan, S.Y., et al. (1950) *Science* 111, 85–90.

Franco, C.M.M., Mukhopadhyay, T., Ganguli, B.N., Rupp, R.H., and Felhaber, H.W. (1987) European patent no. 87101796.8.

Franklin, T.J., and Snow, G.A. (eds.) (1989) *Biochemistry of Antimicrobial Action*, Chapman & Hall, London.

Frere, J.M., Klein, D., and Ghuysen, J.M. (1980) *Antimicrob. Agents Chemother.* 18, 506–516.

Fukui, S., and Beppu, T. (1986) *Misaengul. Kwa. Parhyo.* 10, 8 and *Chem. Abstr.* 105, 224444x.

Ganguli, B.N. (1989) in *Bioactive Metabolites from Microorganisms* (Bushell, M.E., and Grafe, U., eds.), p. 27, Elsevier, Amsterdam.

Goldman, R., Kohlbrenner, W., Lartey, P., and Pernet, A. (1987) *Nature* 329, 162–164.

Goldman, R.C., and Kadam, S.K. (1989) *Antimicrob. Agents Chemother.* 33, 1058–1066.

Goldstein, B.P., Sela, E., Gastaldo, L., et al. (1987) *Antimicrob. Agents Chemother.* 31, 1961–1972.

Gryczan, T.J., Israeli-Reches, M., and Dubnau, D. (1984) *Mol. Gen. Genet.* 194, 357–366.

Hammond, S.M., Claesson, A., Jansson, A.M., et al. (1987) *Nature* 327, 730–734.

Harada, S., Tsubotani, S., Hida, T., Ono, H., and Okazaki, H. (1986) *Tetrahedron Lett.* 27, 6229–6231.

Harned, R.L., Hidy, P.H., and LaBaw, E.K. (1955) *Antibiot. Chemother.* 5, 204–208.

Harris, C.M., Kopecka, C., and Harris, T.M. (1985) *J. Antibiotics* 38, 51–56.

Hata, T., Sano, Y., Ohki, N., et al. (1953) *J. Antibiotics* 6A, 87–94.

Hotta, K., Takahashi, A., Okami, Y., and Umezawa, H. (1983) *J. Antibiotics* 36, 1789–1792.

Hudson, H., Lindberg, A.A., and Stocker, B.A.D. (1978) *J. Gen. Microbiol.* 109, 97–104.

Huisman, O., D'Ari, R., and Gottesman, S. (1984) *Proc. Natl. Acad. Sci. USA* 81, 4490–4494.

Imada, A., Kitano, K., Kintaka, K., Muroi, M., and Asai, M. (1981) *Nature* 289, 590–594.

Imada, A., Kintaka, K., Nakao, M., and Shinagawa, S. (1982) *J. Antibiotics* 35, 1400–1406.

Imada, A., and Okazaki, H. (1987) *Antibiotic Research and Biotechnology* (Cape, R.E., Goldberg, M.I., Hata, T., and Maeda, D., eds.), p. 3, Japan Antibiotics Research Association, Tokyo.

Imanaka, H. (1985) *Recent Adv. Chemother. Proc. 14th Int. Congr. Chemother.*, San Diego, CA, p. 15, (Ishigami, J., ed.), University Tokyo Press, Tokyo.

Jackson, M. (1988) in *Developments in Industrial Microbiology*, no. 29 (Pierce, G., ed.), p. 267, Elsevier Science Publishers, Amsterdam.

Jann, K. (1983) in *Bacterial Lipopolysaccharides: Structures Synthesis and Biological Activities* (Anderson, L., and Unger, F., eds.), p. 171, American Chemical Society Press, Washington, DC.

Janz, S., Wolff, G., and Storch, H. (1988) *Zentralbl. Mikrobiol.* 143, 645–651.

Johnson, J.R., Woodward, R.B., and Robinson, R. (1949) in *The Chemistry of Penicillin* (Clarke, H.T., Johnson, J.R., and Robinson, R., eds.), p. 440, Princeton University Press, Princeton, NJ.

Kadam, S.K. (1989) *J. Bacteriol.* 171, 4518–4520.

Kadam, S.K., Doran, C.C., and Goldman, R.C. (1989) *Can. J. Microbiol.* 35, 646–652.

Kadam, S.K., Rehemtulla, A., and Sanderson, K.E. (1985) *J. Bacteriol.* 161, 277–285.

Kahan, J.S., Kahan, F.M., Goegelman, R., et al. (1979) *J. Antibiotics* 32, 1–8.

Kelly, J.A., Knox, J.R., Zhao, H., Frere, J.M., and Ghuysen, J.M. (1989) *J. Mol. Biol.* 109, 281–288.

Kumagai, H., Tomoda, H., and Omura, S. (1990) *J. Antibiotics* 43, 397–408.

Kurath, P., Rosenbrook, W., Dunnigan, D.A., et al. (1984) *J. Antibiotics* 37, 1130–1142.

Lakey, J.H., Maget-Dana, R., and Ptak, M. (1989) *Biochim. Biophys. Acta* 985, 60–68.

McGuire, J.H., Bunch, R.L., Anderson, R.C., et al. (1952) *Antibiot. Chemother.* 2, 281–289.

Nara, T., Yamamoto, M., Kawamoto, I., et al. (1977) *J. Antibiotics* 30, 533–539.

Naveh, A., Potasman, I., Bassan, H., and Ulitzur, S. (1984) *J. Appl. Bacteriol.* 56, 457–462.

Newton, G.G.F., and Abraham, E.P. (1955) *Nature* 175, 548–551.

Nisbet, L.J., and Porter, N. (1989) in *Microbial Products: New Approaches* (Baumberg, S., Hunter, I., and Rohdes, M.J., eds.), p. 309, Cambridge University Press, Cambridge, England.

Nolan, R.D., and Cross, T. (1988) in *Actinomycetes in Biotechnology* (Goodfellow, M., Williams, S.T., and Mordarski, M., eds.), p. 1, Academic Press, London.

Numata, K., Yamamoto, H., Hatori, M., Miyaki, T., and Kawaguchi, H. (1986) *J. Antibiotics* 39, 994–998.

Okami, Y., and Hotta, K. (1988) in *Actinomycetes in Biotechnology* (Goodfellow, M., Williams, S.T., and Mordarski, M., eds.), p. 33, Academic Press, London.

Okazaki, H. (1987) in *Frontiers of Antibiotic Research* (Umezawa, H., ed.), p. 303, Academic Press, Japan.

Omura, S. (1986) *Microbiol. Rev.* 50, 259–264.

Omura, S., Murata, M., Kimura, D., et al. (1985) *J. Antibiotics* 38, 1016–1022.

Omura, S., Tanaka, H., Oiwa, R., et al. (1979) *J. Antibiotics* 32, 978–982.

O'Sullivan, J., McCullough, J.E., Johnson, J.H., et al. (1990) *J. Antibiotics* 43, 913–922.

O'Sullivan, J., McCullough, J.E., Tymiak, A.A., et al. (1988) *J. Antibiotics* 41, 1740–1744.

Parenti, R., Ciabatti, R., Kettenring, J., and Cavalleri, B. (1990) in *Abstracts 9th Int. Symp. Future Trends in Chemother.* p. 180, Geneva, Switzerland, March 26–28.

Parker, C.T., Kloser, A.W., Schnaitman, C.A., et al. (1992) *J. Bacteriol.* 174, 2525–2530.

Patel, M.V. (1985) *J. Antibiotics* 38, 527–534.

Perlman, D. (1977) *J. Antibiotics* 30(suppl.), S133.

Rake, J.B., Gerber, R., Metha, J.J., et al. (1986) *J. Antibiotics* 39, 58–66.

Reading, D., and Cole, M. (1977) *Antimicrob. Agents Chemother.* 11, 852–861.

Saitoh, D., Tsukanawa, M., Tomita, K., et al. (1988) *J. Antibiotics* 41, 855–891.

Schats, A., and Waksman, S.A. (1944) *Proc. Soc. Exp. Biol. Med.* 57, 244.

Schindler, P.W., Konig, W., Chatterjee, S., and Ganguli, B.N. (1986) *J. Antibiotics* 39, 53–60.

Sensi, P., Margalith, P., and Timbal, M.T. (1959) *Il Farmaco, Ed. Sci.* 14, 146.

Sitrin, R.D., Chan, G.W., Dingerdissen, J.J., et al. (1985) *J. Antibiotics* 38, 561–570.

Sobin, B.A., English, A.R., and Celmer, W.O. (1955) in *Antibiotics Ann.–1954/1955* (Welch, H., and Mari-Ibannez, F., eds.), p. 827–830, Medical Encyclopedia Inc., New York.

Sutcliffe, J.S. (1988) *Annual Reports in Medicinal Chemistry,* Academic Press Inc., San Diego, CA.

Sykes, R.B., Cimarusti, C.M., Bonner, D.P., et al. (1981) *Nature* 291, 489–492.

Thompson, J., Schmidt, F., and Cundliffe, E. (1982) *J. Biol. Chem.* 257, 7915–7921.

Tsuno, T., Ikeda, C., Numata, K., et al. (1986) *J. Antibiotics* 39, 1001–1006.

Ulitzur, S. (1986) *Methods Enzymol.* 133, 275–289.

Ulitzur, S., and Weiser, I. (1981) *Proc. Natl. Acad. Sci. USA* 78, 3338–3342.

Umezawa, H., Ueda, M., Maeda, K., et al. (1957) *J. Antibiotics* 10A, 181–189.

Vicario, P.P., Green, B.G., and Katzen, H. (1987) *J. Antibiotics* 40, 209–216.

Waksman, S.A., and Lechevalier, H.A. (1949) *Science* 109, 305–316.

Weisblum, B. (1983) in *Gene Function in Prokaryotes* (Beckwith, J., Davies, J., and Gallant, J.A., eds.), pp. 91–121, Cold Spring Harbor Laboratory Press, Cold Spring Harbor, NY.

Williams, D.H., and Waltho, J.P. (1988) *Biochem. Pharmacol.* 37, 133–138.

Yao, R.C., and Mahoney, D.F. (1984) *J. Antibiotics* 38, 58–66.

Yao, R.C., and Mahoney, D.F. (1989) *Appl. Environ. Microbiol.* 55, 1507–1512.

Oncogene Function
Inhibitors of Microbial Origin

Kazuo Umezawa

Genetic alteration of normal genes involved in cell growth regulation can induce neoplastic transformation. These normal genes are called proto-oncogenes and the activated genes are oncogenes. The oncogenes are transcribed and translated into their protein products, which have various functions that induce neoplastic transformation. These oncogene functions include acting as a modified growth factor receptor (*erbB, neu, fms*), as a growth factor itself (*sis, hst*), as a tyrosine kinase (*src* oncogene group, *abl*), and acting as a G protein to activate phosphatidylinositol turnover (*ras*).

Streptomyces and other microorganisms produce antibiotics, anticancer agents, and enzyme inhibitors as secondary metabolites. Therefore, microorganisms are a treasury of organic compounds that have various structures and biological activities. Isolation of the specific bioactive products in culture broths depends on the testing method. Since oncogene theory has been extensively developed in the field of carcinogenesis research, we have screened oncogene function inhibitors as a new group of microbial secondary metabolites.

In this chapter, we describe isolation and activity of these inhibitors. Inhibitors from other laboratories, such as genistein, herbimycin, and 2-demethylsteffimycin D, are also included.

9.1 TYROSINE KINASE INHIBITORS

9.1.1 Erbstatin

The binding of epidermal growth factor (EGF) to its receptor, which has an associated tyrosine kinase activity, stimulates the phosphorylation of a tyrosine residue in the EGF receptor, and the phosphorylated receptor transfers the terminal phosphate of ATP to the tyrosine residue of other proteins. Culture filtrates of *Streptomyces* were screened for inhibitors of tyrosine kinase, using the membrane fraction of human epidermoid carcinoma cell line A431. This cell line contains large amounts of EGF receptor on the membrane. One strain produces a novel compound (Figure 9–1) that inhibits tyrosine kinase, and was named erbstatin (Umezawa et al. 1986). Taxonomic studies indicated that the strain producing erbstatin was closely related to *Streptomyces viridosporus*.

Erbstatin in culture filtrates (5 1) was extracted with 5 1 of butyl acetate, and the extract was concentrated under reduced pressure to give a yellowish powder (370 mg). The dried material was dissolved in CHCl₃ and subjected to silicic acid column chromatography. After washing with CHCl₃-MeOH (100:2), the active fraction was eluted with CHCl₃-MeOH (100:5) and concentrated in vacuo to give a yellow powder (110 mg). Erbstatin was dissolved in MeOH-CHCl₃ (1:1) and kept at 4°C. Sixty milligrams of crude crystals were obtained, which were recrystallized from MeOH-CHCl₃. Erbstatin is

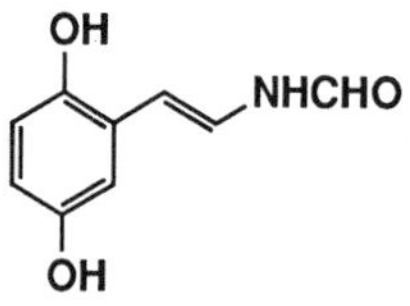

erbstatin

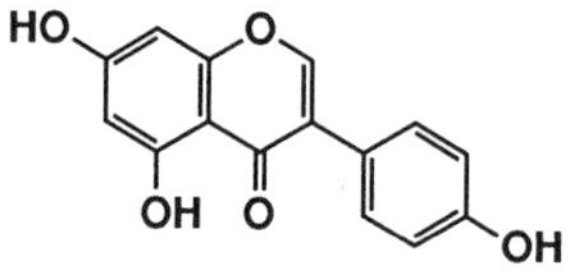

genistein

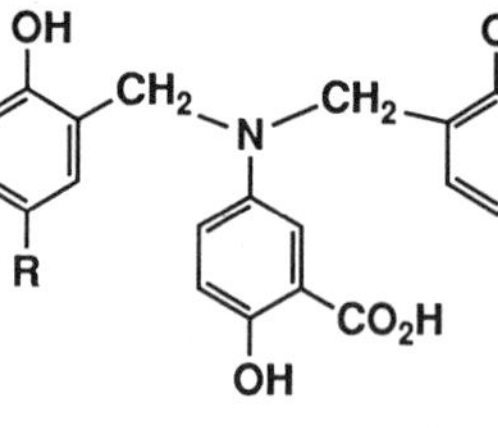

lavendustin A: R=OH
          B: R=H

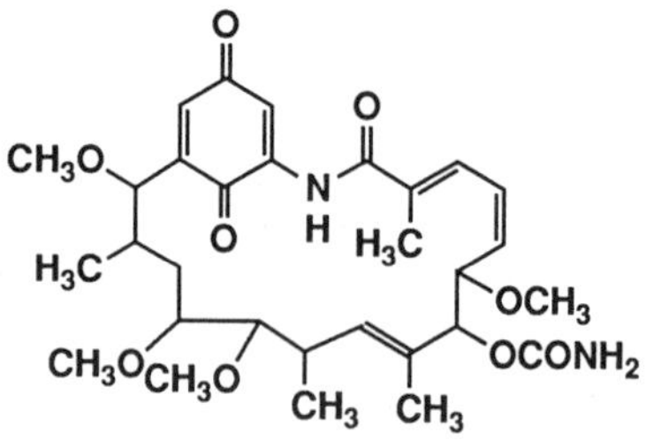

herbimycin A

FIGURE 9–1 Tyrosine kinase inhibitors.

soluble in MeOH, ethanol, and acetone, slightly soluble in $CHCl_3$ and EtOAc, and insoluble in H_2O and *n*-hexane.

The 50% inhibition concentration of erbstatin against tyrosine kinase was 0.55 μg/ml, when it was examined as follows: The reaction mixture contained 1 mM $MnCl_2$, 100 ng EGF, 40 μg protein of A431 membrane fraction, 75 μg of albumin, 3 μg of histone and HEPES (*N*-2-hydroxyethylpiperazine-*N′*-2-ethanesulfonic acid) buffer (20 mM, pH 7.4) in a final volume of 50 μl. The reaction tubes were placed on ice and incubated for 10 minutes in the presence or absence of erbstatin. The reaction was initiated by the addition of labeled ATP (10 μl) and the incubation was continued for 30 minutes at 0°C. Aliquots of 50 μl were pipetted onto Whatman 3MM filter paper and put immediately into a beaker of cold 10% TCA containing 0.01 M sodium pyrophosphate. The filter papers were washed extensively with TCA solution containing 0.01 M sodium pyrophosphate at room temperature, extracted with ethanol and ether, and then dried. Radioactivity was measured by a scintillation counter.

The inhibition of tyrosine kinase by erbstatin was further confirmed by sodium dodecyl sulfate (SDS)-gel electrophoretic analysis of the A431 membrane components. Erbstatin did not inhibit protein kinase A or C.

Erbstatin was chemically synthesized by several groups. One of the effective syntheses includes the reaction of diethyl(isocyanomethyl)phosphonate with dihydroxybenzaldehyde by a modified Schollkopf's procedure (Isshiki et al. 1987a) as shown in Figure 9–2. This procedure was used for the preparation of various erbstatin analogues (Isshiki et al. 1987b).

Erbstatin is easily decomposed in calf serum and in cell culture medium in the presence of oxygen and ferric ion (Imoto et al. 1987a). We isolated 2,5-dihydroxyphenylketoaldehyde by performing an overnight incubation of erbstatin with ferric chloride. The decomposition of erbstatin in calf serum was prevented by the addition of foroxymithine, a ferric ion chelator isolated from *Streptomyces*. We have synthesized methyl 2,5-dihydroxycinnamate (2,5-MeC) as a stable analog of erbstatin (Umezawa et al. 1990). This analog is about four times more stable than erbstatin in calf serum, and its inhibition of EGF receptor tyrosine kinase in vitro and in situ is comparable to that of erbstatin.

Erbstatin and 2,5-MeC delayed the onset of EGF-induced DNA synthesis in quiescent normal kidney cells without suppressing total DNA synthesis (Umezawa et al. 1990). Thus, it is indicated that tyrosine kinase activity is essential for EGF-induced DNA synthesis. Phospholipase C-γ is known to be tyrosine-phosphorylated by EGF or platelet-derived growth factor receptor. However, the effect of tyrosine phosphorylation on the enzyme activity was not understood. We have shown that erbstatin inhibits EGF-induced activation of phospholipase C in A431 cells both in vitro and in situ (Imoto et al. 1990). The cytoplasmic phospholipase C activity was enhanced by EGF, which was inhibited by erbstatin. Therefore, it was suggested that tyrosine phosphorylation of phospholipase C activates the enzymic activity. Erbstatin induces normal phenotypes in temperature-sensitive Rous sarcoma virus-

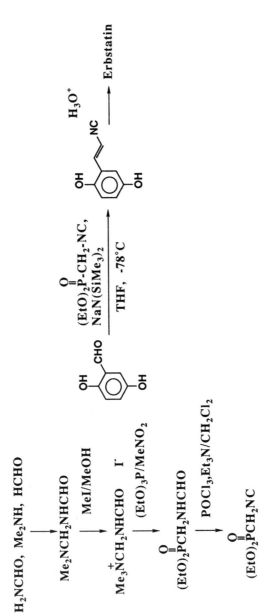

FIGURE 9–2 Synthesis of erbstatin.

infected normal rat kidney (RSV^{ts}-NRK) cells (Umezawa et al. 1991). Erbstatin inhibits the growth of human breast carcinoma MCF-7 and esophageal carcinoma EH-4 in nude mice when intraperitoneally administered in combination with foroxymithine (Toi et al. 1990). 2,5-MeC showed antitumor activity on MCF-7 cells without foroxymithine. Naccache et al. found that erbstatin inhibited human neutrophil responsiveness to chemotactic factors (Naccache et al. 1990).

9.1.2 Genistein

Genistein (see Figure 9–1), a known plant isoflavonoid, was isolated from *Pseudomonas* as an inhibitor of EGF receptor tyrosine kinase (Ogawara et al. 1986). It shows an IC$_{50}$ for the tyrosine kinase of the EGF receptor (A431 cell membrane) at 0.7 μg/ml, and that of RSV-3Y1 cells at 8.0 μg/ml. Genistein does not inhibit protein kinase A or C. The inhibition was competitive with respect to ATP and noncompetitive to a phosphate acceptor, histone H2B (Akiyama et al. 1987). Orobol, a closely related flavonoid from *Streptomyces*, is as potent as genistein is as a tyrosine kinase inhibitor (Umezawa et al. 1986).

9.1.3 Lavendustin

We have screened *Streptomyces* culture filtrates for more potent tyrosine kinase inhibitors in a new tyrosine kinase assay system containing an A431 membrane fraction as the enzyme and the tridecapeptide RRLIEDAE-YAARG as a substrate. As a result, we have isolated two compounds of novel structure from a *Streptomyces* strain. We named them lavendustin A and lavendustin B (Onoda et al. 1989).

For isolation of lavendustin A, fermentation broth filtrate (20 l) was extracted with butyl acetate after adjustment of the pH to 2.4. The extract was concentrated in vacuo to give an oily residue (7.05 g) that was applied to a silica gel column. The column was washed with CHCl$_3$-MeOH (100:2); the active compounds were then eluted with CHCl$_3$-MeOH (3:1). The active fractions were concentrated (2.1 g) and charged on a column of Sephadex LH-20, which was eluted with MeOH. After evaporation of the active fractions, the residue (1.03 g) was further purified twice by high pressure liquid chromatography (HPLC) with 15% MeCN/H$_2$O. The active fractions were collected and extracted with butyl acetate, and the extract was concentrated in vacuo to give a mixture of lavendustins (105.2 mg). This mixture was applied to preparative silica gel TLC [CHCl$_3$-MeOH (2:1)] to obtain lavendustin A (37.0 mg) and lavendustin B (29.7 mg) (see Figure 9–1). These compounds were soluble in MeOH, EtOH, Me$_2$CO, and Me$_2$SO but insoluble in CHCl$_3$, H$_2$O, and Et$_2$O. Taxonomic features indicated that the producing strain was related to *Streptomyces griseolavendus*.

As shown in Table 9–1, lavendustin A inhibited EGF receptor tyrosine kinase with an IC$_{50}$ of 4.4 ng/ml, which was about 50 times more inhibitory

TABLE 9–1 Inhibition of Kinases by Lavendustins

Inhibitor	Tyrosine Kinase[1]	IC_{50} ($\mu g/ml$)		
		Kinase A	Kinase C	PI Kinase[2]
Lavendustin A	0.0044	>100	>100	6.4
Lavendustin B	0.49	—	—	—
Erbstatin	0.20	100	100	25.0

[1] The tridecapeptide was used as a substrate.
[2] PI, phosphatidylinositol kinase.

than erbstatin. It did not inhibit protein kinase C or A but weakly inhibited phosphatidylinositol kinase. Lavendustin B showed much weaker activity than lavendustin A.

The EGF-receptor-associated tyrosine kinase was reported to act in the sequential ordered bi-bi mechanism, in which the peptide comes first and ATP comes second to the enzyme active site (Erneux et al. 1983). Kinetic studies using Lineweaver-Burk plotting indicated that lavendustin A inhibits the tyrosine kinase competitively with ATP, and noncompetitively with the peptide substrate. Thus, the mechanism of inhibition is different from that of erbstatin (Imoto et al. 1987b), which competes with the peptide, but is similar to that of orobol and genistein.

The total synthesis of lavendustin A was accomplished by using successive reductive alkylation of aminohydroxybenzoic acids and hydroxybenzalde-hydes, as shown in Figure 9–3. The synthetic lavendustin A gave exactly the

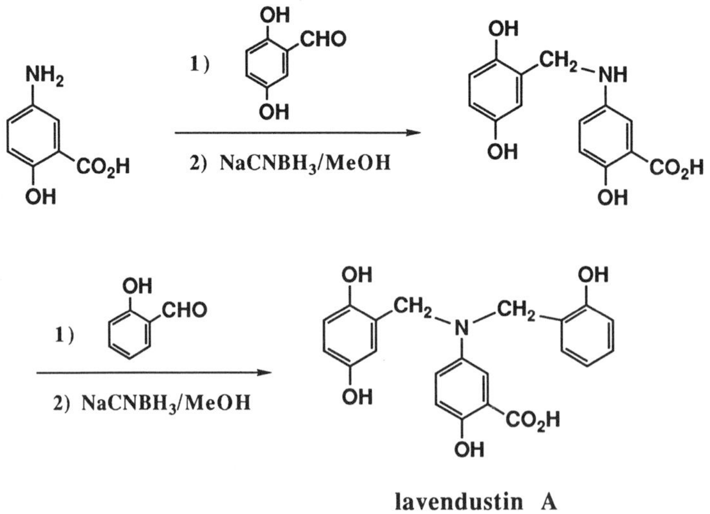

lavendustin A

FIGURE 9–3 Synthesis of lavendustin A.

same results as the natural product in terms of R_f value, spectral data, and inhibition of tyrosine kinase.

Although lavendustin A is a potent inhibitor of tyrosine kinase in vitro, it did not show any biological effect in cultured cells. Assuming that the molecule is so polar that it can not pass through the plasma membrane, we therefore prepared lavendustin A methyl ester. The methyl ester derivative showed weaker inhibitory effect on tyrosine kinase, but inhibited EGF receptor autophosphorylation and internalization in cultured A431 cells (Onoda et al. 1990).

9.1.4 Herbimycin

Herbimycin was isolated as an ansamycin antibiotic having herbicidal activity (Omura et al. 1979). Recently, it was again isolated from *Streptomyces* as a compound that altered the transformed cell morphology into the normal cell morphology in RSV^ts-NRK cells (Uehara et al. 1985a).

The addition of 0.5 μg/ml of herbimycin at the permissive temperature induced a morphological change of the cells back to their normal morphology within 2 days. We also found that the level of fibronectin mRNA was elevated by herbimycin in RSV^ts-NRK cells at the permissive temperature (Umezawa et al. 1987). Previously, herbimycin was considered to inhibit *src* oncogene functions by enhancing degradation of the p60^src protein. Recently, it was suggested that herbimycin irreversibly inhibits tyrosine kinase of the *src* product by binding to the SH group around the active site (Uehara et al. 1989). It does not inhibit EGF receptor tyrosine kinase.

9.2 INHIBITORS OF *ras* ONCOGENE FUNCTIONS

9.2.1 Oxanosine

Oxanosine (Figure 9–4) was isolated as an antibiotic from *Streptomyces* (Shimada et al. 1981). It is a novel, unusual nucleoside and shows weak antibac-

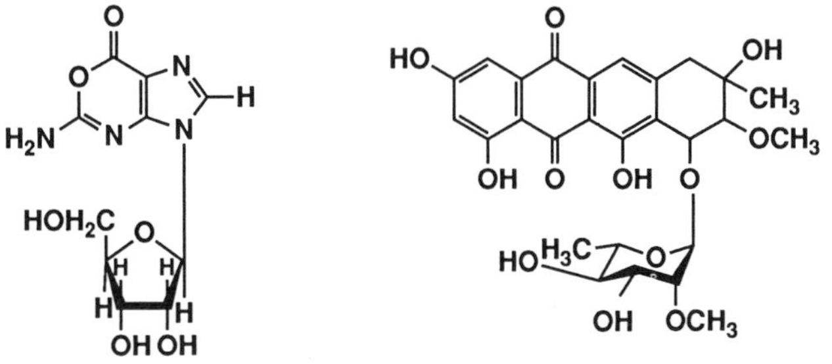

oxanosine

2-demethylsteffimycin D

FIGURE 9–4 *ras* oncogene function inhibitors.

terial activity that is antagonized by guanine and guanosine. Oxanosine also has a weak antitumor effect on L1210 mouse leukemia.

Oxanosine was found to alter tumor cell morphology into the normal morphology in temperature sensitive Kirsten sarcoma virus-infected rat kidney (K-ras^{ts}-NRK) cells (Itoh et al. 1989). The addition of 2 μg/ml of oxanosine for 2 days at the permissive temperature changed the tumor cell morphology into the normal cell morphology.

In Northern blotting analysis, oxanosine increased fibronectin mRNA expression. Using the Western blotting technique, we analyzed the cellular levels of fibronectin. Incubation of K-ras^{ts}-NRK cells at the permissive temperature with 2 μg/ml of oxanosine for 48 hours increased cellular fibronectin content dramatically to the level of normal cells. The fibronectin content in the cells at the nonpermissive temperature did not change in the presence of oxanosine.

For the expression of *ras* oncogene function, the p21^{ras} protein must bind to GTP. In the metabolic pathway of GMP synthesis, IMP is converted to XMP by IMP dehydrogenase, and XMP is converted to GMP by GMP synthetase. Oxanosine itself does not inhibit IMP dehydrogenase, but oxanosine 5'-monophosphate inhibits IMP dehydrogenase almost competitively with the substrate (Uehara et al. 1985b). Therefore, it is likely that oxanosine is phosphorylated in the cells and thereby inhibits IMP dehydrogenase to reduce GTP synthesis. The intracellular levels of GTP, GDP, and GMP are actually lowered by oxanosine, while the level of ATP is not affected. Mycophenolic acid, an IMP dehydrogenase inhibitor, also induces normal morphology in K-ras^{ts}-NRK cells (Itoh et al. 1989).

9.2.2   2-Demethylsteffimycin D

A novel anthracycline was isolated from *Streptomyces* in the course of screening for *ras* function inhibitors (Tsuchiya et al. 1991). 2-Demethylsteffimycin D (see Figure 9–4) at 3 μg/ml induces normal morphology in K-*ras*-NRK cells in 2 days. It does not alter the morphology of RSV-NRK cells. Since the yield of 2-demethylsteffimycin D by the producing strain was very poor, the mechanism of action has not been elucidated yet. Other anthracyclines, such as steffimycin, steffimycin B, C, or D, aclacinomycin, daunomycin, and adriamycin, did not induce normal morphology in K-*ras*-NRK cells. 2-Demethylsteffimycin D inhibits the growth of K-ras^{ts}-NRK cells at a lower concentration at the permissive temperature (IC$_{50}$, 0.73 μg/ml) than at the nonpermissive temperature (IC$_{50}$, 5.8 μg/ml).

9.3   INHIBITORS OF PHOSPHATIDYLINOSITOL TURNOVER

9.3.1   Psi-tectorigenin

Many oncogenes such as *ras, src, sis, fms,* and *fes* are reported to enhance cellular phosphatidylinositol turnover. Hydrolytic breakdown of PIP$_2$ by phospholipase C generates two second messengers, diacylglycerol and inositol triphosphate. The former activates protein kinase C, and the latter mobilizes calcium ion from the endoplasmic reticulum. Therefore, we screened culture filtrates of microorganisms for inhibitors of phosphatidylinositol turnover and thus isolated psi-tectorigenin (Imoto et al. 1988).

Phosphatidylinositol turnover was assayed by incorporation of labeled inositol into phospholipids. A431 cells were preincubated in HEPES-buffered saline containing [^{3}H]inositol at 37°C for 30 minutes. A test chemical and EGF were then added, and the incubation was continued for 60 minutes. Subsequently, 10% trichloroacetic acid containing 0.01 M sodium pyrophosphate was added, and the acid-insoluble fraction was scraped from the dish in H_2O. The lipid was extracted by the addition of $CHCl_3$ and CH_3OH (1:1) and [^{3}H]inositol-labeled lipids were counted by liquid scintillation spectrophotometry.

A culture filtrate from a *Nocardiopsis* strain found in a river near Shanghai showed strong inhibition against phosphatidylinositol turnover. The active principle was purified and its structure was found to be identical with that of psi-tectorigenin (Figure 9–5) by nuclear magnetic resonance (NMR). Psi-

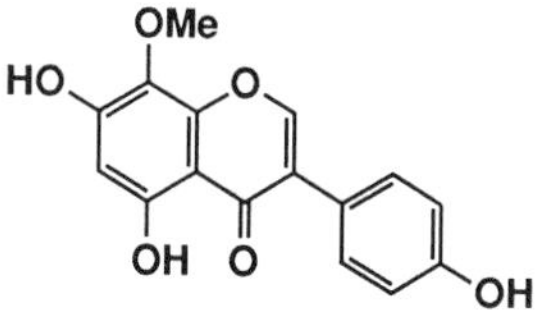

psi-tectorigenin

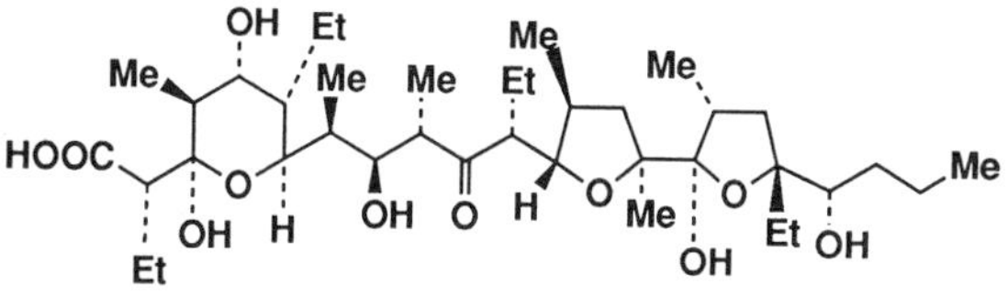

inostamycin

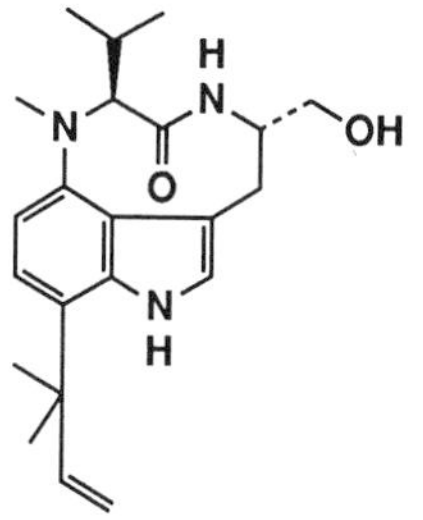

pendolmycin

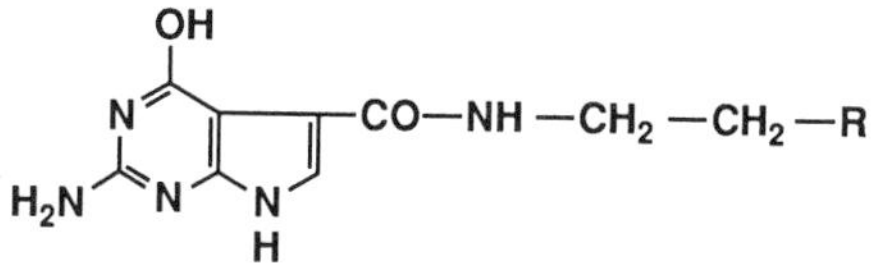

echiguanine A: $R{=}C(=NH){-}NH_2$

B: $R{=}CH_2{-}NH_2$

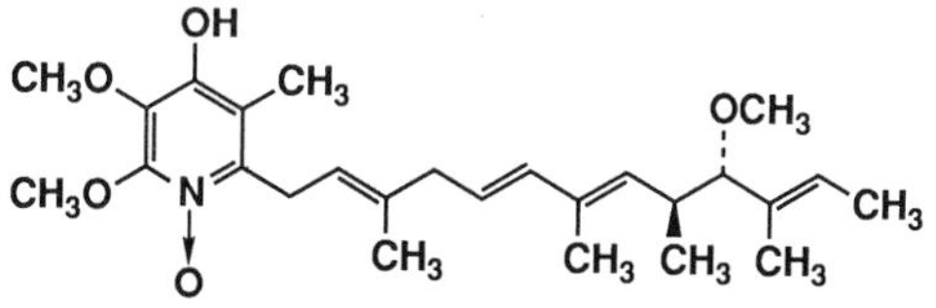

piericidin B_1 N-oxide

FIGURE 9–5 Phosphatidylinositol turnover inhibitors.

tectorigenin inhibited EGF-induced inositol incorporation with an IC_{50} of about 1 μg/ml. The inhibitory activity of psi-tectorigenin was several times stronger than that of orobol or genistein, related isoflavonoids that inhibit tyrosine kinase. Psi-tectorigenin, orobol, and genistein all showed similar in vitro inhibitory activity against EGF receptor tyrosine kinase, with an IC_{50} of about 0.1 μg/ml. However, only psi-tectorigenin did not inhibit EGF receptor tyrosine kinase at 50 μg/ml in situ.

Phospholipase C is a rate-limiting enzyme of phosphatidylinositol turnover. Psi-tectorigenin does not inhibit phospholipase C, but we recently found that it inhibits activation of the enzyme by EGF (Imoto et al. 1991a). The phospholipase C activity in homogenate prepared from EGF-treated A431 cells was fourfold higher than that from untreated cells, and EGF failed to activate the enzyme in the presence of psi-tectorigenin.

9.3.2 Inostamycin

By continuing the screening procedure, we found a novel inhibitor of phosphatidylinositol turnover, one that we named inostamycin (Imoto et al. 1991c).

For isolation of inostamycin, the culture broth (5 l) was centrifuged at 5000 rpm for 10 minutes, and the mycelia was extracted with 1.0 l of acetone. The acetone extract was concentrated in vacuo and combined with the supernatant, then extracted with 3.5 l of EtOAc. This extract was concentrated in vacuo, applied onto a silica gel column (50 g), and eluted with $CHCl_3$-MeOH (100:1). The active fraction was precipitated with MeCN, and the precipitate was further purified by centrifugation partition chromatography. The active fraction was crystallized from hexane/CH_2Cl_2 solution, and the crystals were washed with 1 N HCl, followed by 1 N NaOH and concentrated NaCl solution to obtain the sodium salt. The active component was recrystallized from the same solution to give 112.4 mg of inostamycin. Inostamycin was soluble in MeOH, $CHCl_3$, and Me_2CO and insoluble in hexane and H_2O.

The molecular formula of inostamycin sodium salt was deduced to be $C_{38}H_{67}O_{11}Na$ by mass spectrum and elemental analysis. The tentative structure of inostamycin was determined by ^1H- and ^{13}C-NMR spectroscopy. To confirm the proposed structure and elucidate its stereochemistry, crystallographic analysis was carried out. The relative structure is shown in Figure 9–5.

Thus, inostamycin was shown to be a novel polyether belonging to the polyether antibiotic group, which includes lysocellin, X-14873A, and related compounds from *Streptomyces*. Inostamycin inhibited EGF-induced inositol incorporation into inositol lipids with an IC_{50} of about 0.5 μg/ml in the A431 cell assay system. Lysocellin, which has a closely related structure, also inhibited phosphatidylinositol turnover; however, monensin, a polyether antibiotic having a considerably different structure, did not inhibit phosphatidylinositol turnover.

Inostamycin did not inhibit EGF receptor tyrosine kinase, phospholipase C, or phosphatidylinositol kinase but did inhibit CDP-DG: inositol transferase. In the absence of Ca^{2+} ions, EGF induces rapid rounding of A431 cells. Inostamycin and psi-tectorigenin inhibited this rounding, and phosphatidyl-inositol turnover was suggested to be involved in its mechanism (Imoto et al. 1991b). Inostamycin also showed antitumor activity on Ehrlich tumors in mice. Unexpectedly, inostamycin overcame multidrug resistance in human carcinoma KB cells (Kawada et al. 1991a, 1991b).

9.3.3 Pendolmycin

Pendolmycin (see Figure 9–5) is a novel indole alkaloid, which was isolated as an inhibitor of phosphatidylinositol turnover from *Nocardiopsis* (Yamashita et al. 1988). However, after its structure determination, it was shown to be closely related to known tumor promoters such as teleocidin and lyngbyatoxin. In fact, pendolmycin inhibits phorbol ester binding to the cell surface receptor, activates phospholipase A_2 in cell culture (Umezawa et al. 1989), is a potent activator of protein kinase C, and shows tumor-promoting activity in mice (Nishiwaki et al. 1991). Pendolmycin might suppress EGF receptor tyrosine kinase by phosphorylation, resulting in inhibition of EGF-induced phosphati-dylinositol turnover in A431 cells (Friedman et al. 1984).

Pendolmycin was isolated from the strain that produced psi-tectorigenin. The broth (36 l) filtrate was extracted with an equal volume of EtOAc, and the mycelial cake was extracted with Me_2CO. The Me_2CO extract was dried, dissolved in H_2O, and extracted with EtOAc. The EtOAc extracts were combined and concentrated in vacuo. The dried material was applied to a silica gel column, and the active fraction was eluted with $CHCl_3$-MeOH (100:1). The eluate was dried to give a yellow powder (81.6 mg), which was dissolved in MeOH and applied to a Toyopearl HW-40 column to obtain purified material (40.8 mg). Further purification was carried out using reversed-phase HPLC with 50% $MeCN/H_2O$ as eluent, yielding 8.8 mg of pendolmycin.

9.3.4 Piericidin B₁ N-oxide

Piericidins are insecticidal compounds isolated from mycelia of *Streptomyces mobaraensis* and *S. pactum*. They are toxic to several species of insects, aphids, and mites. Piericidin A has been shown to block electron transport between NADH dehydrogenase and coenzyme Q. In the course of our screening phosphatidylinositol turnover inhibitors, we isolated a novel antibiotic, piericidin B_1 N-oxide, from *Streptomyces* (Nishioka et al. 1991a).

The fermentation broth (6 l) was filtered, and the mycelia were extracted with acetone. After removal of the acetone, the extract was combined with the filtrate, and the mixture was extracted with EtOAc. The EtOAc extract was concentrated in vacuo to give an oily material (2.3 g) that was mixed with silica gel and applied to a silica gel column. The column was washed with $CHCl_3$ and eluted with a mixture of $CHCl_3$ and MeOH (10:1). After evaporation to dryness, the residue (350 mg) was partitioned in a solvent system $CHCl_3$-MeOH-H_2O (5:6:4) by centrifugal partition chromatography in which the lower portion was stationary. The combined active fractions were chromatographed on Sephadex LH-20 with EtOAc. The crude material (90 mg) was further purified by preparative HPLC to give 53 mg of piericidin B_1 N-oxide.

Piericidin B_1 N-oxide showed antibacterial activity against Gram-positive and Gram-negative bacteria and fungi, while piericidin B_1, a known piericidin, did not. Piericidin B_1 N-oxide inhibited the phosphatidylinositol turnover with an IC_{50} of 1.5 µg/ml, while piericidin B_1 showed weaker activity (IC_{50}, 4.0 µg/ml). The mechanism of inhibition has not been determined.

9.3.5 Echiguanine

Phosphatidylinositol kinase is involved in the phosphatidylinositol turnover pathway and may be important for the regulation of phosphatidylinositol 4,5–bisphosphate levels. Therefore, we have screened inhibitors of phosphatidylinositol kinase from microbial secondary metabolites and isolated 2,3-dihydroxybenzoic acid (Nishioka et al. 1989) and toyocamycin (Nishioka et al. 1990). Recently, we isolated novel and potent inhibitors of phosphatidylinositol kinase from a *Streptomyces* strain and named them echiguanine A and B (see Figure 9–5) (Nishioka et al. 1991b).

For the assay of phosphatidylinositol kinase, A431 cell membrane and [γ-^{32}P]ATP with or without inhibitor were mixed in 20 mM HEPES buffer (pH 7.2). The reaction mixture was incubated for 20 minutes at 20°C and stopped by the addition of $CHCl_3$, MeOH, and 1 N HCl (4:1:2). After vigorous vortexing, the lower phase was applied to a silica gel column (1 ml) for the separation of phospholipid and unreacted [γ-^{32}P]ATP (Nishioka et al. 1989). The phosphorylated lipid then obtained was eluted with $CHCl_3$, MeOH, and 4 N NH_4OH (9:7:2) and the radioactivity was quantified by a liquid scintillation counter.

For the isolation of echiguanine A and B, the broth filtrate (27 l) was adsorbed on Diaion HP-20 resin, the resin was washed with distilled water, and the active components were eluted with 50% aqueous methanol. The eluate was concentrated in vacuo, and the concentrate was passed through a column of reversed phase silica gel. The column was successively washed with distilled water and aqueous methanol. After evaporation to dryness, the residue was dissolved in distilled water and applied to a CM-Sephadex column eluting with a gradient of 0.4 to 1.0 M NaCl. The combined eluates were subjected to a column of Diaion CHP-20 and eluted with 0.005 N HCl-50% aqueous methanol. After neutralization of the eluate by Amberlite IR-45 (OH⁻) anion exchange resin, the active fraction was concentrated in vacuo to give a mixture of echiguanine A and B (73 mg). The mixture was further partitioned in the solvent system EtOAc-BuOH-water (4:7:10) using centrifugal partition chromatography to give purified echiguanine A and B at 28.6 mg and 6.7 mg, respectively.

Echiguanine A and B inhibited phosphatidylinositol kinase of the A431 cell membrane with IC_{50}s of 0.04 and 0.11 µg/ml, respectively. Structurally related compounds such as 7-deazaguanine showed only weak activity (IC_{50}, 35 µg/ml), and adenine, hypoxanthine, guanine, and isoguanine did not inhibit the enzyme. Echiguanine A and B showed no antimicrobial activity. The LD_{50}s of echiguanine A and B, when administered intravenously to mice, were >100 mg/kg.

Echiguanine A is the most potent phosphatidylinositol kinase inhibitor discovered. It is stronger than toyocamycin (IC_{50}, 3.3 µg/ml), 2,3-dihydroxybenzaldehyde (IC_{50}, 0.45 µg/ml), quercetin (IC_{50}, 1.8 µg/ml), or orobol (IC_{50}, 0.25 µg/ml).

9.4 FUTURE DIRECTIONS

Various oncogene function inhibitors, including novel compounds, have been isolated from microbial secondary metabolites. As oncogene research progresses, new assay targets will be available, and more interesting bioactive metabolites will be discovered. The functions of nuclear oncogenes, such as *fos* and *myc*, may be possible targets for inhibitor screening in the future.

Several oncogenes such as *ras, erb B*, and *erb B2* (*neu*) are considered to be involved in human carcinogenesis. Erbstatin and inostamycin show antitumor activity in animal models. More effective, stable, or less toxic analogs may become cancer chemotherapeutic agents that act by this new mechanism. We are also screening new structures from nature for clinically useful compounds.

Oncogene function inhibitors usually suppress proto-oncogene product activity. Proto-oncogene products are considered to be essential for regulating cell growth and differentiation. Therefore, oncogene function inhibitors will be used more widely for the mechanistic study of normal cell behavior.

REFERENCES

Akiyama, T., Ishida, J., Nakagawa, S., et al. (1987) *J. Biol. Chem.* 262, 5592–5595.

Ebata, E., Kasahara, H., Sekine, K., and Inoue, Y. (1975) *J. Antibiotics* 28, 118–121.

Erneux, C., Cohen, S., and Garbers, D.L. (1983) *J. Biol. Chem.* 258, 4137–4142.

Friedman, B., Frackelton, A.R., Jr., Ross, A.H., et al. (1984) *Proc. Natl. Acad. Sci. USA* 81, 3034–3038.

Imoto, M., Umezawa, K., Komuro, K., et al. (1987a) *Jpn. J. Cancer Res.* (*Gann*) 78, 329–332.

Imoto, M., Umezawa, K., Isshiki, K., et al. (1987b) *J. Antibiotics* 40, 1471–1473.

Imoto, M., Yamashita, T., Kurasawa, S., et al. (1988) *FEBS Lett.* 230, 43–46.

Imoto, M., Shimura, N., Ui, H., and Umezawa, K. (1990) *Biochem. Biophys. Res. Commun.* 173, 208–211.

Imoto, M., Shimura, N., and Umezawa, K. (1991a) *J. Antibiotics* 44, 915–917.

Imoto, M., Johtoh, N., and Umezawa, K. (1991b) *J. Cell. Pharm.* 2, 49–53.

Imoto, M., Umezawa, K., Takahashi, Y., et al. (1991c) *J. Nat. Prod.* 53, 825–829.

Isshiki, K., Imoto, M., Takeuchi, T., et al. (1987a) *J. Antibiotics* 40, 1207–1208.

Isshiki, K., Imoto, M., Sawa, T., et al. (1987b) *J. Antibiotics* 40, 1209–1210.

Itoh, O., Kuroiwa, S., Atsumi, S., et al. (1989) *Cancer Res.* 49, 996–1000.

Kawada, M., Imoto, M., and Umezawa, K. (1991a) *J. Cell Pharmacol.* 2, 138–142.

Kawada, M., and Umezawa, K. (1991b) *Jpn. J. Cancer Res.* (*Gann*) 82, 1160–1164.

Naccache, P.H., Gilbert, C., Caon, A.C., et al. (1990) *Blood* 76, 2098–2104.

Nishioka, H., Imoto, M., Sawa, T., et al. (1989) *J. Antibiotics* 42, 823–825.

Nishioka, H., Sawa, T., Hamada, M., et al. (1990) *J. Antibiotics* 43, 1586–1589.

Nishioka, H., Sawa, T., Isshiki, K., et al. (1991a) *J. Antibiotics,* 44, 1283–1285.

Nishioka, H., Sawa, T., Nakamura, H., et al. (1991b) *J. Nat. Prod.* 54, 1321–1325.

Nishiwaki, S., Fujiki, H., Yoshizawa, S., et al. (1991) *Jpn. J. Cancer Res.* (*Gann*) 82, 779–783.

Ogawara, H., Akiyama, T., Ishida, J., Watanabe, S., and Suzuki, K. (1986) *J. Antibiotics* 39, 606–608.

Omura, S., Nakagawa, A., and Sadakane, N. (1979) *Tetrahedron Lett.* 1979, 4323–4326.

Onoda, T., Iinuma, H., Sasaki, Y., et al. (1989) *J. Nat. Prod.* 52, 1252–1257.

Onoda, T., Isshiki, K., Takeuchi, T., Tatsuta, K., and Umezawa, K. (1990) *Drugs Expl. Clin. Res.* 16, 249–253.

Shimada, N., Yagisawa, N., Naganawa, H., et al. (1981) *J. Antibiotics* 34, 1216–1218.

Toi, M., Mukaida, H., Wada, T., et al. (1990) *Eur. J. Cancer* 26, 722–724.

Tsuchiya, K. S., Moriya, Y., Kawai, H., et al. (1991) *J. Antibiotics* 44, 344–348.

Uehara, Y., Hori, M., Takeuchi, T., and Umezawa, H. (1985a) *Jpn. J. Cancer Res.* (*Gann*) 76, 672–675.

Uehara, Y., Hasegawa, M., Hori, M., and Umezawa, H. (1985b) *Cancer Res.* 45, 5230–5234.

Uehara, Y., Fukazawa, H., Murakami, Y., and Mizuno, S. (1989) *Biochem. Biophys. Res. Commun.* 163, 803–809.

Umezawa, H., Imoto, M., Sawa, T., et al. (1986) *J. Antibiotics* 39, 170–173.

Umezawa, K., Atsumi, S., Matsushima, T., and Takeuchi, T. (1987) *Experientia* 43, 614–616.

Umezawa, K., Imoto, M., Yamashita, T., et al. (1989) *Jpn. J. Cancer Res.* (*Gann*) 80, 15–18.

Umezawa, K., Hori, T., Tajima, H., et al. (1990) *FEBS Lett.* 260, 198–200.

Umezawa, K., Tanaka, K., Hori, T., et al. (1991) *FEBS Lett.* 279, 132–136.

Yamashita, T., Imoto, M., Isshiki, K., et al. (1988) *J. Nat. Prod.* 51, 1184–1187.

Screening Methodologies for the Discovery of Novel Cytotoxic Antitumor Agents

Anna M. Casazza
Byron H. Long

The discovery of novel antitumor agents has relied heavily in the past on screening procedures applied to materials from natural sources. A recent review reported that approximately half of the anticancer drugs that are commercially available or in late stages of clinical trials were discovered by systematic screening of synthetic compounds and natural products (plant extracts and microbial fermentation broths) (Sikic 1991). Currently, the antitumor natural products successfully used in the clinic and discovered through this process are: daunorubicin and doxorubicin (Adriamycin), mitomycin C, bleomycin, vincristine and vinblastine, actinomycin D, plicamycin, streptozotocin, and more recently, paclitaxel (Taxol®). In addition, novel natural products have been the substrate for chemical modifications, resulting in semisynthetic compounds highly successful for the clinical treatment of cancer, such as etoposide (VP-16), teniposide (VM-26), and several anthracyclines (4'-epidoxorubicin and 4-demethoxydaunorubicin).

Despite these successes, the treatment of cancer patients remains one of the major clinical problems. In particular, patients with solid tumors, such as lung and colon carcinomas, account for a high percentage of cancer cases, and are, in most cases, resistant to the activity of the known anticancer drugs. This has stimulated the search for novel screening assays that could lead to the discovery of compounds with novel mechanisms of action and potential activity against drug-insensitive tumors. This approach has been prompted by novel discoveries in the biology of the cancer cell, and several targets have now been selected based on the mechanisms that regulate cell growth.

In this review we will focus on current methodologies developed for the discovery of novel *cytotoxic* antitumor agents.

The experimental model is the cornerstone of pharmacological research, and in the search for antitumor agents, the selection of suitable models to serve as candidates for clinical trials has always been a critical problem. The largest body of data in this field has been accumulated at the National Cancer Institute (NCI), USA. Initiated in 1937, and reorganized several times (Schepartz 1977; Boyd 1989), the primary screen for anticancer agents at the NCI has undergone several modifications. Originally, very sensitive in vivo experimental tumor models (sarcoma 180, carcinoma 755, leukemia L1210) were used, because all the clinically active compounds were active in these systems (Gellhorn and Hirschberg 1955). Sarcoma 180 and carcinoma 755 were subsequently replaced with the Walker carcinosarcoma. It was later recognized that such in vivo models were overpredictive (identifying compounds that were not confirmed to be active in other tumor models), and in 1975 they were substituted by the mouse leukemia P388, which was used until 1985 as the initial "Stage I" prescreen. Today this model is still a standard tumor model in most laboratories. Concurrently, cell culture and other in vitro assays were being developed, and a body of in vitro-in vivo correlations were being accumulated. The NCI moved, therefore, to a screening flowchart that included in vitro cytotoxicity on KB human carcinoma cells, followed by an in vivo assay against the mouse P388 leukemia, and by the "advanced" evaluation of the activity against a panel of mouse and human tumors transplanted in conventional and nude mice, respectively. It was through this approach that an interesting novel natural compound from plants, paclitaxel, was discovered (Wani et al. 1971). Currently, paclitaxel (Taxol®) is undergoing intensive clinical investigations and shows promising activity against human carcinomas. It is evident that the use of an in vitro assay facilitates discovery of novel natural products if the test can be adapted to follow the activity through the fractionation procedure. In 1985 the NCI, in order to increase the predictivity of in vitro assays for clinical activity, initiated the development of a new screening procedure based on in vitro assay of a large number of human tumor cell lines. This methodology is currently under validation for use in natural antitumor agents discovery programs.

In parallel, several other laboratories have approached the discovery of novel antitumor compounds by introducing a large number of screening assays. This chapter will address this large body of experimental approaches and emphasize mechanistically designed assays, as opposed to empirical approaches. It is our belief that a mechanistic approach is now possible and can open new avenues in the discovery of novel anticancer agents.

10.1 IN VITRO BIOCHEMICAL ASSAYS

10.1.1 Effects on Nucleic Acids

10.1.1.1 Binding to DNA. Several antitumor agents exert their cytotoxic effect because of interaction with cellular DNA, and in vitro techniques that measure this interaction can be used for antitumor drug screening.

The identification of DNA binding compounds can be accomplished by observing altered migration of supercoiled, covalently closed, circular DNA (scccDNA) in agarose gels. This approach provides several advantages for identifying DNA binding agents. First, this is a sensitive method of screening for active agents and, second, preliminary characterization of the type of DNA binding can be accomplished by carefully studying the actual binding effects of a given component upon the migration of scccDNA in agarose gels (Mong et al. 1979a). In this regard, it is possible to distinguish nonintercalative from intercalative binding to DNA by the extent of unwinding of supercoiled DNA brought about by the intercalation process, as analyzed by agarose gel electrophoresis (Waring 1970; Mong et al. 1979a). Furthermore, this DNA binding can be quantified fluorometrically by measuring the displacement of ethidium bromide from DNA in solution (Baguley et al. 1981). However, the amount of time and effort required to prepare, load, and run gels and the fact that results are not readily quantitative represent major disadvantages of this agarose gel-based approach for high throughput screen.

One simple procedure is to measure the ability of a molecule to bind to DNA by size-exclusion high pressure liquid chromatography (HPLC) measurements on a DNA-compound mixture. This technique can be modified to include an analytical column, where DNA elutes as an unretained peak and DNA-binding drugs would cause a reduction of the peak. This method is rather new for application to crude extracts, and the technique will have to be appropriately validated before extending its use to large mass screenings (McPherson and Pezzuto 1983).

10.1.1.2 DNA Strand Scission. DNA strand scission assays using scccDNA from bacteriophage phiX174 or from phage PM2 (Mong et al. 1979a) has been used by several laboratories and also seems suitable for use with crude extracts. As with DNA binding agents, scccDNA can be used to further analyze the DNA breaking activity of active agents such as tallysomycin, bleomycin, auromomycin, and hedamycin (Paoletti et al. 1971; Mirabelli et al. 1980; Suzuki et al. 1979; Strong and Crooke 1978; Lloyd et al. 1978; Mong et al. 1979b). DNA breakage activity on scccDNA can be quantified using a very sensitive fluorometric method based upon the difference in the quantity of ethidium bromide binding to scccDNA and relaxed DNA caused by the introduction into the DNA of either single- or double-strand breaks (Paoletti et al. 1971). Again, this agarose gel-based assay suffers from the same disadvantages for high throughput screening as described for DNA binding systems.

Alkaline elution techniques developed in the laboratory of Dr. K. Kohn provide a very sensitive method for demonstrating DNA break formation in intact cells (Kohn et al. 1981; Bradley and Kohn 1979). However, this method cannot be considered for use in screening for DNA breaking agents because of the requirement that each sample be slowly pumped overnight through a filter. An adaptation of the classical alkaline elution procedure has been made whereby DNA samples are eluted from filters by gravity over a 2-hour period, thus eliminating the requirement for pumping samples through filters (Hincks and Coulombe 1986). It is likely that this procedure could be adapted for automation through the use of microtiter plates, filters, and the appropriate processing equipment.

10.1.1.3 DNA Cross-Linking Formation. The procedure of Morgan and Pulleyblank (1974) may be adapted to identify and quantify the presence of DNA cross-linking agents in a large number of samples. This procedure also relies upon the difference in fluorescence of ethidium bromide in the presence of native duplex DNA where intercalation is favored, or in the presence of denatured DNA where intercalation is much less likely to occur. DNA containing interstrand cross-links will rapidly renature by the process of "snap back" following exposure to denaturing conditions such as heat when conditions favor hybridization of complementary DNA strands, whereas denatured duplex DNA lacking interstrand cross-links will renature very slowly under the same conditions. It is likely that such an assay can be adapted to microtiter plates and fluorescent readers.

10.1.2 Inhibition of Topoisomerases I and II

Several antitumor compounds that have proven successful in the clinic, such as etoposide, teniposide, and possibly the anthracyclines, are considered to exert their cytotoxic effects through interaction with topoisomerase II (Long and Brattain 1984). The camptothecins, the only known inhibitors of topoisomerase I (Hsiang et al. 1985), are now being developed in the clinic because of their promising antitumor activity in experimental tumor models. The search for novel topoisomerase I and topoisomerase II inhibitors can therefore be a successful approach for the discovery of novel anticancer agents.

The most direct approach for identifying the presence of topoisomerase I and II active agents in a natural extract consists of screen systems designed to measure activity against a purified or partially purified enzyme in vitro. In addition, one could consider cytotoxicity screens based upon the differential cytotoxic effects produced against cancer cells, which contain more or less of a given topoisomerase.

In vitro assays can be divided into those that identify agents capable of inhibiting the catalytic activity of either topoisomerase or those agents capable of stabilizing the normally transient covalent intermediate that forms between either enzyme in the process of performing its catalytic activity. In vivo assays have the advantage of being able to identify both types of activities (if the results are evaluated properly) and of not requiring the preparation of large quantities of purified enzyme or radiolabeled DNA.

When addressing the issue of whether inhibition of catalytic activity or intermediate stabilization should be the desired activity of a topoisomerase-targeted anticancer agent, consideration should be given to the fact that essentially all topoisomerase-active anticancer agents known today appear to produce their effect by intermediate stabilization rather than by inhibition of catalytic activity. Camptothecin and its analogs stabilize the covalent intermediate formed between topoisomerase I and its DNA substrate (Hsiang et al. 1985). Although the actual cytotoxic event is not yet known, recent evidence suggests that the toxic event requires DNA synthesis, since inhibition of DNA synthesis almost completely protects cells from the cytotoxic effects of camptothecin (Hsiang et al. 1989; Holm et al. 1989; D'Arpa et al. 1990). All the known topoisomerase II active agents also stabilize the covalent DNA-enzyme intermediate (Nelson et al. 1984; Ross et al. 1984; Minocha and Long 1984a, 1984b; Long and Brattain 1984; Chen et al. 1984; Tewey et al. 1984a, 1984b) except for fostriecin, which appears to bind to the enzyme and thus prevents the enzyme from interacting with the DNA (Bortizki et al. 1986, 1988). Again, the mechanism by which the presence of stabilized topoisomerase II-DNA intermediates causes cell death is not known, but it is unlikely that a requirement for DNA synthesis plays a role (Holm et al. 1989; D'Arpa et al. 1990). One possibility is that subsequent DNA deletion and recombination events may occur (Bae et al. 1991; Charron and Hancock 1991; Long and Stringfellow 1988; Long 1987).

Microtiter assays that measure the inhibition of topoisomerase I and II using labeled DNA have been described and can be easily set up. This type of assay is particularly useful for testing topoisomerase I.

An additional test for topoisomerase I, developed by Jaxel et al. (1989), employs P^{32}-labeled, closed-circular pBR322 DNA and commercial topoisomerase I. The samples are electrophoresed, and the DNA portions in the gel are transferred to a Nytran membrane and exposed to Kodak XAR x-ray films. A similar test can be used for topoisomerase II, with the addition of a topoisomerase II stop mix to halt the reaction.

10.1.2.1 In vitro Inhibition of the Catalytic Activity of Topoisomerase II.

The assays described in this section represent means by which inhibition of the catalytic activity of DNA strand passage can be qualitatively or quantitatively assessed. Such assays would be specific for topoisomerase II and would not be influenced by topoisomerase I or DNases unless stated otherwise.

DNA Catenation. Type II topoisomerases convert scccDNA into catenated rings by DNA strand passage in the presence of a DNA aggregating agent such as histone H1 or polyamines and ATP (Hsieh and Brutlag 1980; Goto and Wang 1982; Shelton et al. 1983). Substrate depletion and catenated DNA production can be evaluated by agarose gel electrophoresis. Such assays are only semiquantitative, have low sample capacity because of the agarose gel electrophoresis step, and are labor intensive. The problems of quantification and labor requirements could be overcome through the use of radiolabeled DNA and by substituting differential centrifugation for gel electrophoresis as a means of separating the substrate from the products. Under appropriate conditions where enough enzyme activity is added to almost completely convert the scccDNA to the catenated product, inhibition would be reflected by increased levels of uncatenated DNA.

DNA Decatenation. Decatenation is similar to catenation except that topoisomerase II converts the naturally catenated kinetoplast DNA (kDNA) into covalently closed, circular DNA (cccDNA) in the presence of ATP but in the absence of DNA aggregating agents (Miller et al. 1981; Marini et al. 1980). This assay is the most popular system currently in use for the discovery of topoisomerase II active agents, but suffers from the same disadvantages as described above for catenation reactions and also has the disadvantage that histones contaminating the enzyme preparations in high enough concentrations would favor catenation rather than decatenation. Although kDNA is not commercially available, trypanosomes are easy to grow and the kDNA is readily prepared in large quantities. Two new assays were described recently that avoid the use of agarose gel electrophoresis to resolve the reactants and products. Radiolabeled kDNA prepared from trypanosomes grown in medium containing radiolabeled thymidine may be either entrapped in loose agarose gels in microtiter wells where topoisomerase II activity frees the cccDNA to diffuse from the gel into the reaction mixture above (Muller et al. 1989) or sedimented by medium speed centrifugation to separate the kDNA from the freed cccDNA. In either assay, inhibition would be reflected by a decrease in radioactivity from untreated control levels.

DNA Unknotting. This assay utilizes the strand passing activity of topoisomerase II to unknot a naturally occurring knotted, covalently closed, circular DNA (kcccDNA) found in immature bacteriophage P4 heads (Liu et al. 1980, 1981) and is similar to and has all the advantages of the above-described assays. In addition, any influence from contaminating histones in the purified topoisomerase II preparation would be clearly discernable as a catenated product upon agarose electrophoresis. The unknotting process is a more quantitative reaction and was recently used to define kinetic parameters for the topoisomerase II reaction and its inhibition by different agents (Hofmann et al. 1990) but can only be evaluated by the use of agarose gel electrophoresis and, thus, is not amenable to automation.

DNA Relaxation.   The relaxation of supercoils in scccDNA˙ is a well known reaction of topoisomerase II (Goto and Wang 1982; Miller et al. 1981; Hsieh and Brutlag 1980), but this reaction is also accomplished by topoisomerase I, and the substrate and products can only be efficiently separated by agarose gel electrophoresis. Topoisomerase I is one of the most likely contaminants in the purification of topoisomerase II. This is the least desirable in vitro assay for use in high throughput screening.

10.1.2.2 In Vitro Inhibition of the Catalytic Activity of Purified Topoisomerase I.

The only assay suitable for assaying topoisomerase I activity and its inhibition by agents under evaluation is the DNA relaxation reaction, which has the same disadvantages as described for use in identifying topoisomerase II inhibitors. This reaction is, however, useful for follow-up studies of topoisomerase I inhibitors. No other biochemical screens are currently available for use in the screening of topoisomerase I inhibitors.

10.1.2.3 In Vitro Intermediate Stabilization Assays Utilizing Purified Topoisomerase II.

These assays rely upon the unique ability of type II topoisomerases to catalytically pass a DNA strand through an adjacent strand of DNA by first forming a transient double-strand break in one of the DNA strands through which the other DNA strand will pass, and then religating this break. In the process of forming the break, the enzyme becomes covalently bound to the 5' terminus through a phosphodiester linkage to a tyrosine in the active site of the enzyme (Hsieh 1983; Sander and Hsieh 1983; Liu et al. 1983).

Linearization of scccDNA.   All topoisomerase II active agents displaying the ability to stabilize the normally transient enzyme-DNA intermediate will cause the conversion of scccDNA (Form I) into linearized (Form III) DNA under appropriate conditions. The incubation of topoisomerase II with DNA in the presence of an agent that stabilizes the enzyme-DNA intermediate results in the rapid establishment of an equilibrium between the ternary complex composed of DNA, enzyme, and drug and the same components that exist free in the reaction mixture. This stabilized intermediate, experimentally referred to as the "cleavable complex," can be transformed into a double-strand DNA break with a 4 base overhang of the 5'-terminus of each strand at the break site upon treatment with the protein denaturant sodium dodecyl sulfate (SDS)

(Liu et al. 1983; Hsieh 1983) and digestive removal of the topoisomerase II covalently bound to the 5′-terminus (Liu et al. 1983; Hsieh 1983; Sander and Hsieh 1983; Marshall and Ralph 1982). It is imperative that this reaction mixture be maintained at the temperature of the reaction, normally 30° to 37°C to prevent reversal of this equilibrium at cooler temperatures (Liu et al. 1983; Sander and Hsieh 1983; Hsieh 1983).

This assay requires agarose gel electrophoresis to separate Form III from Form I and the other DNA forms produced in the reaction, and consumes substantially larger quantities of purified topoisomerase II than do inhibition studies. Furthermore, samples incubated with DNA in the absence of enzyme must be included to verify that any linearized DNA produced was not the result of endonucleases. It is possible to quantify the results, but quantification requires scanning of photographic negatives. This assay would be suitable for screening large numbers of samples on a qualitative basis by visually observing the presence of linearized DNA in the agarose gels and for follow-up studies of active agents in detail but would be expensive in time, labor costs, and the consumption of purified enzyme.

A closely related assay uses ^{32}P-labeled DNA fragments of a given size defined by restriction enzyme cleavage from larger DNA sources. These fragments, labeled only on the 3′ or 5′ strand at one end, are incubated with topoisomerase II in the presence or absence of active agents, then subjected to agarose gel electrophoresis to define the DNA cleavage specificity of the agent (Chen et al. 1984; Tewey et al. 1984a, 1984b; Nelson et al. 1984). This procedure is highly labor intensive and would be used only to further define the characteristics of the active agent and its interaction with topoisomerase II and DNA.

Potassium SDS Precipitation of DNA-Protein Complexes. This procedure, which was developed independently in the laboratories of Dr. M. Muller (Muller 1983; Trask et al. 1984) and Dr. L. Liu (Liu et al. 1983), is based on the ability of many molecules of SDS to become intertwined with the polypeptide chains of a protein through hydrophobic interactions as the means by which this detergent denatures proteins. Once bound to protein molecules in this manner, the addition of potassium ions to this solution of SDS-denatured proteins and free SDS results in the precipitation of not only the free SDS as insoluble potassium SDS salts but also all proteins containing SDS. If any of these precipitated proteins are topoisomerases existing as covalent stabilized intermediates with DNA, then the covalently bound DNA will also be precipitated. This reaction could be conducted in microtiter wells and K-SDS precipitates would be sedimented by centrifugation of microtiter plates. Thus, the use of radiolabeled DNA in the reaction mixture would provide a convenient method for identifying and quantifying topoisomerase II active agents. The only disadvantages of such an assay would be the large amount of purified topoisomerase required and the preparation and use of radiolabeled DNA.

10.1.2.4 In Vitro Intermediate Stabilization Assays Using Purified Topoisomerase I
Like topoisomerase II, topoisomerase I forms a transient, covalent intermediate between the DNA substrate and the enzyme. This intermediate differs from that formed by topoisomerase II in that only a single DNA strand is broken at any one site and the enzyme is covalently bound to the 3′ terminus at the DNA break site (Hsiang et al. 1985). The only well-characterized topoisomerase I active agents, camptothecin and its derivatives, stabilize this intermediate (Hsiang et al. 1985).

Nicking of scccDNA. The introduction of a one DNA single-strand break causes scccDNA to convert from the supercoiled (Form I) to the nicked, relaxed DNA (Form II), which migrates more slowly than the supercoiled (Form I), or linear (Form III) topologically isomeric forms when they are subjected to agarose gel electrophoresis in the presence of the DNA intercalator ethidium bromide. When scccDNA is incubated with topoisomerase I in the presence of an intermediate stabilizing agent such as camptothecin, an equilibrium is rapidly established between a ternary complex consisting of an enzyme, DNA, and compound and those components existing free in the reaction mixture. The addition of SDS prevents reversal of the DNA-enzyme complex by denaturing the enzyme, which is then removed from the DNA by addition of proteinase K to reveal a frank single-strand DNA break. This break can be visualized by agarose gel electrophoresis in the presence of ethidium bromide (Hsiang et al. 1985). As with the equivalent topoisomerase II assay, samples must also be incubated with DNA in the absence of enzyme to verify that any observed DNA nicking activity is not the result of endonuclease activity. Again, this is a labor-intensive gel-based assay that is not amenable to automation and has the added disadvantage that highly pure scccDNA (as free from contaminating Form II DNA as possible) must be used in order to reduce background signal levels.

Potassium SDS Precipitation of DNA-Protein Complexes. As described in Section 10.1.2.3, the detergent SDS strongly binds to any protein and, thus, denatures that protein. The addition of potassium ions to the mixture causes the precipitation of free and protein-bound SDS. This precipitative action would also cause the precipitation of any radiolabeled DNA covalently bound to topoisomerase I. The advantages and disadvantages of this method are the same as those described above for the equivalent topoisomerase II assay.

10.1.2.5 Assays for Agents Active against Topoisomerase II in the Cell.
As described in Section 10.1.2, in vivo assays, in some cases, have the advantage over in vitro assays of being able to identify both topoisomerase I and II active agents and do not require the preparation of large quantities of purified enzyme or radiolabeled DNA, since these components are already present or can be readily produced in cells.

Alkaline DNA Elution Techniques.   Alkaline elution is a very sensitive method to demonstrate DNA break formation in intact cells (Kohn et al. 1981; Bradley and Kohn 1979). In fact, it is through the use of alkaline elution technology that topoisomerase II was identified as the target for VP-16 and VM-26, adriamycin, mAMSA (Amsacrine, 4'-(9-acridinylamino)-methane-sulfon-*m*-anisidide), and ellipticine due to the DNA breaks produced in cells that reflect the stabilized topoisomerase-DNA intermediates. However, this method cannot be considered for use in screening for DNA breaking agents because of the amount of processing required for each sample and the limited number of samples that can be run in each experiment, because each sample must be slowly pumped overnight through a filter. An adaptation of the classical alkaline elution procedure has been made whereby DNA samples are eluted from filters by gravity over a 2-hour period, thus eliminating the requirement for pumping samples through a filter (Hincks and Coulombe 1986). It is likely that this procedure could be adapted for automation through the use of microtiter plates and filters and the appropriate processing equipment.

Potassium SDS Precipitation of DNA-Protein Complexes in Cells.   The procedure of Rowe et al. (1986) describing the K-SDS precipitation assay for the presence of enzyme-DNA complexes in intact cells provides a relatively sensitive microtiter method for identifying both topoisomerase I and II active agents capable of stabilizing the enzyme-DNA intermediate. This procedure involves exposing cells containing tritium-labeled DNA to test samples for 1 hour and lysing the cells with hot SDS solution. The SDS denatures any covalently bound topoisomerases and prevents dissociation of the enzyme from DNA. After shearing the DNA by vortexing or passing through a syringe needle to reduce the viscosity, KCl is added to precipitate SDS as insoluble K-SDS salt and to precipitate all proteins, including topoisomerases covalently bound to DNA. In this process, only radiolabeled DNA covalently bound to protein will precipitate. The precipitates are washed several times in high KCl buffer by heating at 65°C to dissolve the pellets and then cooling on ice to reprecipitate the K-SDS salts before counting the radioactivity in the pellets.

This procedure offers the advantages that topoisomerases do not have to be purified, that activities against both topoisomerase I and II will be detected, and that the assay is capable of being automated and is relatively inexpensive. Follow-up studies conducted by agarose gel electrophoresis would be able to confirm activities and distinguish between topoisomerase I and II active agents. The only disadvantage is that the DNA shearing must be well controlled in order to maintain a high degree of sensitivity.

10.1.3   Tubulin Binding Assays

The positive clinical results recently obtained in patients treated with paclitaxel have redirected the attention of investigators toward cytotoxic compounds that interact with tubulin. Paclitaxel (Taxol®) was identified in 1971 in extracts of the bark of the Western Yew (*Taxus brevifolia*) using the in vitro cytotoxicity assay on KB cells (Wani et al. 1971). Paclitaxel had antitumor activity against P388 leukemia and some human tumor xenografts, and was therefore introduced into clinical trials. However, the development of paclitaxel proceeded very slowly due to the scarcity of bulk compound. It was only after the discovery that this compound acted through a completely novel mechanism (Schiff et al. 1979) that efforts were intensified to obtain adequate supplies to complete Phases I and II. Clinical results now indicate that paclitaxel has impressive activity in refractory ovarian carcinoma patients (McGuire et al. 1989) and substantial activity in the treatment of breast carcinoma (Holmes et al. 1991). Paclitaxel acts by promoting the polymerization of α and β tubulin to form abnormally stable microtubules. In this respect, paclitaxel differs from the other known tubulin-interacting agents, such as vincristine, vinblastine (also established anticancer agents), and colchicine, which inhibit tubulin polymerization. Cells treated with paclitaxel either do not enter mitosis, or are arrested during mitosis because of the lack of spindle formation.

Screening methodologies capable of identifying novel chemotypes with tubulin-interacting properties are therefore of high interest. Technically, several methodologies have been reported in the literature based on a measurement of the modification of UV light absorption caused by diffusion by the microtubules in the presence or absence of microtubule proteins (Schiff et al. 1979). This assay is being used in several laboratories to investigate paclitaxel analogs using tubulin from mammalian organs (generally brain); however, reports on the use of similar assays to identify novel chemotypes have not yet appeared in the literature.

10.2 MAMMALIAN CELL-BASED ASSAYS

10.2.1 Cell Cytotoxicity Assays

The selection of novel chemotypes through the measurement of cytotoxicity against tumor cells in vitro has been employed for several years, and adopted for large-scale screening by institutions like the NCI, therefore contributing to large diffusion of such methodologies. The most commonly used cell lines were derived from human carcinomas, such as HeLa or KB (Geran et al. 1972; Shoemaker et al. 1983). Several novel antitumor agents have been isolated using this approach, including natural products, because the cell cytotoxicity assay can be readily adapted to support bioassay-guided fractionation.

10.2.2 Differential Cytotoxicity Assays

In an effort to increase the cytotoxic selectivity of compounds for cancer cells, several approaches have been developed based on a comparison of normal and tumor cell cytotoxicity, or on a comparison of activity in proliferating and quiescent cells. Several cell types or cell lines have been employed as being representative of normal cells, such as bone marrow cells (in an attempt to discover cytotoxic non-myelosuppressive agents), and human fibroblasts derived from human foreskin and maintained at early passages. However, no compound has been discovered using this approach that could not have been discovered using cytotoxicity assays alone. Generally, it is recognized that the comparison of cytotoxicity against tumor cells with that against normal cells can be used as a secondary assay, rather than as a primary "discovery" assay. More interesting are tests developed in the last 10 years to select compounds that show specific activity against selected tumor cell lines in vitro, which are described below.

10.2.2.1 Disk Diffusion Soft Agar Assay. The first design and application on a relatively large scale of an assay based on selective cytotoxicity against a specific tumor type was initiated by T.H. Corbett (Corbett 1984; Corbett et al. 1986). In addition to using two different tumor cell lines, this assay was original in that it was based on the evaluation of the inhibition of colony formation by tumor cells in soft agar, a property that seems to be more related to the cancer cells' malignancy than growth on a plastic surface in a liquid medium. In addition, compounds were not tested at various dilutions, but absorbed onto paper disks located on the agar, and zones of inhibition of colony formation were evaluated using an automated device, thus rendering this assay amenable to large throughput screening. In the original design of this assay, materials are evaluated against two different tumors at the same time, in the same disk in soft agar, where colonies formed are morphologically distinguishable: a leukemia (L1210) and a drug-insensitive solid tumor, such as a pancreatic or a colon carcinoma. The evaluation is based not on cytotoxicity per se, but on specific cytotoxicity against the solid tumor cells, where the leukemia cells serve as controls. The assay was subsequently modified to use one cell type per dish.

A typical agent identified using this approach is flavone acetic acid, an agent having cellular selectivity against solid tumors in vitro and in vivo. However, this compound did not show activity in the clinic, probably due to metabolic factors.

This approach can be employed for tumor cells of various origin, including selectivity against hormone-dependent and hormone-independent tumor cell lines, or against tumors of similar origin but with different biological properties, such as expression of growth factors and/or their receptors.

10.2.2.2 The NCI Cell Panel. Existing anticancer agents were selected on the basis of leukemia in vivo assays, which could explain the limited spectrum of clinical activity. In 1984 the NCI initiated the design and development of an in vitro *disease-oriented,* preclinical, anticancer drug discovery strategy. Due to the impracticality of in vivo screening with adequate capacity and specific tumor type representation, the NCI decided to rely on an in vitro panel of 60 human tumor cell lines. The cell lines panel represented seven cancer types of clinical relevance: lung, colon, melanoma, renal, ovarian, brain, and leukemia (Boyd 1989). The development of such a panel was initiated in 1985 and is now complete. The cell lines have been selected, adapted to grow in vitro in a single growth medium, and characterized to meet quality assurance criteria and for other biological parameters. A microtiter assay was developed in which cellular growth was initially assessed using a metabolic assay, based on the cellular reduction of a colorless tetrazolium salt (MTT or XTT) to yield a colored formazan derivative (which can be measured by an automated colorimeter) in proportion to viable cell number. During the evolution of the large-scale screen, however, technical limitations of these assays became apparent and an alternative approach was developed in which cell growth is measured with sulphorodamine B, a dye that binds to basic amino acids of cellular macromolecules (Skehan et al. 1990). The full screen was validated at the end of 1989 and the NCI is currently testing at a rate of 400 materials per week against the 60-line panel, using five sample dilutions in duplicate, and 2 days of exposure. The minimum amount of compound required is 40 ml, containing the material at two times the highest concentration desired. The screening of each material generates a voluminous amount of data, because for each cell line three major parameters are evaluated: GI_{50} (as well as GI_{10} and GI_{90}), TGI, and LC_{50} (50% lethal concentration), which represent the material concentration resulting in 50% (10%, 90%): growth inhibition, total growth inhibition, and 50% reduction in the cell number at the end of the incubation period compared to that at the beginning of the experiment (Monks et al. 1991).

During the validation of the cell panel using previously known agents, and using a computer program called COMPARE, the observation was made that antitumor drugs have a profile ("fingerprint") of cytotoxicity. This profile is similar for compounds that have similar mechanisms of action. This observation allows prioritization of compounds acting through particularly interesting mechanisms and possibly the identification of compounds active through novel mechanisms.

The NCI is planning to complement the in vitro panel with an in vivo panel of the corresponding tumors in order to validate the antitumor activity of selected compounds. The complete in vitro-in vivo panel will hopefully be available to the scientific community.

10.2.2.3 The Experience at Bristol Myers Squibb.
In an effort to increase the probability of discovering compounds active against the most drug resistant solid tumors, we set up a small cell line panel constituted of colon and lung carcinoma cell lines, and used it for the primary screening in our fer-

mentation program (Catino et al. 1985). After several years of experience, we realized that novel agents were being identified that could have been discovered using only one of the cell lines, the HCT-116 human carcinoma line. We therefore proceeded by using only this cell line in our primary screen, and several novel antibiotics have been discovered using this approach, the most recent being kedarcidin (Crosswell et al. 1990a). In addition, a study was conducted to evaluate the correlation between the results obtained with the cell lines panel in vitro, and the results obtained in vivo in the P388 leukemia model (Rose et al. 1988). Each material, regardless of its in vitro cytotoxicity, was evaluated in vivo, and a highly significant relationship between these two assays was demonstrated for each cell line. When the cell lines were compared, HCT-116 provided a small advantage in predicting for P388 activity. Therefore, our in vitro screen against the human colon carcinoma cell line HCT-116 could be followed by an in vivo assay using the P388 leukemia model, which is convenient and easy to use. To evaluate if materials active in vitro on the HCT-116 cell line, but not in vivo against the P388 leukemia line, could be specifically active against the colon carcinoma cell line, we developed the corresponding in vivo tumor model. This model involves transplantation of the tumor subcutaneously or under the renal capsule (subrenal capsule assay) (Basler et al. 1986). Currently, our natural products program uses these models for the discovery of novel cytotoxic antitumor agents. Interesting antitumor compounds have been discovered from fermentation broths and, more recently, from marine organisms through a National Cooperative Natural Products Drug Discovery Group (Staley and Clardy 1993).

10.2.2.4 Differential Response between Sensitive and Resistant Cell Pairs.

Resistance to antitumor agents is the cause of insensitivity of several tumor types to drug treatment and relapse after chemotherapeutic regimens. The biochemical mechanism of resistance depends on the mechanism of action of the drug to which resistance is expressed. However, a single mechanism can cause resistance to several classes of drugs, resistance to one drug can be sustained by several different mechanisms, and resistance to several agents can coexist in the same cell. One of the most explored mechanisms is that of multidrug resistance (MDR), which is expressed in cells treated with most lipophilic antitumor antibiotics (except bleomycin) and therefore of interest in relation to natural products (Moscow and Cowan 1988). MDR is related to the overexpression or mutation of a 170-kilodalton (kDa) transmembrane protein (P-glycoprotein) that is involved in xenobiotics extrusion from the cell (Roninson 1987).

Another mechanism of MDR, also called "atypical MDR," has been related to reduced levels of activity of topoisomerase II or mutation in topoisomerase II (Danks et al. 1987). Many laboratories have now reported that resistance to the topoisomerase II active drugs VP-16 and VM-26 (De

Jong et al. 1990; Hong et al. 1990; Potmesil et al. 1988), mAMSA (Charcosset et al. 1988; Pommier et al. 1986; LeFevre et al. 1991; Potmesil et al. 1988), anthracyclines (De Jong et al. 1990; Deffie et al. 1989; Potmesil et al. 1988; Friche et al. 1991), and other agents (Charcosset et al. 1988) and resistance to the topoisomerase I active compound camptothecin (Sugimoto et al. 1990; Eng et al. 1990), can be related to lower enzyme levels and/or activities. In addition, yeasts carrying a *rad52* DNA repair mutation are hypersensitive to camptothecin, whereas yeasts lacking the yeast topoisomerase I are completely resistant to this drug (Eng et al. 1988; Nitiss and Wang 1988, 1991; Bjornsti et al. 1989). However, the insertion of the cDNA of human topoisomerase I restores camptothecin sensitivity (Bjornsti et al. 1989; Nitiss and Wang 1991). To date, the human topoisomerase II cDNA has not been successfully cloned into a yeast or bacteria and expressed.

The use of sensitive and resistant pairs of a given cell line, where the mechanism of resistance to a topoisomerase active agent has been clearly shown to relate exclusively to the level of topoisomerase activity, offers a valuable, sensitive, and specific means of identifying topoisomerase-active agents. An ideal topoisomerase I screening system would be the insertion of the human topoisomerase I cDNA into a genetically permeabilized bacteria or yeast strain carrying the *rad52* DNA repair mutation and lacking the yeast topoisomerase I gene. The expression of the human enzyme would be under the control of an inducible promoter so that the same strain could be used under uninduced and induced conditions. This situation would represent a truly isogenic pair except for the inducible presence or absence of human topoisomerase I.

A similar approach can be applied to P388 leukemia cells selected for camptothecin resistance, which possess essentially no topoisomerase I (Eng et al. 1990). Thus, the parental cells would be sensitive and the camptothecin-resistant cells would be resistant to topoisomerase I active compounds that produce cytotoxicity by a mechanism similar to that of camptothecin.

Unfortunately, similar model systems do not exist for identifying topoisomerase II active agents because of the current inability to express a functioning eukaryote topoisomerase II cDNA in other cells, such as yeast or bacteria. A solution to this problem could be the use of a well-characterized, resistant eukaryote cell line expressing low levels of endogenous topoisomerase II, to be compared with its parental cell line. One possible example of such a cell line is the human colon carcinoma line HCT116 and its acquired VP-16-resistant counterpart, HCT116(VP)35, which expresses about half the parental levels of topoisomerase II mRNA and activity, yet has essentially no expression of P-glycoprotein, and normal glutathione levels (Long et al. 1991). Other interesting cell pairs reported in the literature include KB human cells and their resistant variants KB/VCR (which represents the mdr mechanism), KB/1c, and KB/1d (which are characterized by decreased levels of topoisomerase II and decreased intracellular drug accumulation) (Ferguson et al. 1988).

Compounds that interact with topoisomerase II can either inhibit the catalytic activity of the enzyme or stabilize the enzyme-DNA intermediate complex. A drug that inhibits the catalytic activity would be active in cell lines resistant to etoposide because of the reduced levels of topoisomerase II. Therefore, the use of these panels can help in understanding the mechanism of activity of the compound tested.

In our own experience, we continue to perform the primary screening against the human HCT-116 colon carcinoma cell line, as described in Section 2.2.3, but have introduced two of its sublines as secondary screen: the HCT-116/VP 35 subline, resistant because of low levels of topoisomerase II, and the HCT-116/VM46 MDR subline. This allows us to prioritize the chemical fractionation of extracts that show activity in the resistant lines.

10.2.3 Reversal of Drug Resistance

In addition to the use of resistant cells in cell line panels, described earlier (Section 2.2.4), there is an interest in the discovery of agents that do not have antitumor activity by themselves, but are able to reverse drug resistance or restore sensitivity to anticancer agents. Such agents, called resistance revertants, can be identified using combination assays with known agents in cell lines that have a selected established mechanism of resistance. Since one of the most common mechanisms of drug resistance is related to decreased intracellular accumulation of the cytotoxic agent, rapid tests have been developed to measure radiolabeled drug accumulation in cell lines pairs, sensitive and mdr-resistant.

10.3 IN VIVO ANIMAL MODELS

In pharmacology, any type of in vitro assay should be followed by appropriate in vivo assays to verify if compounds active in the artificial conditions of the in vitro assay are able to maintain their activity when injected in animals and subjected to the variables related to distribution and metabolism. In cancer, this is of even greater importance due to the fact that several compounds able to inhibit tumor cell growth in vitro, if devoid of the necessary specificity, will result in toxicity in the absence of any therapeutic effect. The use of in vivo models for primary antitumor screening has been discontinued in most laboratories, due to the high cost and impracticality for a high volume screen and, therefore, will not be discussed in detail in this chapter. However, and it cannot be stressed enough: in vivo assays must be present in a flow-chart for selection of novel antitumor agents for clinical trials. Below we outline a few fundamental questions, and provide some comments addressing these questions.

Are in vitro assays the most appropriate primary assays to discover antitumor agents, or do we need to use in vitro assays? This question has been animately debated for many years. Even if it is realized that in vivo assays are more complete, offer a large variety of information, and are more predictive of the clinical antitumor activity than any in vitro assay, because of the high cost, most laboratories now rely on in vitro prescreens similar to those described above and use in vivo assays as secondary tests to prioritize among discovered leads.

Among the in vivo assays to use as secondary screens, which models are the most appropriate? Most laboratories were using the models adopted by the NCI until the question arose: "Is the P388 tumor model no longer adequate as a drug discovery model?" (Corbett et al. 1987). We have summarized the changes implemented at the NCI. Many laboratories have introduced other tumor models in the screening process.

Are human tumor xenografts more predictive of the clinical activity than mouse tumors? There is no confirmed evidence that this is the case. In our experience at Bristol Myers Squibb, we have had the opportunity to evaluate in vivo fermentation extracts tested in the in vitro cell panel described above. Using two mouse tumor models, the P388 leukemia and the B16 melanoma (intraperitoneal and subcutaneous), an attempt was made to identify materials specifically active against solid tumors (Rose et al. 1988). This exercise was very useful for the evaluation of the predictability of our in vitro assays towards in vivo effectiveness. It was shown that the prescreen in the cell panel followed by the in vivo evaluation did not allow us to select compounds specifically active against the solid tumors. On the basis of these results, we have continued to use the P388 tumor model as the first in vivo assay as an indicator of the toxicity and to allow us to test in vivo a large number of compounds in modified Phase I-II preclinical evaluation. Positive materials are then evaluated in vivo in a panel of solid tumors of various origin. Because of the interest in discovering materials specifically active against human solid tumors, materials negative in vivo in the P388 leukemia model are subsequently tested against the HCT 116 tumor in nude mice, in a subrenal capsule assay (Basler et al. 1986; Crosswell et al. 1990b).

10.4 CONCLUSIONS

We have summarized the methodology to discover novel cytotoxic antitumor agents and the innovative changes that have occurred in the last 10 years. The development of novel assays based on newly identified targets in the cancer cell and the rationale drug design to affect relevant targets will surely enlarge and improve the armamentarium available to clinicians to treat cancer patients.

Some of the most relevant screens described above, such as the NCI cell panel, are complex, require large resources, and are not feasible for small organizations. This panel may complement the traditional evaluation of novel anticancer compounds, and have a role in further selection of compounds screened using other approaches and in the selection of the most promising analog derived from known antitumor agents.

REFERENCES

Bae, Y.-S., Chiba, M., Ohira, M., and Ikeda, H. (1991) *Gene* 101, 285–289.
Baguley, B.C., Denny, W.A., Atwell, G.J., and Cain, B.F. (1981) *J. Med. Chem.* 24, 170–177.
Basler, G.A., Schurig, J.E., Henderson, A.J., et al. (1986) *Proc. Am. Assoc. Cancer Res.* 27, 410 (abstract).
Bjornsti, M.-A., Benedetti, P., Viglianti, G.A., and Wang, J.C. (1989) *Cancer Res.* 49, 6318–6323.
Boritzki, T.J., Hann, F.S., Fry, D.W., et al. (1986) *Proc. Am. Assoc. Cancer Res.* 27, 276 (abstract).
Boritzki, T.J., Wolfard, T.S., Besserer, J.A., Jackson, R.C., and Fry, D.W. (1988) *Biochem. Pharmacol.* 37, 4063–4068.
Boyd, R.M. (1989) *Principles Practices Oncol.* 3(10), 1–12.
Bradley, M.O., and Kohn, K.W. (1979) *Nucleic Acids Res.* 7, 793–804.
Catino, J.J., Francher, D.M., Edinger, K.J., and Stringfellow, D.A. (1985) *Cancer Chemother. Pharmacol.* 15, 240–243.
Charcosset, J.-Y., Saucier, J.-M., and Jacquemin-Sablon, A. (1988) *Biochem. Pharmacol.* 37, 2145–2149.
Charron, M., and Hancock, R. (1991) *Chromosoma* 100, 97–102.
Chen, G.L., Yang, L., Rowe, T.C., et al. (1984) *J. Biol. Chem.* 259, 13560–13566.
Corbett, T.H. (1984) *Proc. Am. Assoc. Cancer Res.* 25, 325 (abstract).
Corbett, T.H., Valeriote, F.A., and Baker, L.H. (1987) *Invest. New Drugs* 5, 3–20.
Corbett, T.H., Wozniak, A., Gerpheide, S., Hanka, L. (1986) in *14th Int. Congr. Chemother.* (Hanka, L.J., Kondo, T., and White, R.J., eds.), pp. 5–14, University of Tokyo Press, Tokyo.
Crosswell, A.R., Rose, W.C., Clark, J.L., et al. (1990a) *Proc. Am. Assoc. Cancer Res.* 31, 416 (abstract).
Crosswell, A.R., Honyotsky, S.M., Robinson, C.E. (1990b) *J. Clin. Res. Clin. Oncol.* 116(suppl.), 450.
Danks, M.K., Yalowich, J.C., and Beck, W.T. (1987) *Cancer Res.* 47, 1297–1301.
D'Arpa, P., Beardmore, C., and Liu, L.F. (1990) *Cancer Res.* 50, 6919–6924.
De Jong, S., Zijlstra, J.G., De Vries, E.G.E., and Mulder, N.H. (1990) *Cancer Res.* 50, 304–309.
Deffie, A.M., Batra, J.K., and Goldenberg, G.J. (1989) *Cancer Res.* 49, 58–62.
DeJong, S., Zijlstra, J.G., DeVries, E.G.E., and Mulder, N.H. (1990) *Cancer Res.* 50, 304–309.
Eng, W.-K., Faucette, L., Johnson, R.K., and Sternglanz, R. (1988) *Mol. Pharmacol.* 34, 755–760.
Eng, W.-K., McCabe, F.L., Tan, K.B., et al. (1990) *Mol. Pharmacol.* 38, 471–480.
Ferguson, P.J., Fisher, M.H., Stephenson, J., et al. (1988) *Cancer Res.* 48(21), 5956–5964.
Friche, E., Danks, M.K., Schmidt, C.A., and Beck, W.T. (1991) *Cancer Res.* 51, 4213–4218.
Gellhorn, A., and Hirschberg, E. (1955) *Cancer Res.* 3(suppl.), 1–113.

Geran, R.I., Greenberg, N.H., MacDonald, M.M., Schumacher, A.M., and Abbott, B.J. (1972) *Cancer Chemother. Rep.* Part 3, 3, 1–103.

Goto, T., and Wang, J.C. (1982) *J. Biol. Chem.* 257(10), 5866–5872.

Hincks, J.R., and Coulombe, R.A., Jr. (1986) *Biochem. Biophys. Res. Commun.* 137, 1006–1014.

Hofmann, G.A., Mirabelli, C.K., and Drake, F.H. (1990) *Anticancer Drug Design* 5, 273–282.

Holm, C., Covey, J.M., Kerrigan, D., and Pommier, Y. (1989) *Cancer Res.* 49, 6365–6368.

Holmes, F.A., Walterts, R.S., Theriault, R.L., et al. (1991) *J. Natl. Cancer Inst.* 83, 1797–1805.

Hong, J.H., Okada, K., Kusano, T., et al. (1990) *Biomed. Pharmacother.* 44, 41–45.

Hsiang, Y.-H., Hertzberg, R., Hecht, S., and Liu, L.F. (1985) *J. Biol. Chem.* 260, 14873–14878.

Hsiang, Y.-H., Lihou, M.G., and Liu, L.F. (1989) *Cancer Res.* 49, 5077–5082.

Hsieh, T.-S. (1983) *Methods Enzymol.* 100, 161–170.

Hsieh, T.-S., and Brutlag, D. (1980) *Cell* 21(1), 115–125.

Jaxel, C., Kohn, K.W., Wani, M.C., et al. (1989) *Cancer Res.* 49, 1465.

Kohn, K.W., Ewig, R.A.G., Erickson, L.C., and Zwelling, L.A. (1981) in *DNA Repair: A Laboratory Manual of Research Procedures,* vol. 1 (Friedberg, E.C., and Hanawalt, P.C., eds.), pp. 379–401, Marcel Dekker, New York.

LeFevre, D., Riou, J.-F., Ahomadegbe, J.C., et al. (1991) *Biochem. Pharmacol.* 41, 1967–1979.

Liu, L.F., Liu, C.-C., and Alberts, B.M. (1980) *Cell* 19(3), 697–707.

Liu, L.F., Davis, J.L., and Calendar, R. (1981) *Nucleic Acids Res.* 9, 3979–3989.

Liu, L.F., Rowe, T.C., Yang, L., Tewey, K.M., and Chen, G.L. (1983) *J. Biol. Chem.* 258, 15365–15370.

Lloyd, R.S., Haidle, C.W., and Robberson, D.L. (1978) *Biochemistry* 17, 1890–1896.

Long, B.H. (1987) *NCI Monogr.* 4, 123–127.

Long, B.H., and Brattain, M.G. (1984) in *Etoposide (VP-16): Current Status and New Developments* (Issell, B.F., Muggia, F.M., and Carter, S.K., eds.), pp. 63–86, Academic Press, New York.

Long, B.H., and Stringfellow, D.A. (1988) in *Advances in Enzyme Regulation,* vol. 27 (Weber, G., ed.), pp. 223–256, Pergamon Press, New York.

Long, B.H., Wang, L., Lorico, A., et al. (1991) *Cancer Res.* 51(19), 5275–5283.

Marini, J.C., Miller, K.G., and Englund, P.T. (1980) *J. Biol. Chem.* 255, 4976–4979.

Marshall, B., and Ralph, R.K. (1982) *FEBS Lett.* 145, 187–190.

McGuire, W.P., Rowinsky, E.K., Rosenshein, N.B., et al. (1989) *Ann. Intern. Med.* 111, 273–279.

McPherson, D.D., and Pezzuto, J.M. (1983) *J. Chromatogr.* 281, 348–354.

Miller, K.G., Liu, L.F., and Englund, P.T. (1981) *J. Biol. Chem.* 256, 9334–9339.

Minocha, A., and Long, B.H. (1984a) *Fed. Proc.* 43, 1542.

Minocha, A., and Long, B.H. (1984b) *Biochem. Biophys. Res. Commun.* 122, 165–170.

Mirabelli, C.K., Huang, C.-H., and Crooke, S.T. (1980) *Cancer Res.* 40, 4173–4177.

Mong, S., Strong, J.E., Bush, J.A., and Crooke, S.T. (1979a) *Antimicrob. Agents Chemother.* 16, 398–405.

Mong, S., Strong, J.E., and Crooke, S.T. (1979b) *Biochem. Biophys. Res. Commun.* 88, 237–243.

Monks, A., Scudiero, D., Skehan, P., et al. (1991) *J. Natl. Cancer Inst.* 83(11), 757–766.

Morgan, A.R., and Pulleyblank, D.E. (1974) *Biochem. Biophys. Res. Commun.* 61, 396–403.

Moscow, J.A., and Cowan, K.H. (1988) *J. Natl. Cancer. Inst.* 80, 14–20.

Muller, M.T. (1983) *Biochem. Biophys. Res. Commun.* 114, 99–106.

Muller, M.T., Helal, K., Soisson, S., and Spitzner, J.R. (1989) *Nucleic Acids Res.* 17, 9499.

Nelson, E.M., Tewey, K.M., and Liu, L.F. (1984) *Proc. Natl. Acad. Sci. USA* 81, 1361–1365.

Nitiss, J., and Wang, J.D., (1988) *Proc. Natl. Acad. Sci. USA* 85, 7501–7505.

Nitiss, J., and Wang, J.C. (1991) in *DNA Topoisomerases and Cancer* (Potmesil, M., and Kohn, K., eds.), pp. 77–91, Oxford University Press, London.

Paoletti, C., Lepecq, J.B., and Lehman, I.R. (1971) *J. Mol. Biol.* 55, 75–100.

Pommier, Y., Kerrigan, D., Schwartz, R.E., Swack, J.A., and McCurdy, A. (1986) *Cancer Res.* 46, 3075–3081.

Potmesil, M., Hsiang, Y.-H., Liu, L.F., et al. (1988) *Cancer Res.* 48, 3537–3543.

Roninson, I.B. (1987) *Clin. Physiol. Biochem.* 5, 140–151.

Rose, W.C., Schurig, J.E., and Meeker, J.B. (1988) *Anticancer Res.* 8, 355–368.

Ross, W., Rowe, T., Glisson, B., Yalowich, J., and Liu, L. (1984) *Cancer Res.* 44, 5857–5860.

Rowe, T.C., Chen, G.L., Hsiang, Y.-H., and Liu, L.F. (1986) *Cancer Res.* 46, 2021–2026.

Sander, M., and Hsieh, T.-S. (1983) *J. Biol. Chem.* 258, 8421–8428.

Schepartz, S. (1977) *Jpn. J. Antib.* 30(suppl.), 35–40.

Shelton, E.R., Osheroff, N., and Brutlag, D.L. (1983) *J. Biol. Chem.* 258(15), 9530–9535.

Schiff, P.B., Font, J., and Horwitz, S.B. (1979) *Nature* 277, 665–667.

Shoemaker, R.H., Abbott, B.J., MacDonald, M.M., et al. (1983) *Cancer Treat. Rep.* 67, 97.

Sikic, B.I. (1991) *J. Natl. Cancer Inst.* 83(11), 738–740.

Skehan, P., Storeng, R., Scudiero, D.A., et al. (1990) *J. Natl. Cancer Inst.* 82, 1107–1112.

Staley, A.L., and Clardy, J. (1993) Marine invertebrates and microbes as sources of potential antitumor compounds. in *Anticancer Drug Discovery and Development,* Kluwer Academic Publishers, Boston (in press).

Strong, J.E., and Crooke, S.T. (1978) *Cancer Res.* 38, 3322–3326.

Sugimoto, Y., Tsukahara, S., Oh-hara, T., Isoe, T., and Tsuruo, T. (1990) *Cancer Res.* 50, 6925–6930.

Suzuki, H., Nishimura, T., and Tanaka, N. (1979) *Cancer Res.* 39, 2787–2791.

Tewey, K.M., Chen, G.L., Nelson, E.M., and Liu, L.F. (1984a) *J. Biol. Chem.* 259, 9182–9187.

Tewey, K.M., Rowe, T.C., Yang, L., Halligan, B.D., and Liu, L.F. (1984b) *Science* 226, 466–468.

Trask, D.K., DiDonato, J.A., and Muller, M.T. (1984) *EMBO J.* 3, 671–676.

Waring, M. (1970) *J. Mol. Biol.* 54, 247–279.

Wani, M.C., Taylor, H.L., Wall, M.E., Coggon, P., and McPhail, A.T. (1971) *J. Am. Chem. Soc.* 93, 2325–2327.

Hypocholesterolemic Agents

Akira Endo

The recent introduction of a unique class of cholesterol-lowering drugs offers new promise for the treatment of hypercholesterolemia. These drugs are inhibitors of 3-hydroxy-3-methylglutaryl coenzyme A (HMG-CoA) reductase, the major rate-controlling enzyme in the synthesis of cholesterol. Low doses of these drugs reduce plasma cholesterol levels markedly, and they have not been found to produce serious side effects. If reductase inhibitors prove to be free of long-term adverse effects, they will be widely used for treating hypercholesterolemia. This review will examine discoveries and biological activities of HMG-CoA reductase inhibitors.

11.1 BACKGROUND OF APPROACH

Clofibrate and its derivatives have been widely used throughout the world as hypocholesterolemic agents. These drugs have multiple effects on lipid metabolism, and their actions at the biochemical level are not well understood. The fibrates reduce plasma cholesterol levels only moderately, i.e., 10% to 20% (Brown et al. 1986; Frick et al. 1987). Another widely used drug is probucol: this drug lowers plasma cholesterol levels to an extent similar to that achieved with the fibrates. Its principal action is to reduce the level of plasma low-density lipoproteins (LDL), although it also reduces the levels of high density lipoproteins (HDL), an effect that may be deleterious (Kesäniemi and Grundy 1984). Recent studies show that probucol retards the development of atherosclerosis in rabbits with genetic hypercholesterolemia by means of mechanisms that are not related to its cholesterol-lowering action (Kita et al. 1987). The bile acid sequestrants (cholestyramine and cholestipol) lower the level of the plasma LDL by enhancing hepatic conversion of cholesterol into bile acids, whereas niacin appears to suppress the hepatic synthesis of lipoprotein. Both sequestrants and niacin reduce plasma cholesterol levels by 15% to 30%. Although these drugs can be highly effective, they unfortunately are not tolerated by all patients. Finally, the various available drugs can be used in combination to produce a greater reduction of plasma cholesterol levels than can be achieved by treatment with one agent. A combination of bile acid sequestrants and niacin is particularly potent; in patients with severe hypercholesterolemia, this regimen can reduce cholesterol levels by 30% to 40% (Kane et al. 1981; Blankenhorn et al. 1987).

In humans, the greater part of the cholesterol in the body was believed to be synthesized de novo, mostly in the liver. This led to the search for a new class of drugs to inhibit cholesterol biosynthesis as a means of lowering plasma cholesterol levels. Thus, Endo and co-workers began to search for microbial products in 1971 that inhibited in vitro sterol synthesis in a cell-free system of rat liver (Endo 1985, 1988).

11.2 ASSAY OF CHOLESTEROL BIOSYNTHESIS

Over 20 enzymes are known to be involved in cholesterol biosynthesis. Of these enzymes, HMG-CoA reductase, which catalyzes the reduction of HMG-CoA to mevalonate, is the major rate-limiting enzyme in the pathway (Figure 11–1). This enzyme, therefore, is a prime target for pharmacological intervention. In addition to reductase, several other enzymes, which include HMG-CoA synthase (Tanaka et al. 1982; Bergstrom et al. 1984), squalene synthetase (Faust et al. 1979), and squalene epoxidase (Eilenberg and Scheater 1987), are reported to be important sites of control in cholesterol production. An agent that inhibits either one of these enzymes can be evaluated by testing

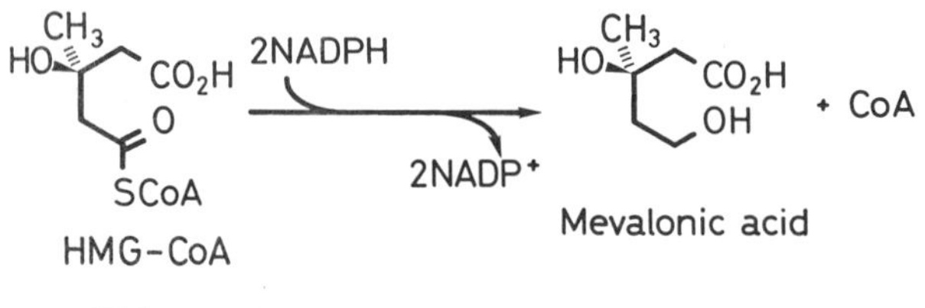

FIGURE 11–1 HMG-CoA reductase reaction.

for its inhibitory effect on the sterol synthesis from a labeled precursor or on the individual enzyme.

11.2.1 Incorporation of Labeled Precursors into Sterols

A sensitive and simple assay of in vitro cholesterol synthesis was described by Kuroda and Endo (1977), in which the labeled precursors ([14C]acetate, [14C]HMG-CoA, and [14C]mevalonate) are incubated with rat liver enzyme preparations. The synthesized sterols are saponified and extracted with light petroleum. Sterols are isolated from the extracts as digitonide and counted. The rat liver enzyme preparations are stable at −80°C and retain their catalytic activity for at least 4 months, unless freezing and thawing are repeated, which causes a loss in activity. By using this procedure, one can test 50 to 100 culture broths at a time for their activity to inhibit sterol synthesis.

Many cultured mammalian cell lines can be utilized to determine sterol synthesis from labeled precursors (Brown et al. 1978; Kaneko et al. 1978; Hasumi et al. 1985). Cells grown in monolayer or in suspension are incubated with [14C]acetate or [14C]mevalonate and saponified, followed by isolating and counting the digitonin-precipitable [14C]-labeled sterols formed. This procedure is widely used by many groups to determine sterol synthesis but is more laborious than the cell-free system described earlier.

11.2.2 Assay for Specific Enzymes

HMG-CoA reductase is assayed by using either isolated microsomes of mammalian cells or a solubilized and purified enzyme preparation using [14C]HMG-CoA as substrate (Endo et al. 1976a; Brown et al. 1978). Cytoplasmic HMG-CoA synthase is prepared from rat livers and assayed by the procedure of Clinkenbeard et al. (1975). Squalene synthetase and squalene epoxidase are assayed by using rat liver microsomes as described by Cohen et al. (1986) and Yamamoto and Bloch (1970), respectively. Acetoacetyl-CoA thiolase, the first enzyme in cholesterol synthetic pathway, can be assayed using either a spectrophotometric or a radioactive assay (Vagelos and Alberts 1960; Clinkenbeard et al. 1973).

11.3 COMPACTIN-RELATED COMPOUNDS

11.3.1 Isolation

Compactin (ML-236B) was the first competitive inhibitor of HMG-CoA reductase discovered. This compound was reported as an antibiotic produced by *Penicillium brevicompactum* (Brown et al. 1976) and as an inhibitor of cholesterol biosynthesis (called ML-236B) isolated from *Penicillium citrinum* (Endo et al. 1976b) (Figure 11–2). Compactin is produced by several other species of fungi (Doss et al. 1986; Endo et al. 1986a). Along with compactin,

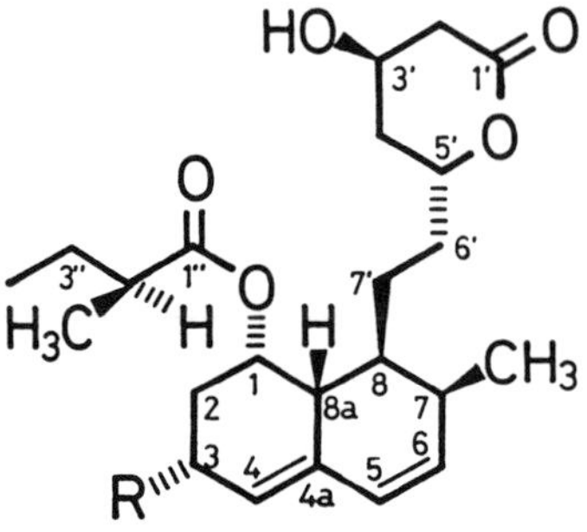

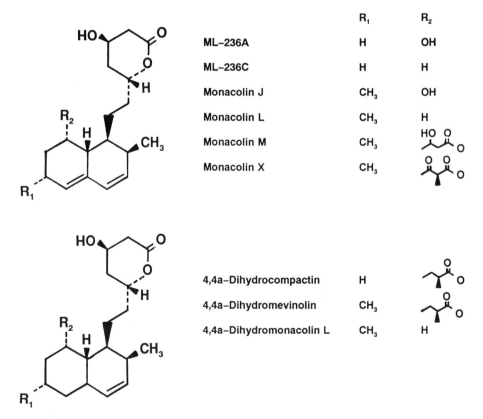

FIGURE 11–2 Compactin (ML-236B) and monacolin K (mevinolin).

R=H, Compactin (ML-236B)
R=CH₃, Monacolin K (mevinolin)

	R₁	R₂
ML–236A	H	OH
ML–236C	H	H
Monacolin J	CH₃	OH
Monacolin L	CH₃	H
Monacolin M	CH₃	(HO O ester)
Monacolin X	CH₃	(diketo ester)

	R₁	R₂
4,4a–Dihydrocompactin	H	(2-methylbutyryl ester)
4,4a–Dihydromevinolin	CH₃	(2-methylbutyryl ester)
4,4a–Dihydromonacolin L	CH₃	H

FIGURE 11–3 Minor metabolites related to compactin and monacolin K.

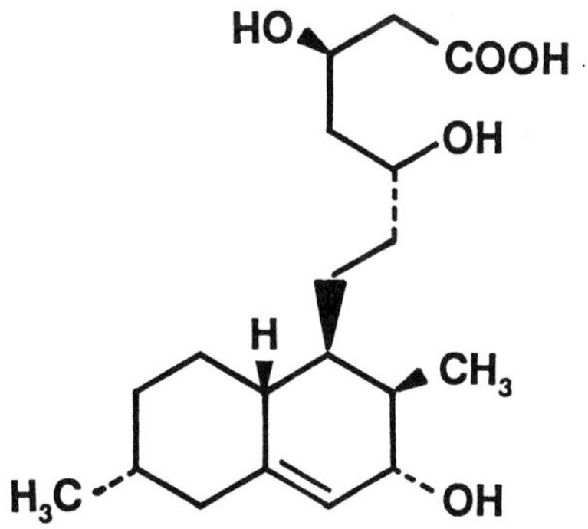

FIGURE   11–4 6α-Hydroxy-4,6-dihydro-monacolin L acid.

ML-236A, ML-236C, and 4,4a-dihydrocompactin, which are structurally analogous to compactin, are isolated from *P. citrinum* (Endo et al. 1976b; Lam et al. 1981) (Figure 11–3). Monacolin K (mevinolin, lovastatin) is the major active compound produced by *Monascus ruber* (Endo 1979, 1980) and *Aspergillus terreus* (Alberts et al. 1980) (see Figure 11–2). In addition, monacolin J, L, M, and X, dihydromonacolin L (see Figure 11–3), and 6α-hydroxy-4,6-dihydromonacolin L acid (Figure 11–4) are produced by *M. ruber* (Endo et al. 1985a, 1985b, 1986b; Nakamura et al. 1990). 4-4a-Dihydromevinolin is isolated from *A. terreus* (Albers-Schönberg et al. 1981) (see Figure 11–3).

11.3.2 Microbial Modification

Compactin is converted to 3β-hydroxycompactin (pravastatin) by *Mucor hiemalis* and *Nocardia* sp. (Serizawa et al. 1983a, 1983d) (Figure 11–5). Compactin is also hydroxylated to 3α-hydroxycompactin by *Syncephalastrum nigricans* (Serizawa et al. 1983b). 6-Hydroxy-*iso*-compactin is derived from compactin by *Absidia coerulea* (Serizawa et al. 1983c) (see Figure 11–5). 8a-β-Hydroxycompactin and 8a-β-hydroxymonacolin K are obtained from compactin and monacolin K, respectively, by growing *Schizophyllum commune* (Yamashita et al. 1985). 5′-Phosphocompactin acid and 5′-phosphomonacolin K are produced by several fungal strains (Endo et al. 1985d) (see Figure 11–5). The 2(S)-methylbutyryl ester of compactin and monacolin K can be hydrolyzed to ML-236A and monacolin J, respectively, by the action of a carboxylesterase of the fungus *Emericella unguis* (Komagata et al. 1986; Murakawa et al. 1987). A series of the side chain ester analogs of monacolin K (mevinolin) are derived from the silyl ether of monacolin J, which includes simvastatin (2″-methylmonacolin K) (Hoffman et al. 1986a) (Figure 11–6). This compound, which is more active than mevinolin, can be converted to L-669,262 (*iso*-simvastatin-6-one) (see Figure 11–6) (Joshua et al. 1991) by a *Nocardia autotrophica* species. This transformation product is six to seven times as active as simvastatin.

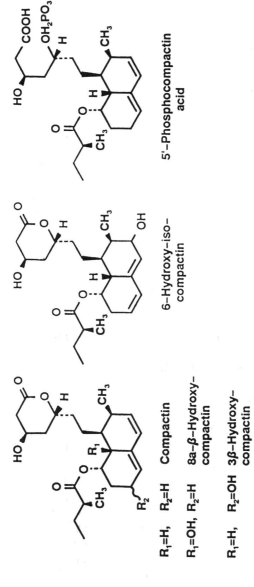

FIGURE 11–5 Compactin analogs produced by microbial conversion.

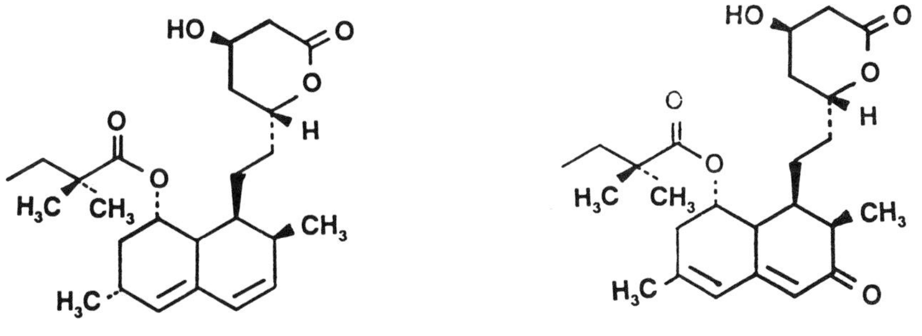

FIGURE 11-6 Simvastatin (2″-methylmonacolin K and L-669,262 (*iso*-simvastatin-6-one).

11.3.3 Biosynthesis of Compactin-Related Agents

Monacolins isolated from *M. ruber* include 4,4a-dihydromonacolin L, 6α-hydroxy-4,6-dihydromonacolin L, monacolin L, monacolin J, and monacolin K (see Figures 11–2, 11–3, and 11–4). All these compounds are active in the inhibition of HMG-CoA reductase. 4,4a-Dihydromonacolin L is converted to 6α-hydroxy-4,6-dihydromonacolin L by *M. ruber* in the presence of molecular oxygen (Figure 11–7) (Nakamura et al. 1990). The latter can be spontaneously dehydrated to monacolin L, although conversion of exogenously added 6α-hydroxy-4,6-dihydromonacolin L to monacolin L by *M. ruber* has not been successful. Monacolin L is hydroxylated to monacolin J by the action of a monooxygenase (Komagata et al. 1989). The end-product, monacolin K, can be derived from monacolin J by cells of both *M. ruber* and *Paecilomyces viridis* (Kimura et al. 1990). The latter strain is a producer of compactin (Endo et al. 1986a). One possible mechanism for this conversion is the esterification of monacolin J with α-methylbutyryl-CoA to monacolin K. These results are summarized in Figure 11–7. (For involvement of acetate and methionine see the following paragraph.)

[13C]Acetate, methyl-[13C]methionine, and 18O2 are incorporated into compactin and monacolin K in cultures of *P. citrinum, M. ruber,* and *A. terreus,* and the 1H and 13C nuclear magnetic resonance (NMR) spectra of the two products were fully assigned by combined spectral analysis. Both compounds are formed by the head-to-tail coupling of two polyketide chains (4-carbon and 18-carbon) of acetate units. The 4-carbon chain has one methionine derived methyl group, and a methyl group at C–3 in the bicyclic ring system of monacolin K is also derived from methionine (Figure 11–8) (Chan et al. 1983; Moore et al. 1985; Endo et al. 1985c).

The above data suggest involvement of a biological Diels-Alder cyclization to generate the correct ring stereochemistry in a single step. Such ring forming processes have been suggested for a number of reduced polyketide

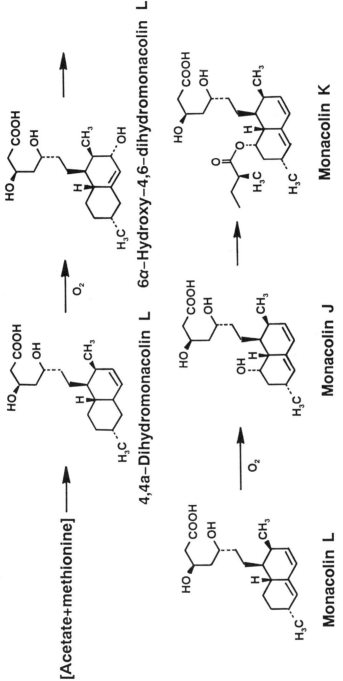

FIGURE 11–7 Biosynthesis of monacolins (shown in acid form).

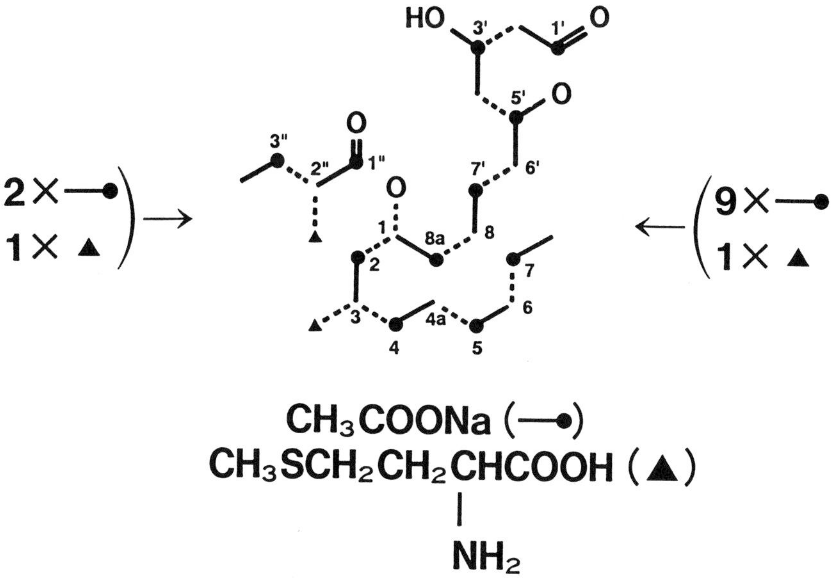

FIGURE 11–8 Biosynthesis of monacolin K (mevinolin) as studied using [^{13}C]acetate and [^{13}C]methionine. Reproduced by permission of *The Journal of Antibiotics* (Endo et al. 1985c).

metabolites. However, no enzyme catalyzed Diels-Alder reaction has yet been demonstrated.

11.3.4 Inhibition of HMG-CoA Reductase

The inhibition of HMG-CoA reductase by compactin analogs is reversible (Endo et al. 1976a; Tanzawa and Endo, 1979). As can be expected from the structure of their acid forms (Figure 11–9), the inhibition by these compounds is competitive with respect to the substrate HMG-CoA (Endo et al. 1976a). The lactone form of these compounds is inactive and in vivo is hydrolyzed to the acid form, which is the principal active form of these compounds. Most compactin analogs reported inhibit HMG-CoA reductase of rat liver at 0.2 to 1.0 nM (Endo 1980, 1985; Hoffman et al. 1986a; Joshua et al. 1991), whereas under the same conditions, the Km value for HMG-CoA is ~10 μM. Thus, the affinity of HMG-CoA reductase for most compactin analogs is 10,000-fold or more than its affinity for the natural substrate, HMG-CoA.

Compactin analogs do not affect other enzymes involved in cholesterol biosynthesis (Endo et al. 1977). Studies on compactin using cultured cells and intact animals suggest that reductase is the only enzyme that is inhibited by compactin analogs.

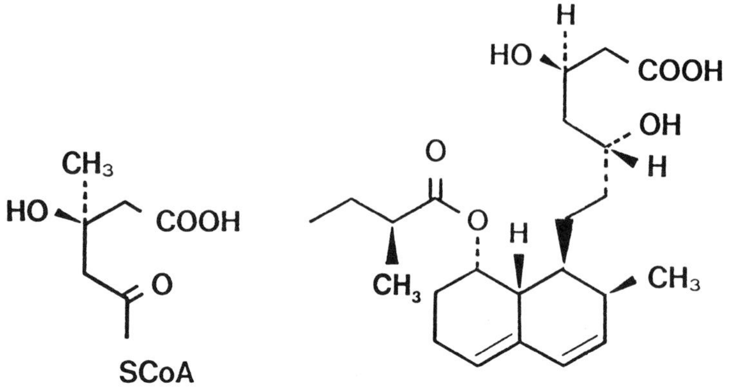

FIGURE 11-9 HMG-CoA (A) and compactin (acid form) (B).

11.3.5 Structure-Activity Relationship

The molecule of compactin analogs is composed of four portions (see Figure 11-9): (1) the lactone or the 3',5'-dihydroxypentanoic acid; (2) the moiety bridging the lactone and the lipophilic groups; (3) the bicyclic decaline ring; and (4) the side chain ester. Structural similarity between HMG-CoA and compactin analogs suggests that the active center of these compounds in the inhibition of HMG-CoA reductase is at the 3',5'-dihydroxypentanoic acid moiety of the molecules. This hypothesis is supported by the data that inhibitory activity of compactin is reduced to 1/100 or less by methylation or acetylation of the hydroxy group at C-3' and/or C-5' (Endo 1985) and that 5'-phosphocompactin acid is one-tenth of compactin in the inhibitory activity (Endo et al. 1985d). The hydroxy group at C-3' of the 3',5'-dihydroxypentanoic acid portion is the *trans*-isomer and much more active than the *cis*-isomer (Stokker et al. 1986; Hoffman et al. 1986b). The addition of a methyl group to the C-3' position to give the *trans*-form yields a compound that more closely resembles the HMG portion of HMG-CoA, but does not stimulate activity (Hoffman et al. 1986b). Replacement of the carboxyl group of compactin with a carboxamide group abolishes activity. Increasing the length of the bridge to three carbons reduces activity (Stokker et al. 1986).

The hydrophilic decaline ring of compactin analogs is essential for inhibitory activity. This is shown by the data that HMG is more than 10^{-6}-fold less active than compactin (Brown et al. 1978). 4,4a-Dihydrocompactin, 4,4a-dihydromevinolin, and 4,4a-dihydromonacolin L (see Figure 11-3) are comparable in activity to compactin, monacolin K, and monacolin L, respectively (Lam et al. 1981; Albers-Schönberg et al. 1981; Endo et al. 1985a). The addition of a methyl group to C-3 of the bicyclic ring of compactin to give monacolin K doubles activity (Endo 1980; Alberts et al. 1980), while the hydroxyl group at this position, which gives pravastatin, slightly reduces activity (Tsujita et al. 1986). L-669,262 (*iso*-simvastatin-6-one) (see Figure 11–6), derived from simvastatin, is six to seven times as active as simvastatin in inhibiting reductase activity (Joshua et al. 1991). Activity is reduced one-third by the hydroxylation at C–8a (Yamashita et al. 1985).

Compactin analogs, which lack the side chain ester (Endo 1985), are far less active than those that have it. The introduction of an additional aliphatic group on the C–2″ of the side chain increases potency. Simvastatin, which has an additional methyl group at this position, has about 2.5 times the inhibitory activity of mevinolin (Hoffman et al. 1986a). Increasing the length of the side chain and the removal of the terminal methyl group lowers the inhibitory activity. Stereochemistry at the C–2″ of the side chain is not crucial because the natural product (mevinolin) and its diastereomer are equally potent (Hoffman et al. 1986a).

11.3.6 Inhibition of Cholesterol Synthesis

Inhibition of HMG-CoA reductase could theoretically impair the formation of mevalonate and all the metabolic products derived from mevalonate: cholesterol, dolichol, isopentenyl adenine, ubiquinone, and prenylated protein (Goldstein and Brown 1990; Glomset et al. 1990). In turn, shortage of these metabolites may cause a variety of biological and pharmacological events.

Compactin analogs inhibit cholesterol biosynthesis in a variety of cultured animal and human cells at nanomolar concentrations (Endo 1985). In cultured human skin fibroblasts, inhibition of sterol synthesis from [^{14}C]acetate by compactin is 50% at 1 nM, and 100% at 10 μM, respectively (Kaneko et al. 1978). Under these conditions, sterol synthesis from [^{14}C]mevalonate and fatty acid synthesis from [^{14}C]acetate are not significantly affected.

At higher concentrations where sterol synthesis is reduced by over 90%, compactin analogs inhibit cell growth as well (Kaneko et al. 1978; Goldstein et al. 1979; Ryan et al. 1981). This inhibition can be overcome and cells can grow normally if mevalonate is added to the culture medium, indicating that these compounds are specific inhibitors of HMG-CoA reductase.

When administered orally to mice and rats, compactin analogs are absorbed into the liver, the major site of cholesterogenesis, as compared with other organs. Thus, when these compounds are given orally to rats, sterol synthesis in vivo in the liver is most strongly inhibited as compared with the inhibition in other organs (ileum, adrenal, kidney, lung, spleen, and testis) (Endo et al. 1977). Inhibition of cholesterol synthesis after a single dose of mevinolin was 50% in rats at 46 μg/kg (Alberts et al. 1980).

In a double-blind placebo-controlled single dose study, mevinolin (100 mg) produced a 78% reduction in plasma mevalonate concentrations between 2 and 6 hours after administration in healthy volunteers (Parker et al. 1984). Plasma mevalonate concentrations were directly correlated with the rate of whole body cholesterol synthesis. Using the sterol balance method to estimate cholesterol synthesis, Grundy and Bilheimer (1984) found mevinolin decreased output of neutral steroids in patients with familial hypercholesterolemia (FH).

11.3.7 Hypocholesterolemic Activity

Compactin analogs fail to reduce plasma cholesterol levels in normal rats and mice (Endo et al. 1979). Compactin is, however, effective in rats treated with Triton WR-1339, a hyperlipidemic detergent (Endo et al. 1979; Kuroda et al. 1977). In normal rats, compactin causes a decrease in fecal excretion of bile acids and in the activity of hepatic cholesterol 7α-hydroxylase, the rate-limiting enzyme in bile acid synthesis. In addition, compactin causes a marked increase in hepatic HMG-CoA reductase activity (Endo et al. 1979). These unexpected changes in the activity of the two microsomal enzymes may, in part, explain the ineffectiveness of compactin.

In normolipidemic dogs, compactin analogs produce a rapid reduction of plasma cholesterol levels at a dose of 10 mg/kg or less (Tsujita et al. 1979; Alberts et al. 1980). Triglyceride levels are not consistently changed. The LDL-cholesterol fraction, which is responsible for atherosclerosis, is preferentially lowered. Fecal excretion of bile acids is, unlike in the rats, not affected or slightly elevated by these inhibitors (Tsujita et al. 1979). In dogs and miniature pigs, a combination of colestipol (bile acid sequestrant) in addition to mevinolin produces greater reduction in LDL cholesterol than occurs with either drug alone (Huff et al. 1985).

In healthy volunteers, compactin analogs produce a rapid and profound cholesterol-lowering effect. Yamamoto et al. (1980) first studied the efficacy of compactin in patients with FH. In this study, plasma cholesterol was reduced by 27% at 60 to 100 mg/day after 4 to 8 weeks of treatment. In a detailed study with heterozygous FH, decreases in LDL-cholesterol levels were 29% at 30 to 60 mg daily dosages (Mabuchi et al. 1981). HDL levels were not reduced but, rather, slightly elevated. Combined treatment with a bile acid-binding resin (cholestyramine) with compactin produces a much greater decrease in LDL-cholesterol levels than does either drug alone (Mabuchi et al. 1983). When patients with heterozygous FH were treated with cholestyramine alone (12 g daily), LDL cholesterol was lowered by 28%, and the addition of compactin (30 mg daily) produced a 53% reduction. Similar cholesterol-lowering effects in humans were obtained with mevinolin and pravastatin (Grundy 1988).

11.3.8 Mechanism for Lowering LDL Cholesterol

There could be two possible mechanisms for lowering LDL-cholesterol levels by an HMG-CoA reductase inhibitor, namely, decreasing LDL synthesis and increasing its clearance from the circulation.

In dogs, the lowering of LDL-cholesterol levels occurs in two ways; by an inhibition of LDL synthesis, and through enhancing the receptor-mediated LDL catabolism in the liver (Kovanen et al. 1981). In humans, however, the relative importance of these two mechanisms in lowering plasma LDL-cholesterol varies among patients, depending on whether they have familial or non-familial hypercholesterolemia. Mevinolin treatment primarily enhances receptor-mediated LDL catabolism in patients with heterozygous FH (Bilheimer et al. 1983), whereas reduced LDL synthesis, rather than increased LDL catabolism, is the predominant factor in patients with non-familial hypercholesterolemia (Grundy et al. 1985).

11.4   OTHER INHIBITORS OF CHOLESTEROL BIOSYNTHESIS

Recently, streptenols, which are not related to compactin analogs, have been isolated from a *Streptomyces* sp. as potent inhibitors of HMG-CoA reductase (Figure 11–10). These compounds inhibit reductase activity of Hep G2 cells at 10^{-9} M, a value comparable to those for compactin analogs (Zeeck et al. 1990). Xerulin analogs isolated from *Xerula melanotricha* are active in the inhibition of sterol synthesis from [^{14}C]acetate at 0.1 μg/ml (Kuhnt et al. 1990) (Figure 11–11). These compounds do not inhibit the incorporation of [^{14}C]mevalonate into sterols.

A novel triyne carbonate was isolated from an Actinomyces culture (L-660,631) (Kempf et al. 1988) or from *Microbispora* (EV-22) (Wright et al. 1988) as an antifungal agent (Figure 11–12). This compound, which inhibits

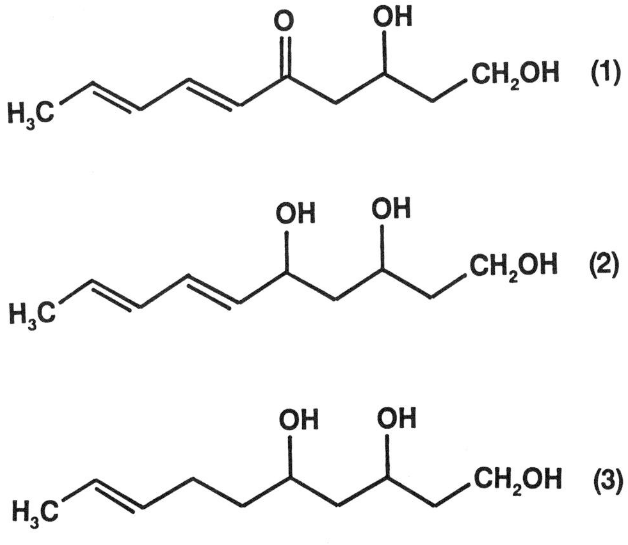

FIGURE 11–10 Streptenols.

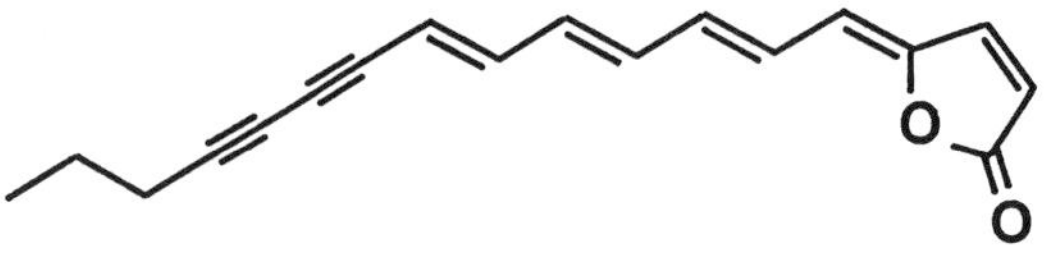

A

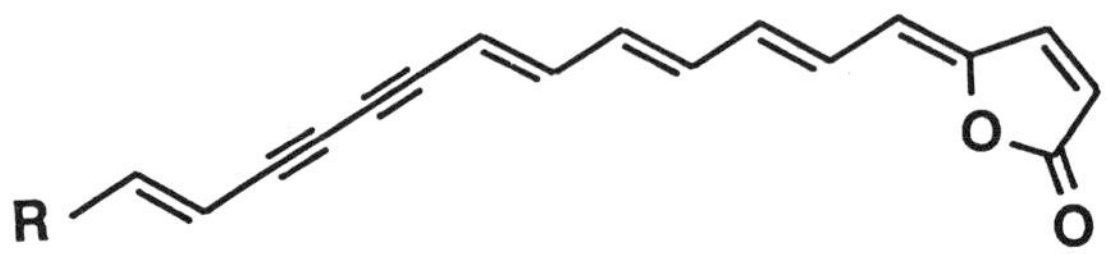

B R=CH₃
C R=COOH

FIGURE 11–11 Dihydroxerulin (A), xerulin (B) and xerulinic acid (C).

sterol synthesis in *Candida albicans* at minimal inhibitory concentrations (MIC) (Onishi et al. 1988), is a potent inhibitor of rat liver cytosolic aceto-acetyl-CoA thiolase, an enzyme that catalyzes the first step in cholesterol synthetic pathway (Greenspan et al. 1989).

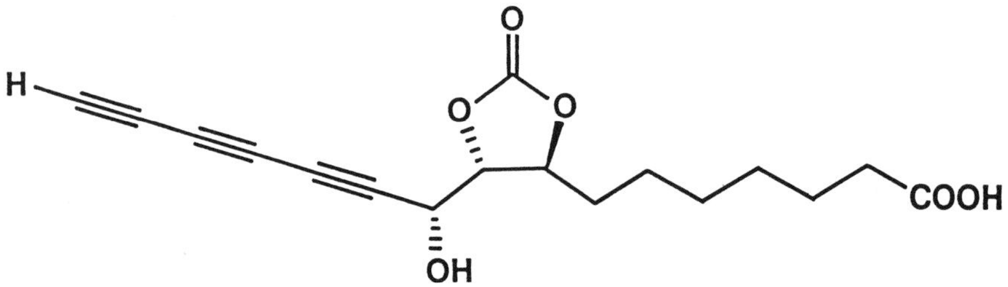

A

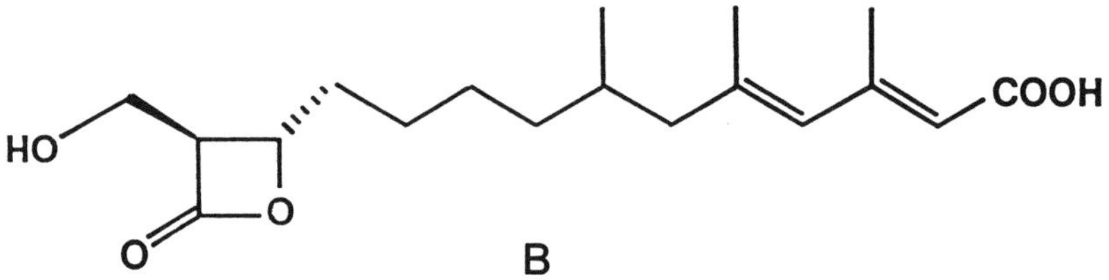

B

FIGURE 11–12 L-660,631(EV-22) (A) and 1233A (L-659,699) (B).

HMG-CoA synthase is an enzyme involved in an early step of cholesterol synthetic pathway. A specific inhibitor of HMG-CoA synthase, 1233A (L-659,699), was reported by two groups in 1987 (Omura et al. 1987; Greenspan et al. 1987) (Figure 11–12). This compound was originally isolated from *Cephalosporium* sp. as an antibacterial product (Aldridge et al. 1971). Inhibitory activity of 1233A is specific for HMG-CoA synthase, and the enzymes HMG-CoA reductase, β-ketoacyl-CoA thiolase, acetoacetyl-CoA synthetase,

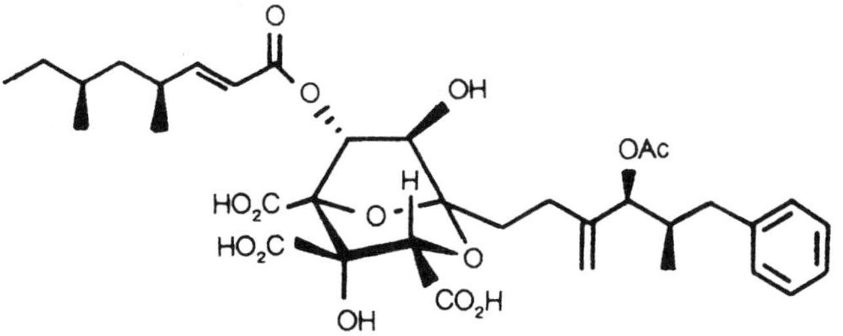

Squalestatin 1 (zaragozic acid A)

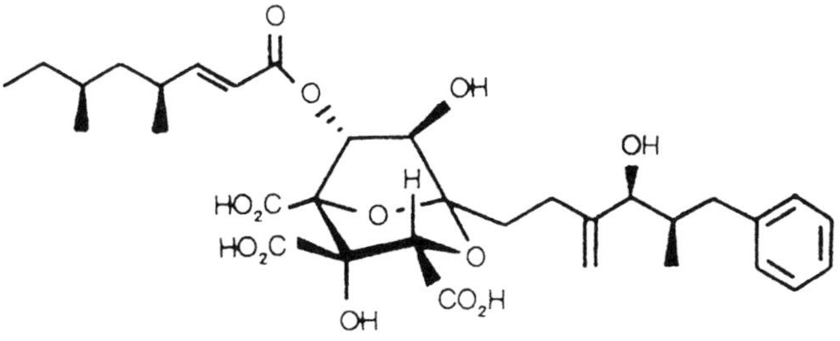

Squalestatin 2

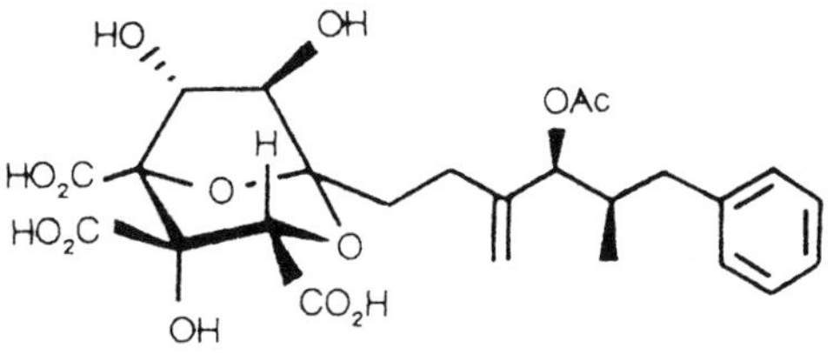

Squalestatin 3

FIGURE 11–13 Squalestatins and zaragozic acids. (*Continued on next page.*)

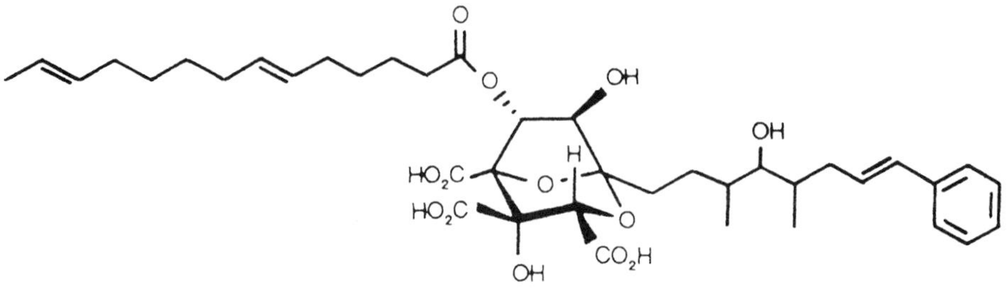

Zaragozic acid B

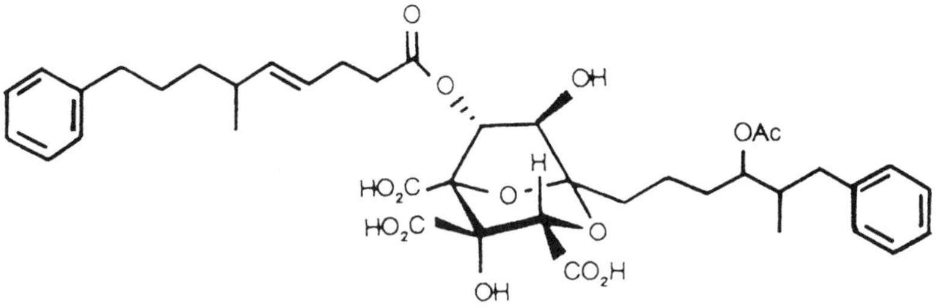

Zaragozic acid C

FIGURE 11–13 (continued)

and fatty acid synthase are not inhibited. In cultured Hep G2 cells, this compound inhibits the incorporation of [^{14}C]acetate into sterols with an IC$_{50}$ of 6 μM, while incorporation of [^{14}C]mevalonate into sterols in these cells is not affected. A number of analogs of 1233A were described as inhibitors of HMG-CoA synthase (Greenspan et al. 1987).

Squalene synthetase is a microsomal enzyme that catalyzes the reductive dimerization of farnesyl pyrophosphate to squalene. Selective inhibition of this enzyme should not directly interfere with essential side pathways to the non-sterol isoprene products including dolichol, coenzyme Q, heme A, isopentenyl tRNA, and isoprenylated proteins. Isoprenyl phosphinyl formates and their esters have recently been synthesized as squalene synthetase inhibitor (Biller et al. 1991). Inhibition of microsomal squalene synthetase by these compounds is competitive with respect to the substrate farnesyl pyrophosphate (Ki = 2.6 to 10 μM).

Recently, a group of potent inhibitors of squalene synthetase, the squalestatins, was isolated from *Phoma* sp. by Glaxo's group (Dowson et al. 1992; Sidebottom et al. 1992). Similar metabolites, named zaragozic acids, were independently isolated from *Sporomiella intermedia* and *Leptodontium elatius,* respectively (Wilson et al. 1992; Defresne et al. 1992; Bergstrom et al. 1993; Hensens et al. 1993) (Figure 11–13). Squalestatin 1 was also isolated from a strain of *Setosphaeria khartoumensis* (Hasumi et al. 1993). Zaragozic acids are competitive inhibitors of rat liver squalene synthetase with apparent Ki values of 21 to 78 pM (Bergstrom et al. 1993). In marmosets, squalestatin 1 significantly lowers serum cholesterol at an oral dose of 10 mg/kg per day (Baxter et al. 1992).

Squalene epoxidase, which catalyzes the conversion of squalene to 2,3-oxidosqualene, is another important enzyme in cholesterol synthetic pathway. A potent inhibitor of mammalian squalene synthetase (NB-598) has been chemically synthesized, which shows a potent hypocholesterolemic activity in dogs (Horie et al. 1990). However, no squalene epoxidase inhibitor has been isolated from microorganisms.

REFERENCES

Albers-Schönberg, G., Joshua, H., Lopez, M.B., et al. (1981) *J. Antibiotics* 34, 507–512.

Alberts, A.W., Chen, J., Kuron, G., et al. (1980) *Proc. Natl. Acad. Sci. USA* 77, 3957–3961.

Aldridge, D.C., Gile, D., and Turner, W.B. (1971) *J. Chem. Soc.* 3888–3891.

Baxter, A., Fitzgerald, B.J., Hutson, J.L., et al. (1992) *J. Biol. Chem.* 267, 11705–11708.

Bergstrom, J.D., Kurtz, M.M., Rew, D.J., et al. (1993) *Proc. Natl. Acad. Sci. USA* 90, 80–84.

Bergstrom, J.F., Wong, G.A., Edwards, P.A., and Edmond, J. (1984) *J. Biol. Chem.* 259, 14548–14553.

Bilheimer, D.W., Grundy, S.M., Brown, M.S., and Goldstein, J.L. (1983) *Proc. Natl. Acad. Sci. USA* 80, 4124–4128.

Biller, S.A., Forster, C., Gordon, E.M., et al. (1991) *J. Med. Chem.* 34, 1914–1916.

Blankenhorn, D.H., Nessim, S.A., Johnson, R.L., et al. (1987) *J. Am. Med. Assoc.* 257, 3233–3240.

Brown, A.G., Smale, T.C., King, T.J., Hasenkamp, R., and Thompson, R.H. (1976) in *J. Chem. Soc., Perktin Trans.* 1, 1165–1170.

Brown, M.S., Faust, J.R., Goldstein, J.L., Kaneko, I., and Endo, A. (1978) *J. Biol. Chem.* 253, 1121–1128.

Brown, W.V., Dujovne, C.A., Farquhar, J.W., et al. (1986) *Arteriosclerosis* 6, 670–678.

Chan, J.K., Moore, R.N., Nakashima, T.T., and Vederas, J.C. (1983) *J. Am. Chem. Soc.* 105, 3334–3336.

Clinkenbeard, K.D., Reed, W.D., Mooney, R.A., and Rane, M.D. (1975) *J. Biol. Chem.* 250, 3108–3116.

Clinkenbeard, K.D., Sugiyama, T., Moss, J., Reed, W.D., and Lane, M.D. (1973) *J. Biol. Chem.* 248, 2275–2284.

Cohen, L.H., Griffioen, A.M., Wanders, R.J.A., et al. (1986) *Biochem. Biophys. Res. Commun.* 138, 335–341.

Doss, S.L., Chu, C.K., Mesbak, M.K., et al. (1986) *J. Nat. Prod.* 49, 357–358.

Dowson, M.J., Farthing, J.E., Marshall, P.S., et al. (1992) *J. Antibiotics* 45, 639–647.

Dufresne, C., Wilson, K.E., Zink, D., et al. (1992) *Tetrahedron* 48, 10221–10226.

Eilenberg, H., and Schechter, I. (1987) *J. Lipid Res.* 28, 1398–1404.

Endo, A. (1979) *J. Antibiot.* 32, 852–854.

Endo, A. (1980) *J. Antibiot.* 33, 334–336.

Endo, A. (1985) *J. Med. Chem.* 28, 401–405.

Endo, A. (1988) *Klin. Wochenschr.* 66, 421–427.

Endo, A., Hasumi, K., Nakamura, T., Kunishima, M., and Masuda, M. (1985a) *J. Antibiotics* 38, 321–327.

Endo, A., Hasumi, K., and Negishi, S. (1985b) *J. Antibiotics* 38, 420–422.

Endo, A., Hasumi, K., Yamada, A., Shimada, R., and Takeshima, H. (1986a) *J. Antibiotics* 39, 1609–1610.

Endo, A., Komagata, D., and Shimada, H. (1986b) *J. Antibiotics* 39, 1670–1673.

Endo, A., Kuroda, M., and Tanzawa, K. (1976a) *FEBS Lett.* 72, 323–326.

Endo, A., Kuroda, M., and Tsujita, Y. (1976b) *J. Antibiotics* 29, 1346–1348.

Endo, A., Negishi, Y., Iwashita, T., Mizukawa, K., and Hirama, M. (1985c) *J. Antibiotics* 38, 444–448.

Endo, A., Tsujita, Y., Kuroda, M., and Tanzawa, K. (1977) *Eur. J. Biochem.* 77, 31–36.

Endo, A., Tsujita, Y., Kuroda, M., and Tanzawa, K. (1979) *Biochim. Biophys. Acta* 575, 266–276.

Endo, A., Yamashita, H., Naoki, H., Iwashita, T., and Mizukawa, Y. (1985d) *J. Antibiotics* 38, 328–332.

Faust, J.R., Goldstein, J.L., and Brown, M.S. (1979) *Proc. Natl. Acad. Sci. USA* 76, 5018–5022.

Frick, M.H., Elo, O., Haapa, K., et al. (1987) *N. Engl. J. Med.* 317, 1237–1245.

Glomset, J.A., Gelf, M.H., and Farnsworth, C.C. (1990) *Trend Biochem. Sci.* 15, 139–142.

Goldstein, J.L., and Brown, M.S. (1990) *Nature* 343, 425–430.

Goldstein, J.L., Helgeson, J.A.S., and Brown, M.S. (1979) *J. Biol. Chem.* 254, 5403–5409.

Greenspan, M.D., Yudokovitz, J.B., Chen, J.S., et al. (1989) *Biochem. Biophys. Res. Commun.* 163, 548–553.

Greenspan, M.D., Yudokovitz, J.B., Lo, C-Y.L., et al. (1987) *Proc. Natl. Acad. Sci. USA* 84, 7488–7492.

Grundy, S.M. (1988) *N. Engl. J. Med.* 319, 24–33.

Grundy, S.M., and Bilheimer, D.W. (1984) *Proc. Natl. Acad. Sci. USA* 81, 2538–2542.

Grundy, S.M., Vega, G.L., and Bilheimer, D.W. (1985) *Ann. Intern. Med.* 103, 339–343.

Hasumi, K., Otsuki, R., and Endo, A. (1985) *J. Biochem.* 98, 319–325.

Hasumi, K., Tachikawa, K., Sakai, K., et al. (1993) *J. Antibiotics* 46, 689–691.

Hensens, O.D., Dufresne, C., Liesch, J.M., et al. (1993) *Tetrahed. Lett.* 48, 10221–10226.

Hoffman, W.F., Alberts, A.W., Anderson, P.S., et al. (1986a) *J. Med. Chem.* 29, 849–852.

Hoffman, W.F., Alberts, A.W., Cragoe, E.J., et al. (1986b) *J. Med. Chem.* 29, 159–169.

Horie, M., Tsuchiya, Y., Hayashi, M., et al. (1990) *J. Biol. Chem.* 265, 18075–18078.

Huff, M.W., Telford, D.E., Woodcroft, K., and Strong, W.L.P. (1985) *J. Lipid Res.* 26, 1175–1186.

Joshua, H., Schwartz, M.S., and Wilson, K.E. (1991) *J. Antibiotics* 44, 366–370.

Kane, J.P., Malloy, M.J., Tun, P., et al. (1981) *N. Engl. J. Med.* 304, 251–258.

Kaneko, I., Hazama-Shimada, Y., and Endo, A. (1978) *Eur. J. Biochem.* 87, 313–321.

Kempf, A.J., Hensens, O.D., Schwartz, R.E., et al. (1988) in *Antifungal Drugs, Annals of the New York Academy of Science,* vol. 544, (St. Georgiev, V., ed.), pp. 183, New York Academy of Science, New York.

Kesäniemi, Y.A., and Grundy, S.M. (1984) *J. Lipid Res.* (1984) 25, 780–790.

Kimura, K., Komagata, D., Murakawa, S., and Endo, A. (1990) *J. Antibiotics* 43, 1621–1622.

Kita, T., Nagano, Y., Yokode, M., et al. (1987) *Proc. Natl. Acad. Sci. USA* 84, 5928–5931.

Komagata, D., Shimada, H., Murakawa, S., and Endo, A. (1989) *J. Antibiotics* 42, 407–412.

Komagata, D., Yamashita, H., and Endo, A. (1986) *J. Antibiotics* 39, 1574–1577.

Kovanen, P.T., Bilheimer, D.W., Goldstein, J.L., Jaramillo, J.J., and Brown, M.S. (1981) *Proc. Natl. Acad. Sci. USA* 78, 1194–1198.

Kuhnt, D., Anke, T., Besl, H., et al. (1990) *J. Antibiotics* 43, 1413–1420.

Kuroda, M., and Endo, A. (1977) *Biochim. Biophys. Acta* 486, 70–81.

Kuroda, M., Tanzawa, K., Tsujita, Y., and Endo, A. (1977) *Biochim. Biophys. Acta* 489, 119–125.

Lam, T.Y.K., Gullo, V.P., Goegelman, R.T., et al. (1981) *J. Antibiotics* 34, 614–616.

Mabuchi, H., Haba, T., Tatami, R., et al. (1981) *N. Engl. J. Med.* 305, 478–482.

Mabuchi, H., Sakai, T., Sakai, Y., et al. (1983) *N. Engl. J. Med.* 308, 609–613.

Moore, R.N., Bigam, G., Chan, J.K., et al. (1985) *J. Am. Chem. Soc.* 107, 3694–3701.

Murakawa, S., Nakamura, T., Komagata, D., Sunagawa, E., and Endo, A. (1987) *Agric. Biol. Chem.* 51, 1879–1884.

Nakamura, T., Komagata, D., Murakawa, S., Sakai, K., and Endo, A. (1990) *J. Antibiotics* 43, 1597–1600.

Omura, S., Tomoda, H., Kumagai, H., et al. (1987) *J. Antibiotics* 40, 1356–1357.

Onishi, J.C., Abruzzo, G.K., Fromtling, R.A., et al. (1988) in *Antifungal Drugs, Annals of the New York Academy of Science,* vol. 544 (St. Georgiev, V., ed.), pp. 229, New York Academy of Science, New York.

Parker, T.S., MacNamara, J., Brown, C.D., et al. (1984) *J. Clin. Invest.* 74, 795–804.

Ryan, J., Hardeman, E.C., Endo, A., and Simoni, R.D. (1981) *J. Biol. Chem.* 256, 6762–6769.

Serizawa, N., Nakagawa, K., Hamano, K., et al. (1983a) *J. Antibiotics* 36, 604–607.

Serizawa, N., Nakagawa, K., Hamano, K., et al. (1983b) *J. Antibiotics* 36, 608–610.

Serizawa, N., Nakagawa, K., Tsujita, Y., et al. (1983c) *J. Antibiotics* 36, 918–920.

Serizawa, N., Serizawa, S., Nakagawa, K., et al. (1983d) *J. Antibiotics* 36, 887–891.

Sidebottom, P.J., Highcock, R.M., Lane, S.J., et al. (1992) *J. Antibiotics* 45, 648–658.

Stokker, G.E., Alberts, A.W., Anderson, P.S., et al. (1986) *J. Med. Chem.* 29, 170–181.

Tanaka, R.D., Edwards, P., Lam, S.F., Knoppel, E.M., and Fogelman, A.M. (1982) *J. Lipid Res.* 23, 1026–1031.

Tanzawa, K., and Endo, A. (1979) *Eur. J. Biochem.* 98, 195–201.

Tsujita, Y., Kuroda, M., Shimada, Y., et al. (1986) *Biochim. Biophys. Acta* 877, 50–60.

Tsujita, Y., Kuroda, M., Tanzawa, K., Kitano, N., and Endo, A. (1979) *Atherosclerosis* 32, 307–313.

Vagelos, P.R., and Alberts, A.W. (1960) *Anal. Biochem.* 1, 8–16.

Wilson, K.E., Burk, R.M., Biftu, T., et al. (1992) *J. Org. Chem.* 57, 7151–7158.

Wright, J.J., Puar, M.S., Pramanik, B., and Fishman, A. (1988) *J. Chem. Soc. Commun.* 413–414.

Yamamoto, A., Sudo, H., and Endo, A. (1980) *Atherosclerosis* 35, 259–266.

Yamamoto, S., and Bloch, K. (1970) *J. Biol. Chem.* 245, 1670–1674.

Yamashita, H., Tsubokawa, S., and Endo, A. (1985) *J. Antibiotics* 38, 605–609.

Zeeck, A., Grbley, S., Wink, J., Hamman, P., and Giani, C. (1990) Japanese Patent (Kokai) Heisei 2-261388.

Immunomodulators

Masakuni Okuhara
Toru Kino

Ever since the "Golden Era" of antibiotic research, it has been known that many antibiotics possess other pharmacological activities. In the immunology field, K.M. Stevens reported in 1953 that several antibiotics, including penicillin, streptomycin, and aureomycin had the ability of suppressing the immune response to a soluble protein antigen in rabbits (Stevens 1953). As for the immune-stimulating activity of microbial metabolites, bacterial extracts have been utilized for nonspecific immunostimulation in cancer patients (Fudenberg and Whitten 1984). Among the natural products that have immunological activities, however, the first compound that has demonstrated dramatic effectiveness in clinical trials is cyclosporin A.

The fact that immunological activities have been found in a variety of microbial metabolites does mean that the immunology is a promising target area for successful natural product screening. Further exploration of microbial metabolite activities will lead to useful immunomodulators for the treatment of autoimmune diseases and the prevention of allograft rejection after transplantation. In this chapter, the assay methods for immunomodulators, the discovery of several agents, and future trends of discovery research will be described.

12.1 SCREENING FOR IMMUNOSUPPRESSANTS

Ever since Edward Jenner introduced vaccination with cowpox as protection against smallpox nearly 200 years ago, humankind has been successfully controlling certain pathogenic diseases with the help of protective immunity. The living body, after exposure to foreign substances such as invading microbes, enhances its own specific protective ability against a given antigen. Several decades ago, the mechanism of immune response was considered so complicated that immunity was taken as an enigmatic phenomenon. In recent decades however, our understanding of "immunity" has grown enormously and nowadays, researchers can focus on the regulatory mechanisms of the immune response.

There are several types of cells responsible for specialized functions in the immune system, that act through hormone-like substances, much as in the nervous system. Since the late 1970s, when antigen-specific T cells were cloned, it was found that these cells could be cultured indefinitely in the presence of immunological hormone interleukin 2 (IL-2). It has been demonstrated again and again that this interleukin plays a crucial role in the immune response (Gillis and Smith 1977; Smith 1988). On the one hand, the immune system protects the body against diseases from invasion by foreign organisms, but on the other hand, it can attack a life-saving organ transplant or mistakenly respond to the body's own molecules that, in turn, results in autoimmune diseases such as multiple sclerosis, rheumatoid arthritis, and juvenile diabetes. A large accumulation of data shows that activated T cells are associated directly with damage to the tissues during graft rejection and autoimmune diseases (Rosenberg et al. 1987; Eden et al. 1985; Cohen et al. 1985). This finding rationalizes the idea that activated T cells can be treated as a pathogen, similar to pathogenic microbes in an infectious disease, and this is leading to new therapeutic approaches for abnormal immune reactions.

12.1.1 Screening for Inhibitors of IL-2 Production

As mentioned above, depressed function of activated T cells can lead to the suppression of an abnormal immune response. One of the targets for the depression of cell function is IL-2. Selective inhibition of IL-2 production could inhibit T cell growth, which in turn, would result in suppression of the immune response. Several screening methods are used in the search for inhibitors of IL-2 production. This section describes two approaches. One method uses the mixed lymphocyte reaction (MLR) (Morgan et al. 1976), and the other uses concanavalin A (Con A)-induced T cell activation (Larsson and Coutinho 1979). Additionally, a method is described by which detection of nonspecific cytotoxic substances can be excluded.

12.1.1.1 An Assay Method Using MLR. Major histocompatibility complex (MHC)-associated antigens were originally detected by transplantation in vivo. Initially, the assumed role of MHC was that animals that differed at this locus would rapidly reject each other's tissue grafts. Now, MHC is known to be a cluster of genes important in immune recognition and the signaling between the cells of the immune system (Bach et al. 1976; Todd et al. 1988). When two types of leukocytes differing in MHC are co-cultured in vitro, they recognize each other's MHC-associated antigens, and T cells are stimulated to transform and proliferate. The mechanism of this blastogenesis is understood as follows. In mixed leukocyte cultures, cytotoxic T cell precursors (CTLp) primarily recognize MHC-encoded Class-I antigens by means of T cell receptors. This recognition of the Class-I antigens transforms the CTLp into activated T cells that express IL-2 receptors on their cell surface. In contrast, helper T cell precursors respond to Class-II alloantigens and are activated to become helper T cells that produce various lymphokines, including IL-2. Both types of lymphocytes use IL-2 and proliferate. The degree of proliferation can be determined by measuring [^{3}H]-thymidine incorporation or through direct microscopical observation. One can search for compounds that suppress the production of IL-2 by testing whether the test samples inhibit the T cell proliferation or not.

Preparation of Spleen Cells.   Mice were killed by cervical dislocation. Their spleens were removed and placed in a petri dish containing phosphate buffer (PBS) supplemented with 5% fetal calf serum (PBS/5% FCS, FCS; Gibco, Paisley, Renfrewshire). The spleens were teased carefully into a single cell suspension using slide glass techniques.

Erythrocytes were lysed by a 3-minute exposure to ammonium chloride buffer and the lysate was centrifuged. The precipitate was washed twice with PBS/5% FCS and the cell suspension was filtered through nylon mesh. Finally, the cells were resuspended in complete RPMI1640 culture medium consisting of 10% FCS, 2 mM L-glutamine, 5×10^{-5} M 2-mercaptoethanol (2ME), 50 U/ml of penicillin, and 50 μg/ml of streptomycin. The cells were counted in a hemocytometer. The viability was judged by trypan blue exclusion and was always greater than 95%.

Mixed Lymphocyte Cultures (MLC).   Spleen cells were prepared from two strains of mice that had different MHC antigens. For example, BALB/c (H-2^d haplotype) and C57BL/6 (H-2^b haplotype) mice can be used. Splenocytes from the two strains of mice were co-cultured in 96-well microtiter plates (4×10^5 cells/well in 200 μl complete RPMI1640 medium, Sumitomo Bakelite Co., Ltd., Tokyo) in the presence of serial dilutions of test samples. The culture plates were incubated at 37°C in a humidified CO_2 incubator for 3 days. The degree of suppression of the MLR in the presence of test samples can be determined by microscopical observation of the lymphocyte proliferation (Figure 12–1).

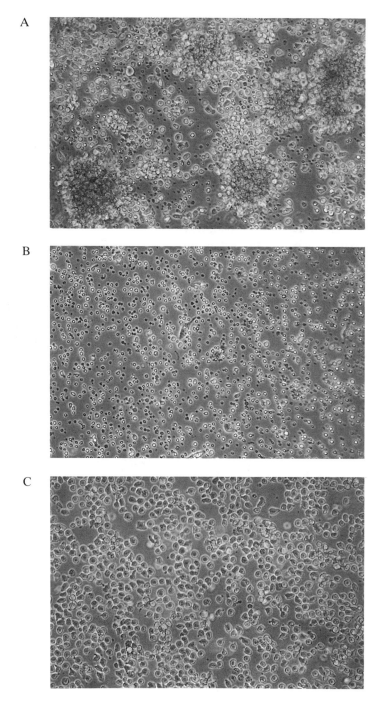

FIGURE 12–1 Effect of FK506 on the blastogenesis in MLC (A, control, and B, FK506, 1 ng/ml) and on the growth of mouse lymphoma EL-4 (C, EL-4, FK506, 1000 ng/ml).

12.1.1.2   An Assay Method using Concanavalin A-Induced Blastogenesis.   There are a series of agents called mitogens that activate T and B cells. T cells activated by Con A, the most extensively characterized T cell mitogen, produce various lymphokines, including IL-2. Stimulated T cells that express IL-2 receptors on their surfaces respond to IL-2 and proliferate. One can select compounds that suppress the production of IL-2 using this reaction. Splenocytes (2×10^5 cells/well in 100 μl, complete RPMI1640 medium including 5 μg/ml Con A) were cultured in 96-well microtiter plates in the presence of serial dilutions of the test samples. The cells were incubated for 3 days at 37°C in a humidified CO_2 incubator. The degree of suppression on Con A-induced T cell proliferation can be determined by microscopically observing lymphocyte growth.

12.1.1.3   Exclusion of Nonspecific Cytotoxic Interference.   Fermentation broths usually contain many substances that inhibit the growth of cells nonspecifically. These interferences make it difficult to detect substances that are specific lymphocyte growth inhibitors. This problem has to be overcome to avoid false positive results. To select such specific inhibitors more efficiently, the cytotoxicity of the sample is tested. The broths, including the active substances, can be selected by comparing their potency for specific suppression with the nonspecific cytotoxicity. Mouse lymphoma EL-4 is an appropriate cell line for this objective because it does not require IL-2 for growth.

12.1.1.4   Screening Procedure.   A set of two, 96-well microtiter plates is prepared as described previously. The fermentation samples can be tested as culture filtrates or acetone extracts from the mycelium of the microorganisms. One plate is used to test for the suppressive effect on T cell proliferation and the other is used to test for the nonspecific cytotoxicity against mouse lymphoma EL-4. Both plates are cultured in a CO_2 incubator. After 3 days of incubation, the plates are studied under a microscope and the samples showing lymphocyte proliferation inhibition at a lower concentration than necessary for the growth of EL-4 cells are selected for purification of the active principle.

12.1.1.5   Discovery of FK506.   After screening thousands of fermentation broths using the blastogenetic response of mouse lymphocytes, a potent immunosuppressant was discovered (Kino et al. 1987a). The active compound, FK506, was found in the broth of an actinomycete isolated from a soil sample collected near Tsukuba city, Japan, and was designated as *Streptomyces tsukubaensis* no. 9993. The fermentation broth inhibited the proliferation of activated T cells at a 1000-fold dilution but the filtrate did not inhibit the growth of EL-4 lymphoma at a tenfold dilution.

The active principle in the cultured broth was purified and FK506 was isolated as colorless prisms. FK506 is a neutral macrolide, which has a hemiketal-masked α, β-diketoamide functionality incorporated in a 23-membered ring (Figure 12–2). During the course of its isolation, five minor components

FK506

FR900525

FR901154

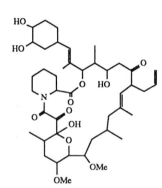

FR901155

FR901156

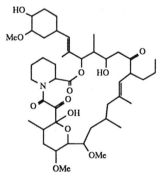

FIGURE 12–2 FK506 and its derivatives. Reproduced by permission of *The Journal of Antibiotics* from Kino et al. (1987a) for FK506, and Hatanaka et al. (1988) for FR900520, FR900523, and FR900525. (*Continued on next page.*)

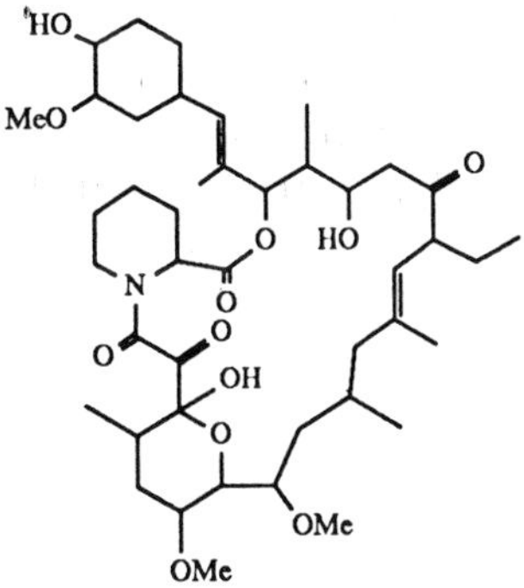

FR900520 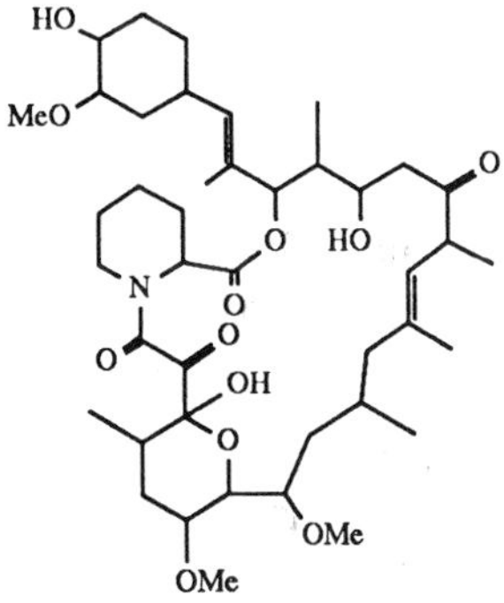   FR900523

FIGURE 12-2 (continued)

(FR900525, FR901154, FR901155, FR901156, and FR900520), were detected and purified (Hatanaka et al. 1989; Okuhara et al. 1990). Later, the FK506-related compounds FR900520 and FR900523 were found in the broth of *Streptomyces hygroscopicus* subsp. *yakushimaensis* no. 7238, a species from a soil sample originating in Yakushima, Japan (Hatanaka et al. 1988). The structure of these compounds are shown in Figure 12-2.

12.1.1.6 Immunosuppressive Effect of FK506 In Vitro. Originally discovered as an agent suppressing the blastogenetic response of mice lymphocytes, FK506 not only suppresses the maturation and proliferation of lymphocytes in the MLR or the response to Con A, but also suppresses various T cell-related immune responses. For example, generation of cytotoxic T cells, production of T cell-derived soluble mediators such as IL-2, IL-3, and IFN-gamma, and expression of the IL-2 receptor are suppressed by this agent (Kino et al. 1987b). Cyclosporin A (CsA), a well known fungal metabolite extracted from *Tolypocladium inflatum,* has similar effects. However, FK506 suppresses in vitro immune models at about 100 times lower concentrations than CsA. On the other hand, FK506 is nontoxic to IL-2 independent mouse lymphoma EL-4 and BW5147 at concentrations below 1000 nM and, moreover, the effect of FK506 on the formation of granulocyte/macrophage colonies by murine bone marrow cells cultured in semisolid agar is negligible. The murine, in vitro experiments demonstrate that FK506 specifically inhibits early events of the activated "T cell cascade" and minimally suppresses myelopoiesis and erythropoiesis (Tocci et al. 1989; Kino et al. 1987b).

12.1.1.7 Immunosuppressive Effects of FK506 In Vivo.

Humoral and Cellular Immunity. Immune response is generally classified into humoral and cellular immunity. The first acts by means of antibodies whereas the latter acts mainly through lymphocytes and macrophages. FK506 inhibits both of these immunities. This was deduced from the findings that FK506 strongly suppressed antibody production against sheep red blood cells (SRBC) as well as the delayed type hypersensitivity (DTH) response to methylated bovine serum albumin (MBSA) in mice. FK506 is about tenfold more potent in suppressing the antibody production and threefold more potent in suppressing DTH responses than CsA (Kino et al. 1987a).

Graft versus Host Reaction (GvHR).   When a recipient cannot reject a graft, the immunocompetent donor cells (e.g., from a bone marrow graft) recognize the recipient's MHC antigens and react against the recipient's tissues. FK506 inhibited GvHR that was induced by the injection of spleen cells from C57BL/6 mice into the footpads of BDF_1 mice. The effective dose of FK506 was about sixfold lower than that of CsA (Kino et al. 1987a).

Animal Model of Organ Transplantations.   Ochiai et al. studied extensively the effect of FK506 on organ transplantations in animal models. In these studies, FK506 prolonged the survival time of grafts for cardiac and skin transplants in rats (Ochiai et al. 1987c; Inamura et al. 1988c), in kidney transplantation in dogs (Ochiai et al. 1987a), and in mouse skin xenografts to rats (Ochiai et al. 1987b).

1. Rat Skin Allograft

Rat skin allografting is well suited for testing the in vivo effectiveness of immunosuppressants at the early stage of evaluation because the technique is not difficult to master.

Ear skin grafts from F344 strain rats (RT-1^{lvl}, donor) were transplanted to the lateral thorax of WKA strain rats (RT^k, recipient). The rejection symptoms appeared in 6 to 7 days and quickly culminated in complete graft necrosis. Intramuscular administration of FK506 (5 days per week starting on the day of grafting, for 2 weeks after grafting) prolonged the acceptance time of F344 skin allografts in WKA rats. One milligram per kilogram and 3.2 mg/kg of FK506 prolonged median allograft survival time to 27 and 43 days, respectively. Similar results were obtained with CsA at higher doses, i.e., 32 mg/kg or more. When WKA recipients of F344 skin grafts were treated with FK506 at 3.2 mg/kg for 5 days per week for 2 weeks and were given subsequent maintenance doses of 0.32 or 3.2 mg/kg twice per week for 120 days, all the animals retained healthy grafts as long as the treatment was continued (Inamura et al. 1988c).

2. Rat Cardiac Allograft

Heterotopic cardiac allografting was performed under ether anesthesia by cuff anastomosis of the donor aorta and pulmonary artery to the recipient's common-carotid artery and external jugular vein, respectively. Rejection occurred, the blood did not flow in the coronary artery of the graft, and finally, the heartbeat ceased. The median graft survival time of F344 cardiac allografts in untreated control WKA rats was 6 days with a range of 5 to 8 days. Engrafted rats were administered 1.0 mg/kg/day of FK506 orally. The schedule of administration was 5 days per week from the day of grafting for 2 weeks. In this system, all grafts survived indefinitely. A similar prolongation was obtained using oral administration of CsA at a dose of 10 mg/kg/day. Donor-specific suppressor cells may play a role in the maintenance of FK506-induced long-term graft acceptance (Ochiai et al. 1987d).

3. Canine Renal Allograft

Kidneys from dogs were transplanted into unrelated dogs whose own kidneys were removed. The dog could live with the renal allograft, but after about 1 week, signs of rejection began to appear, and after 2 weeks or so, the obtained graft malfunctioned and caused the death of the animals. However, when daily treatment with FK506 was started from the day of allografting, the median survival time was 61 days using a dose of 0.08 mg/kg/day. On this regimen, one-half of the recipients survived more than 250 days with a dose of 0.16 mg/kg/day, intramuscularly (i.m.). Oral treatment with 1.0 mg/kg/day enabled all the recipient dogs to survive for more than 140 days (Ochiai et al. 1987a). This improvement of survival time of skin, heart, and kidney transplants made it reasonable to expect good results with such transplantation models as the liver (Todo et al. 1987) and cornea (Kobayashi et al. 1989). Many investigators have shown that FK506 also prolongs the survival time of several other organs.

Autoimmune Animal Models. The immune system specifically recognizes foreign antigens and distinguishes them from its own antigens. Autoregulatory failure causes the immune system to react against its own antigens and to destroy the body's tissues. Autoantibodies and autoreactive T cells are now proven to be actively involved in the manifestation of autoimmune diseases. A variety of autoimmune diseases occur in humans and the corresponding autoimmune diseases have been produced or found in animals. The effects of FK506 on the development of these diseases have been examined in a wide range of animal models.

Type II collagen-induced arthritis rats manifest several immune abnormalities that are similar to human rheumatoid arthritis. Therefore, the model has been widely used to detect therapeutic agents that might exhibit antiarthritic activity in humans. Rats immunized with type II collagen were treated for 12 days with different doses of FK506. The treatment was begun on the day of immunization and was continued for 12 days. Doses of 0.32 mg/kg or higher suppressed arthritis significantly (Inamura et al. 1988a).

MRL/lpr mice (8 weeks old) and NZBxNZW F_1 mice (12 weeks old), which are recognized as important animal models of human systemic lupus erythematosus (SLE), were injected intraperitoneally (i.p.) with FK506 (2.5 mg/kg, 3 times per week for 40 weeks). The results showed that the drug prolonged the lifespan and reduced proteinurea (Takabayashi et al. 1989).

FK506 was given orally to spontaneous diabetes BB rats in doses of 1 mg/kg or 2 mg/kg between 30 and 120 days of age. In these animals, FK506 could delay the onset of diabetes in direct proportion to the dose and prolong the lifespan (Murase et al. 1990). FK506 also prevented experimental allergic encephalomyelitis (Inamura et al. 1988b) and experimental autoimmune uveo-retinitis (Kawashima et al. 1988).

Several research groups have been investigating the mechanism of immunosuppressive action of FK506 (Tocci et al. 1989; Schreiber 1991). This agent, similar to CsA, inhibits T cell proliferation, lymphokine production, and lymphokine gene expression. With these parameters of T cell activation, FK506 is 10- to 100-fold more potent than CsA. Both FK506 and CsA appear to inhibit the TCR-CD3-mediated signal transduction pathway by inhibiting the induction as well as the function of such nuclear transcription factors as the nuclear factor for activated T cells (NF-AT) and AP-1 (Granelli-Piperno et al. 1990; Bierer et al. 1990a; Mattila et al. 1990). FK506 and CsA bind to FK-BP and cyclophilin, respectively. These cytoplasmic receptors, called immunophilins, have a peptidyl-prolyl *cis-trans* isomerase activity (PPIase), which FK506 and CsA inhibit in vitro. However, the exact involvement of PPIase in immunosuppression remains to be clarified (Bierer et al. 1990b). FK506 was discovered by a rational approach: inhibition of proliferation of activated T cells, which was demonstrated to be a pathogenic principle of allograft rejection and the development of autoimmune diseases. This agent would not only provide a breakthrough in the management of clinical transplants, but also could be a useful research tool for studying the signal transduction mechanism in T cells, just as the discovery of penicillin resulted in the elucidation of the structure of bacterial cell wall and of its biosynthesis.

12.1.2 Screening for Inducers of Ia Receptor Expression

The immune response starts with the recognition of an antigen by cell-surface receptors of B and T cells. However, these two types of cells recognize antigens in different ways. T cells, in contrast to B cells, do not recognize a native, unprocessed antigen. They can only recognize a complex of an antigen-derived peptide and the Ia receptor on the surface of the antigen presenting cell (APC) (Grey et al. 1989; Unanue and Allen 1987; Sette et al. 1987). The process of T cell activation involves several steps. First, the antigen is captured by an APC and internalized by endocytosis. Then, the antigen is degraded into peptides and bind to the Ia receptors within the cell. Next, the antigen-derived peptide–Ia complex moves to the cell surface and the antigen is presented to a helper T cell. The interaction between the helper T cell receptor and the peptide–Ia receptor complex leads to the activation of the T cells that, in turn, results in the secretion of interleukins. The Ia molecule is a protein expressed on the APC surface. Since the amount of expressed Ia molecule on the APC surface is proportionate to the level of immune response, we may be able to manipulate the immune system for clinical purposes by using agents that control Ia molecule expression (Unanue and Allen 1987; Sette et al. 1987). In this connection, IFN-gamma, a potent inducer of Ia expression, has been reported to be effective for treating autoimmune diseases in humans and animals (Pernice et al. 1989; Jacob et al. 1989). This fact may demonstrate, at least in part, the validity of the above-described hypothesis. Therefore, IFN-gamma-like substances, which can positively modulate Ia expression on APCs, could be expected to have a therapeutic potential in autoimmune diseases.

12.1.2.1 An Assay Method for Ia Expression. An enzyme-linked immunosorbent assay (ELISA) was used to screen for compounds that enhance Ia expression. The amount of Ia molecule expressed on macrophages was detected using horseradish peroxidase (HRP) conjugated anti-mouse I-A^k monoclonal antibody. Serial twofold dilutions of test samples were prepared in 96-well microtiter plates in a volume of 50 μl, which were seeded with 50 μl of peritoneal exudate cells (PEC) (1×10^5 cells/well). The samples were then incubated for 3 days at 37°C in a humidified CO_2 incubator. The plates were aspirated, 100 μl of HRP conjugated anti-mouse I-A^k monoclonal antibody diluted in Dulbecco's modified Eagle's medium (DMEM) containing 5% FCS was added, and the mixture was incubated for 2 hours at room temperature. After incubation and washing (three times with 1 volume PBS containing 1% FCS), 200 μl of substrate solution (2.5 mg/ml o-phenylenediamine dihydrochloride and 0.018% H_2O_2 in 0.1 M citrate-phosphate buffer, pH 4.5 was added. The samples were incubated for another 30 minutes (room temperature, in the dark) and the reaction was stopped by adding 50 μl of 1 N HCl. The absorbance was read at 492 nm by a dual wavelength automated plate reader, with the reference wavelength set at 630 nm (Kaizu et al. 1985).

HRP Conjugated Anti-Mouse I-A^k Monoclonal Antibody. Conjugation of HRP (POD type IV, Sigma Chemical Co., St. Louis, MO) to monoclonal antibody was performed as described by Nakane and co-workers (Nakane and Kawaoi 1974).

Purification of Monoclonal Antibody. Anti-I-A^k monoclonal antibody producing hybridoma, 10-3.6.2 was purchased from American Type Culture Collection, Rockville, MD. The hybridoma cells were cultured in DMEM supplemented with 10% FCS, 4.5 mg/ml glucose, 50 U/ml penicillin, and 50 μg/ml streptomycin. Ascites fluids containing high-titer monoclonal antibody were produced by injecting hybridoma i.p. into BALB nu/nu mice primed with 2,6,10,14-tetramethylpentadecane (Wako Pure Chemical Industries Ltd., Osaka, Japan). Large amounts of monoclonal antibodies were purified from ascites fluids by ammonium sulfate precipitation and diethylaminoethyl (DEAE)-cellulose chromatography using a DE52-cellulose column (Whatman Chemical Separation Ltd., Maidstone, Kent, U.K.).

Preparation of Peritoneal Exudate Cells (PECs). Lipopolysaccharide, a major outer membrane component of Gram-negative bacteria, has been shown to have Ia-inducing activity both in vitro and in vivo (Wentworth and Ziegler 1987). Therefore, the C3H/HeJ mouse strain, a hyporesponder to lipopolysaccharide, was used to prepare PECs. This strain was used in order to avoid detection of bacterial endotoxins. C3H/HeJ mice were injected i.p. with 2 ml of 3% Brewer's thioglycollate medium (BBL, Cockeysville, MD). Four days after the injection, thioglycollate-induced peritoneal exudate cells (TG-PEC) were harvested by peritoneal lavage with 3 ml of ice-cold DMEM containing 20 U heparin/ml. The obtained cell suspension was centrifuged at 1000 rpm for 5 minutes and resuspended at 2×10^6 cells/ml in DMEM supplemented with 10% FCS and 50 µM of 2ME.

12.1.2.2 Ia Detection of Macrophages Stimulated with IFN-Gamma. For the control experiment, macrophages were stimulated with IFN-gamma and analyzed for Ia expression, because IFN-gamma is known to induce Ia production on macrophages (Smith et al. 1990). The dose-response of Ia expression versus concentration of IFN-gamma is illustrated in Figure 12–3.

12.1.2.3 Discovery of WS9482 (Kifunensine). WS9482 was produced in fermentation from a new species of the genus *Kitasatosporia*, *K. kifunense* sp. nov. (Figure 12–4). WS9482 significantly increased the Ia expression of

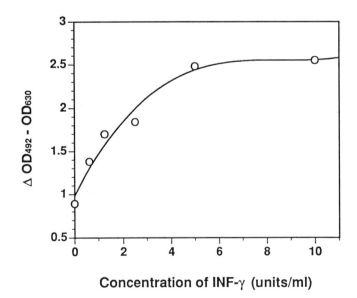

FIGURE 12–3 Dose response for Ia induction by IFN-γ in vitro.

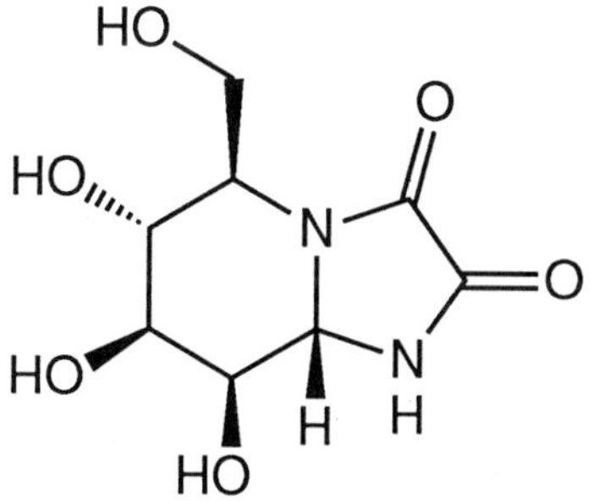

FIGURE 12–4 WS9482 (kifunensine). Reproduced with permission from Kayakiri et al. (1990).

peritoneal macrophages both in vitro and in vivo (Okuhara 1990). In the in vitro system, WS9482 dose-dependently increased the expression of Ia antigen starting with a concentration of 1.0 ng/ml and reaching maximum expression at around 100 ng/ml. At the latter concentration, the activity was about twice that of the control. Expression of Ia antigen in vivo was also significantly enhanced in a macrophage population by WS9482.

Treatment of C3H/HeJ mice with WS9482 resulted in a dose-dependent expression of the antigen, where subcutaneous injection with 100 mg/kg of WS9482 increased Ia antigen by three to four times that of the saline control. Kinetic studies on Ia induction in mice showed a maximum threefold increase over the control on day 1, followed by a rapid decline of Ia expression on days 2 and 3.

Effect of WS9482 on Rat Models of Autoimmune Diseases. Delayed-type hypersensitivity (DTH) reactions are one of the kinds of reactions that are caused by sensitized T cells after contact with an antigen. WS9482 showed a potent suppressive effect of DTH reactions caused by SRBC. The administration of WS9482 on days 0, 1, 2, 5, and 6 profoundly reduced DTH to SRBC when the animals were challenged with SRBC on day 6 and the DTH was measured on day 7. The ED_{50} was 18 and 22 mg/kg by subcutaneous and oral administration, respectively. Since WS9482 suppressed the T cell-involving immune response, this agent was expected to show suppressive properties on collagen-induced and adjuvant-induced arthritis in rats. Lewis rats immunized with collagen on day 0 and treated subcutaneously with 1.0, 3.0, and 10 mg/kg of WS9482 beginning on day 0 for 3 weeks showed a significant decreased prevalence and incidence of arthritis compared to the vehicle control. In the rats given WS9482, the DTH and antibody titers to collagen were lower than those values of untreated control on day 21. WS9482 also suppressed the development of adjuvant arthritis in rats.

Effect of WS9482 on Skin Allograft Survival in Rats. Ear skin grafts of F344 rats were transplanted on the lateral thorax of WKA strain rats. Transplanted grafts to untreated control rats were rejected in 6.0 days. WS9482 given subcutaneously in doses of 3, 10, and 30 mg/kg, 5 days per week beginning on day 0 increased the mean allograft survival time to 7.0, 12.5, and 24.0 days, respectively, although four out of five recipients given 30 mg/kg died on day 11, probably owing to the drug's toxicity.

12.1.2.4 Discovery of WS4490 (FR900483). WF4490 was produced by a fungus, *Nectria lucida* F-4490. The structure of WF4490 is (3R,4R,5R)-3,4-dihydroxy-5-hydroxymethyl-1-pyrroline (Figure 12–5). Addition of WF4490 to peritoneal macrophages also resulted in a significant rise of Ia antigen expression, with a maximum, 1.7-fold increase (100 ng/ml) over the control (T. Kaizu, M. Hashimoto, M. Okamoto, and M. Okuhara, unpublished observations). Both orally and subcutaneously, WF4490 decreased the DTH response in rats. This compound also inhibited chronic inflammatory responses in the collagen-induced arthritis model (Figure 12–6).

Our working hypothesis was that low molecular weight natural products showing enhancing activity of Ia expression could be useful immunomodulators for the treatment of chronic inflammatory diseases. We based our hypothesis on the fact that IFN-gamma, a potent inducer of Ia antigen expression, is therapeutically effective for adjuvant arthritis in rats (Jacob et al. 1989) and juvenile rheumatoid arthritis in humans (Pernice et al. 1989). As a result, we discovered two immunomodulators, WS9482 and WF4490. Both of these compounds in in vivo and in vitro models, potently enhanced Ia antigen expression in macrophages. As expected, both compounds inhibited the development of autoimmune diseases and prolonged the survival time of skin allografts in rat models. Whether there is a relationship between the enhancement of Ia expression and the immunosuppressive activity is not clear, but the injection of these compounds in mice caused a maximum increase of Ia antigen on day 1, followed by a rapid disappearance of antigen expression, and their suppressive effect on DTH reactions to collagen declined shortly after the drugs were stopped. These results may suggest that the Ia inducing activity of these compounds is associated with their suppressive effects on immune response. As stated earlier, screening of MLR inhibitors is a rational approach for the search for immunosuppressants. However, in the case of Ia inducers, WS9482 and WF4490 did not inhibit, but stimulated the MLR reaction. In this respect, they behaved similar to IFN-gamma (Siegel 1988).

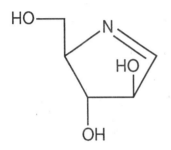

FIGURE 12–5 WF4490 (FR900483). Reproduced by permission of *The Journal of Antibiotics* from Shibata et al. (1988).

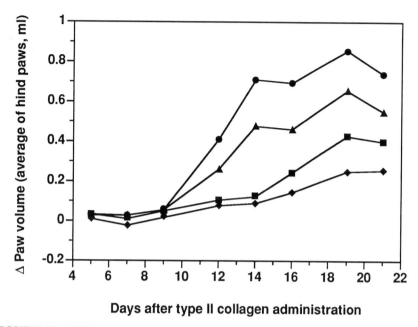

Days after type II collagen administration

FIGURE 12–6 Effect of WF4490 on the induction of type II collagen-induced arthritis in rats. Control, —●—; 10 mg/kg, —▲—; 32 mg/kg, —■—; and 100 mg/kg, —◆—. On day 0, Lewis rats in groups of 10 were immunized intradermally with an emulsion of bovine type II collagen and Freund's incomplete adjuvant. The rats were examined over the course of 3 weeks to record the edema volume of the hind paws. The drug was administered orally every day from day 0 to 21.

The immunosuppressive mechanisms of compounds with stimulating activity of MLR is still poorly understood. However, one can speculate that these Ia inducers stimulate the APCs (macrophages), which in turn generate suppressor T cells (Usui et al. 1984; Baxevanis et al. 1990). It has been shown that treatment of arthritic rats with WS9482 reduced the ratio of $CD4^+$ to $CD8^+$ (helper to suppressor) T cell populations. As an alternative explanation, it has been suggested that WS9482 and WF4490 enhance the expression of the Ia antigen not only in macrophages but also in B cells. Macrophages having increased Ia molecules enhance the MLR, while B cells with enhanced expression of Ia antigens cause the paralysis of helper T cells, which in turn leads to tolerance (E. Shevach, personal communication; Weaver and Unanue 1990).

12.2 SCREENING FOR IMMUNOSTIMULANTS

Application of microbial products with immunostimulating activity to clinical situations dates back to the nineteenth century, when William B. Coley used bacterial extracts for the treatment of tumors. After that, it was discovered that Bacillus Calmette-Guerin (BCG), a viable, attenuated strain of *Mycobacterium bovis,* could inhibit tumor growth when injected intralesionally. Preliminary trials of BCG have been conducted in malignant melanoma (Morton et al. 1970), lymphocytic leukemia (Mathe et al. 1976), lung cancer (Holmes et al. 1977), and other tumors. Several immunostimulants of microbial origin have been developed and are used in Japan in combination with chemotherapy, radiotherapy, and/or surgery for the treatment of patients who have tumors. However, most of the immunostimulants have unidentified structures and appear to lack in the degree of efficacy that would allow their adoption as useful clinical agents. Compounds with clearly defined structure and greater activity would be more promising for the treatment not only of tumors, but also infectious diseases caused by fungi, parasites, and viruses.

In the course of the screening for new antibiotics, we have noticed that some antibiotics show more potent antibacterial efficacy in vivo than was expected from their in vitro activity. We have pursued this observation and tried to find low molecular weight microbial products with protective efficacy in experimental animals, despite an apparent lack of antimicrobial activity in vivo.

12.2.1 An Assay Method for Immunostimulants

Since substances that have the properties described in the preceding section would have the property of enhancing the host's defensive ability, the protocol of the screening was designed as follows. Fermentation samples showing no in vitro antimicrobial activity were injected subcutaneously to mice 6, 5, 4, and 1 days before challenge with the infecting strain. On day 0, the mice were challenged with *Escherichia coli* and 2 days later surviving animals were counted. In this screening system, groups of three to five male ICR strain mice were inoculated i.p. with 4.8×10^7 colony forming units (cfu)/mouse of *E. coli* no. 22, i.e., the minimum lethal dose.

12.2.2 FK156 and Its Chemical Derivative FK565

Protective activity against *E. coli* infection in mice was detected in the culture filtrates of *Streptomyces olivaceogriseus* and *Streptomyces violaceus* (Gotoh et al. 1982a, 1982b). The active principle was isolated and the structure was confirmed to be an acyl tetrapeptide consisting of D-lactic acid, L-alanine, D-glutamic acid, L-meso-DAP, and L-glycine (Figure 12–7; Kawai et al. 1982). FK156 was chemically synthesized and a number of derivatives were prepared. Among these derivatives, FK565 was the most potent compound in its protecting activity against *E. coli* infections. Besides their anti-infectious activities, FK565 and its parent compound have various kinds of immunological activities both in vitro and in vivo. These include several activities outlined in Table 12–1. Of these, the anti-infectious, antitumor, and antiviral properties will be described in more detail.

```
            OH          CH₃
            |           |      (D)
A    CH₃CH–CONH–CH–CONH–CH–COOH
       (D)         (L)    |
                         (CH₂)₂  (L)
                          |
                          CONH–CH–CONH–CH₂–COOH
                               |
                              (CH₂)₃
                               |
                          NH₂–CH–COOH
                               (D)


                              (D)
B              CH₃(CH₂)₅CONH–CH–COOH
                              |
                             (CH₂)₂            CH₃
                              |      (L)       |
                              CONH–CH–CONH–CH–COOH
                                   |          (D)
                                  (CH₂)₃
                                   |
                              NH₂–CH–COOH
                                   (D)
```

FIGURE 12–7 FK156 (A) and FK565 (B). Reproduced by permission of *The Journal of Antibiotics* from Mine et al. (1983b).

12.2.2.1 Host Resistance Enhancing Activity against Microbial Infection.
FK156 and FK565 significantly enhanced the host's defensive ability of mice against microbial infection with various microorganisms (Mine et al. 1983b). In the experiment, both FK156 and FK565 potently suppressed the effects of challenge with various extracellular and facultative intracellular organisms. FK156 was only effective when given parenterally but FK565 was effective through both the parenteral and oral routes. Because microbial infections are frequent in immunocompromised patients and cause serious problems, it is important to find immunostimulants that are effective against infections in immunosuppressed hosts. FK156 and FK565 were evaluated in models of mice immunosuppressed with various immunosuppressive agents and were found to significantly restore their defensive mechanisms against microbial invasion (Yokota et al. 1983). These compounds also enhanced the therapeutic

TABLE 12–1 Immunological Properties of FK156 and Its Derivative FK565

Enhance RES function, humoral antibody production, and DTH reactions
Activate macrophages for tumor cell cytotoxicity
Activate NK cells against tumor cells
Augment IL-1 release from macrophages
Induce colony-stimulating factor
Protect against bacterial, fungi, and viral infection
Regress pulmonary metastases

effect of such antibiotics as ticarcillin or gentamicin in immunosuppressed mice. The host defense mechanisms of FK156 and FK565 were investigated in mice and it was found that these compounds increase phagocytes in both the peripheral blood and the peritoneal cavity. Phagocyte function such as chemotaxis, phagocytosis, and killing activities were also enhanced in drug-treated mice (Mine et al. 1983a).

12.2.2.2 Tumor Growth Inhibiting Activity. FK156 and FK565, which had been demonstrated to have diverse and highly potent immunostimulatory activities, were therefore expected to have therapeutic effects against tumor growth. In general, from the studies on the therapeutic effects of immuno-modulators on transplantable tumors, we know that tumors inoculated in allogeneic animals are more sensitive to immunomodulators than tumors in syngeneic ones, and highly immunogenic tumors also are more sensitive to drugs than weakly immunogenic ones. Moreover, the site of administration of the immunomodulators influences the host-mediated suppressive activity in solid tumors. Compounds injected directly into tumors lead to a higher rate of tumor growth inhibition than when the compounds are administered systemically. Taken together, weakly immunogenic tumor-syngeneic mouse systems, such as P388 leukemia-DBA/2 mice would probably be a more realistic model than highly immunogenic tumor-syngeneic mouse or allogeneic mouse systems for the development of the clinically useful immunotherapy treatments. Systemic treatment is also a more realistic model than local injections.

FK156 and FK565 were evaluated for therapeutic activity in the P388-DBA/2 model by both intralesional and systemic injection (Izumi et al. 1983). It was found that both compounds, when injected directly into solid P388 tumors that had been established 7 days after tumor inoculation, caused a significant decrease in the tumor size. Systemic treatment also produced significant therapeutic effect in this tumor model. Although regression of the primary tumor was observed in these protocols, the lifespan was not prolonged. However, consecutive administration of FK156, 4 times per week starting on day 1, resulted in a considerable increase in the median survival time.

12.2.2.3 Antiviral Activity of FK565. The antiviral activity of FK565 was investigated because several immunomodulators with macrophage-stimulating activity are known to increase host resistance against viral infections. Macrophages from mice pretreated with FK565 inhibited the multiplication of herpes simplex virus type 1 (HSV-1) and type 2 (HSV-2) in monkey Vero cells, and that of vesicular stomatitis virus (VSV) in murine L929 cells. Subcutaneous administration of FK565 to mice, 6 mg/kg once per week for 3 weeks after infection, significantly increased survival time (Oku et al. 1986).

Generally, in vivo screening with live animals has several disadvantages not encountered with in vitro screening. In vivo screening is usually less sensitive to biological activities so that only compounds with highly potent activity are detected. In vivo screening is also laborious and time-consuming, and moreover, adds significant time to the isolation of the active principles. The advantages of the system are that active compounds inevitably respond in animal models, and the possibility exists that in vivo screening could detect new substances having unknown mechanisms of action. The latter is possible because in vivo screening is not target oriented, like enzyme inhibitors or receptor antagonists. Another potential advantage would be that the compounds detected by in vivo screening will be possibly novel ones because very few researchers will undertake such a "crazy screening program."

12.3 PERSPECTIVES: NEW TARGETS FOR IMMUNOMODULATORS

The increasing understanding of basic processes of both normal and abnormal immune responses offers new insights of and approaches for the screening of new compounds for immunotherapy. This has been supported by parallel advances in instrumentation and analytical methodologies obtained in the last two decades. In the field of immunology, perhaps the most applicable advance for new screening methods concerns cell-cell interactions. The following discussion will briefly review possible new targets for screening for immunomodulators.

12.3.1 Antigen-APC-T Cell Hybridoma System

The antigen-APC-T cell hybridoma system serves as a good in vitro model for immune response. In the body, which is immunized with an antigen, the only detectable immune reaction is against that antigen specifically. In the T cell hybridoma system, the T cells produce IL-2 in an antigen specific-APC dependent manner (Kaye et al. 1984). This in vitro system is applicable for the screening of compounds that inhibit steps in the immune response such as antigen uptake, processing, presentation, and/or T cell activation. T cell hybridomas are established by polyethylene glycol-induced fusion of spleen cells with the T cell hybridoma line BW5147. Spleen cells are obtained from mice (DBA/2) immunized with ovalbumin (OVA, an antigen) and complete Freund's adjuvant. The cells are then restimulated in vitro for 3 days with OVA (100 µg/ml) before cell fusion. After fusion, the cells are cultured in selective medium and growing hybrids are screened according to their reactivity to OVA by assaying production of IL-2.

12.3.1.1 An Assay Method. The experiment is carried out in 96-well microtiter plates. Serial twofold dilutions (50 µl) of test samples are seeded with

50 μl of hybridoma cells (5×10^4 cells/well) and 50 μl of syngeneic PEC (macrophages, 2.5×10^5 cells/well). The cultures are incubated in the presence of OVA (100 μl/ml). After 24 hours, 50 μl of supernatant medium is transferred to microculture wells containing 10^4 CTLL-2 cells, a cytotoxic T-lymphocyte line (established by Baker et al. 1979). After incubation for 24 hours the presence of IL-2 is assessed by measuring the [^{3}H]-thymidine incorporation by CTLL-2 cells.

12.3.1.2 Microbial Products with Inhibiting Activity of Antigen Presentation.

There are several compounds of microbial origin that inhibit processes involved in antigen presentation. Cerulenin is an antibiotic produced by a *Streptomyces* sp. The mechanism of its antimicrobial activity is inhibition of fatty acid and sterol biosynthesis (Matsumae et al. 1963; Ohmura 1976). In the antigen-APC-T cell hybridoma system, cerulenin treatment inhibits the ability of APC to present the given antigen to T cells. However, when APC are allowed to take up and process antigens before drug treatment, inhibition of presentation does not occur. This suggests cerulenin inhibits an antigen processing step in which a native nonimmunogenic protein is converted into an immunogenic form (Falo et al. 1987).

Brefeldin A was discovered as an antifungal agent (Hayashi et al. 1974). This bicyclic antibiotic was isolated from the fermentation broth of a fungus identified as *Phyllosticta medicaginis*. Brefeldin A specifically inhibits the presentation of endogenous antigens for MHC class-I as well as class-II restricted T cells (Nuchtern et al. 1989; Adorini et al. 1990; Yewdell and Bennink 1989).

Various kinds of microbial protease inhibitors, including leupeptin, antipain, chymostatin, elastatinal, pepstatin, phosphoramidon, amastatin, and bestatin, have been demonstrated to inhibit antigen presentation in a highly selective manner. The site of action was suggested to be the step leading to the antigen processing by the APCs (Puri and Factorovich 1988). The idea that the protease inhibitory activity of these inhibitors is associated with the inhibition of the antigen processing is easy to assume but still remains to be elucidated. Since most of the known microbial protease inhibitors are non-cytotoxic and of low molecular weight, protease inhibiting activity may be a promising target for immunotherapeutic agents. However, the effective doses of the above mentioned protease inhibitors for inhibition of the antigen presentation are rather high, 100 to 200 μg/ml. More potent activity (50% inhibitory doses of less than 1.0 μg/ml) would be necessary for therapeutic efficacy.

D.A. Palay and co-workers observed that cyclosporin inhibited macrophage-mediated antigen presentation in vitro at a dose less than 1.0 μg/ml. Although the significance of this effect is not known, it may be involved in at least part of the drug's immunosuppressive properties (Palay et al. 1986).

As stated earlier, the antigen receptor of T cells recognizes processed antigens in association with an Ia antigen. When the processed antigen and the Ia molecule are recognized by a T cell receptor, these two components of the ligand are considered to physically interact on the surface of APC before T cell recognition. Substitute peptides, in which amino acid sequences of epitope are chemically modified, have been suggested to bind to the Ia molecule of the APC and form a complex that can no longer be recognized by T cells. Such "blocking peptides" also have been demonstrated to effectively inhibit the T cell responses when injected into mice (Adorini et al. 1988; Sette et al. 1989). However, owing to their structure, these peptides would be unstable in vivo and also would be poorly absorbed by oral administration. Screening for nonpeptide natural products is a promising approach toward the prevention and possibly also for the treatment of autoimmune diseases. In the future, when disease-associated antigens and disease-related human leukocyte antigens (HLA) class-II molecules are identified more precisely, screening systems for competitors of such molecules could be established using hybridoma cells.

12.3.2 Adhesion Receptors

A specific immune response is initiated by activation of antigen-specific receptors on T cells and B cells. However, recent findings have demonstrated that antigen-independent interactions between accessory molecules (adhesion receptors) play a critical role in cell-cell interactions in the generation of the immune response (Springer 1990). A variety of adhesion receptors have been identified so far. They can be divided into three families according to their structural and functional similarities: (1) the immunoglobulin superfamily, which includes the antigen-specific receptors of T cells and B cells; (2) the integrin family, which plays a role in the regulation of adhesion and migration; and (3) the selectin family, which helps regulate leukocyte binding to endothelium. Because of the involvement of adhesion receptors in the immune response, inhibition of the expression of the adhesion receptors or antagonism of the corresponding ligands by receptor antagonists would provide a therapeutic approach for reducing pathological conditions. Several recent studies have suggested that blocking the functions of adhesion receptors could lead to the inhibition of disease manifestation in animal experiments and clinical trials (Fischer et al. 1986; Price et al. 1987; Wegner et al. 1990).

LFA-1 is a member of the integrin family, and is expressed on the surface of T and B cells, NK cells, monocytes, macrophages, and granulocytes. Monoclonal antibodies to LFA-1 inhibit a wide variety of adherence-related cellular functions, such as T cell-mediated killing, antigen nonspecific natural killing, antigen presentation, and the binding of tumor cells by activated macrophages. This antibody also has the ability to inhibit immune responses in vivo. Fischer et al. reported that the administration of anti-human LFA-1 monoclonal antibody facilitated the engraftment of T cell-depleted HLA mismatched bone marrow transplantation in children (Fischer et al. 1986).

CD11B/CD18 (LFA-1) is also important in normal neutrophil functions. Murine monoclonal antibody to CD18 is a potent inhibitor of human neutrophil functions in vitro. The antibody also inhibited neutrophil function in an inflammatory model using rabbits (Price et al. 1987).

ICAM-1 is a member of the immunoglobulin superfamily and has been shown to mediate the adhesion of lymphocytes and neutrophils to endothelial cells. Barton et al. (1989) reported that pretreatment with antibody to ICAM-1 prevented granulocyte infiltration into the lung in a phorbol ester-induced inflammatory model in rabbits. Wegner et al. (1990) also showed that a monoclonal antibody to ICAM-1 attenuated eosinophil infiltration and the induction of airway hyperresponsiveness in vivo.

The above-mentioned in vivo effectiveness of monoclonal antibodies to adhesion receptors suggests that substances having a similar mode of action might serve as useful anti-inflammatory and antigraft rejection agents. A specific inhibitor of the adhesion receptors' function has yet to be discovered from microorganisms; however, there are several findings that give hints for the establishment of in vitro screening systems for such inhibitors. Two examples are described here. Homogeneous cell populations such as T, B, or NK cell lines are stimulated by treatment with phorbol 12-myristate-13 acetate (PMA) and adhere to one another to form multicellular clusters. The clusters in this assay are completely inhibited by monoclonal antibodies to LFA-1 (Van Kooyk et al. 1989).

ICAM-1 expression on cultured epidermal keratinocyte is increased by treatment with IFN-gamma. T lymphoblasts are strongly adherent to IFN-gamma-treated epidermal cell monolayers through ICAM-1–LFA-1 interactions. Adhesion of T lymphoblasts to the monolayers can be measured using [^{51}Cr]-labeled lymphocytes (Dustin et al. 1988).

12.3.3 Enzymes on the Surface of Leukocytes

It is now known that many of the differentiation antigens on the surface of leukocytes are either enzymes in themselves or are associated with enzymes. CD75 is a 5'-ectonucleotidase that is expressed on subpopulations of human T and B cells. This enzyme may play an important part in immune regulation because its activity is markedly lower in lymphocytes from patients with several immunodeficiency diseases (Thompson et al. 1987).

CD10, a common acute lymphocytic leukemia antigen (CALLA), has been widely used as a diagnostic marker of common ALL. Recently, Letarte et al. reported that the amino acid sequence deduced from cDNA sequence coding for CD10 was identical to that of neutral endopeptidase (NEP, enkephalinase). Several studies indicated that this protein may be important in neutrophil function (Letarte et al. 1988). CD45 is a protein tyrosine phosphatase (PTP), and is thought to participate in signal transduction events involving the dephosphorylation of target proteins in T cells (Tonks et al. 1988; Kiener and Mittler 1989).

Tyrosine protein kinase (TPK), pp56[lck] is associated with CD4 and CD8 molecules and is believed to carry out important physiological functions in T cell responses (Veillette et al. 1988; Rudd et al. 1988).

Specific inhibition (or in some cases, activation) of the enzymatic activity of these receptor molecules could modulate the leukocyte functions and contribute to reversing the biochemical reactions that occur during abnormal immune responses. Another interesting enzyme for immunosuppressants is T cell heparanase. This enzyme is produced by activated, but not resting, T cells and is considered to play an important role in causing autoimmune diseases (Naparstek et al. 1984). Lider et al. (1989) reported that low-dose heparin could prevent the development of autoimmune diseases and allograft rejection in animal models and also showed that heparins inhibit the activity of heparanase in vitro.

In the past, the ability of microbes to produce immunological activities was mostly overlooked. However, the development of CsA and FK506 as immunosuppressants in organ transplantation has changed the situation dramatically. Today we can predict that more potent as well as less toxic immunomodulators will be discovered from microbial products.

REFERENCES

Adorini, L., Cardinaux, F., Falcioni, F., and Muller, S., et al. (1988) *Nature* 334, 623–625.

Adorini, L., Ullrich, S.J., Appella, E., and Fuchs, S. (1990) *Nature* 346, 63–66.

Bach, F.H., Back, M.L., and Sondel, P.M. (1976) *Nature* 259, 273–281.

Baker, P.E., Gillis, S., and Smith, K.A. (1979) *J. Exp. Med.* 149, 273–278.

Barton, R.W., Rothlein, R., Ksiazek, J., and Kennedy, C. (1989) *J. Immunol.* 143(4), 1278–1282.

Baxevanis, C.N., Reclos, G.L., and Papamichail, M. (1990) *J. Immunol.* 144(11), 4166–4170.

Bierer, B.E., Mattila, P.S., and Standaert, R.F., et al. (1990a) *Proc. Natl. Acad. Sci. USA* 87, 9231–9235.

Bierer, B.E., Somers, P.K., Wandless, T.J., Burakoff, S.J., and Schreiber, S.L. (1990b) *Science* 250, 556–559.

Cohen, I.R., Holoshitz, J., Eden, W.V., and Frenkel, A. (1985) *Arthritis Rheum.* 28, 841–845.

Dustin, M.L., Singer, K.H., Tuck, D.T., and Springer, T.A. (1988) *J. Exp. Med.* 167, 1323–1340.

Eden, V.W., Holoshitz, J., and Nevo, Z. (1985) *Proc. Natl. Acad. Sci. USA* 82, 5117–5120.

Falo, L.D., Benacerraf, B., Rothstein, L., and Rock, K.L. (1987) *J. Immunol.* 139(12), 3918–3923.

Fischer, A., Griscelli, C., and Blanche, S., et al. (1986) *Lancet* 2, 1058–1061.

Fudenberg, H.H., and Whitten, H.D. (1984) *Annul. Rev. Pharmacol. Toxicol.* 24, 147–174.

Gillis, S., and Smith, K.A. (1977) *Nature* 268, 154–156.

Gotoh, T., Nakahara, K., Iwami, M., Aoki, H., and Imanaka, H. (1982a) *J. Antibiotics* 35(10), 1280–1285.

Gotoh, T., Nakahara, K., and Nishiura, T., et al. (1982b) *J. Antibiotics* 35(10), 1286–1292.

Granelli-Piperno, A., Nolan, P., Inaba, K., and Stainman, R.M. (1990) *J. Exp. Med.* 172, 1869–1872.

Grey, H.M., Sette, A.S., and Buus, S. (1989) *Sci. Am.* 261, 38–46.

Hatanaka, H., Iwami, M., Kino, T., Goto, T., and Okuhara, M. (1988) *J. Antibiotics* 41(11), 1586–1591.

Hatanaka, H., Kino, T., and Asano, M., et al. (1989) *J. Antibiotics* 42(4), 620–622.

Hayashi, T., Takatsuki, A., and Tamura, G. (1974) *J. Antibiotics* 27, 65–72.

Holmes, E.C., Ramming, K.P., and Mink, J., et al. (1977) *Lancet* 2, 586–587.

Inamura, N., Hashimoto, M., and Nakahara, K., et al. (1988a) *Clin. Immunol. Immunopathol.* 46, 82–90.

Inamura, N., Hashimoto, M., and Nakahara, K., et al. (1988b) *Int. J. Immunopharmacol.* 10(8), 991–995.

Inamura, N., Nakahara, K., and Kino, T., et al. (1988c) *Transplantation* 45(1), 206–209.

Izumi, S., Nakahara, K., and Gotoh, T., et al. (1983) *J. Antibiotics* 36(5), 566–574.

Jacob, C.O., Holoshitz, J., Meide, P.V.D., Strober, S., and MacDevitt, H.O. (1989) *J. Immunol.* 142, 1500–1505.

Kaizu, T., Kojima, K., and Iwasaki, K. (1985) *Thrombosis Res.* 40, 91–99.

Kawai, Y., Nakahara, K., and Gotoh, T., et al. (1982) *J. Antibiotics* 35(10), 1293–1299.

Kawashima, H., Fujino, Y., and Mochizuki, M. (1988) *Invest. Ophthalmol. Visual Sci.* 29(8), 1265–1271.

Kayakiri, H., Oku, T., Hashimoto, M., et al. (1990) *Chem. Pharm. Bull.* 38(1), 293–295.

Kaye, J., Gillis, S., and Mizel, B.S., et al. (1984) *J. Immunol.* 133(3), 1339–1345.

Kiener, P.A., and Mittler, R.S. (1989) *J. Immunol.* 143(1), 23–28.

Kino, T., Hatanaka, H., and Hashimoto, M., et al. (1987a) *J. Antibiotics* 40(9), 1249–1255.

Kino, T., Hatanaka, H., and Miyata, S., et al. (1987b) *J. Antibiotics* 40(9), 1256–1265.

Kobayashi, C., Kanai, A., Nakajima, A., and Okumura, K. (1989) *Transplant. Proc.* 21(1), 3156–3158.

Larsson, E., and Coutinho, A. (1979) *Nature* 280, 239–241.

Letarte, M., Vera, S., and Tran, R., et al. (1988) *J. Exp. Med.* 168, 1247–1253.

Lider, O., Baharav, E., and Mekori, Y.A. (1989) *J. Clin. Invest.* 83, 752–756.

Mathe, G., Vassal, F., Delgado, M., et al. (1976) *Cancer Immunol. Immunother.* 1, 77–87.

Matsumae, A., Kamio, Y., and Hata, T. (1963) *J. Antibiotics* 16, 236–238.

Mattila, P.S., Ullman, K.S., and Fiering, S., et al. (1990) *EMBO J* 9(13), 4425–4433.

Mine, Y., Watanabe, Y., and Tawara, S., et al. (1983a) *J. Antibiotics* 36(8), 1059–1066.

Mine, Y., Yokota, Y., and Wakai, Y., et al. (1983b) *J. Antibiotics* 36(8), 1045–1050.

Morgan, D.A., Ruscetti, F.W., and Gallo, R. (1976) *Science* 193, 1007–1008.

Morton, D.L., Eilber, F.R., and Joseph, W.L., et al. (1970) *Ann. Surg.* 172, 740–743.

Murase, N., Lieberman, I., and Nalesnik, M., et al. (1990) *Lancet* 336(8711), 373–374.

Nakane, P.K., and Kawaoi, A. (1974) *J. Histochem. Cytochem.* 22, 1084–1091.

Naparstek, Y., Cohen, I.R., Fuks, Z., and Vlodavsky, I. (1984) *Nature* 310, 241–244.

Nuchtern, J.G., Bonifacino, J.S., Biddison, W.E., and Klausner, R.D. (1989) *Nature* 339, 223–226.

Ochiai, T., Nagata, M., and Nakajima, K., et al. (1987a) *Transplant. Proc.* 19(5 Suppl. 6), 53–56.

Ochiai, T., Nagata, M., and Nakajima, K., et al. (1987b) *Transplant. Proc.* 19 (5, Suppl. 6), 84–86.

Ochiai, T., Nakajima, K., Nagata, K., et al. (1987c) *Transplant. Proc.* 19(1), 1284–1286.

Ochiai, T., Nakajima, K., and Nagata, M., et al. (1987d) *Transplantation* 44, 734–738.

Ohmura, S. (1976) *Bacteriol. Rev.* 40, 681–697.

Oku, T., Imanishi, J., and Kishida, T. (1986) *Antiviral Res.* 6, 233–239.

Okuhara, M. (1990) in *Second International Conference on the Biotechnology of Microbial Products: Novel Pharmacological and Agrobiological Activities*, October 14–17, 1990, Sarasota, FL.

Okuhara, M., Goto, T., and Hatanaka, H., et al. (1990) European Patent no. 353678.

Palay, D.A., Cluff, C.W., Westworth, P.A., and Ziegler, H.K. (1986) *J. Immunol.* 136(12), 4348–4353.

Pernice, W., Schuchmann, L., and Dippell, J., et al. (1989) *Arthritis Rheum.* 32(5), 643–646.

Price, T.H., Beatty, P.G., and Corpuz, S.R. (1987) *J. Immunol.* 139(12), 4174–4177.

Puri, J., and Factorovich, Y. (1988) *J. Immunol.* 141(10), 3313–3317.

Rosenberg, A.S., Mizuochi, T., Sharrow, S.O., and Singer, A. (1987) *J. Exp. Med.* 165, 1296–1315.

Rudd, C.E., Trevillyan, J.M., Dasgupta, J.D., Wong, L.L., and Schlossman, S.F. (1988) *Proc. Natl. Acad. Sci. USA* 85, 5190–5194.

Schreiber, S.L. (1991) *Science* 251, 283–287.

Sette, A., Adorini, L., and Appella, E., et al. (1989) *J. Immunol.* 143(10), 3289–3294.

Sette, A., Buus, S., and Colon, S., et al. (1987) *Nature* 328, 395–399.

Shibata, T., Nakayama, O., Okuhara, M., et al. (1988) *J. Antibiot.* 41(3), 296–301.

Siegel, J.P. (1988) *Cell Immunol.* 111, 461–472.

Smith, K.A. (1988) *Science* 240, 1169–1176.

Smith, M.R., Muegge, K., and Keller, J.R., et al. (1990) *J. Immunol.* 144(5), 1777–1782.

Springer, T.A. (1990) *Nature* 346, 425–434.

Stevens, K.M. (1953) *J. Immunol.* 71, 119–124.

Takabayashi, K., Koike, T., and Kurasawa, K., et al. (1989) *Clin. Immunol. Immunopathol.* 51, 110–117.

Thompson, L.F., Ruedi, J.M., Low, M.G., and Clement, L.T. (1987) *J. Immunol.* 139(12), 4042–4048.

Tocci, M.J., Matkovich, D.A., and Collier, K.A., et al. (1989) *J. Immunol.* 143(2), 718–726.

Todd, J.A., Acha-Orbea, H., and Bell, J.I., et al. (1988) *Science* 240, 1003–1009.

Todo, S., Podesta, L., and Kahn, D. (1987) *Transplant. Proc.* 19(5, Suppl. 6), 64–67.

Tonks, N.K., Charbonneau, H., Diltz, C.D., Fischer, E.H., and Walsh, K.A. (1988) *Biochemistry* 27(24), 8695–8701.

Unanue, E.R., and Allen, P.M. (1987) *Science* 236, 551–557.

Usui, M., Aoki, I., Sunshine, G.H., and Dolf, M.E. (1984) *J. Immunol.* 132(4), 1728–1734.

Van Kooyk, Y., Kemenade, P., van de Wiel-van., and Weder, P., et al. (1989) *Nature* 342, 811–813.

Veillette, A., Bookman, M.A., Horak, E.M., and Bolen, J.B. (1988) *Cell* 55, 301–308.

Weaver, C.T., and Unanue, E.R. (1990) *Immunology Today* 11(2), 49–55.

Wegner, C.D., Gundel, R.H., and Reilly, P., et al. (1990) *Science* 247, 456–459.

Wentworth, P.A., and Ziegler, H.K. (1987) *J. Immunol.* 138(10), 3167–3173.

Yewdell, J.W., and Bennink, J.R. (1989) *Science* 244, 1072–1075.

Yokota, Y., Mine, Y., Wakai, Y., Watanabe, Y., and Nishida, M. (1983) *J. Antibiot.* 36(8), 1051–1058.

Isolation and Structure Determination of Natural Products

13

Isolation and Purification of Secondary Metabolites

James B. McAlpine
Jill E. Hochlowski

Secondary metabolites, by virtue of their nature as idiosyncratic and of no vital significance to the producing organism, are often produced in very low yields by the original wild type strains. Indeed, as modern highly specific and sensitive screens to detect bioactive molecules have become available and as researchers have looked at ways to concentrate extracts for primary screening, novel metabolites have been discovered from fermentations in which they were produced in levels as low as 1 µg/l. Calicheamicin, the potent diynene antitumor agent from *Micromonospora echinospora* subsp. *calichensis* was detected in a modified λ prophage induction assay at this level (Fantini et al. 1986). As submerged fermentations of actinomycetes or other microbes typically contain upwards of 10 g/l of total solids at harvest, the isolation chemist is faced with the task of isolating and purifying one part in 10^7, a feat that is only made more difficult if it is not achieved in reasonable yield. Moreover, unlike the medicinal chemist, who usually concentrates on a series of compounds of similar chemical and physical properties and, hence, is able to master the limited number of separation techniques applicable to the specific chemotype, the natural product chemist must be prepared to deal with molecules of the whole spectrum of bioactive metabolites. These can vary in hydro- and lipophilicity, charge, solubility, and size. In the immunosuppressant field, for example, microbial products range from the extremely lipophilic, hexane soluble, cyclic undecapeptides, the cyclosporins, to the very water soluble, basic, spermidine derivative, 15-deoxyspergualin. Yet, perhaps the greatest challenge is supplied by the fact that many secondary metabolites are produced as families of closely related congeners, which often differ in ways that impart negligible changes to polarity or other chromatographic properties. Hence, the natural product chemist must have mastered a wide range of techniques and know when and how to apply them to effect crude and fine separations to obtain a pure product. The task has been made more feasible by advances in two areas. Modern spectral methods of structure determination (see Chapter 14) have dramatically reduced the amount of material required to elucidate a novel structure. In our laboratory in recent years, seldom has the chemist isolated 20 mg of a novel metabolite before the structure has been, at least, tentatively assigned. This is in contrast to the situation prior to Fourier transformed nuclear magnetic resonance (NMR) when several grams of a compound of modest structural complexity was the normal requirement. The second area of major advance is that of separation science itself, where several new chromatographic techniques and a wide array of solid supports aimed at specific types of separations have been developed and commercialized.

The isolation and identification of new secondary metabolites for the sake of discovering novel structure per se is no longer an activity for which there is much support, either in academic or industrial laboratories. Most natural product chemistry is directed at some biological screen and the purification process tends to follow a bioactivity-directed fractionation. This gives the chemist a tool for learning something of the nature of the metabolite prior to its isolation. Thus, after the initial identification of activity in a crude extract or fermentation broth, by quantitatively following this activity, albeit in some arbitrary and undefined units, the chemist can explore the stability, lipophilicity, ionizability, and chromatographic mobility of the active principle through the bioassay. The initial steps of a purification process usually involve many-fold purifications through simple extractions, either solid-liquid or liquid-liquid extractions, or triturations.

Sequential trituration of a solid residue with solvent systems of increasing polarity or sequential liquid-liquid partitioning was made popular several years ago by Kupchan, who frequently partitioned a crude extract with a series of solvent systems. A typical example is the concentration of the antitumor activity of an alcoholic extract of *Brucea antidysenterica* by sequential partitioning into chloroform-water, 10% aqueous methanol-petroleum ether, 20% aqueous methanol-carbon tetrachloride, and finally 40% aqueous methanol-chloroform (Kupchan et al. 1973). The final chloroform extract contained the active principle bruceantin, which was purified by silica gel chromatography (13.1).

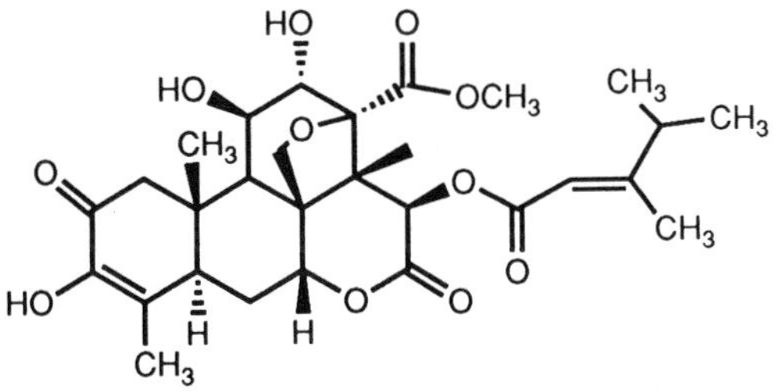

BRUCEANTIN (13.1)

The more hydrophilic metabolites may be candidates for ion exchange chromatography, reversed phase silica gel chromatography, or size exclusion chromatography on polysaccharide resins. The more lipophilic metabolites can be further purified by chromatography on normal phase silica gel, florisil, alumina, or lipophilic size exclusion resins such as Sephadex LH-20. They may also be candidates for a variety of high-speed countercurrent techniques or chromatography on polystyrene resins.

This chapter aims at reviewing current methods of separation science applicable to the isolation of secondary metabolites by surveying examples from recent literature.

13.1 PARTITION CHROMATOGRAPHY

Within the last decade, perhaps no separation technique has changed so significantly as countercurrent chromatography. More striking is the fact that most of this change has been inspired by the work of one man, Yochiro Ito. Working initially in Japan and later at the National Institutes of Health in Bethesda, Yochiro has spent a productive career studying the behavior of two-phase systems under a tremendous variety of different force fields and constraints. His work has led to the commercialization of a number of different instruments that have revolutionized this methodology. Previous techniques, such as countercurrent distribution and droplet countercurrent chromatography, required days for a single separation and suffered the further problem that the instruments required very static environments and were particularly tedious to set up and clean. Modern high speed countercurrent instruments allow for a complete chromatography, including set-up and clean-up, to be completed in 3 to 6 hours.

13.1.1 Droplet Countercurrent Chromatography

Droplet countercurrent chromatography is still used in some laboratories because of the mild conditions employed and the fine separations that can be obtained between closely related congeners.

13.1.1.1 Stilbene Glucosides. A chloroform/methanol/water (7:13:8) system was used in a descending mode DCCC to effect a separation of the α-glucosidase inhibitors, 4'-*O*-methylpiceid (13.2A) and rhapontin (13.2B) (Kubo et al. 1991). The dried powder obtained from a methanolic extract of kelembak, the dried root of *Rheum palmatum*, was extracted sequentially with hexane, chloroform, ethyl acetate, and water. The ethyl acetate fraction gave a yellow solid on concentration, 1 g of which was chromatographed with this DCCC system. The flow rate was 21 ml/hour and fractions were collected hourly. 4'-*O*-Methylpiceid was eluted in fractions 28 to 35 and rhapontin in fractions 87 to 141. Each was recrystallized from methylene chloride/ethanol mixtures.

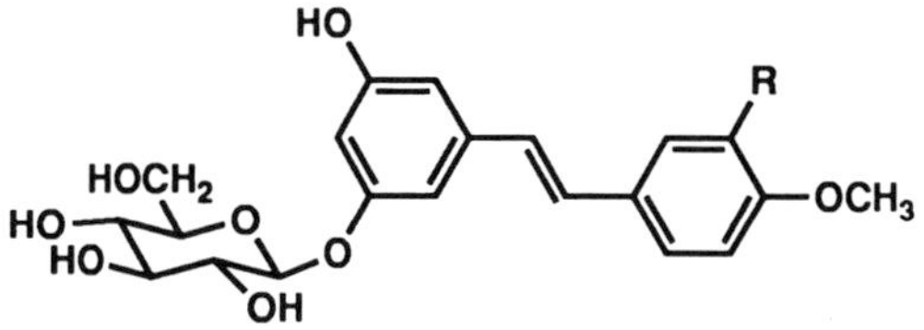

4'-O-METHYLPICEID R = H   A
RHAPONTIN R = OH   B

(13.2)

13.1.1.2 Anthraquinone-Anthrone-*C*-Glycosides. This technique was used by Parilli and co-workers to achieve a crude separation of the ethereal extract of defatted tubers of *Asphodelus ramosus* from which they obtained several novel *C*-glycosides of an anthraquinone-anthrone nucleus (13.3). A Rikakikai 300 tube instrument was used with the lower phase of a chloroform, methanol, water (4:4:3) solvent system as the mobile phase and a flow rate of 1 ml/hour. Sixty percent of the 3-g load for this chromatography was eluted with the solvent front and contained a mixture of glycosides that were subsequently separated by silica gel chromatography (Adinolfi et al. 1989). These glycosides from the initial fraction from the DCCC are characterized by having only three hydroxyl groups on the sugar moiety and attached carbon, e.g., R = H and R' = rhamnosyl, xylosyl, arabinosyl, or quinovosyl (Adinolfi et al. 1991). Later fractions yielded compounds in which four hydroxyls were

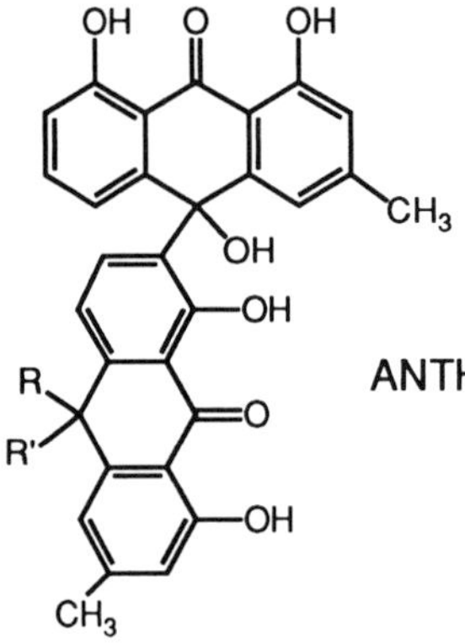

ANTHRAQUINONE-ANTHRONE-*C*-GLYCOSIDES

(13.3)

present in this part of the molecule, e.g., R = OH and R' = xylosyl or R = H and R' = glucosyl (Adinolfi et al. 1989). Finally a much more polar disaccharide, R = OH, R' = β-D-glucosyl(1 → 4)-C-β-D-glucosyl, was recovered by high pressure liquid chromatography (HPLC) of the extruded stationary phase (Lanzetta et al. 1990).

13.1.2 High Speed Countercurrent Chromatography (HSCC)

The most commonly used of the Ito instruments is the horizontal multilayered synchronous coil planet centrifuge. It offers the chemist a number of very desirable features. A complete chromatography can be effected in a matter of a few hours. None of the instability problems occasionally associated with solid supports arise, and a near quantitative recovery is virtually guaranteed. Moreover, the method has a very high resolution power and often baseline separations can be obtained of congeners of almost identical polarity. The instrument relies on the difference between the Archimedian screw forces experienced by each phase of a two-phase mixture, to retain one phase against the introduction of the other. The vectors arising from a planetary spin motion appear to enhance the mixing and separating of the phases and result in high chromatographic efficiency. An attractive feature of the synchronous planetary motion in the absence of a need for mechanical seals. Recently, vertical multilayered synchronous planet coil centrifuges have become commercially available in which two or three coils are connected in series and positioned such that they counterbalance one another. This has the double advantage of increasing the overall capacity of the instruments and removing the need to manually adjust a counterbalance to the particular solvent system being used. Another HSCC instrument, the centrifugal partition chromatograph, functions rather like a high speed droplet countercurrent. It does this by using centrifugal force in place of gravity, and is particularly useful with some viscous, butanol-containing systems that give poor retention of stationary phase in the coil planet centrifuge at room temperature. A centrifugal partition chromatography instrument is marketed by Sanki company.

13.1.2.1 Deoxyerythromycins. The synchronous multilayered planet coil centrifuge has been used to effect the separation of several erythromycin analogs (13.4), virtually indistinguishable by thin layer chromatography (TLC). From recombinant inactivation of the *EryH* gene of *Saccharapolyspora erythraea*, a mutant was obtained that, under slightly different fermentation conditions, produced a total of eight closed related 6-deoxyerythromycins. Each compound could be obtained in analytical purity by chromatography of a basic extract in only two chromatographies; a "partition" LH-20 column followed by a HSCC. At harvest, the fermentation broth was adjusted to pH 9 and extracted with ethyl acetate. The concentrated extract was partitioned

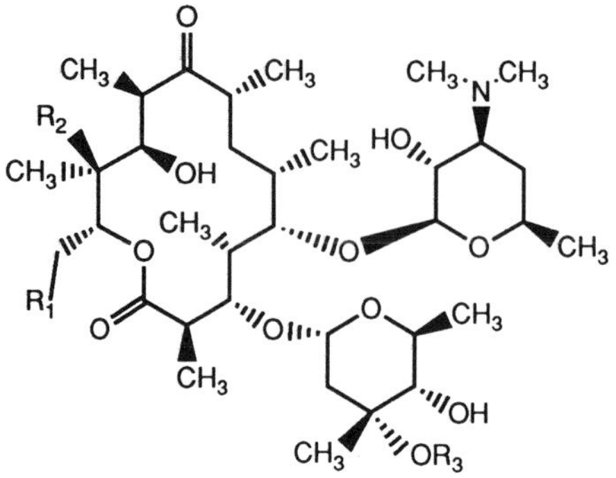

ERYTHROMYCIN	R1	R2	R3
6-DEOXY A	CH3	OH	CH3
6-DEOXY B	CH3	H	CH3
6-DEOXY C	CH3	OH	H
6-DEOXY D	CH3	H	H
15-NOR-6-DEOXY A	H	OH	CH3
15-NOR-6-DEOXY B	H	H	CH3
15-NOR-6-DEOXY C	H	OH	H
15-NOR-6-DEOXY D	H	H	H

(13.4)

between methanol and heptane and the methanol solubles were chromato-graphed over Sephadex LH-20 in chloroform-heptane-ethanol (10:10:1). The erythromycin derivatives were collected in two fractions from this column. The earlier eluting, more hydrophilic group was subjected to HSCC on an Ito Horizontal Multi-layered Coil Planet Centrifuge in a carbon tetrachloride-methanol-0.01 M aqueous potassium phosphate, pH 7 buffer (1:1:1) system. This chromatography was run at 800 rpm with the upper phase mobile and the tail as inlet. With a 3 ml/minute flow-rate this gave approximately 80% retention of stationary phase. The elution of components was in the order: 15-nor-6-deoxyerythromycin C, 6-deoxyerythromycin C, 15-nor-6-deoxye-rythromycin A, 15-nor-6-deoxyerythromycin D, and, lastly, 6-deoxyerythro-mycin A. The more lipophilic erythromycins from the LH-20 column were separated using another countercurrent chromatography identical to this one except that the lower phase was maintained as the stationary phase by re-versing the direction of spin of the column. Slightly greater (85%) stationary phase retention was achieved and the components were eluted in the order 6-deoxyerythromycin B, 15-nor-6-deoxyerythromycin B and 6-deoxyerythro-mycin D. Noteworthy is the fact that the eight compounds consisted of 6-deoxyerythromycins A, B, C, and D and their 15-nor counterparts, i.e., four pairs of compounds differing only as aliphatic homologs of highly polar com-plex molecules with molecular weight of approximately 700 daltons (Weber et al. 1991).

13.1.2.2 Squalestatins. Several squalene synthase inhibitors containing the same 2,6-dioxobicyclo-[3.2.1]octane tricarboxylic acid pharmacophore have recently been isolated from a variety of different fungal species (Bartizal et al. 1991a, 1991b; Bergstrom et al. 1991). Another family of these, the squa-lestatins (13.5), was obtained from a *Phoma* sp., purified, and three closely related congeners were separated by HSCC (Dawson et al. 1992). A pH 2.8 ethyl acetate extract of the harvest fermentation broth was extracted into 1% sodium bicarbonate solution and, after pH adjustment to 2.8, reextracted

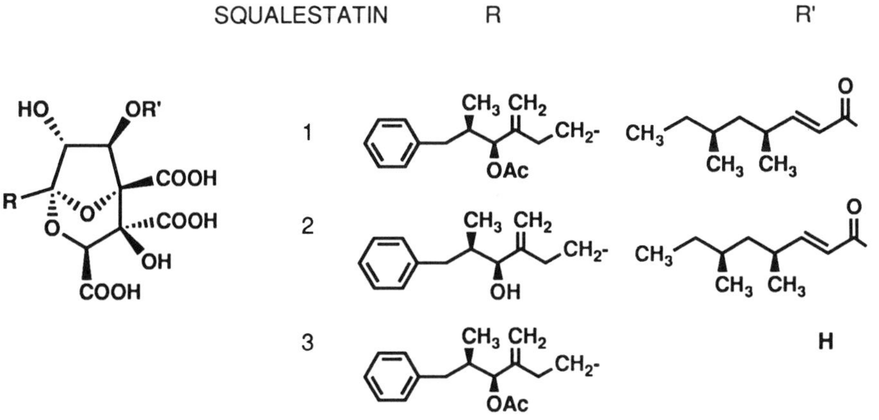

(13.5)

into ethyl acetate. This simple extraction procedure gave 500 mg of mixed squalestatins, >20% pure, from 4 l of fermentation broth. These were separated and purified by chromatography on an Ito Planet Coil Centrifuge, capacity 380 ml, operating at 800 rpm in "reverse" spin and the "tail" as inlet. The solvent system was ethyl acetate, hexane, methanol, 0.01 N aqueous sulfuric acid (6:5:5:6) with the lower layer as stationary. Fractions were collected and, after 1:5000-fold dilution, monitored for squalene synthase inhibition. Base-line separation was obtained from the three squalestatins despite the fact that squalestatins 2 and 3 were eluted from the column by displacing the stationary phase with methanol. These last two compounds were each further purified by rechromatography in a similar system but with the component ratios changed to 3:2:2:3 for squalestatin 2 and 6:1:1:6 for squalestatin 3.

13.1.3 Diol Column Chromatography

The bonded-phase silica gel packing diol ($SiOCH_2CH_2CH_2OCH_2CHOHCH_2$-OH) can be used as a partition chromatography medium as well as an absorbent (Rasmussen and Scherr 1987). This packing material can be used to replace other partition chromatography supports such as Sephadex LH-20, offering the advantage of rigidity. Diol bonded-phase silica gel is available from Analytichem, San Fernando, CA.

13.1.3.1 Coloradocin. These same authors used diol as a partition support for the isolation of the antibiotic coloradocin (13.6) (Rasmussen et al. 1987). An amberlite XAD-2 extract from the Actinoplanete-producing culture was first separated by Sephadex LH-20 chromatography eluted with methanol to give 19 g of material from a 49.8-l fermentation. This bioactive material was

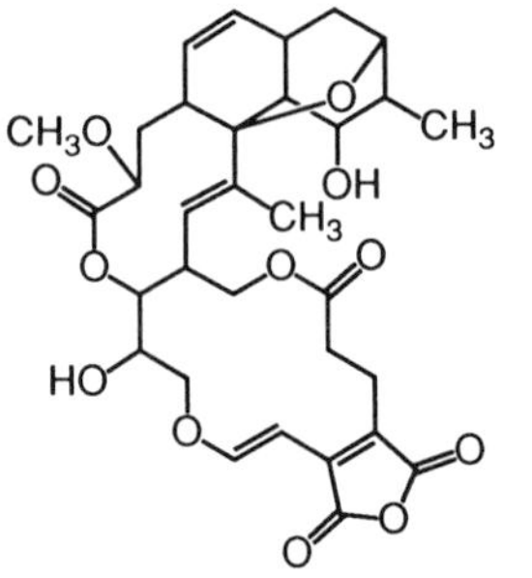

COLORADOCIN

(13.6)

then chromatographed in two successive partition chromatographies over diol-bonded silica gel (Sepralyte, 40 μm packing). The first column was equilibrated with the lower phase of a CCl_4-$CHCl_3$-MeOH-H_2O (5:5:8:2) solvent system and eluted with the upper layer of the same system. Active material from this column was then chromatographed over a second diol-bonded silica gel column equilibrated with the lower layer of a CCl_4-MeOH-H_2O (5:4:1) solvent system and eluted with the upper layer of this system to yield 4.7 g of crude material. This material was separated on a final C-18 bonded silica gel, reversed phase column followed by countercurrent chromatography in an Ito Multi-layered Coil Planet Centrifuge to yield pure coloradocin.

13.1.3.2 Benzanthrins. Diol-bonded-phase silica gel has also been employed as a key step in the isolation of the antibiotics benzanthrin A (13.7A) and B (13.7B) (Rasmussen et al. 1986). A methylene chloride extract of the producing culture *Nocardia lurida* was first separated on Sephadex LH-20

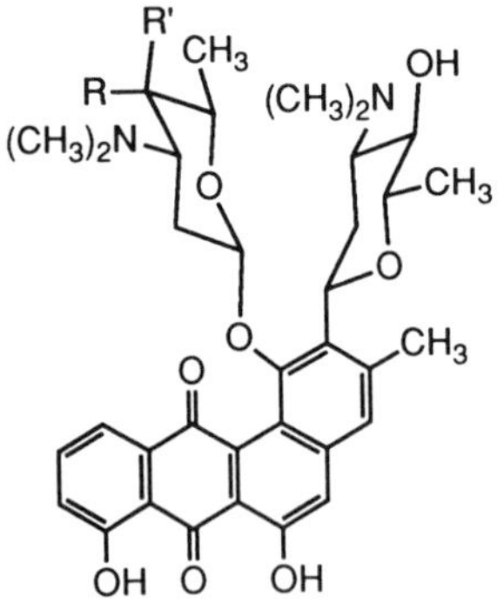

BENZANTHRIN A, R = OH, R' = H **A**
BENZANTHRIN B, R = H, R' = OH **B**

(13.7)

into two active bands by elution with methanol. Each of these was chromatographed over a diol column equilibrated with the lower layer and eluted with the upper layer of a CCl_4-$CHCl_3$-MeOH-H_2O (5:5:8:2) solvent system. The loading ratio was 1 g of sample load per 80 g of packing. The two active fractions applied to diol weighed 4.7 and 1.3 g and were reduced to weights of 620 and 232 mg, respectively. A final countercurrent chromatography step, again in a CCl_4-$CHCl_3$-MeOH-H_2O (this time at ratios of 4:1:4:1) solvent system, yielded 265 mg of benzanthrin A and 43 mg of benzanthrin B.

13.2 ADSORPTION CHROMATOGRAPHY

13.2.1 Polymeric Adsorbents

Polymeric resins of various composition for use in chromatographic separation are available. These are most commonly polymers of aromatics such as styrene and divinylbenzene (and their co-polymers) as well as substituted analogs. Acrylic acid ester polymers are also available. Different grades of these are produced that vary in pore volume and surface area. These resins are employed as chromatography packings and, additionally and more typically, are used to make batch extracts of organic materials from aqueous solutions such as fermentation broths. Mitsubishi Kasei sells a number of polymeric resins, including the Diaion and Sepabead series, and Rohm and Haas sells a comparable Amberlite XAD series.

13.2.1.1 Mycophenolic Acid. Amberlite XAD-2 (polystyrene) has been employed by the authors in the isolation of mycophenolic acid (13.8) both as a column chromatography packing and as an extraction resin. Whole broth from the producing fungal culture was combined with XAD-2 (10 l of broth with 1 l of resin), the mixture was stirred for 4 hours, and the solids were collected by centrifugation. This solid cake was then washed with water and eluted with methanol. Additional partition and size exclusion chromatography steps yielded a semipurified oil that was applied to an XAD-2 column and eluted with a step gradient of aqueous methanol. Material eluting in 75% methanol was subjected to a final purification step of countercurrent chromatography to yield pure mycophenolic acid (recovered yield of 16 mg/l).

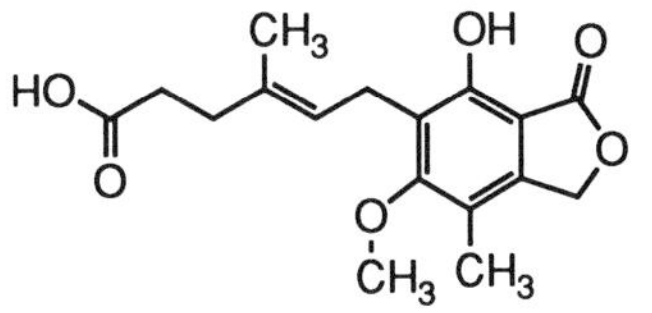

MYCOPHENOLIC ACID

(13.8)

13.2.1.2 Antiarones J and K. Two novel phenolics have been isolated on an XAD-2 column eluted with organic solvents (Hano et al. 1991). A methanol extract (150 g) from the tree bark of *Antiaris toxicaria* was applied to an XAD-2 column that was eluted sequentially with n-hexane, benzene, diethyl ether, and acetone. Material eluting in diethyl ether (25 g) was further chromatographed on silica gel to yield antiarones J (13.9A) (270 mg) and K (13.9B) (29 mg).

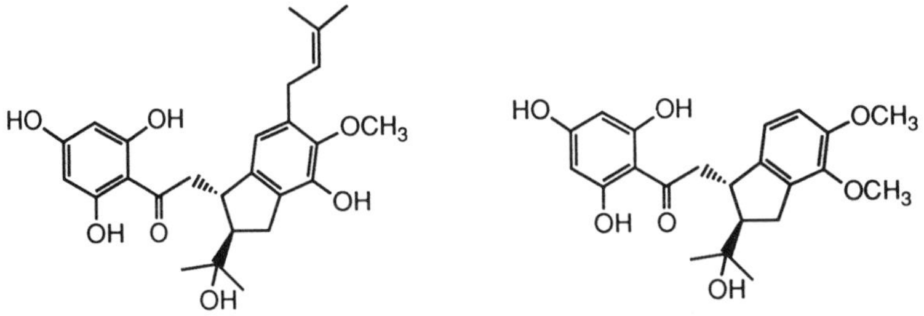
HO
OH
OCH3
OH
OH O
OH
HO
OH
OCH3
OCH3
OH O
OH

ANTIARONE J — A   ANTIARONE K — B (13.9)

13.2.1.3 **Pumilacidin.** A complex of peptide antivirals have been isolated from *Bacillus pumilus* with Diaion HP-20 (a polystyrene-divinylbenzene co-polymer) (Naruse et al. 1990). An ethyl acetate-precipitated, butanol extract was treated batchwise with resin that was then packed into a column and eluted with water, 50% aqueous methanol, and finally 80% aqueous acetone. The pumilacidin complex (13.10) was eluted with the last solvent. Diaion HP-20 is used in a later step as a desalting medium for the purified components.

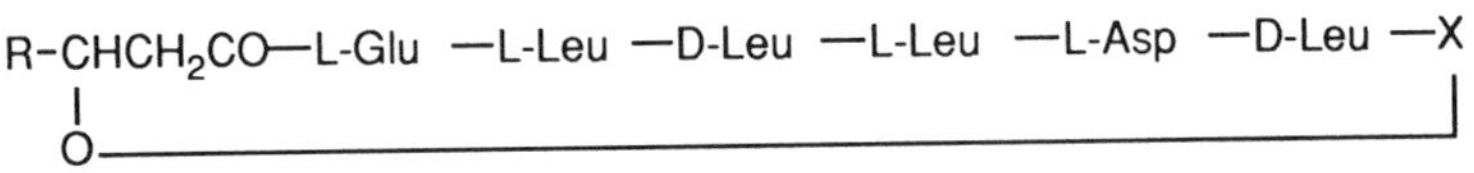
R-CHCH2CO—L-Glu —L-Leu —D-Leu —L-Leu —L-Asp —D-Leu —X
O

PUMILACIDIN	R	X
A	CH_3-CH(CH_3)-(CH_2)$_9$- or CH_3-CH_2-CH(CH_3)-(CH_2)$_8$-	L-Ile
B	CH_3-CH(CH_3)-(CH_2)$_9$- or CH_3-CH_2-CH(CH_3)-(CH_2)$_8$-	L-Val
C	CH_3-CH(CH_3)-(CH_2)$_{11}$- or CH_3-CH_2-CH(CH_3)-(CH_2)$_{10}$-	L-Ile
D	CH_3-CH(CH_3)-(CH_2)$_{11}$- or CH_3-CH_2-CH(CH_3)-(CH_2)$_{10}$-	L-Val
E	CH_3-CH(CH_3)-(CH_2)$_{10}$-	L-Ile
F	CH_3-CH(CH_3)-(CH_2)$_{10}$-	L-Val
G	CH_3-(CH_2)$_{12}$-	L-Val

(13.10)

After reversed phase chromatography separation of the pumilacidins in an acetonitrile-phosphate buffer solvent system, the aqueous concentrates of the individual components are desalted by passing over an HP-20 column.

13.2.2 Diatomaceous Earth

Diatomaceous earth (diatomaceous silica, celite, celatom, Fuller's earth, infusorial earth), composed of aluminum magnesium silicates (primarily SiO_2), has been used as a filter agent as well as a chromatographic absorbent. It is available from a variety of suppliers in different forms varying by particle size, exact mineral content, and purity.

13.2.2.1 Taxol. Celite-454 (Aldrich) was used as a chromatographic absorbent for the isolation of the chemotherapeutic agent, taxol (13.11) (Witherup et al. 1990). The methylene chloride soluble portion of a methylene chloride/methanol extract of *Taxus media* cv. Hicksii was applied to a bed of celite and flash chromatographed sequentially with hexane, methylene chloride, ethyl acetate, and methanol. Taxol was contained in the material

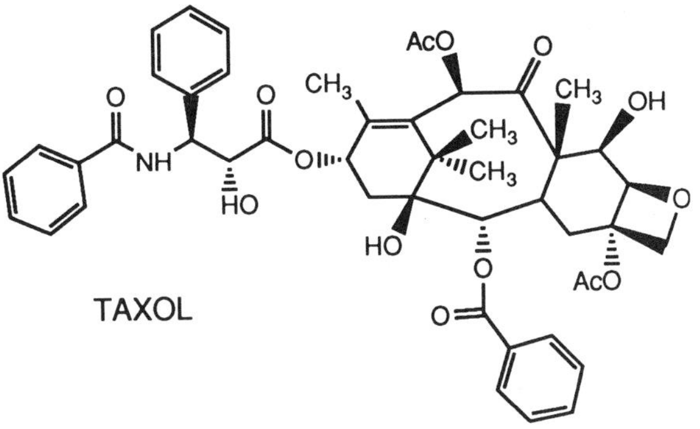

TAXOL

(13.11)

that eluted with methylene chloride, and this isolation step achieved a reduction in weight from the original 29 g of extract to 4.6 g eluting from the celite bed. This extract was further fractionated by silica gel and phenyl bonded-phase HPLC chromatography to yield pure taxol. A final weight of taxol isolated from this extract of 530 mg indicates that the Celite-454 flash chromatography yielded material of 11.5% taxol content from a crude extract containing 1.83% taxol.

13.2.2.2 Flavonoids. Celite was also used as an initial isolation step in the purification of three novel flavonoids (Ngadjui et al. 1991). Twigs from *Hoslundia opposita* Vahl were extracted with methanol, precipitated with acetone, and treated with activated carbon. Chromatography was then carried out using celite eluted with acetone to give material that, upon further separation by silica gel chromatography and prep TLC, yielded the three novel compounds hoslundin (13.12A), hoslundal (13.12B), and hoslunddiol (13.12C).

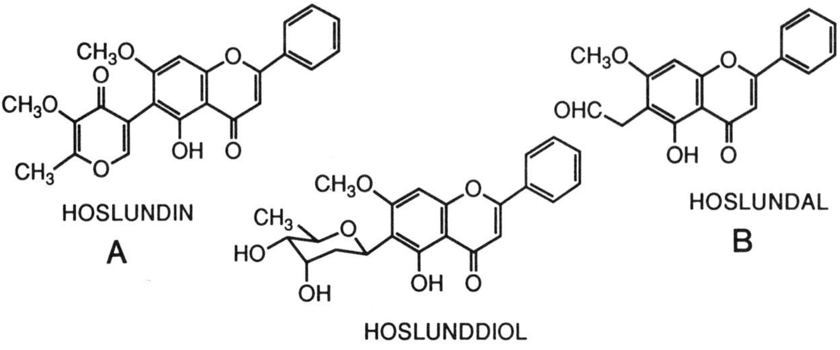

HOSLUNDIN

A

HOSLUNDAL

B

HOSLUNDDIOL

C

(13.12)

13.2.3 Alumina

Alumina chromatography packings (Al_2O_3) are available in acidic, basic, and neutral forms. Different mesh sizes of the alumina chromatography packings as well as variances of water content and, hence, activity allow a wide range of selectivities.

13.2.3.1 Sophazrine. A series of three successive alumina chromatographies was employed to purify the novel plant alkaloid sophazrine (13.13) (Rahman et al. 1991). After an initial ethanol extraction of the leafy shoots of *Sophora griffithii*, a preparation containing 80 g of crude alkaloids is obtained by partitioning between chloroform and water at various pHs. This crude material was then applied to an alumina column eluted sequentially with

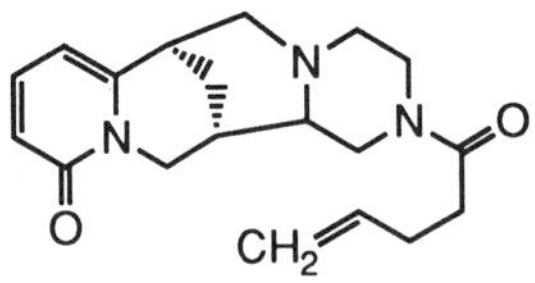

SOPHAZRINE

(13.13)

petroleum ether-chloroform, chloroform, acetone-chloroform, acetone, and acetone-methanol. Material eluting from this column in chloroform-acetone mixtures was then rechromatographed over a second alumina column, employing TLC grade alumina, and eluted isocratically with petroleum ether-chloroform (1:2). A final alumina chromatography on a chromatotron instrument (see Section 13.2.5.3) eluted with chloroform afforded 20 mg of pure sophazrine.

13.2.3.2 Isodelectine. Alumina chromatography was also employed as the final purification step in the isolation of the norditerpenoid, isodelectine (13.14), from *Delphinium vestitum* (Desai et al. 1990). A crude basic residue was prepared via an aqueous ethanol extract of the dried plant material. This residue was further separated by ion exchange chromatography on a Dowex 50W-X8. A sample of this material was then chromatographed across an

ISODELECTINE

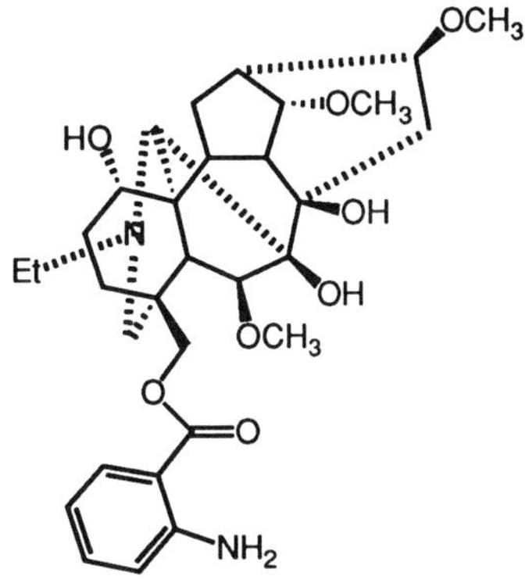

(13.14)

alumina rotor eluted isocratically with hexane-diethyl ether (3:1) on a chromatotron (see Section 13.2.5.3). This step reduced the weight of crude basic extract from 285 to 56 mg of material that was further separated by prep TLC on alumina plates eluted with diethyl ether in order to yield 24 mg of pure isodelectine as well as smaller quantities of two related norditerpene alkaloids.

13.2.4 Florisil

Florisil®, a trademark of the Floridin company, is generally used as a generic name for magnesium silicate chromatographic packing material. Distributors in the United States include Aldrich, Fisher, and U.S. Silica, which sell various mesh ranges of adsorbent.

13.2.4.1 Taxinine M. Florisil was used in the isolation of taxinine M (13.15), a congener of the antitumor agent taxol, from the bark of *Taxus brevifolia* (Beutler et al. 1991). The methanol extract from this tree bark after partition between ethyl acetate and water and trituration with hexane and acetone yielded 26 g of mixed taxanes. This was chromatographed over florisil that was eluted with a step gradient of acetone in hexane. Material containing both taxol and taxinine M eluted between 45% and 75% acetone, and this step achieved a reduction in weight to 7 g. Final purification was achieved by silica gel column chromatography and CN bonded-phase reversed phase HPLC.

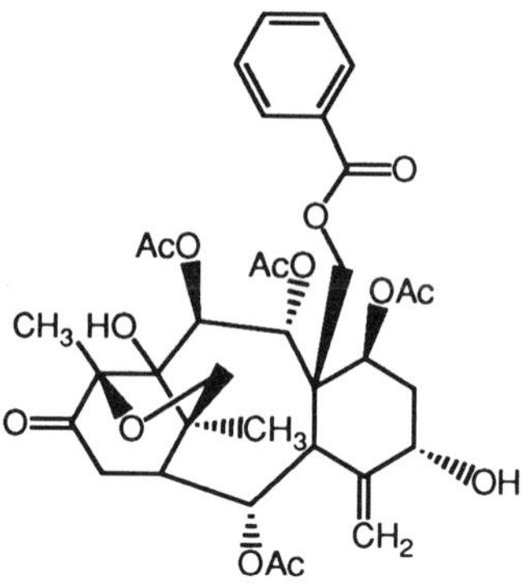

TAXININE M.

(13.15)

13.2.4.2 Aphidicolin. The authors have employed florisil as the final purification step in the isolation of the DNA replication inhibitor, aphidicolin (13.16), from an unidentified fungal culture. A series of partitions and triturations of the original extract from this culture was chromatographed over Sephadex LH-20 eluted with methanol. Active fractions from the LH-20

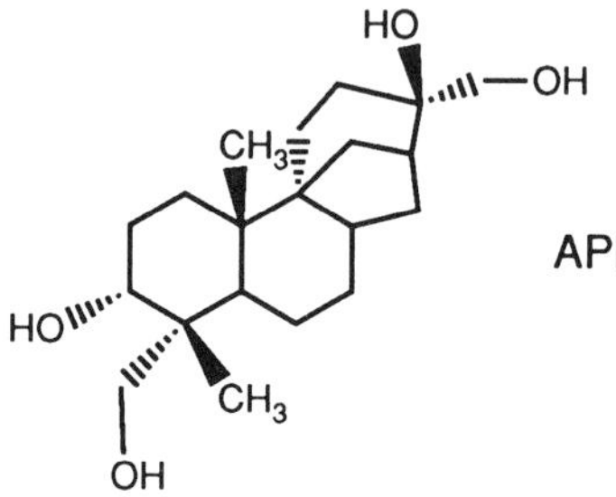

APHIDICOLIN

(13.16)

column were combined and the resultant oil was deposited onto a small amount of florisil by evaporation from methanol. This was then loaded onto a florisil column eluted with a step gradient of ethyl acetate with increasing concentrations of methanol. Pure aphidicolin eluted from this column with 25% methanol in ethyl acetate.

13.2.4.3 Pyranocoumarins. A comparison has been made between florisil and silica gel packings for the separation of plant pyranocoumarins (13.17) (Glowniak 1991). Several useful elution solvent systems are included. Retention values for various compounds on florisil in solvent systems of dichloromethane with diisopropyl ether, heptane, acetonitrile, and ethyl acetate are reported.

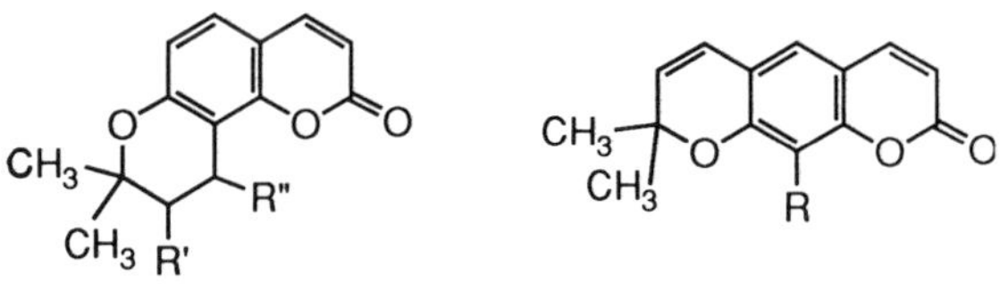

PYRANOCOUMARIN	R	R'	R"
SAMIDIN		OC(O)CH=C(CH3)2	OC(O)CH3
ISOSAMIDIN		OC(O)CH3	OC(O)CH=C(CH3)2
VISNADIN		OC(O)CH(CH3)C2H	OC(O)CH3
		5	
DIHYDROSESELIN		H	H
PTERIXIN		OC(O)CH3	OC(O)C(CH3)=CHC
			H3
DISENECIOYL-*cis*-		OC(O)CH=C(CH3)2	OC(O)CH=C(CH3)2
KHELLACTONE			
cis-KHELLACTONE		OH	OH
trans-KHELLACTONE		OH	OH
XANTHYLETIN	H		
LUVANGETIN	OCH3		

$$(13.17)$$

13.2.5 Silica Gel

The high resolving power of silica gel as a solid adsorbent support makes this by far the most widely used chromatographic medium, especially for the separation of lipophilic compounds. A survey of papers in the *Journal of Natural Products* from January 1990 to December 1991 shows that silica gel chromatography was used at some point in the separation and purification of over 90% of the new agents reported! Silica gel comes in a wide variety of mesh sizes and activities; the finer the mesh size, the greater the resolving power. Very high resolution columns can be obtained by packing with TLC-grade silica gel.

13.2.5.1 Column Silica Gel Chromatography.

Flavonoids.   Silica gel was used to isolate a series of 12 new cytotoxic flavonoids (13.18) from the ether layer of a partitioned methanol extract of the dried roots of *Muntinga calabura* (Kaneda et al. 1991). The ether solubles were first chromatographed over silica gel in a chloroform/methanol step gradient and separated into eight mixed fractions. These were then rechromatographed over silica gel columns eluted with increasing concentrations of acetone in toluene or with ethyl acetate. The polarity of the second chromatography that was chosen, corresponded to the elution from the first. Five

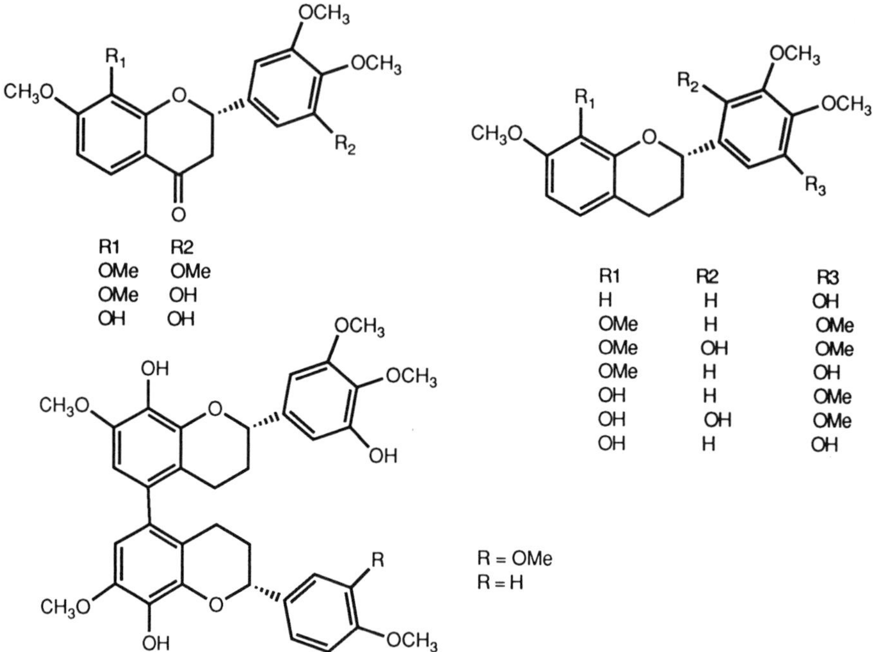

R1	R2
OMe	OMe
OMe	OH
OH	OH

R1	R2	R3
H	H	OH
OMe	H	OMe
OMe	OH	OMe
OMe	H	OH
OH	H	OMe
OH	OH	OMe
OH	H	OH

R = OMe
R = H

FLAVONOIDS FROM *M. CALABURA*

(13.18)

of the twelve compounds were obtained pure, directly from the second chromatography, whereas the other seven required a recrystallization from methanol.

Lupin Alkaloids.   Perhaps one of the most striking features of silica gel is the wide range of polarities that can be handled efficiently on this adsorptive support. Alkaloids can be sharply eluted by solvents systems containing either ammonium hydroxide or an organic base. A series of lupin alkaloids (13.19) were separated from the basic portion of the 75% ethanol extract of viable seeds of *Lupinus termis* using silica gel column chromatography with basic elution solvents (Mohamed et al. 1991). An initial large column (7 × 150 cm) containing 1 kg of Merck Kieselgel 60F254 was loaded with 27 g of crude basics and eluted with a step gradient of methanol in methylene chloride containing concentrated ammonium hydroxide at the level of 1 part per 500. The 2% methanol fractions contained 250 mg of a mixture of ($-$)-11,12-*seco*-12,13-didehydromultiflorine and the angelic and tiglic esters of 13α-hydroxy-lupanine. The ($-$)-11,12-*seco*-12,13-didehydromultiflorine (200 mg) was separated from the two esters by a second silica gel column eluted with cyclohexane-diethylamine 9:1. The 8% methanol fractions from the large column gave 600 mg of pure ($-$)-albine then a mixture of ($-$)-Δ^5-dehydroal-

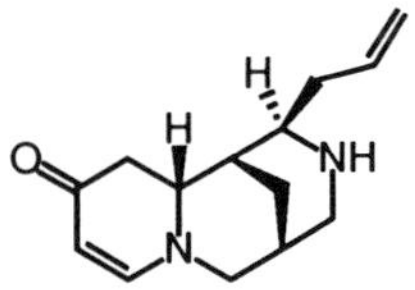

(-)-ALBINE

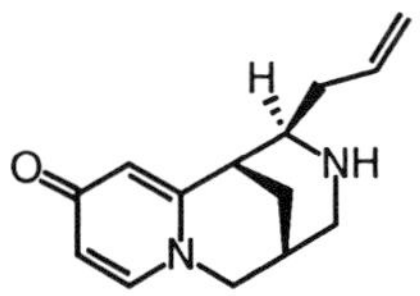

(-)-Δ^5-DEHYDROALBINE

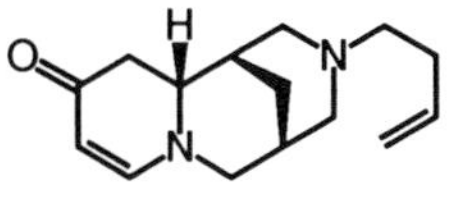

(-)-11,12-*seco*-12,13-DI-
DEHYDROMULTIFLORINE

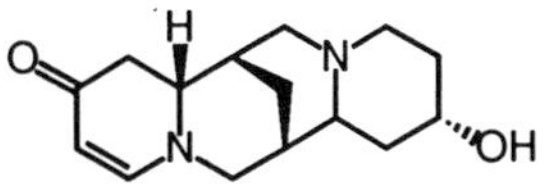

(-)-13α-HYDROXYMULTIFLORINE

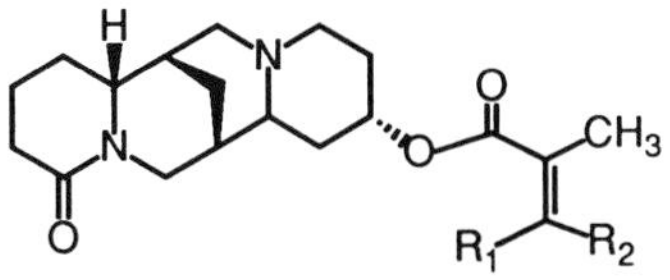

(-)-13α-ANGELOYLOXYLUPANINE R$_1$ = CH$_3$, R$_2$ = H
(-)-13α-TIGLOYLOXYLUPANINE R$_1$ = H, R$_2$ = CH$_3$

(13.19)

bine and (−)-13α-hydroxymultiflorine. This mixture was separated by chromatography on another silica gel column eluted initially with cyclohexane-diethylamine 7:3 to give pure (−)-13α-hydroxymultiflorine and subsequently washed with methanol to give the (−)-Δ^5-dehydroalbine.

Lipid Olefins. Highly lipophilic aliphatics with major contributions to their polarity from olefinic bonds can have their adsorptive properties enhanced by silver nitrate impregnation of silica gel. This technique is often referred to as argentation chromatography and can be employed both in column and thin layer formats. In a search for sex attractants of the Guam brown tree snake, six long chain (C$_{35}$ to C$_{37}$) methyl ketodienes with the two separated olefins at various positions in the chain were isolated (Murata et al. 1991). Hexane washes of the skin of nine snakes yielded 547 mg of crude lipids that were initially chromatographed over alumina to give 14 mg of methyl ketones. This was chromatographed over 4.3 g of Analtech silica gel, impregnated with 10% silver nitrate. The column was developed with a gradient of diethyl ether (2% to 8%) in hexane and only three fractions were collected. The first was essentially saturated and monounsaturated methylketones while the methylketodienes were retained. The second fraction contained three dienes in which the olefin closest to the ketone was separated from it by only three methylene groups. Most strongly retained were dienes, in which this separation between the ketone and the closest olefin, consisted of five methylene groups.

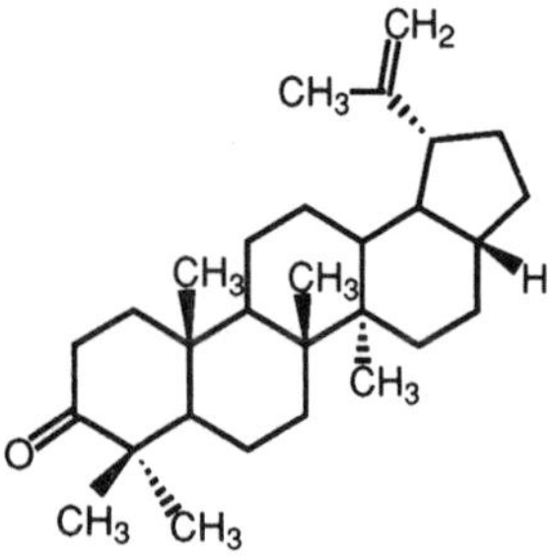

3-OXO-28-NORLUP-20(29)-ENE

A

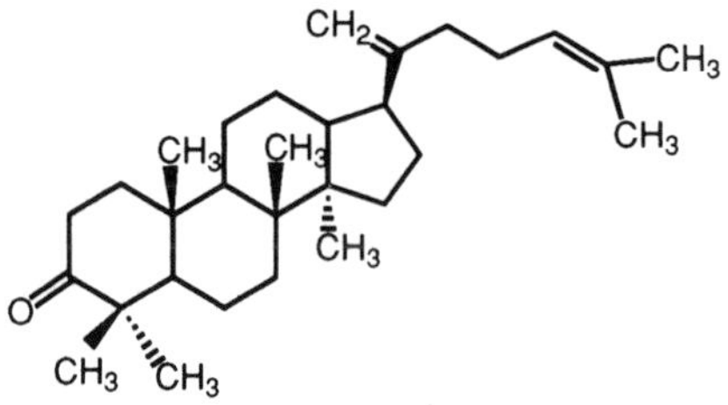

3-OXO-DAMMARA-20(21),24-DIENE

B

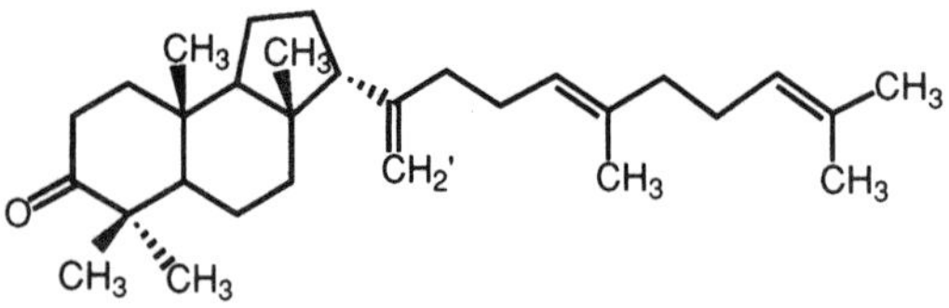

3-OXO-MALABARICA-14(26),17E,21-TRIENE

C

(13.20)

Triterpenoids. Perhaps a more classic example of argentation chromatography is provided by its use in the triterpene field to separate, otherwise closely related, mono, di, and triene congeners (Marner et al. 1991). A mixture containing ketones with various degrees of unsaturation was obtained as the first fraction from a petrol-diethyl ether gradient-developed silica gel chromatography of the neutrals from gum mastic. This mixture was further chromatographed on silica gel impregnated with 10% silver nitrate to give pure samples of 3-oxo-28-norlup-20(29)-ene (13.20A), 3-oxo-dammara-20(21),24-diene (13.20B), and 3-oxo-malabarica-14(26),17E,21-triene (13.20C).

Cyclic Peptides. The very fine separations that can be achieved on analytical TLC are often lost when apparently similar systems are transferred to preparative column systems. This can usually be attributed to the different activity of the coarser silica commonly used for column chromatography. TLC grade silica gel (250 μm) is expensive but it can be used in columns. It packs tightly, which results in slower flow rates but it does give separations that approximate those seen on plates. Ganguli's group used a column (3 × 85 cm) of TLC-grade silica gel to purify deoxymulundocandin (13.21B) from a crude anti-

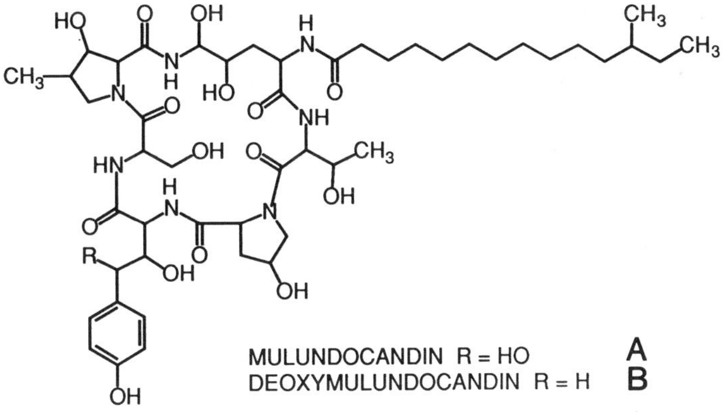

MULUNDOCANDIN R = HO **A**
DEOXYMULUNDOCANDIN R = H **B**

(13.21)

biotic extract containing mulundocandin (13.21A) (Mukhopadhyay et al. 1992). The concentrate had been obtained from an acetone extract of 33.5 kg of mycelia of *Aspergillus sydowii* by partitioning between water and ethyl acetate, triturating the concentrated ethyl acetate extract with acetonitrile and extracting the remaining solids with methanol. This procedure reduced the weight from 253 g of ethyl acetate extract to 6 g of methanol solubles. The silica gel column was developed under pressure with ethyl acetate-propanol initially in the ratio 5:2 and finally at 5:3. This procedure gave 85 mg of pure deoxymulundocandin.

13.2.5.2 Preparative Thin Layer Silica Gel Chromatography.

Viranamycins A and B.   Seto's group has employed preparative TLC in the isolation of the novel 18-membered macrolide antitumor agents viranamycin A (13.22A) and viranamycin B (13.22B) (Hayakawa et al. 1991). The activity from an acetone extract of the mycelial cake of the streptomycete-producing organism was extracted into ethyl acetate at pH 2. This ethyl acetate-soluble

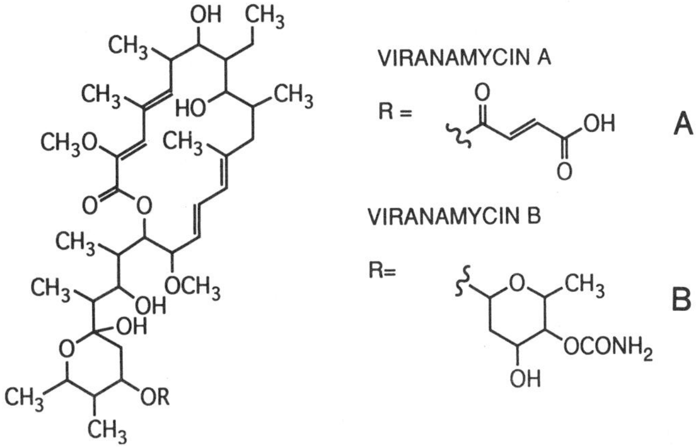

VIRANAMYCIN A

R =

VIRANAMYCIN B

R=

(13.22)

material was then chromatographed over a silica gel column developed with a chloroform/methanol step gradient. The two active components were separated on this initial silica gel column. Material containing viranamycin A was subjected to preparative TLC on a 500-μm thick Merck silica gel plate developed with chloroform-methanol-28% ammonia (8:4:1). The active band from this plate was then subjected to a second preparative TLC. This was developed with chloroform-methanol (5:1). A final purification step on Sephadex LH-20 eluted with chloroform-methanol yielded 2 mg of viranamycin A from a 5-l fermentation. A preparative TLC developed with chloroform-methanol (1:1) was also employed in the purification of viranamycin B. A Sephadex LH-20 column was again used as the final purification step to give 20 mg of viranamycin B.

MH-031. TLC was used in the purification of a small γ-lactone with hepatoprotective activity, MH-031 (13.23), isolated from *Streptomyces rishiriensis* (Itoh et al. 1991). The filtered mycelial cake from 25 l of fermentation broth was extracted with 80% aqueous acetone and the extract was concentrated and added to the culture filtrate. The active principle was adsorbed onto Diaion HP-20 resin, which was washed with 50% aqueous methanol before being eluted with 100% methanol. This eluate was concentrated and then extracted with ethyl acetate at pH 4 to give the crude hepatoprotective agent.

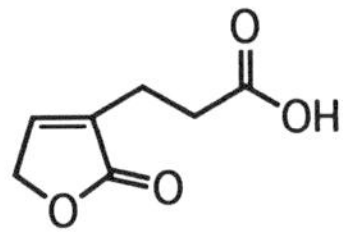

MH-031   (13.23)

Chromatography over a silica gel column developed in a chloroform-methanol step gradient afforded a 15-fold purification. A further 15-fold purification was effected by preparative TLC on a Merck Kieselgel plate developed with chloroform-methanol-acetic acid (100:50:1, active band Rf between 0.14 and 0.25) followed by a final recrystallization from hexane-ethyl acetate to yield 76 mg of pure MH-031.

13.2.5.3 Radial Chromatography (the Chromatotron).

A modification of preparative TLC is represented by the Chromatotron instrument, available from Harrison Research, Palo Alto, CA. This system, patented by Shuyen Harrison, consists of a circular, motor-driven wheel onto which a rotor coated with an absorbent, such as silica gel, is placed (Harrison 1979). A sample applied to the center of a plate will travel under centrifugal acceleration to the perimeter, where bands can be collected as they elute from the plate. Prepoured silica gel and alumina plates for the instrument are available from Analtech in thicknesses varying from 1000 to 4000 μm. Plates can also be poured onto glass blanks with any of a number of absorbents. The instrument is fitted with a quartz lid to allow for the observation of UV, visible, or fluorescent materials on plates with an incorporated UV indicator.

Sesquiterpenes. The sponge sesquiterpenes, furospinulosin 1 (13.24A) and 22-deoxyvariabilin (13.24B) were separated on a chromatotron (Kernan et al. 1991). Seven grams of dried sponge (*Thorecta* sp.) was extracted with methanol and the concentrate from this extract partitioned between methylene chloride and water. Methylene chloride layers (240 mg) were separated

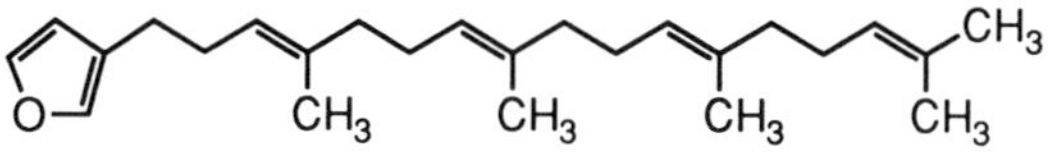

FUROSPINULOSIN 1 A

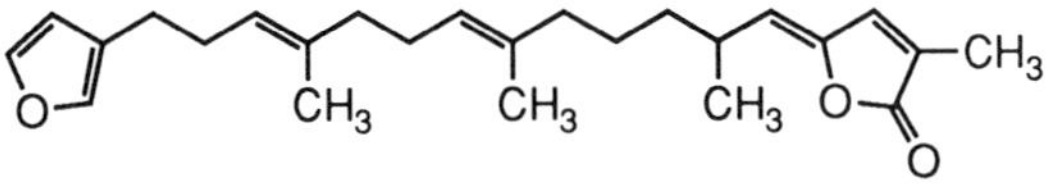

22-DEOXYVARIABILIN B

(13.24)

first on a silica gel column eluted with ethyl acetate-hexane (1:1) and subsequently on the chromatotron with a 1 mm thick silica gel coated rotor eluted sequentially with ethyl acetate-hexane (1:1) and (5:1) to yield furospinulosin 1 (10 mg) and 22-deoxyvariabilin (2 mg).

Ryanoids. Radial chromatography was a key step in the isolation and purification of the ryanoid, ryanodyl-3-(pyridine-3-carboxylate) (13.25) from the stemwood of *Ryania speciosa*, an insecticidal plant from South America (Jeffries et al. 1991). An initial wet chloroform extract was partitioned between ethyl acetate-water and the ethyl acetate-soluble material was filtered through

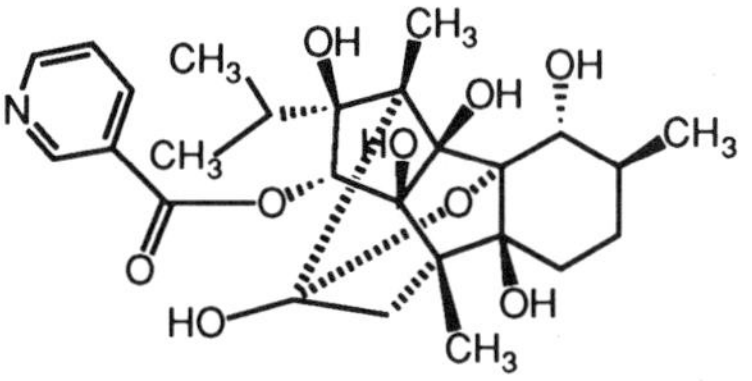

RYANODYL-3-(PYRIDINE-3-CARBOXYLATE)

(13.25)

silica gel in chloroform-methanol (6:1) with 2% aqueous methylamine to remove the more polar components. This material was then applied batchwise to a chromatotron with a 2000-μm silica gel rotor and eluted with chloroform/methanol mixtures (20:6 through 6:1) containing 2% aqueous methylamine. A final HPLC step was required to obtain pure ryanodyl 3-(pyridine-3-carboxylate).

Anthrones. The chromatotron was also used for the isolation of monoamino oxidase inhibitors from *Gentiana detonsa* (Hostettmann et al. 1980). A 400-mg sample of the chloroform extract from this plant was applied to a 2-mm silica gel rotor. Samples were monitored by UV detection as well as TLC on alumina plates of fractions eluted. The chromatotron plate was developed in chloroform, which initially eluted various fats and pigments from the extract followed by pure decussatin (13.26A) (7 mg) and gentiacaulein (13.26B) (12 mg). The entire chromatographic run required only 20 minutes to complete.

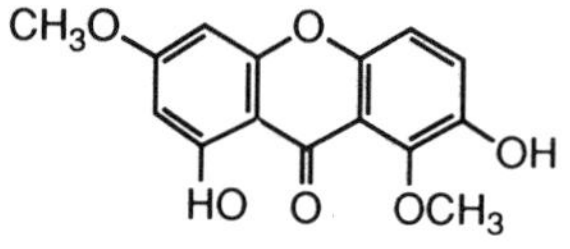

DECUSSATIN

A

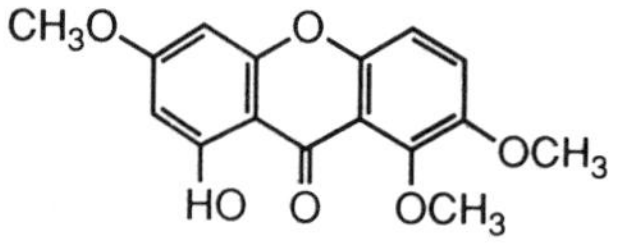

GENTIACAULEIN

B (13.26)

Furanocoumarins. Furanocoumarins (13.27) have been purified from *Heracleum sphondylium* by using a chromatotron for the final purification step (Erdelmeier et al. 1985). A chloroform extract of the plant roots was first separated by alumina column chromatography to separate the furanocoumarin fraction mixture. This mixture was then separated by sequential chromatography steps on 2-mm thick silica gel rotors. The first separation was carried out with an eluting solvent of hexane-tetrahydrofuran (7:3), which separated the furanocoumarin mixture into two groups of compounds. The first group was rechromatographed on the chromatotron in a solvent system of hexane-chloroform-tetrahydrofuran (63:16:1) to yield pure pimpinellin and isobergapten. The second fraction was rechromatographed in the same solvent system to yield pure bergapten and a third furanocoumarin mixture. This third mixture was subjected to another chromatography on the chromatotron

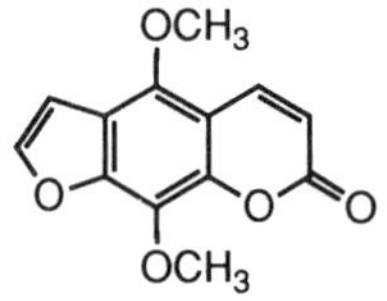

ISOPIMPINELLIN

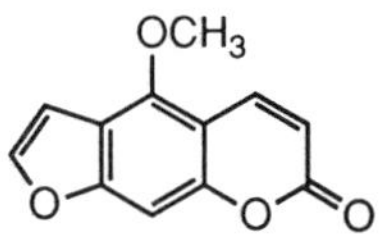

BERGAPTEN

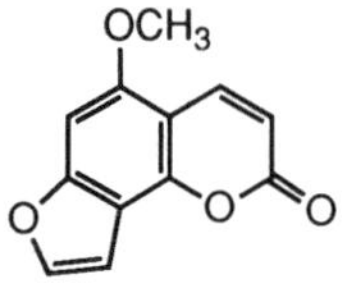

ISOBERGAPTEN 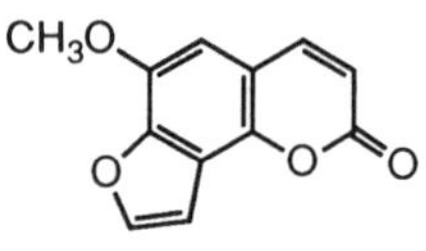   SPHONDIN 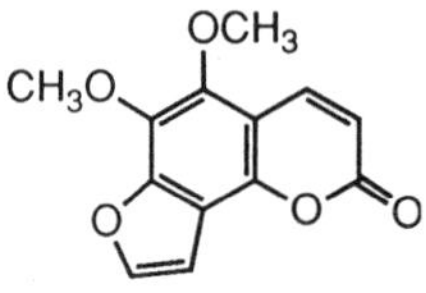   PIMPINELLIN   (13.27)

in which the plate was developed with straight methylene chloride. This gave pure sphondin and isopimpinellin.

Neolignans. McLaughlin's group used silica gel chromatography both as columns and as thin layers on the chromatotron to separate a group of cytotoxic neolignans (13.28) from the Peruvian shrub, *Endlicheria dysodantha* (Ma et al. 1991). A 95% ethanolic extract of the powdered roots was partitioned between methylene chloride and water, and the organic layer was concentrated to give the crude extract. Simple adsorbtion chromatography on silica gel using gradients of methanol, methylene chloride, and hexane gave two novel neolignans dysodanthins A and B and megaphone acetate as colorless oils and a known analog of burchellin as white crystals.

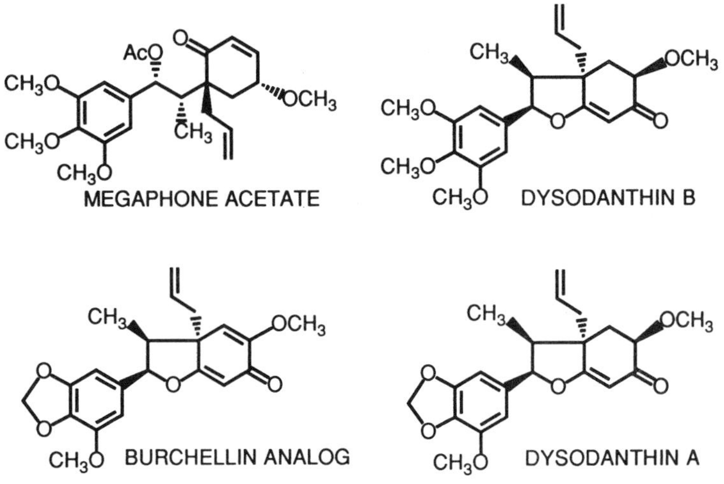

MEGAPHONE ACETATE   DYSODANTHIN B

BURCHELLIN ANALOG   DYSODANTHIN A   (13.28)

13.3 SIZE EXCLUSION CHROMATOGRAPHY

13.3.1 Hydrophilic Resins

Fractogel® TSK, available from EM Science, Gibbstown, NJ and TSK-Gel®, available from TosoHaas, Philadelphia, are the most commonly used media for size exclusion chromatography of smaller natural products. These are hydrophilic vinyl polymer gels that can be packed for medium pressure chromatography. Each is available in different size exclusion limits and particle sizes. Aqueous buffer is the most commonly used elution solvent, but organic solvents may also be employed.

13.3.1.1 Pacidamycins. Fractogel was employed for the purification of the antipseudomonal pacidamycins (13.29) from *Streptomyces coeruleorubidus* (Chen et al. 1989). An acidic butanol extract of an Amberlite XAD-2 prep-

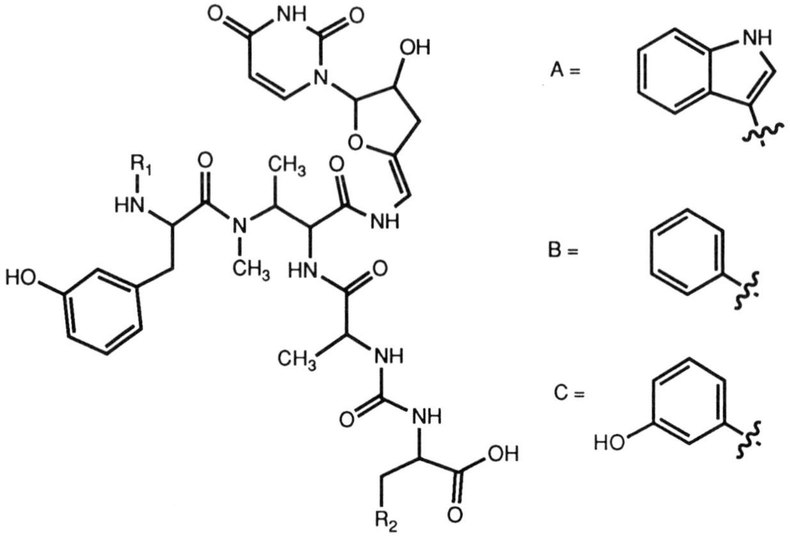

PACIDAMYCIN	R₁	R₂
1	Ala	A
2	Ala	B
3	Ala	C
4	H	A
5	H	B

$$(13.29)$$

aration of the culture was concentrated to yield a precipitate. This precipitate was chromatographed over Fractogel® TSK HW-40(S) eluted with 50 mM ammonium acetate adjusted to pH 8 with NH_4OH. The authors monitored the elution of individual components via UV detection at 254 nm. Pacidamycins 1 through 5 were obtained in pure form from this column.

13.3.1.2 Peptolides. Aqueous size exclusion chromatography was employed as the final step in the isolation of the peptolide aladapcin (13.30) from the fermentation broth of a *Nocardia* species (Shiraishi et al. 1990). Twelve thousand liters of fermentation broth of this soil isolate were subjected to a series of ion exchange chromatographies to yield approximately 1 g of crude powder. This powder was applied to a Toyopearl HW-40(F) column eluted with water to yield 5.4 mg of pure aladapcin.

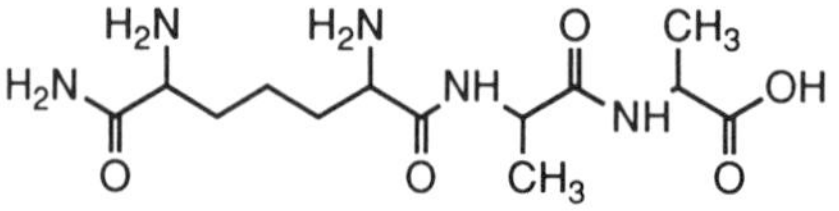

ALADAPCIN

$$(13.30)$$

13.3.1.3 Dioxindoles. Organic solvent elution from TSK-gel was used as an intermediate step in the isolation of dioxindoles from cabbage (Monde et al. 1991). Initial silica gel fractions eluting in $CH_2Cl_2/MeOH$ (49:1) were subjected to gel filtration chromatography on TSK-GEL® HW-40 developed with methanol. Final purification of silica gel HPLC and reversed phase HPLC yielded pure 3-cyanomethyl-3-hydroxyoxindole (13.31A) and dioxibrassinin (13.31B), respectively.

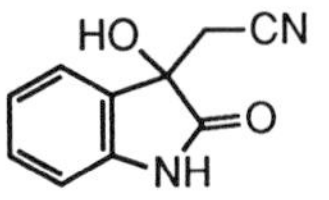

3-CYANOMETHYL-3-HYDROXYOXINDOLE

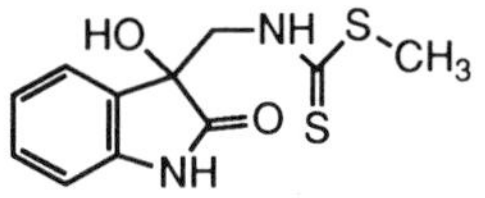

DIOXIBRASSININ

A

B   (13.31)

13.3.1.4 Sesquiterpene Esters. Another example of organic solvent elution from TSK gel was described for the isolation of triptogelin sesquiterpene esters (Takaishi et al. 1991). Extraction of *Tripterygium wilfordii* followed by silica gel and Sephadex LH-20 chromatographies yielded material from which triptogelin E-1 (13.32A) crystallized. The mother liquors from this crystallization were then subjected to chromatography on Toyopearl® HW40 developed with chloroform/methanol (1:1). A final reversed phase HPLC step yielded pure triptogelin G-1 (13.32B).

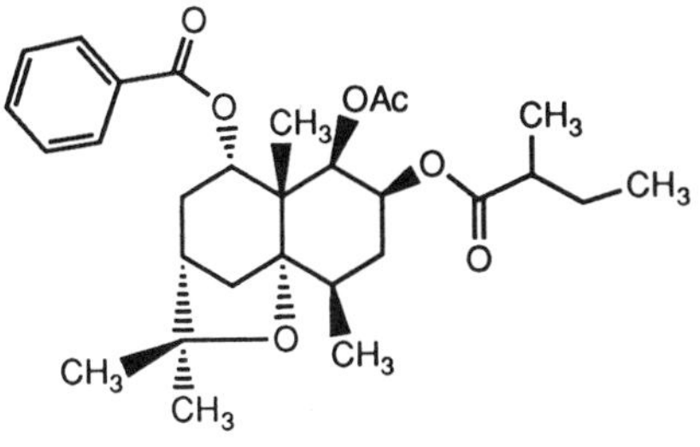

TRIPTOGELIN E-1

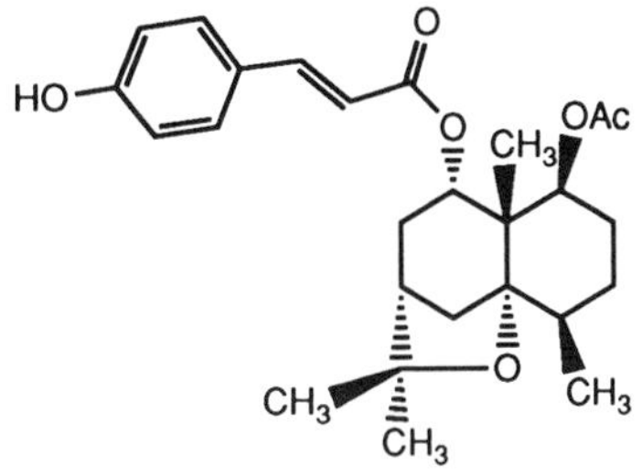

TRIPTOGELIN G-1

A

B

(13.32)

13.3.2 Lipophilic Size-Exclusion Chromatography

Pharmacia supplies a line of lipophilic size-exclusion polysaccharide resins analogous to their G-series of cross-linked dextrans, which are widely used by protein chemists both for preparative separations and as a method of estimating molecular weight. The lipophilic "LH" resins have the hydroxyl groups of the polysaccharide backbone capped with 2-hydroxypropyl groups, thus increasing the ratio of backbone carbons to hydroxyl groups considerably.

LH-20 and LH-60 are the lipophilic versions of G-25 and G-50, respectively. Unlike the G-series resins, they are used with a wide variety of solvents that greatly influence their chromatographic nature. In highly polar solvents, such as dimethyl formamide (DMF), they act essentially by size exclusion and effect separations on the basis of molecular weight. In less polar solvents, other adsorbtive forces come into play and aromatics tend to be strongly held to the resin. In solvents such as chloroform or carbon tetrachloride, with higher specific gravity than the resin, it is necessary to run columns in the ascending mode. To avoid the technical difficulties that this involves, solvent mixtures such as hexane/chloroform can be used as a gravity column to effect similar chromatographies. LH-20 chromatography in methanol is a very common first clean-up step on a crude extract (Brill et al. 1985).

13.3.2.1   2,5-Dihydroxyacetanilide.   A high degree of purification of the D1 selective dopamine receptor ligand, Sch 42029 (13.33) (2,5-dihydroxyacetan-ilide) was achieved using Sephadex LH-20 (Hegde et al. 1991). The broth from the fermentation of an Actinoplanete was extracted with XAD-16, and the resin was washed with water and eluted with 50% methanol. The eluates were concentrated and the resulting crude extract was chromatographed over Sephadex LH-20 in water. The small aromatic ligand is well retained and 5.4 g of extract yielded 135 mg of material of greater than 30% purity. The pure material was obtained from a second chromatography by C-18 reversed phase HPLC.

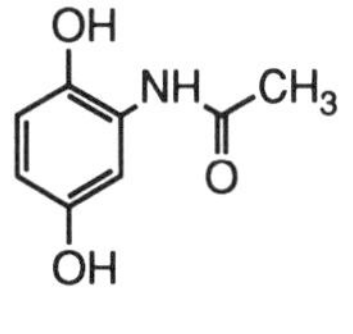

Sch 42029

(13.33)

13.4   REVERSED PHASE CHROMATOGRAPHY

Several bonded-phases, i.e., lipophilic materials covalently bonded to a silica gel matrix, are available for reversed phase chromatography. These include octadecyl (C_{18}), octyl (C_8), butyl (C_4), phenyl (ϕ), and cyanopropyl (CN). By far the most commonly used bonded-phase packing is C_{18}. Reversed phase chromatography is run either in an HPLC modality or as a column chro-matography under slight pressure. Reversed phase packings are available from numerous vendors (including J.T. Baker, Phillipsburg, NJ, and EM Science, Gibbstown, NJ) in particle sizes from 5 μm for analytical HPLC chromatog-raphy to a nominal 40 μm for pressurized column chromatography.

13.4.1 Preparative Reversed Phase Chromatography

13.4.1.1 Tirandalydigin. Bonded-phase C_8 chromatography was used as the final purification step in the isolation of the antibiotic tirandalydigin (13.34) (Brill et al. 1988). A methylene chloride extract of the fermentation broth of this streptomycete was partitioned between hexane and methanol. The methanol layer containing tirandalydigin was first subjected to Sephadex LH-20 column chromatography developed in heptane-chloroform-methanol (1:1:1). Active fractions from this column were then subjected to multiple, subsequent countercurrent chromatographies on an Ito Multi-layered Coil Planet Centrifuge in solvent systems of varying ratios of both chloroform-carbon tetrachloride-methanol-water and hexane-ethyl acetate-methanol-water. Active material from the final countercurrent chromatography was subjected to reversed phase chromatography on a Merck LiCroprep RP-8 column yielding 135 mg of tirandalydigin from 80 l of fermentation broth.

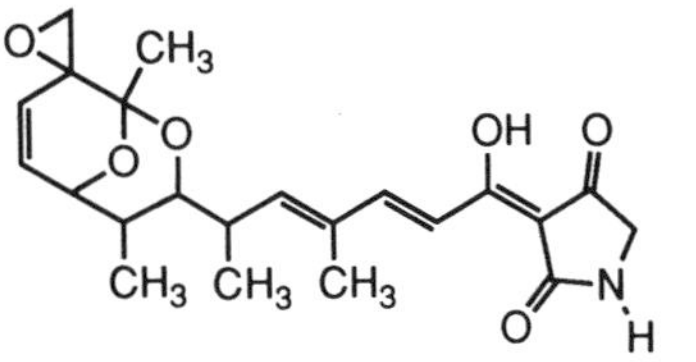

TIRANDALYDIGIN

(13.34)

13.4.1.2 Tiacumicins D and E. The authors have employed bonded-phase C_{18} chromatography as the final step in the separation of the antibiotics tiacumicins D (13.35A) and E (13.35B) (Hochlowski et al. 1987). The acetone-ethyl acetate extract from 4500 l of *Dactylosporangia aurantiacum* subsp. *hamdenensis* was partitioned between first chloroform-methanol-water (1:1:1) and then hexane-methanol (2:1). Active material from these partitions (lower layer from each) was subjected to Sephadex LH-20 column chromatography developed with methylene chloride-methanol (2:1). Active fractions from this column were next subjected to reversed phase "flash chromatography" on Baker C_{18} bonded-phase (in which packing material in a scintered glass funnel is developed via the application of vacuum beneath the packing material) developed with a water-methanol gradient. Active fractions from this column were further subjected to normal phase chromatography on silica gel eluted with a chloroform-methanol gradient. Two active bands eluted from the silica gel column, one of which was resolved over a Baker C_{18} bonded-phase silica gel column into tiacumicins D and E. These metabolites were isolated in quantities of 0.0077 and 0.022 mg/l, respectively, from whole broth.

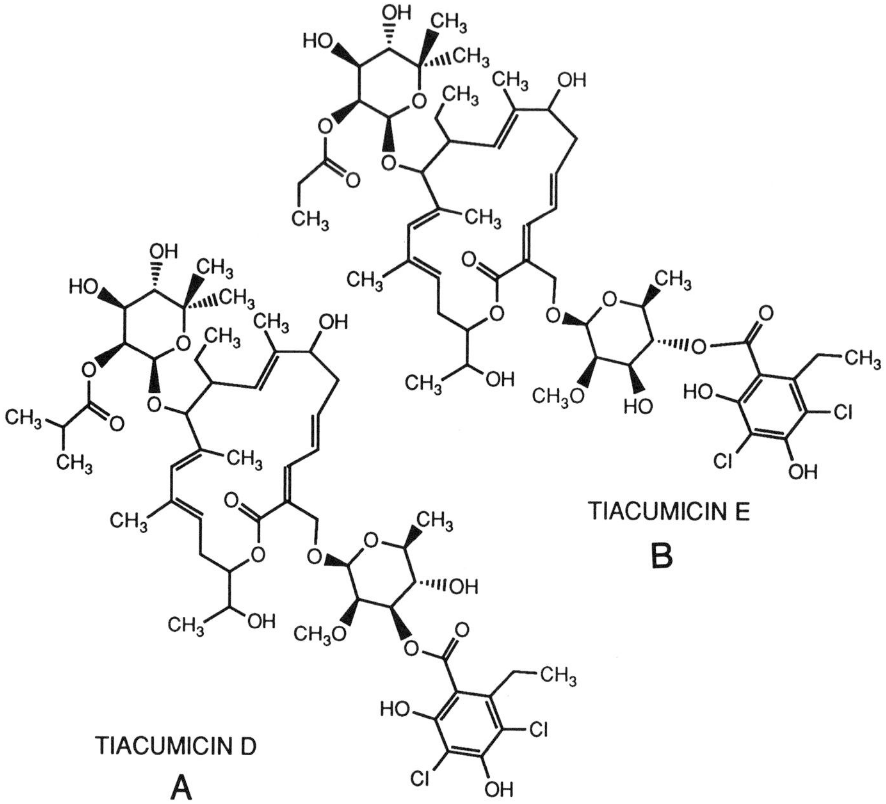

TIACUMICIN E

B

TIACUMICIN D

A

(13.35)

13.4.2 Reversed-Phase HPLC

13.4.2.1 Sesquiterpene Isothiocyanates. The final purification step of the sponge sesquiterpenes 13-isothiocyanatocubebane (13.36A) and 1-isothio-cyanatoaromadendrane (13.36B) by Faulkner's group was a C_{18} reversed phase HPLC step (He et al. 1992). A methanol extract of the frozen sponge *Axinyssa aplysinoides* was partitioned between ethyl acetate and water. The ethyl acetate layer was chromatographed over silica gel developed with a hexane-ethyl acetate gradient. A mixture of the above-mentioned isothio-cyanates was eluted in hexane-EtOAc (99:1) and was subsequently chro-matographed over a Dynamax C_{18} HPLC column developed in MeOH-H_2O (99:1), yielding 13-isothiocyanato-cubebane and 1-isothiocyanato-aromaden-drane in quantities of 7 and 8 mg, respectively, from the 25-g dry weight sponge sample.

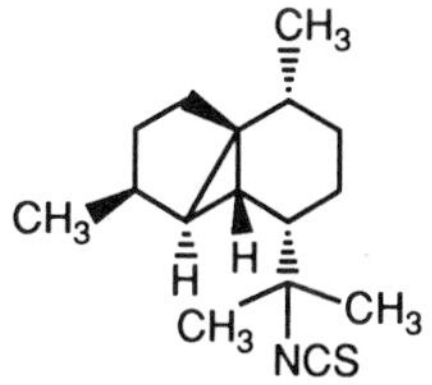

13-ISOTHIOCYANATOCUBEBANE

A

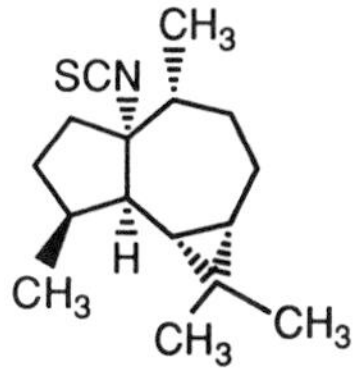

1-ISOTHIOCYANATOAROMADENDRANE

B

(13.36)

13.4.2.2 3α-Hydroxy-3,5-dihydromonacolin L. A novel hydroxymethyl-glutaryl (HMG)-CoA reductase inhibitor of the monacolin family has been isolated from the culture broth of *Monascus ruber* employing reversed phase HPLC as the final step in the isolation (Nakamura et al. 1990). Culture filtrate was extracted with Diaion HP-20 eluted with acetonitrile-water (2:3) and this concentrate was chromatographed over a Diaion HP-20SS column developed with an aqueous acetone gradient. A prepurification step over a Supelclean LC-18 column developed in water yielded material that was purified in small portions (250 mg/run) on an octadecyl silica (ODS) HPLC column developed in a CH_3CN-0.1% H_3PO_4 (35:65) solvent system. Finally, the natural product was desalted by extraction into ethyl acetate and basification to produce 46.8 mg/run of the sodium salt of 3α-hydroxy-3,5-dihydromonacolin L (13.37).

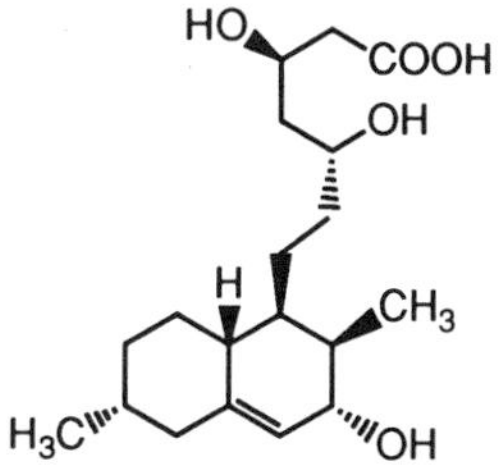

3α-HYDROXY-3,5-DIHYDROMONACOLIN L

(13.37)

13.5 ION EXCHANGE CHROMATOGRAPHY

The basis of ion exchange chromatography is that an ionized compound will be adsorbed onto a charged surface. Thus, ion exchange resins or gels all consist of a rigid matrix covalently attached at a charged (at some pH) functional group.

These are designated "cation exchangers" if the covalently attached group can take on a negative charge and therefore adsorb positively charged (cationic) compounds. The most typical functional groups bound to the matrix are carboxymethyl ($—OCH_2COO—$) and sulphopropyl ($—CH_2CH_2$-$CH_2SO_3—$), the latter being a stronger ion exchanger than the former. The metabolites most commonly purified by cation exchange chromatography contain an amine functionality that, when protonated, can bind to the anionic surface of a cation exchanger.

"Anion exchangers" in contrast, contain a positively charged (at some pH) functional group covalently attached to the matrix. The most common anion exchange charged groups are aminoethyl ($-OCH_2CH_2NH_3^+$), diethylaminoethyl ($-OCH_2CH_2NH^+(CH_2CH_2)_2$) and quaternary aminoethyl ($-OCH_2CH_2N^+(C_2H_5)_2CH_2CH(OH)CH_3$). The relative strength of anion exchangers increases from a 1° to 2° to 3° amine functionality. Metabolites that can be purified by anion exchange chromatography commonly contain a carboxylic acid functionality that, when deprotonated, can bind to the cationic surface of an anion exchanger.

Varying the exchangeable counter-ion associated with the ion exchanger functionality emparts further selectivity for adsorption of compounds. As the ease of displacement of counter-ion decreases, increasingly strong interactions between functionality of the ion exchanger and the functionality of the compound being purified are required for displacement. The usual method of elution of compounds from an ion exchanger employs a gradient pH that passes through the point at which the desired compound is no longer charged and, hence, elutes from the column. High ion strength or increasing ionic strength gradients are also used for desorption.

Matrices available for ion exchange chromatography range from those suitable for bulk extraction of crude materials such as the IRC and IRA series from Rohm and Haas, Philadelphia, or the Dowex series from Dow Chemical, (sold through vendors such as Rio-Rad, Richmond, CA) to numerous column chromatography grade gel such as the Sephadex ion exchangers from Pharmacia, Piscataway, NJ.

13.5.1   Cation Exchange Chromatography

13.5.1.1   Oligopeptides.   The isolation of the oligopeptide antibiotic FR112123 (13.38) from *Streptomyces viridochromogenes* uses cation exchange chromatography in two key steps; extraction from the fermentation broth and final purification (Fujie et al. 1990). The fermentation culture filtrate is first passed over a bed of Diaion SK-1B cation exchange resin that is pre-equilibrated with ammonium ion. The column is then washed with water, followed by elution of the antibiotic with 0.1 N ammonium hydroxide. Two silica gel chromatographies are then performed in butanol-ethanol-chloroform-

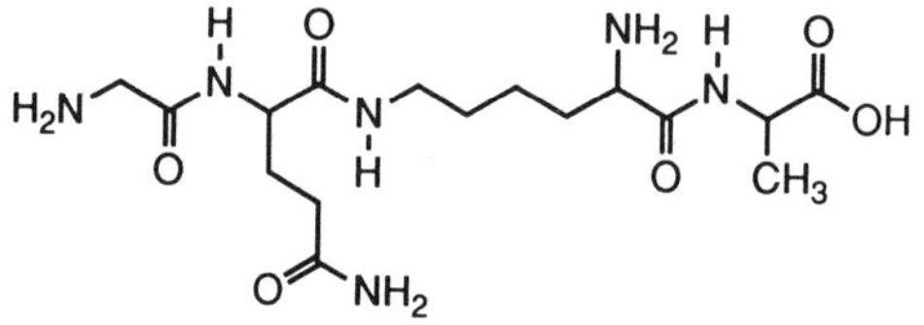

FR112123

(13.38)

aqueous ammonia and butanol-acetic acid-water solvent systems. Active material from the final silica gel column is then chromatographed over a CM-sephadex C-25 column pre-equilibrated with ammonium ion, the column is washed with water, and is developed with 0.02 N ammonium hydroxide to yield pure FR112123.

13.5.1.2 Nicaeensin. Anion exchange chromatography comprised the majority of the isolation of nicaeensin (13.39), an amidinoureido containing metabolite of the red alga, *Schottera nicaeensis* (Chillemi et al. 1990). Algal material, freshly collected, was repeatedly extracted with aqueous methanol (30%), concentrated to an aqueous solution, adsorbed onto Dowex-50W (H⁺ form), and eluted from this cation exchanger with 2 N ammonium hydroxide. Material eluting from this column was then passed over a further cation exchange column (Amberlite IRC-50, H⁺ form) and eluted with 2 N ammonium hydroxide. After one silica gel chromatography column, a final cation exchange step on Dowex-50W, again eluted with 2 N ammonium hydroxide, provided pure nicaeensin after lyophilization.

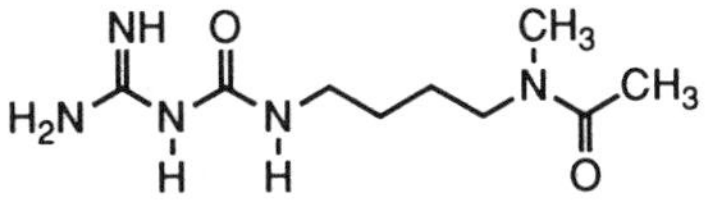

NICAEENSIN

(13.39)

13.5.1.3 FR109615. Both cation and anion exchange steps were used in the isolation of the zwitterionic antifungal compound, FR109615 (13.40) from *Streptomyces setonii* (Iwamoto et al. 1990). Culture filtrate from the fermentation broth of this producing culture was run through a column of Diaion SK-1B (H⁺ form), the column was washed with water, and was developed with 0.5 N ammonium hydroxide. Eluate from this column was then passed through a Dowex IX2 (OH⁻ form) column, washed with water, and developed with 0.05 N acetic acid. These two ion exchange chromatographies reduced 15 l of crude culture filtrate to 3 g of concentrate. A final silica gel column developed in 90% aqueous isopropanol yielded 1.2 g of crude material, which upon recrystallization from hot ethanol gave 800 mg of pure FR109615.

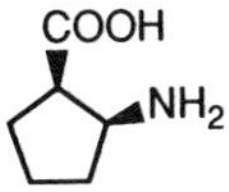

FR109615

(13.40)

13.5.2 Anion Exchange

13.5.2.1 Melanostatin. Advantage is taken of the carboxylic acid functionality of the melanin synthesis inhibitor, melanostatin (13.41), by an anion

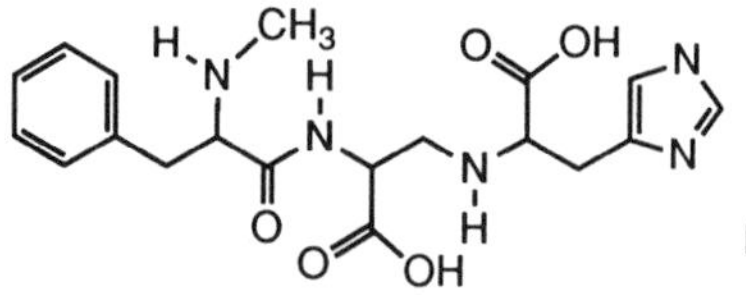

MELANOSTATIN

(13.41)

exchange step in the isolation (Ishihara et al. 1991). Carbon was used to adsorb organics from 200 l of filtered fermentation broth. An aqueous acetone eluate from this carbon batch extraction was concentrated to 1 l of aqueous solution and adsorbed onto 1 l of Amberlite IRC-50 (H⁺ form) that was washed with water and developed with 1.1 l of 2 N ammonium hydroxide. This yielded 18 g of crude material that was further purified by silica gel chromatographies developed in CH_2Cl_2-EtOH-14% NH_4OH (9:14:3) and BuOH-HOAc-H_2O (4:1:1). A final desalting step on Sephadex LH-20 yielded 220 mg of pure melanostatin.

13.5.2.2 Fosfadecin. Three anion exchange steps comprised the entire isolation of the nucleotide antibiotic fosfadecin (13.42) from *Pseudomonas viridiflava* (Katayama et al. 1990). Ninety-eight liters of culture filtrate was adsorbed onto 5 l of Amberlite IRA-402 (Cl⁻ form) that was then developed with 1 M NaCl. The material eluting from this first anion exchanger was then applied to a carbon column and subsequently desorbed with 8% iBuOH-0.02 M NH_4OH. Eluates from the carbon column were then applied to a QAE Sephadex A-25 (Cl⁻ form) column and developed with 0.01 M NaCl. Active fractions from this second anion exchange column were combined and desalted on a carbon column developed with 8% i-BuOH. This carbon column eluate was then applied to another QAE Sephadex A-25 (Cl⁻ form) column developed with 0.09 M NaCl. Active fractions eluting from this third anion exchange column were again desalted on carbon to yield 1.46 g of fosfadecin (as the disodium salt).

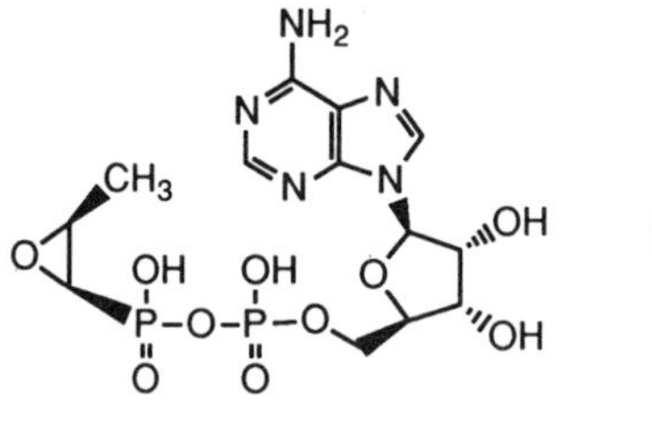

FOSFADECIN

(13.42)

13.5.2.3 Ichangensin Glucoside. The carboxylic acid functionality of ichangensin glucoside (13.43) enabled the use of anion exchange chromatography in the isolation of this novel limonoid from *Citrus junos* (Ozaki et al. 1991). A pectinase digest of ground seeds was centrifuged and the supernate was

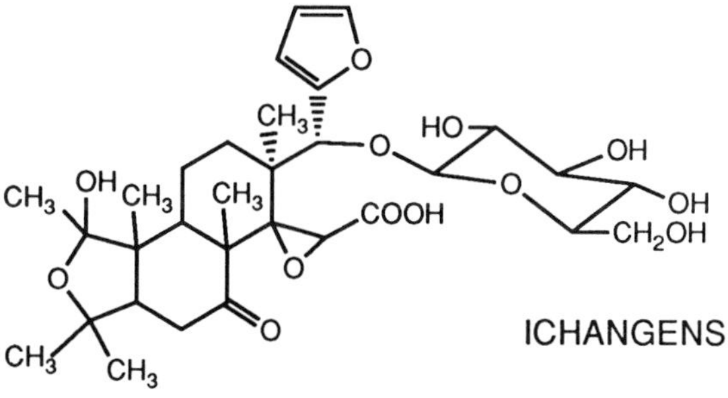

ICHANGENSIN GLUCOSIDE

(13.43)

filtered and then applied to an XAD-2 column. After a water wash, the XAD-2 column was developed with CH_3CN and the dried CH_3CN eluate was applied to a DEAD Sephacel column in pH 6.5 aqueous solution. The DEAD column was washed with water and developed with 0.2 M NaCl. Final purification involved desalting on a bonded-phase C_{18} column and C_{18} reversed phase HPLC chromatography.

13.5.2.4 Dehydropeptidase Inhibitors. Ion exchange chromatography was employed extensively in the purification of the two closely related dehydro-peptidase inhibitors WS1358A1 (13.44A) and WS1358B1 (13.44B), from *Streptomyces parvulus* (Hashimoto et al. 1990). The harvest broth was filtered through diatomaceous earth and the filtrate was adjusted to pH 10 with sodium hydroxide solution. The active principles were adsorbed onto Dowex 1 × 2 in the chloride form and eluted with 0.2 N sodium chloride solution. The solution was adjusted to pH 2 with hydrochloric acid and then desalted with a column of activated charcoal. This column was washed initially with water and then with 25% aqueous methanol before the inhibitors were eluted with 25% aqueous methanol containing 0.5 N ammonium hydroxide. The eluate was concentrated and adjusted to pH 2 before being applied to a column of Diaion SP-207, which was developed with water. The slightly greater lipo-philicity of WS1358A1 enabled it to be retained longer on the column than the B analog and, hence, a separation of the two active principles was achieved at this point. Each required several further chromatographies utilizing DEAE Sephadex A-25, Sephadex G-15, and cellulose or Dowex 1 × 2 packings to obtain pure compounds.

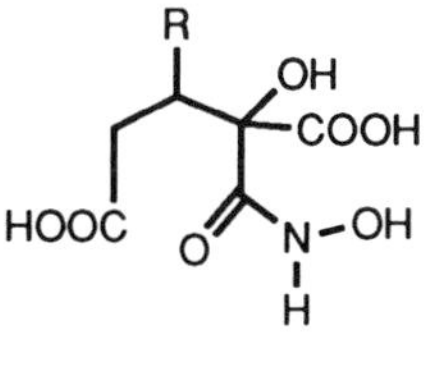

WS1358A1 R = CH_3 A
WS1358B1 R = H B

(13.44)

13.6 AFFINITY CHROMATOGRAPHY

Affinity chromatography refers to types of chromatography on solid supports in which very specific chemical interactions occur between some group or series of groups on the support and a unique class of analytes. The types of interactions that are used are usually of the ligand-receptor type but in some cases the term "affinity chromatography" is used to describe systems that rely on looser and more general binding phenomena. For example, this term was used to describe chromatography of nucleotides on Sepharose impregnated with ferric chloride, where the retention appears to rely on the interaction of the negatively charged free phosphate groups of the nucleotides with the positively charged iron (III) ions (Dobrowlska et al. 1991). Nucleosides, such as cyclic AMP, without a free phosphate group are not retained and are eluted with the void volume, whereas those with a free phosphate group are retained at pH 5.5 and eluted with modest differentiation while the pH of the eluent is increased to 8. True affinity chromatography, relying on a receptor-ligand type interaction is, by its nature, very specific. This has the advantage that the affinity supports can be used to extract very low concentrations of the analyte from complex media with remarkable efficiency. It has the disadvantage that the column support tends to have a very narrow range of applications, i.e., new supports must be designed and prepared for each new application.

13.6.1 Glycopeptides

Perhaps one of the most classic examples of the use of affinity chromatography in the isolation and purification of secondary metabolites is in the area of the antibiotic glycopeptides (13.45) These antibacterials have a biological mode of action that involves complexing to the pentapeptide portion of the peptidoglycan of the bacterial cell wall. In particular, they have been shown to bind to the terminal lysyl-D-alanyl-D-alanine region, which is involved in peptidoglycan cross-linking. This interaction was used in the design of affinity columns with D-alanyl-D-alanine linked to Activated CH-Sepharose 4B (Corti and Cassani 1985). These authors showed that this support would selectively adsorb glycopeptides such as teicoplanin, vancomycin, and ristocetin from complex fermentation beers. A one-step isolation and purification could be achieved by washing the loaded resin at pH 7.4 and subsequently eluting the antibiotic at pH 11. Their resin preparation had a capacity of 6 to 7 μmol glycopeptide/ml of resin.

This type of affinity resin was used to isolate and purify the products of microbial transformations of teicoplanin derivatives (Borghi et al. 1991). The complex of glycopeptides obtained from the affinity resin was then separated by semipreparative HPLC using a HIBAR Lichrosorb RP-18 column eluted with a stepwise gradient of 0.02 M aqueous sodium dihydrogen phosphate with increasing concentrations of acetonitrile.

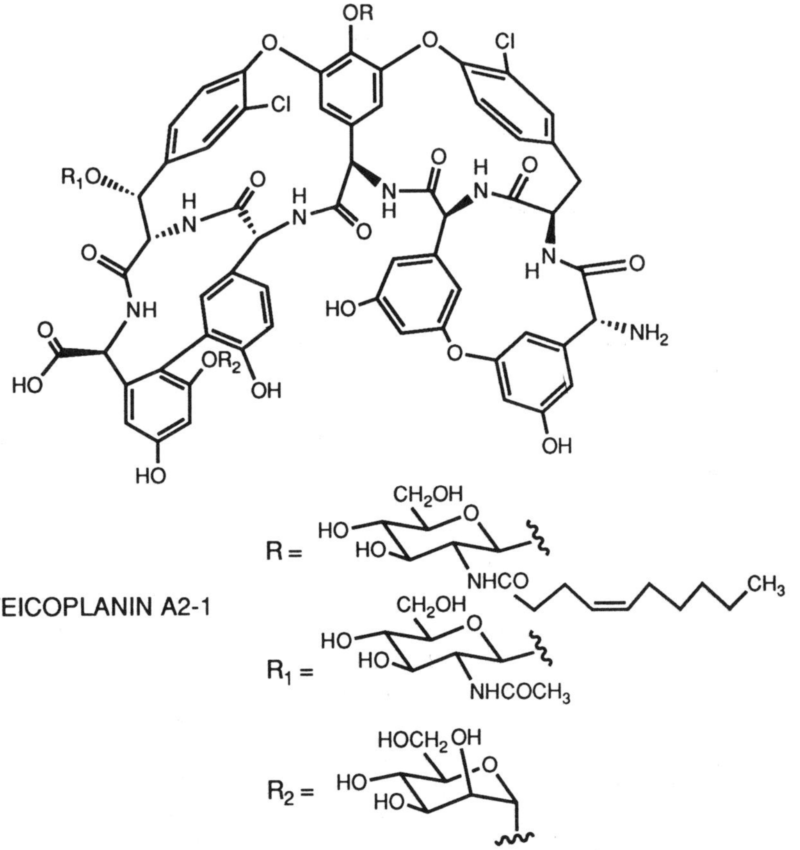

TEICOPLANIN A2-1

(13.45)

13.6.2 DNA Interactive Agents

In the search for novel agents with potential anticancer activity from plants, Pezzuto has argued that predominant modes of action of such compounds will involve binding to DNA and that this interaction could be the basis for both a detection system and a means of isolation (Pezzuto et al. 1991). In a reversed phase HPLC system, they used the ability of such agents, provided that they are retained by the column, to decrease, in a dose dependent manner, the intensity of a DNA peak that would otherwise be unretained. Similar effects were observed on RP-18 TLC plates, where the addition of fagaronine or ethidium bromide significantly retarded the mobility of DNA. This technique was used to examine a series of plant extracts and led to the isolation of the novel macrocyclic alkaloids, the budmunchiamines (13.46), from *Albizia amara* (Mar et al. 1991). DNA-cellulose was used as an affinity resin

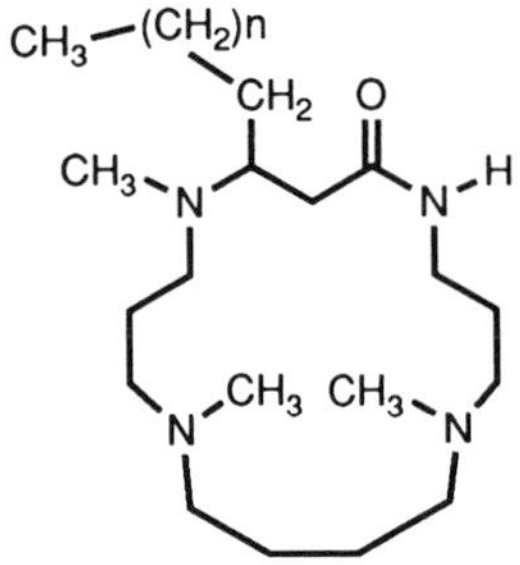

CH$_3$—(CH$_2$)n
CH$_2$ O
CH$_3$—N   N—H
N—CH$_3$ CH$_3$—N

BUDMUNCHIAMINE   A n = 9

B n = 7

C n = 11

(13.46)

to isolate enough of these agents to act as TLC standards to follow a more classical isolation.

13.7 CONCLUSIONS

The need for the isolation chemist, working in the area of secondary metabolites, to tailor his isolation procedures to the particular active principle being sought is a feature of this discipline and injects an element of art into the basic science of separation and purification. It seems likely that, within the foreseeable future, artificial intelligence programs will be developed to deduce quite complex structures from spectral data and, hence, deprive many of us from the great joy and sense of achievement that we derive from those wonderful "jig-saw" puzzles today. By contrast, the position of the isolation chemist seems secure. We are a long way from being able to predict a logical and efficient isolation scheme for an active principle when its bioactivity is first detected in a crude fermentation broth, or plant, or marine organism extract. Here, the trial and error approach reigns supreme. Productive chemists cast their early nets wide; testing as many different conditions as possible for their effect on the stability, adsorption, polarity, and partition distribution of active principles with whatever assays are available. Then, the subsequent trials of the trial and error approach are intuitive rather than random.

The position that normal phase silica gel chromatography holds as the pre-eminent solid support in the isolation of natural products is perhaps surprising, given the large number of other supports and nonadsorptive chromatographic methods that have been developed since the 1970s. Perhaps it reflects the high separatory power of this method and the ease in which lipophilic compounds can be isolated with it.

HSCC has been relatively slow in fulfilling its potential. This may be the result of the reputation for tedium deservedly acquired by countercurrent distribution apparatus and droplet countercurrent devices. It may also be a reflection of the lack of large company promotion for this methodology. Part of the reason for this slow introduction seems to have been the fact that very few of the major graduate schools responsible for training natural product chemists embraced this technology at its inception; hence, its use spread more from the experience of a handful of isolated end-users, rather than as an essential innovation. Most of these graduate schools now have one or more high-speed countercurrent devices in their natural products isolation laboratories and this technique should be in the repertoire of new isolation chemists and should achieve a position of greater prominence.

The explosion in molecular biology techniques and their use by screening groups should generate opportunities for affinity chromatography. As receptors are cloned and produced in milligram quantities for high throughput ligand binding assays, they can also be used on immobile supports as affinity resins to selectively isolate the ligand antagonists being discovered by such a screen. Even in cases where it is impractical to do this on a scale to allow for the spectral identification of the active principle, it should be possible to emulate the work of Pezzuto (Mar et al. 1991) and isolate an analytical sample as a standard to enable a classical isolation followed by TLC or HPLC.

REFERENCES

Adinolfi, M., Corsaro, M.M., Lanzetta, R., Parrilli, M., and Scopa, A. (1989) *Phytochemistry* 28(1), 284–288.

Adinolfi, M., Lanzetta, R., Marciano, C.E., Parrilli, M., and De Giulio, A. (1991) *Tetrahedron* 47(25), 4435–4440.

Bartizal, K.F., and Onishi, J.C. (1991a) U.S. patent no. 5,026,554.

Bartizal, K.F., Rozdilsky, W., and Onishi, J.C. (1991b) U.S. patent no. 5,053,425.

Bergstrom, J.D., Onishi, J.C., Hensens, O.D., et al. (1991) European patent 0 448 393 A1.

Beutler, J.A., Chmurny, G.M., Look, S.A., and Witherup, K.M. (1991) *J. Natural Products* 54(3), 893–897.

Borghi, A., Ferrari, P., Gallo, G.G., et al. (1991) *J. Antibiotics* 44(12), 1444–1451.

Brill, G.M., McAlpine, J.B., and Hochlowski, J.E. (1985) *J. Liquid Chromatogr.* 8(12), 2259–2280.

Brill, G.M., McAlpine, J.B., and Whittern, D. (1988) *J. Antibiotics* 41(1), 36–44.

Chen, R.H., Buko, A.M., Whittern, D.N., and McAlpine, J.B. (1989) *J. Antibiotics* 42(4), 512–520.

Chillemi, R., Morrone, R., Patti, A., Piattelli, M., and Sciuto, S. (1990) *J. Natural Products* 53(5), 1220–1224.

Corti, A., and Cassani, G. (1985) *Appl. Biochem. Biotechnol.* 11, 101–109.

Dawson, M.J., Farthing, J.E., Marshall, P.S., et al. (1992) *J. Antibiotics* 45(5), 639–647.

Desai, H.K., El Sofany, R.H., and Pelletier, S.W. (1990) *J. Natural Products* 53(6), 1606–1608.

Dobrowolska, G., Muszynska, G., and Porath, J. (1991) *J. Chromatogr.* 541, 333–339.

Erdelmeier, C.A.J., Nyiredy, S., and Sticher, O. (1985) *J. High Resolution Chromatogr. Chromatogr. Commun.* 8(3), 132–134.

Fantini, A.A., Korshalla, J.D., Pinho, F., et al. *Abstracts of the 26th Interscience Conference on Antimicrobial Agents and Chemotherapy*, no. 227, New Orleans.

Fujie, A., Iwamoto, T., Shigematsu, N., et al. (1990) *J. Antibiotics* 43(5), 449–455.
Głowniak, K. (1991) *J. Chromatogr.* 552(7), 453–461.
Hano, Y., Mitsui, P., Nomura, T., Kawai, T., and Yoshida, Y. (1991) *J. Natural Products* 54(4), 1049–1055.
Harrison, S. (1979) U.S. patent no. 4,139,458.
Hashimoto, S., Murai, H., Ezaki, M., et al. (1990) *J. Antibiotics* 43(1), 29–37.
Hayakawa, Y., Takaku, K., Furihata, K., Nagai, K., and Seto, H. (1991) *J. Antibiotics* 44(12), 1294–1299.
He, H.-Y., Salvá, J., Catalos, R.F., and Faulkner, J.D. (1992) *J. Org. Chem.* 57(11), 3191–3194.
Hegde, V.R., Patel, M.G., Horan, A.C., et al. (1991) *J. Ind. Microbiol.* 8(3), 187–192.
Hochlowski, J.E., Swanson, S.J., Ranfranz, L.M., et al. (1987) *J. Antibiotics* 40(5), 575–588.
Hostettman, K., Hostettman-Kaldas, M., and Sticher, O. (1980) *J. Chromatogr.* 202, 154–156.
Ishihara, Y., Oka, M., Tsunakawa, M., et al. (1991) *J. Antibiotics* 44(1), 25–32.
Itoh, Y., Shimura, H., Ito, M., et al. (1991) *J. Antibiotics* 44(8), 832–837.
Iwamoto, T., Tsujii, E., Ezaki, M., et al. (1990) *J. Antibiotics* 43(1), 1–7.
Jeffries, P.R., Toia, R.F., and Casida, J.E. (1991) *J. Natural Products* 54(4), 1147–1149.
Kaneda, N., Pezzuto, J.M., Soejarto, D.D., et al. (1991) *J. Natural Products* 54(1), 196–206.
Katayama, N., Tsubotani, S., Nozaki, Y., Harada, S., and Ono, H. (1990) *J. Antibiotics* 43(3), 238–246.
Kernan, M.R., Cambie, R.C., and Berquist, P.R. (1991) *J. Natural Products* 54(1), 265–268.
Kubo, I., Muai, Y., Soediro, I., Soetarno, S., and Sastrodihardjo, S. (1991) *J. Natural Products* 54(4), 1115–1118.
Kupchan, S.M., Britton, R.W., Ziegler, M.F., and Sigel, C.W. (1973) *J. Org. Chem.* 33(1), 178–179.
Lanzetta, R., Parrilli, M., Adinolfi, M., Aquilla, T., and Corsaro, M.M. (1990) *Tetrahedron* 46(4), 1287–1294.
Ma, W.-W., Kozlowski, J.F., and McLaughlin, J.L. (1991) *J. Natural Products* 54(4), 1153–1158.
Mar, W., Tan, G.T., Cordell, G.A., and Pezzuto, J.M. (1991) *J. Natural Products* 54(6), 1531–1542.
Marner, F.-J., Freyer, A., and Lex, J. (1991) *Phytochemistry* 30(11), 3709–3712.
Mohamed, M.H., Saito, K., Kadry, H.A., et al. (1991) *Phytochemistry* 30(9), 3111–3115.
Monde, K., Sasaki, K., Shirata, A., and Takasugi, M. (1991) *Phytochemistry* 30(9), 2915–2917.
Mukhopadhyay, T., Roy, K., Bhat, R.G., et al. (1992) *J. Antibiotics* 45(5), 618–623.
Murata, Y., Yeh, H.J.C., Pannell, L.K., et al. (1991) *J. Natural Products* 54(1), 233–240.
Nakamura, T., Komagata, D., Murakawa, S., Sakai, K., and Endo, A. (1990) *J. Antibiotics* 43(12), 1597–1600.
Naruse, N., Tenmyo, O., Kobaru, S., et al. (1990) *J. Antibiotics* 43(3), 267–280.
Ngadjui, B.T., Ayafor, J.F., Sondengam, B.L., Connolly, J.D., and Rycroft, D.S. (1991) *Tetrahedron* 47(22), 3555–3564.

Ozaki, Y., Miyake, M., Maeda, H., et al. (1991) *Phytochemistry* 30(8), 2659–2661.

Pezzuto, J.M., Che, C.-T., McPherson, D.D., et al. (1991) *J. Natural Products* 54(6), 1522–1530.

Rahman, A.-U., Pervin, A., Choudhary, M.I., Hasan, N., and Sener, B. (1991) *J. Natural Products* 54(4), 929–935.

Rasmussen, R.R., Nuss, M.E., Scherr, M.H., et al. (1986) *J. Antibiotics* 39(11), 1515–1526.

Rasmussen, R.R., and Scherr, M.H. (1987) *J. Chromatogr.* 386, 325–332.

Rasmussen, R.R., Scherr, M.H., Whittern, D.N., Buko, A.M., and McAlpine, J.B. (1987) *J. Antibiotics* 40(10), 1383–1393.

Shiraishi, A., Nakajima, M., Katayama, T., et al. (1990) *J. Antibiotics* 43(6), 634–638.

Takaishi, Y., Tamai, S., Nakano, K., Murakami, K., and Tomimatsu, T. (1991) *Phytochemistry* 30(9), 3027–3031.

Weber, J.M., Leung, J.O., Swanson, S.J., Idler, K.B., and McAlpine, J.B. (1991) *Science* 252, 114–117.

Witherup, K.M., Look, S.A., Stasko, M.W., Ghiorzi, T.J., and Muschik, G.M. (1990) *J. Natural Products* 53(5), 1249–1255.

Modern Spectroscopic Approaches to Structure Elucidation

Otto D. Hensens

The objective of this chapter is to critically review current trends and strategies in the structure determination of natural products using modern spectroscopic methods. Advances, particularly in nuclear magnetic resonance (NMR) spectroscopy, in recent years have been pivotal and have brought about a revolution unprecedented in the structural analysis of organic compounds. It is therefore not uncommon nowadays to find complex natural products structure determinations reported in the literature using predominantly NMR methodology. The whole area of NMR structural analysis continues to benefit significantly from the tremendous strides made in overcoming the resolution

I would like to thank my colleagues, Drs. Mary W. Baum, Robert P. Borris, Jerrold M. Liesch, Steven M. Pitzenberger, Sheo B. Singh, James P. Springer, and Ken Wilson for critically proofreading this manuscript and for their helpful comments and suggestions. Special thanks go to Dr. Gregory Helms for the HMBC experiments on L-156,602 and for useful NMR discussions. I am also indebted to Professors Haruo Seto, Kenneth L. Rinehart, and journal publishers, for permission to reproduce published material and in particular to Professor Seto for providing further details on the structure determination of cyclothiazomycin. I would like to especially express my appreciation to Dr. Georg Albers-Schonberg for the intellectual stimulus, and continued support and encouragement through the years at Merck.

problem in NMR conformational/structural studies of peptides and proteins. This work has recently been the subject of several excellent texts (Wüthrich 1986; Oppenheimer and James 1989a, 1989b) and reviews (Hosur et al. 1988; Kaptein et al. 1988; Prestegard 1988; Bax 1989; Wüthrich 1989, 1990). Advances in radio frequency (rf) and probe technology, in the application of higher magnetic fields, and the ever expanding repertoire of pulse sequences in one-dimensional (1D) and two-dimensional (2D) NMR and in the biological area of three-dimensional (3D) NMR and even four-dimensional (4D) NMR, will inevitably be passed down to the structural organic chemist to allow the resolution of more complex structural problems on increasingly smaller sample quantities. This nondestructive methodology contrasts sharply with chemical degradation studies that so typically have dominated natural product structure determination in the not too distant past.

The tremendous progress in NMR in recent years should not overshadow the very significant developments that have been made in mass spectrometry (MS). MS remains the method of choice for determining molecular formulas and identifying known substances. Where applicable, depending largely upon volatility and fragmentation patterns, MS can be a very powerful tool, as has been demonstrated in the analysis and sequencing of peptides and carbohydrates. In parallel with developments in NMR, the field has benefited greatly from studies of peptides and proteins where the problem of volatility has been paramount. Judging from the recent literature, however, the application of MS techniques to the structure determination of natural products is in general more selective and complementary to NMR, and for these reasons the discussion will essentially be limited to NMR methodology.

The outline of this chapter follows typically the *modus operandi* for structure elucidation of an unknown natural product that has in large part evolved in the author's laboratory during the last decade. In PART A of the structure elucidation process, Section 14.1 deals with the process of establishing an empirical formula, which is seldom straightforward and often involves a combination of MS and NMR techniques for moderate (molecular weight (MW), 500 to 1000) to large (MW > 1000) natural products. The subsequent chronology of steps derives from practical considerations in acquiring the critical NMR data on a small sample in the most expedient manner. For example, $^1H-^1H$ connectivity NMR data are almost always obtained and analyzed before multiple-bond $^1H-^{13}C$ connectivity data because of the relatively more difficult problem of acquiring the latter information. The concept of establishing partial structures from predominantly $^1H-^1H$ 2D-correlation spectroscopy (COSY) NMR data is thus developed in Section 14.1.2, which is followed in PART B by a discussion on the assembly of these subunits into a total structure by predominantly long-range connectivity NMR methodology (Section 14.2). Although they are basically long-range $^1H-^1H$ connectivity experiments, relayed coherence transfer experiments via scalar or isotropic mixing are included in the preceding Section 14.1.2 because they require a contiguous $^1H-^1H$ sequence and do not in general provide connectivity through quaternary centers or heteroatoms. A rather restricted view of long-range methodologies is therefore taken in this chapter.

This arbitrary chronology, involving modest instrumentation by today's standards, has generally suited us well but has already begun to change as the difficulty and time constraints of acquiring critical NMR data, especially long-range ^1H–^{13}C connectivities, rapidly improves. Various data can thus be analyzed simultaneously rather than sequentially. This most recent and less equivocal strategy, which relies on more recent NMR instrumentation using inverse ^1H detection methods, has placed a much greater emphasis on the ^{13}C nucleus in structural analysis despite considerable advances in the detection of direct and relayed ^1H–^1H J coupling information from Hartman-Hahn spin-locking experiments via isotropic mixing. This is expounded in greater detail in Section 14.2.1.2.

It is the application of primarily NMR spectroscopic techniques to structure elucidation of natural products from a wide variety of fermentation, marine, and plant sources that is emphasized in this chapter, and the reader is referred to the original literature given in the references for detailed information on the particular spectroscopic techniques employed. In particular, many excellent NMR texts (Bax 1982; Atta-ur-Rahman 1986, 1989; Wüthrich 1986; Croasmun and Carlson 1987; Derome 1987; Sanders and Hunter 1987; Brey 1988; Martin and Zektzer 1988a; Neuhaus and Williamson 1989; Oppenheimer and James 1989a, 1989b) and reviews (Benn and Günther 1983; Bax and Lerner 1986; Morris 1986; Farrar 1987a, 1987b; Kessler et al. 1988a; Martin and Zektzer 1988b; Sadler 1988; Derome 1989) dealing with 2D NMR and modern pulse sequences have appeared during the last few years and are recommended for further reading. The techniques and examples chosen and highlighted here are meant to be illustrative and by no means exhaustive.

14.1 STRUCTURE ELUCIDATION PROCESS: PART A

The great potential of NMR in organic structural analysis is inherent in the phenomenon of scalar and dipolar coupling between nuclei with spin. Methodology to unravel such coupling in particular as it applies to nuclei of spin 1/2 (e.g., ^1H, ^{13}C, ^{15}N, ^{31}P), has literally exploded since the advent of 2D NMR in 1976 (Aue et al. 1976) following the visionary concept of Jeener in 1971. Spectroscopists, and natural product and organic chemists alike have preoccupied themselves with deciphering ^1H–^1H, ^1H–^{13}C and ^{13}C–^{13}C coupling networks that have either directly or indirectly led to mapping of the carbon skeleton of an organic molecule. Some of the more recent applications and developments in this area will be discussed and the reader is referred in particular to the excellent texts by Derome (1987), Sanders and Hunter (1987), Atta-ur-Rahman (1989), Martin and Zektzer (1988a), and Neuhaus and Williamson (1989) for further details on establishing connectivity based on scalar and dipolar nuclear Overhauser effect (NOE) coupling.

A word needs to be said about the manner in which connectivity data can be conveniently reported, as expressed in structural formulas. Unless otherwise stated, we adopt the way in which some authors have depicted ^1H–^1H connectivities from COSY and relayed coherence transfer pathways, by bold lines (━━━━) and long range ^1H–^1H and ^1H–^{13}C correlations by solid arrows (——→), the tail emanating from the proton resonance in the latter case. Sometimes long-range ^1H–^1H correlations are also embraced by the bold line designation. Where several connectivities are depicted in a given structure, NOE correlations can be distinguished from other long-range correlations by dotted arrows. 1D NOE correlations are depicted by single dotted arrows (----→), the tail emanating from the proton that was saturated and the head signifying the proton resonance experiencing an intensity change. Double dotted arrows (←---→) can be used to distinguish 2D nuclear Overhauser enhancement spectroscopy (NOESY) or rotating frame nuclear Overhauser enhancement spectroscopy (ROESY) correlations or 1D NOE connectivities where the NOE effect was shown to be reciprocated.

Before developing partial structure methodology, some current trends in establishing molecular formulas of natural products will be discussed.

14.1.1   Determination of Molecular Formula

The establishment of the molecular formula is critical in the structure determination process of natural products. High resolution electron impact mass spectrometry (EIMS) has usually provided such unequivocal determinations. More recently, "soft" ionization techniques such as high resolution-fast atom bombardment mass spectrometry (HR-FABMS) methods have considerably extended the applications of high resolution mass spectrometry to higher molecular weight ranges, particularly of nonvolatile, polar compounds (Barber et al. 1981; Tomer 1989). Depending, however, on the particular instrumentation available, the accuracy of these methods does not always define an empirical formula uniquely but provides a range of formulas especially in the molecular weight range above 500. An increasing number of examples have recently appeared in the literature where determining empirical formulas of natural products with molecular weight 500 or more have required an interplay between MS and NMR methods. In certain cases, the time-honored combustion analysis has been required (Iwasaki et al. 1984; Kato et al. 1986; Funaishi et al. 1987; Murata et al. 1987; Helms et al. 1988; Kobayashi et al. 1988; Fusetani et al. 1989a; Moore et al. 1989; Inman et al. 1990; Kettingring et al. 1990; Hensens et al. 1991a). A brief outline of this strategy is provided in the following discussion.

A proton-decoupled ^{13}C NMR spectrum can in principle provide a reliable carbon count of the molecule. However, depending on carbon spin-lattice relaxation times, the flip angle (pulse width) and the acquisition time employed, quaternary carbons can sometimes appear as weak signals that may not be readily distinguished from impurity peaks that are present. In particular, if the peak is weak and not quaternary, it should be regarded as suspect (such as a solvent peak) unless some exchange phenomenon, judging for example from its linewidth, is suspected to be present (e.g., the tautomeric structures of the agglomerins; Terui et al. 1990a). Changing the solvent, pH, temperature, and/or parameter selection (e.g., smaller flip angle with a suitable delay) may be helpful. Demonstrating its long-range connectivity to an assigned proton usually ensures that a quaternary carbon belongs to the molecule (see Section 14.2.1.2). As an example of exchange-broadening, no carbons of the flavensomycinic acid portion of the novel macrolide C-077 (14.1) (O.D. Hensens, unpublished observations) were observed in the ^{13}C NMR spectrum in 20% $CD_3OD/CDCl_3$. All carbons were accounted for,

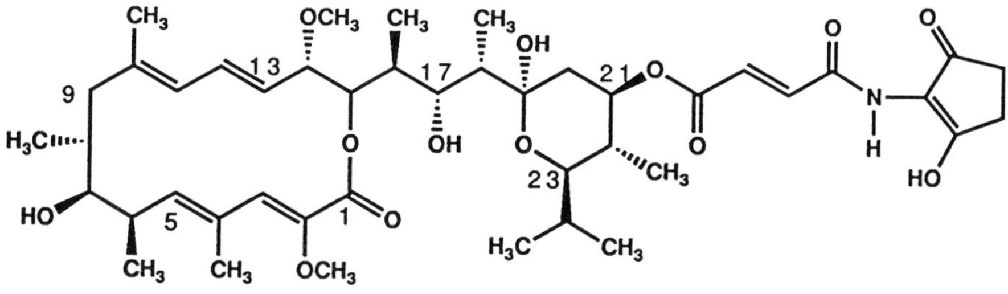

$$(14.1)$$

however, in $CDCl_3$. This was confirmed independently for the identical structures bafilomycin B_1 (Werner et al. 1983, 1984; Baker et al. 1989) and setomycin (Otoguro et al. 1988) as well as for the related macrolide L-155,175 (Goetz et al. 1985). The distinction between the presence of more than one component or a multiplicity of conformations, as in the case of lyngbyatoxin A (Cardellina et al. 1979), rapamycin (Findlay and Radics 1980), teleocidin B, and olivoretins (Sakai et al. 1984), FK-506 (Tanaka et al. 1987, Mierke et al. 1991), ulithiacyclamide (Ishida et al. 1989), and patellin 2 (Zabriskie et al. 1990), can be made in a similar manner by changing the experimental conditions mentioned above.

The number of ^{13}C resonances in a molecule does not immediately infer the correct carbon count and degeneracy in chemical shift position is often the reason. It would therefore appear prudent to record spectra in different solvents under conditions such that the signal intensities accurately reflect the number of carbon atoms. In other cases, the degeneracy may be associated with a symmetry feature of the molecule such as monomeric versus dimeric forms. Performing quantitative ^{13}C measurements, however, is not a trivial task since ^{13}C intensities depend critically on the NOE enhancements observed under proton decoupling, which can vary widely in a given molecule. Because of the necessary long delays (three to five times the longest ^{13}C spin-lattice relaxation time, T_1), using a "gated" pulse sequence without NOE is often impractical for small samples. In practice, carbons with the same multiplicity are compared in the same region of the spectrum with the implicit assumption that comparison is made between carbons having similar T_1s. For example, the ^{13}C NMR spectrum of the antifungal compound L-681,102B (14.2) $C_{25}H_{40}NO_8P$ (O.D. Hensens unpublished observations; Burg et al. 1989, 1991), contains only 23 signals, two of which at 27.1 and 34.2 ppm correspond to two carbon intensities by comparison with five other methylene triplets in the 22 to 41 ppm region, consistent with the presence of a substituted cyclo-hexane. The structure is identical to that recently reported for phospholine (Ozasa et al. 1989) and phoslactomycin B (Fushimi et al. 1989).

$$(14.2)$$

The next step is to determine the proton count. Several methods are currently available to determine carbon multiplicities and, therefore, the carbon-bound proton count. Among these, the *J*-modulated attached-proton-test (APT) (Rabenstein and Nakashima 1979; Le Cocq and Lallemand 1981; Patt and Shoolery 1982; Shoolery 1984; Crews et al. 1985; Madsen et al. 1986; Radeglia and Porzel 1989) and the distortionless enhancement by polarization transfer (DEPT) (Bendall et al. 1982a, 1982b; Doddrell et al. 1982; Pegg and Bendall 1983) sequences appear to be the most often used. For various reasons, DEPT is often preferred for spectral editing because it has reduced dependence on *J*, is not very sensitive to misset pulses or in-homogeneities in the rf, and has definite sensitivity advantages. In addition, one has the ability to pulse faster since the polarization transfer depends on the T_1 of the proton, not the carbon. As a rule, four spectra are acquired with a pulse width of $\pi/4$, $\pi/2$, $\pi/2$, and $3\pi/4$, where the $\pi/2$ experiment is repeated in order to obtain comparable absolute intensities. Clean separation into CH, CH$_2$, and CH$_3$ subspectra can be obtained by linear combinations of the four spectra with computer optimization of the weighting coefficients. Since the experiment is generally optimized to accomplish polarization trans-fer via large one-bond couplings, quaternary carbons disappear from the spectra. It is often sufficient to acquire only two sets of data with pulse widths of $\pi/2$, giving only positively phased CH carbons, and $3\pi/4$, which distin-guishes between inverted methylene carbons and CH and CH$_3$ carbons that have a positive amplitude. The solvent peak is effectively suppressed in all spectra, which may thus advantageously detect nonquaternary carbons obscured by the solvent peak. Some pertinent applications have appeared in the literature (Morris 1986; Komoroski et al. 1986) and the subject was recently reviewed (Sorensen and Jacobsen 1988). In our earlier experience, with the availability of adequate sample quantities (usually >10 mg) and despite the potential for overlapping signals, the acquisition of a conventional "gated" coupled spectrum in combination with a simple APT spectrum using a delay of $1/J$ (e.g., 7 mseconds) has the advantage of simultaneously providing $J_{^{13}C-^{1}H}$ values of many carbons. These carry invaluable information about substituent electronegativity, orientation, and bond strain (Hansen 1981).

A superior methodology for small samples involves a combination of the DEPT sequence and the recent sensitive ^{1}H detected HMQC (heteronuclear multiple-quantum coherence) $^{1}H-^{13}C$ correlation experiment (see Section 14.1.2.2). Without ^{13}C decoupling, one-bond $^{1}H-^{13}C$ coupling constants can be obtained from F_1 (^{13}C) traces of the correlation map (Bax and Subramanian 1986; Cavanagh et al. 1988), which can also overcome the overlap problems associated with 1D gated coupled spectra. The technique was effectively used in the structure determination of the complex macrolide antibiotic desertomycin (Bax et al. 1986).

Having established the number of carbon-bound protons in the molecule, it remains to determine the number of active protons, which is usually less straightforward. ^{1}H NMR methods depend on the assumption that all actives are observable in aprotic solvents. Whereas this usually applies to slowly exchanging amide NH protons, this may or may not be the case for hydroxylic OH, thiol SH, or amine NH protons, depending on strong hydrogen bonding effects, the solution characteristics and experimental conditions. ^{2}H-Exchange experiments are most often used. Saturation transfer experiments can also be used to detect active protons (Von Dreele et al. 1972; Brewster and Hruby 1973; Glickson et al. 1974). In thiostrepton (14.3) for example, irradiation of the residual water peak in $CDCl_3$, resulted in the loss of resonance intensity of all six active OH protons whereas the amide protons were unaffected (Hensens and Albers-Schonberg 1983a).

A method for the detection of active protons using ^{13}C NMR takes advantage of the ^{2}H-isotope induced shift of a ^{13}C resonance on ^{2}H exchange of any active proton in the molecule. 2-Bond (β), 3-bond (γ), and long-range isotope shifts have been observed for a variety of substrates (Hansen 1983, 1988), including carbohydrates (Gagnaire and Vincendon 1977; Pheffer et al. 1979; Bock and Lemieux 1982; Christofides and Davies 1982, 1983a, 1983b; Reuben 1984, 1985b, 1985c; Wise et al. 1984; Hoffman and Davies 1988), alcohols, diols, and polyols (Newmark and Hill 1980; Hansen 1986; Reuben 1986a, 1986b), amines, anilines, and ammonium derivatives (Reuben 1985a, 1987; Nakashima et al. 1986; Yashiro et al. 1986), indoles (Morales-Rios and Joseph-Nathan 1989), nucleosides, and nucleotides (Reuben 1987; Brush et al. 1988), and amino acids and peptides (Feeney et al. 1974; Ladner et al. 1975; Otter et al. 1990). In general, the usually upfield and larger β-shifts can be observed under fast or slow exchange and thus be used for assignment

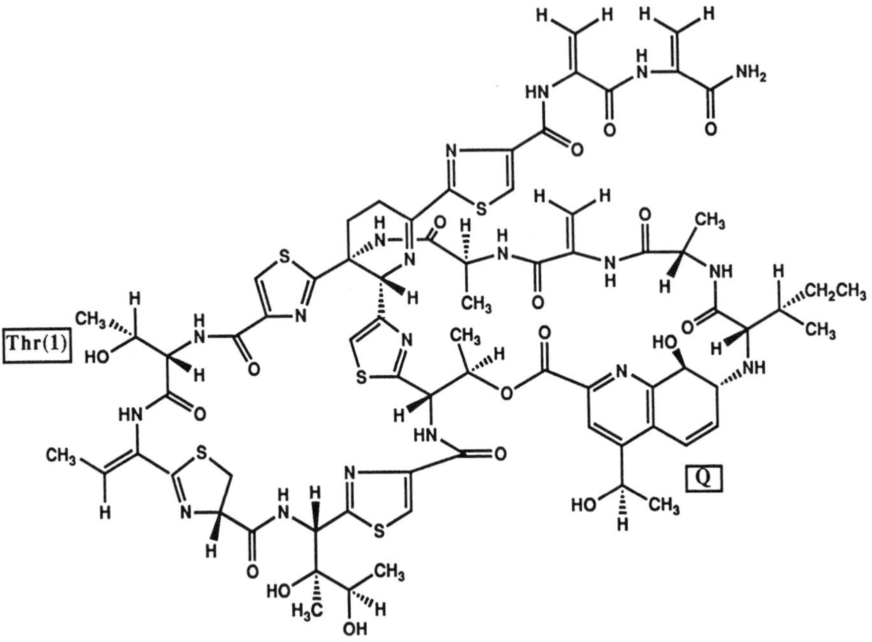

(14.3)

purposes. For instance, slow ^{2}H exchange of the amide NH protons in thio-strepton (14.3) and thiopeptin B_a in 20% $CD_3OD/CDCl_3$ resulted in doubling of the adjacent α- and carbonyl carbons, facilitating their assignment, whereas all alcohol carbons were observed as singlets due to fast exchange (Hensens and Albers-Schonberg 1978, 1983b, 1983c). Insufficient digital resolution of these experiments masked the observance of γ- or long-range effects. That the β-shift is dominant in alcohols has been applied very effectively in distinguishing alcohol from ether or ester linkages in such complex natural products as the polyether yessotoxin (Murata et al. 1987) and the macrolides prorocentrolide (Torigoe et al. 1988), patellazole C (Zabriskie et al. 1988), and tiacumicins (Hochlowski et al. 1987), where solutions in CD_3OD and CD_3OH (or in admixture with a non-protic solvent) were simply compared.

As far as MS methods are concerned for the determination of the number of active protons in a molecule, the formation of trimethylsilyl (TMS) ethers and esters have enjoyed great popularity. MS data for the derivatives are compared with that of the corresponding deuterated d_9-TMS derivatives. The propensity of a functional group in a molecule to silylate depends on a number of factors and the reader is referred to the reviews by Poole (1977) and Anderegg (1988) for further details.

With the molecular weight and ^{1}H and ^{13}C counts in hand, severe restrictions are now placed on the number of possible empirical formulas for a given molecule. This is particularly the case if some knowledge of the elements present has been obtained from either a combustion analysis or by inference from NMR data. The example of the $C5_a$ antagonist L-156,602 (14.4), $C_{38}H_{64}N_8O_{13}$ is illustrative (Hensens et al. 1991a). Positive-ion FABMS initially indicated a molecular weight of 822 (M^+-H_2O) but this was revised to 840 by FABMS in the negative-ion mode, giving by contrast a very strong response. ^{13}C NMR analysis in CD_3COOD indicated a carbon count of 37 with the α-carbon (α–C) of the β-OHLeu residue differentially exchange-broadened whereas 38 were observed in CD_3CN and C_5D_5N. APT and gated

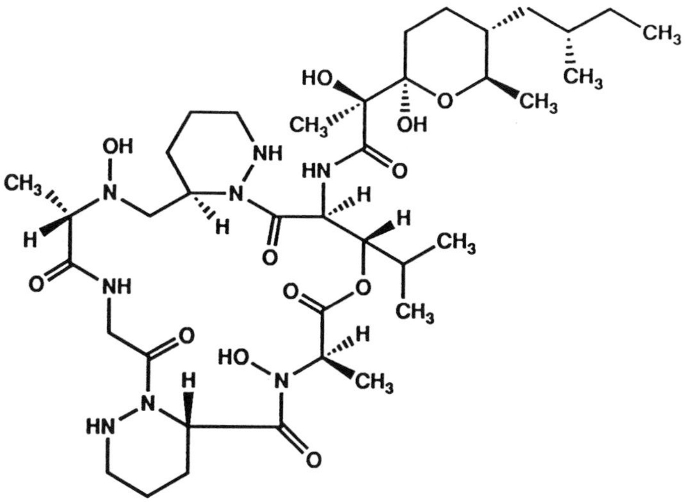

(14.4)

coupled spectra yielded the carbon-bound proton count of 56. The number of eight active protons was determined by ^1H NMR by "titration" of a CD_3CN solution with DMSO-d_6 up to 25% (v/v) and subsequent spiking with CD_3OD. This was established only after it was noted that the somewhat broadened α–H of one of the NOHAla residues also disappeared on addition of CD_3OD. This was consistent with the extensive exchange-broadened signal for the corresponding α–C in the ^{13}C NMR spectrum in CD_3OD (O.D. Hensens, unpublished observations). The NMR data suggested a C/H/N/O compound and from the partial structures, readily favored the empirical $C_{38}H_{64}N_8O_{13}$ over the only other allowed possibility, $C_{38}H_{64}N_{16}O_6$. Subsequent confirmation was obtained by HR-EIMS under fast heating conditions by peak matching on a weak M^+-H_2O peak (found mass to charge ratio (m/z) 822.4493; calculated m/z 822.4487). Within 5 millimass units (mmu) there are 30 reasonable solutions assuming a C/H/N/O compound. Hence, the HRMS data by themselves can only be regarded as consistent with the calculated empirical formula as distinct from rigorously establishing it. This should be kept in mind because journals such as the *Journal of the American Chemical Society* and the *Journal of Organic Chemistry*, for example, set widely varying, acceptable limits in

1991 of ± 3 and ± 13 mmu, respectively, for molecular weights up to 500 and ± 6 and ± 16 mmu, respectively, for molecular weights 500 to 1000.

14.1.2 Determination of Partial Structures

The tremendous success that natural products structure determination has enjoyed in establishing partial structures from ^{1}H–^{1}H connectivity data, whether it is from conventional double irradiation or 2D-COSY-type experiments attests in large part to the existence of a large body of empirical chemical shift and coupling constant data of organic compounds, including natural products as well as the interpretive skills of the natural product chemist. We believe the latter claim is justified because in spite of a large reservoir of information, there is no unambiguous correlation between chemical shift position and functionality (Bhacca and Williams 1964; Szymanski and Yelin 1968; Jackman and Sternhell 1969; Yamaguchi 1970; Batterham 1973; Nakanishi et al. 1974, 1975; Brugel 1979; Sasaki 1985, 1986, 1987). Moreover, despite the dependence of J_{HH} on structure, no absolute or unambiguous correlation exists between the magnitude of the coupling and the number of intervening bonds. Long-range coupling ($^nJ_{HH}$ where n $>$ 3) can generally be observed up to five bonds away (Sternhell 1969; Jackman and Sternhell 1969; Marchand 1982; Platzer et al. 1987) and with some notable exceptions are usually less than 3 Hz, which falls well within the range of many vicinal and some geminal couplings (Bothner-By 1965; Cookson et al. 1966; Bovey 1969). Invaluable as this relatively sensitive methodology has become, it is perhaps fair to say that ^{1}H–^{1}H connectivities provide only an indirect means to establishing the carbon skeleton of an organic molecule.

Included in the development of partial fragments is usually some knowledge of preliminary ^{13}C NMR data discussed in Section 14.1.1 as well as one-bond ^{1}H–^{13}C correlations. One-bond ^{1}H–^{13}C correlation techniques, which by definition are limited to nonquaternary carbons, provide little connectivity data unless used in conjunction with long-range ^{1}H–^{13}C data where such ^{13}C NMR assignments are essential. The most important information the 2D ^{1}H–^{13}C correlation experiment offers is that it can provide a clear distinction between methine and methylene proton positions in crowded regions of a ^{1}H NMR spectrum. This information cannot always be unambiguously obtained from COSY-type experiments and when attempted, is solely inferred from the number and size of the couplings involved, a process that is not unambiguous and difficult in situations where there is significant overlap. Armed, however, with definitive ^{1}H chemical shift positions from ^{1}H–^{13}C correlated spectra, the COSY data become significantly more amenable to interpretation.

With ^{1}H–^{1}H connectivity and ^{13}C multiplicity information in hand, it is often instructive at this stage to first establish consistency between the number and types of protons from ^{1}H and ^{13}C spectra. Various possibilities are then considered in terms of carbon type and combinations of functionalities to account for all heteroatoms, active protons, C—X and C = X bonds as previously illustrated for the ATPase inhibitor L-681,110A$_1$ (fumarate instead of flavensomycinic acid at C21 of (14.1)) (Hensens et al. 1983). The number of rings in the compound can then be determined from the molecular formula and knowledge of the degrees of unsaturation or double-bond equivalents.

14.1.2.1 ^1H–^1H Connectivities.

COSY-Type Experiments. Many versions of the basic COSY experiment have appeared since 1976 when Ernst and co-workers (Aue et al. 1976) translated Jeener's visionary concept of homonuclear proton correlated 2D NMR in 1971 into practice (Martin and Zektzer 1988a). It is evident from a random survey of natural product structure determinations since about 1985, that the basic COSY as well as the double-quantum filtered COSY (DQF-COSY) experiments have been the most popular and that the complementary relayed COSY (RCOSY) and homonuclear Hartman-Hahn spectroscopy (HOHAHA) or total correlation spectroscopy (TOCSY) experiments are becoming increasingly important. The COSY pulse sequence consists of two 90° proton pulses that cause transfer of coherence between scalar coupled spins during an evolution time t_1, incremented to digitize the second frequency domain, followed by an acquisition period

$$90°-t_1-90°-\text{acquire}$$

Many variants of the experiment have been devised to produce a cleaner diagonal and the first such attempt was the COSY-45 where the first 90° pulse in the pulse sequence is substituted by a 45° pulse, reducing the intensity of near-diagonal components. The introduction of phase sensitive spectra provided major advantages in the presentation of COSY spectra (States et al. 1982; Marion and Wüthrich 1983; Keeler and Neuhaus 1985; Nagayama 1986; Wemmer 1989; Traficante 1990a, 1990b; Pelczer 1991) where pure absorption line shapes give improved resolution that is ideal for the extraction of chemical shifts and coupling constants. The other major benefit is the possibility of discriminating between positive and negative intensities, which is especially useful for interpreting cross peaks in such experiments as 2D-NOESY. Many advantages are combined in the phase-sensitive version of the DQF-COSY experiment as discussed below.

Multiple quantum filtration NMR methods were introduced by Ernst and co-workers (Pianti et al. 1982; Mareci 1988; Rance et al. 1989). These techniques selectively remove different spin systems from a COSY plot. During the COSY experiment, the second 90° pulse produces multiple quantum coherence (MQC), which is not detectable. If, however, immediately after the second pulse another pulse is applied, after a minimal fixed delay Δ, the MQC is converted back into observable magnetization.

$$90°-t_1-90°-\Delta-90°-\text{acquire}$$

The order of MQC can be selected according to its coherence transfer pathway by means of appropriate phase cycling. When a DQF is applied, for instance, two-spin systems or higher are selected, filtering out singlets, especially methyl singlets, which are often prevalent in natural products. Because of their intensity they often obscure much weaker proton multiplets near the diagonal; hence, a method for filtering them out is highly desirable. Potential dynamic range problems are also decreased. Moreover, unlike the phase-sensitive COSY experiment that has dispersive diagonal peaks, the phase-sensitive DQF-COSY experiment has a definite advantage in producing well-defined cross and diagonal peaks with near absorption line shapes in both directions (Rance et al. 1983; Morris and Richards 1985). Recently, Derome and Williamson (1990) have recommended a minimum eight-step phase cycling to eliminate rapid-pulsing artifacts where the relaxation delays between successive scans are less than 3 to 4 T_1. Many applications of DQF-COSY experiments to natural products have been reported, including the diterpene aldehyde halitunal (Koehn et al. 1991), several sesquiterpene lactones (Aljancic-Solaja et al. 1988; Fukuyama et al. 1990a), the ursane triterpenoid coleonolic acid (Roy et al. 1990), the aurodox-like antibiotic SB22484 (Ferrari et al. 1990), the polyene macrolide vacidin A (Sowinski et al. 1989), the cyclic peptide anantin (Wyss et al. 1991), the lanthiopeptin PA48009 (Hayashi et al. 1990), and the glycopeptides A40926 (Waltho et al. 1987), UK-68,597 (Skelton et al. 1990), and UK-69, 542 (Skelton et al. 1991).

Relayed Coherence Transfer Experiments.   The popular relayed COSY experiment attributed to Eich et al. (1982) resolves the question of accidental degeneracy in proton chemical shift. The experiment allows magnetization transfer between protons that are both coupled to a common third proton. The pulse sequence for the single relayed coherence transfer experiment (RCOSY) is an extension of the COSY pulse sequence where a third 90° pulse transfers coherence, initially transferred between vicinal protons, to a third neighbor. The pulse sequence takes the form

$$90°-t_1-90°-\tau_m-180°-\tau_m-90°-\text{acquire}$$

where optimization of the coupling constant dependent mixing period τ_m has been worked out by Bax and Drobny (1985). The 180° refocusing pulse removes the dependence on chemical shift coherence transfer and sufficient phase cycling is required to eliminate further transfer beyond the usual 4-bonds (Andersen et al. 1988). Optimization of the relative intensities of the relayed cross peaks was predicted for τ_m settings of $1/(4J_{HH})$, where J_{HH} is the average of the couplings involved in the transfer. In practice we have found values of 0.02 and 0.05 seconds to be satisfactory, which effectively optimize for couplings of 12.5 and 5 Hz, respectively. It should be noted that normal COSY cross peaks appear as well in the 2D plot; hence, a COSY spectrum should always be acquired for comparison.

The pulse sequence was critically put to the test in our recent structure determination of the sesterterpenoid variecolin (14.5) (Hensens et al. 1991b).

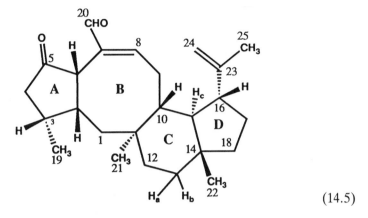

(14.5)

Despite extensive analysis of COSY and direct and long-range ^1H–^{13}C data in CD_3CN, a unique structure was not indicated and the challenge became a choice between the three isomeric structures (14.5), (14.6), and (14.7). This

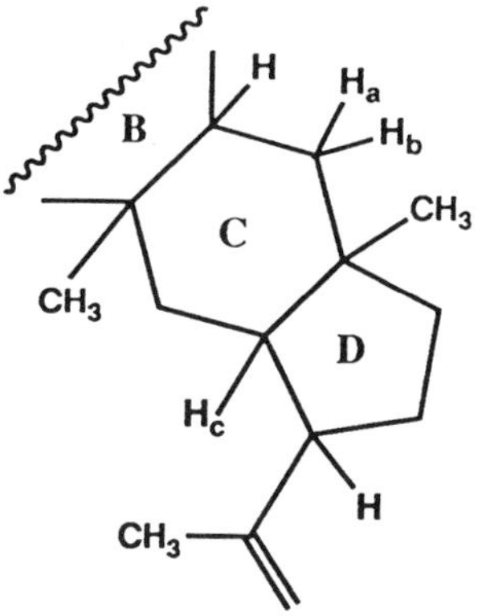

(14.6)

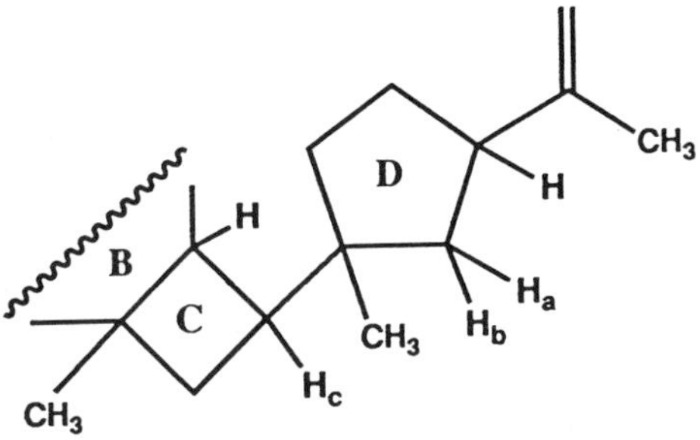

(14.7)

was a direct result of the degeneracy in chemical shift position near δ1.47 of the three protons designated H_a, H_b, and H_c. With a mix time of 0.02 seconds, many long-range correlations were observed (Figure 14–1) but the correlation between H10 and H16 (through H15) in particular, allowed a clear distinction in favor of (14.5) for the structure of variecolin. No correlation between the two protons was observed using a mix time of 0.05, which optimizes for a smaller coupling constant and stresses the need for choosing the correct value for this parameter. This experiment suggested large vicinal couplings between H10 and H15 and between H15 and H16, which was subsequently borne out from analysis of ^1H NMR data in C_6D_6 where H15 was clearly resolved (Hensens et al. 1991b).

Other applications to structure determinations include some cubitane and cembrane diterpenoids (Shin and Fenical 1991), the lantibiotic actagardine (Kettingring et al. 1990), the cyclic peptide anantin (Wyss et al. 1991), the polyether yessotoxin (Murata et al. 1987), and the complex macrolides PA-46101 A and B (Matsumoto et al. 1990).

Care should be exercised in interpretation because coherence transfer can be relayed via long-range couplings (Otter and Kotovych 1986; Kay et al. 1987; Hughes 1988). It is possible to optimize this "anomalous" but useful experiment by extending the RCOSY experiment over 5 bonds by inclusion of a second mixing period and appropriately coined the 2RCOSY experiment. The pulse sequence takes the form

$$90°-t_1-90°-\tau_m-180°-\tau_m-90°-\tau'_m-180°-\tau'_m-90°-\text{acquire}$$

Homans et al. (1984, 1987) have even used the 3RCOSY experiment by including a third mixing time period to obtain 5-bond (H1 → H4) and 6-bond

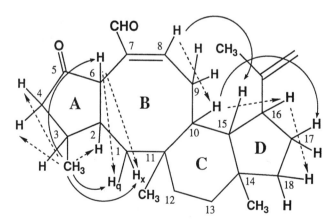

FIGURE 14–1 Single (-·-→) and double (——→) ^1H–^1H relayed connectivities in variecolin (14.5) (Hensens et al. 1991b).

(H1 → H5, H2 → H6) correlations in carbohydrates. In the case of the depsipeptide dolastatin 13, predominantly ^{1}H–^{1}H COSY and ^{1}H–^{1}H single, double, and triple relayed COSY experiments indicated all eight discrete spin-coupled amino acid residues (Pettit et al. 1989). Despite the advantage of the control one has over the coherence-transfer pathway, as demonstrated in peptides and carbohydrates alike, the intensity of the relayed responses suffers from the strong dependence on J and falls off significantly with further mixing periods.

It is apparently for these reasons that the isotropic mixing experiment HOHAHA or TOCSY, which is much less dependent on J (Braunschweiler and Ernst 1983; Davis and Bax 1985), is replacing RCOSY-type experiments in natural product structure determination. The 2D-HOHAHA or TOCSY pulse sequence is similar to that of COSY where the second 90° pulse is replaced by a pulse train such as MLEV-17 (Bax and Davis 1985).

$$90°–t_1–SPINLOCK\ (\tau_m)–acquire$$

Similar to the principle of cross polarization in solids first introduced by Hartman and Hahn, coherence transfer in the rotating frame takes place during the mixing interval or spin-lock, the extent of spin propagation depending on the mixing period τ_m and the magnitude of the coupling between the spins.

Some control over the extent of transfer can be exercised by suitable selection of the mixing time. For small molecules, Martin and Zektzer (1988a) recommend periods of 20 to 50 mseconds for connectivity between protons up to 4 bonds apart, 50 to 90 mseconds for up to 5 or 6 bonds and 100 to 150 mseconds for even longer transfers that can be reasonably interpreted. Caution should be exercised against overreliance in identifying specific resonances in a given spin system, by observing the order in which they appear with increasing mixing time (Cavanagh et al. 1990). The power of the method has been demonstrated in numerous applications reported so far, which include the alkaloids stellettamide A (Hirota et al. 1990) and ptilomycalin A (Kashman et al. 1989), the novel tetramic acid lydicamycin (Hayakawa et al. 1991), the diterpenes oryzalic acid A (Kono et al. 1991), astrogorgiadol and astrogorgin (Fusetani et al. 1989), the anthraquinone glycosides altromycins (Brill et al. 1990), the tannin α-punicalagin (Doig et al. 1990), the depsipeptides variapeptin and citropeptin (Nakagawa et al. 1990), and the cyclic peptides cyanoviridin RR (Kusumi et al. 1987), microviridin (Ishitsuka et al. 1990), and syringostatins A and B (Isogai et al. 1990).

14.1.2.2 ^{1}**H–**^{13}**C One-Bond Connectivities.** The 2D-heteronuclear correlation (HETCOR) experiment, which correlates a ^{13}C nucleus with its attached protons, was introduced by Bodenhausen and Freeman in 1977. This represented a marked improvement over 1D methods available at the time, such as single-frequency-off-resonance-decoupling (SFORD) (Ernst 1966; Reich et al. 1969). However, it is perhaps Freeman and Morris' experiment (1978) and Bax and Morris' modification (1981), allowing phase cycling with quadrature detection in both dimensions, that have been the most popular in their application to natural products. These are basically INEPT-type (insensitive nuclei enhanced by polarization transfer) magnetization transfer experiments where the choice of two $^1J_{CH}$ dependent delays have to be compromised. The first is a defocusing delay that is optimal at $\Delta_1 = 1/(2^1J_{CH})$, whereas optima for the refocusing delay Δ_2 depend on the multiplicity of the carbon. For this reason it is set at $\Delta_2 = 1/(3^1J_{CH})$ where we have used $^1J_{CH}$ values generally in the range of 125 to 150 Hz. In our experience, the sensitivity of the experiment is quite forgiving for methines. Depending on the number of increments required for suitable digitization along the ^1H axis (F_1) and the amount of available spectrometer time (i.e., the number of scans), results can usually be obtained from only a few milligrams of sample at 75 to 125 MHz. The longer T_1s of methyl groups, due primarily to spin rotation, appear to be the main reason for their more difficult detection. It is, however, the methylene carbons that are the most difficult to detect, especially where there is strong coupling. The choice of $\Delta_2 = 1/(3^1J_{CH})$ is almost optimal for a methylene carbon and we have chosen $^1J_{CH}$ values of 125 to 130 Hz to optimize for aliphatic methylenes. It is seldom that complete correlation between both protons and their attached carbon is observed unless at least 20 mg of compound is available, depending on instrumentation. There is therefore ample opportunity for improvement not only in sensitivity but also in achievable digital resolution along both the ^1H (F_1) and ^{13}C (F_2) axes. Since the one-bond HETCOR experiment usually accompanies the long-range version LR-HETCOR, the reader is referred to Section 14.2.1.2 for pertinent examples.

On the sensitivity front, the most important advance has been development of the ^1H detected HMQC experiment via direct coupling (Bax and Subramanian 1986). This experiment requires powerful and stable amplifiers for complete broad-band decoupling of the heteronucleus. Folding of the ^{13}C chemical shift range has therefore been helpful (Bax and Summers 1986; Summers et al. 1986). The challenge of suppressing the enormous ^1H signal not directly attached to ^{13}C has been largely overcome by suitable phase cycling and subtraction routines, as well as the inclusion of a BIRD (bilinear rotation decoupling) pulse applied to ^{13}C at the beginning of the pulse sequence. In contrast to HETCOR, the HMQC experiment is displayed with the ^1H shift along the F_2 axis and that of its directly attached ^{13}C nucleus along F_1. Digitization of the carbon frequency domain therefore results directly from incrementation of the evolution period, which can be a definite drawback for closely spaced ^{13}C NMR signals. Folding of the ^{13}C chemical shift range as noted earlier is therefore often used coupled with linear prediction (Zhu and Bax 1990). Due to the extreme sensitivity of this experiment, there is no reason not to employ a larger number of increments than is usually used in HETCOR with suitable zero-filling. For pertinent examples the reader is referred to references given in Section 14.2.1.2 with the corresponding long-range heteronucleus multiple bond correlation (HMBC) experiment.

14.2 STRUCTURE ELUCIDATION PROCESS: PART B

14.2.1 Determination of Total Structure

It is in the area of long-range correlation methodology that the most progress has been made in assembling partial fragments of a molecule, mainly constructed from ^1H–^1H connectivities, into a total structure. These techniques are particularly powerful in establishing connectivity through nodes of quaternary centers and heteroatoms. In the first category are included techniques that detect long-range interactions between protons that are coupled through either scalar or dipolar (NOE) mechanisms. Both are indirect methodologies since the relationship between connectivity and the number of bonds implicated in the interaction is not unequivocal (see Section 14.1.2.1). In particular, "through-space" NOE correlations have been widely used to bridge isolated spin systems in natural products. To avoid possible ambiguity, but with some notable exceptions, they should be reserved for determining stereochemical conformational assignments wherever possible. It is, however, in the realm of long-range ^1H–^{13}C connectivities that the most dramatic achievements have been realized. The combination of restricted ambiguity and dramatic improvements in sensitivity has projected this methodology into the forefront of modern structural analysis. In particular our survey reveals that the introduction of the ^1H detected HMBC (heteronuclear multiple bond correlation) experiment by Bax and Summers in 1986, has revolutionized natural product structure elucidation and has dramatically shifted the focus from the ^1H to the ^{13}C nucleus. Due to the extreme sensitivity of this experiment, in combination with the corresponding one-bond HMQC experiment, this methodology can now be routinely applied to 1 to 2 mg or submilligram quantities of natural products of moderate size at 400 to 600 MHz. With some exceptions (see Section 14.2.1.2), correlations are limited to 2- or 3-bonds and the data can be analyzed simultaneously with ^1H–^1H connectivities, thus avoiding possible ambiguities inherent in formulating partial structures only from such data (see Section 14.1.2.1). The pronounced shift to the ^{13}C nucleus in structural analysis can be regarded as fundamental, since the analysis now involves manipulation of discrete C, CH, CH_2, and CH_3 building blocks by virtue of multiplicity experiments. In principle, this strategy is inherent to all long-range ^1H–^{13}C shift correlation experiments but because of the dramatic sensitivity of the ^1H detected experiment, this methodology with the requisite instrumentation has become a practical reality. This methodology conceptually approaches the elegant double quantum INADEQUATE experiment (incredible natural abundance double quantum transfer experiment), based on ^{13}C–^{13}C couplings (Bax et al. 1980a, 1981a, 1981b), in terms of its unique and direct mapping capability of the carbon skeleton of an organic molecule. Because of its inherent insensitivity, the experiment does not enjoy widespread usage.

14.2.1.1 ^{1}H–^{1}H Long-Range Connectivities.

J *Coupled Methods.* Whereas the observation of long-range correlations in the COSY experiment is variable, the experiment can be optimized for long-range couplings (LR-COSY) where a fixed delay Δ is inserted at the beginning of the t_1 and acquisition (t_2) periods

$$90°–\Delta–t_1–90°–\Delta–\text{acquire}$$

Values of 0.2 to 0.5 seconds for Δ have, in our experience, proved useful, revealing couplings (<1.0 Hz) that are often difficult to detect by double irradiation experiments. Recent applications of the technique include the diterpene aldehyde halitunal (Koehn et al. 1991), the polyhydroxylated lactone discodermolide (Gunasekera et al. 1990, 1991), a polyhydroxylated piperidine glucoside (Evans et al. 1985), the 18-membered macrolides tiacumicins (Hochlowski et al. 1987), the triterpenoid coleonolic acid (Roy et al. 1990), and the triterpenoid galactosides pouosides A-E (Ksebati et al. 1988).

Despite their restricted use in natural product ^{1}H–^{1}H correlation studies, it has been amply demonstrated that double quantum COSY (DQ-COSY) (Mareci and Freeman 1983) and zero quantum COSY (ZQ-COSY) experiments (Bax et al. 1980b; Pouzard et al. 1981; Blumich 1984; Muller 1984; Hall and Norwood 1987a, 1987b) are particularly useful in the detection of small couplings. The reason for their limited use is not immediately apparent since other advantages of these experiments include the absence of a crowded diagonal and in the case of the ZQ-COSY experiment, intrinsically higher resolution. Barnekow et al. (1989) have successfully applied the ZQ-COSY experiment to the structure determination of some marine sesquiterpenoids not only for long-range but 2- and 3-bond couplings as well. Optimization of the experiment was discussed as well as alternative ways in which the data can be presented graphically. To compensate for their greater dependence on J_{HH}, Martin and Zektzer (1988a) have recommended running two experiments of DQ-COSY with different mixing times, resulting in some editing of cross peaks, which may be particularly desirable for crowded regions. Although somewhat less sensitive, a scheme that purports to create uniform excitation of double quantum coherence largely independent of J, should make this technique more attractive for natural products (Limat et al. 1988). Other applications of DQ-COSY to compound classes include nucleotides (Hore et al. 1982; Hanstock and Lown 1984), oligosaccharides (Dabrowski et al. 1987), glycopeptides (Fesik et al. 1985), and peptides and proteins (Macura et al. 1984; Boyd et al. 1985; Rance et al. 1985; Otting and Wüthrich 1986).

If a long-range rather than vicinal interaction is suspected, the assignment can be aided by well documented geometric constraints such as *W* and *Zig-Zag* configurations for 4- and 5-bonds, respectively, for a variety of systems. This can be particularly diagnostic for the identification of axial methyl singlets in ring systems such as steroids or terpenoids where a *W* coupling path to some axial proton is available (Platzer et al. 1987; Schneider et al. 1985). Examples include the sesterterpenoid variecolin (14.5) (Hensens et al. 1991b), and the triterpenoids galactosides pouosides A–F (Ksebati et al. 1988) and olean-12-ene-3β,22β,24-triol (Steffens et al. 1983).

Allylic and homoallylic couplings in unsaturated or aromatic systems are well documented and may also be used in the construction of partial fragments or in their assembly to a full structure across quaternary or heteroatoms. There are numerous examples in the natural products literature involving couplings to vinyl methyls as well as allylic methylene and methine protons. It is the latter that facilitate connectivity through sp^2 quaternary carbons, as will be illustrated for the triene macrolide phosphate oxydifficidin (14.8) (O.D. Hensens, unpublished observations; Wilson et al. 1987). Partial fragments Me–C4–C13, C14–C15, C17–C23, and C25–C27 were readily deduced

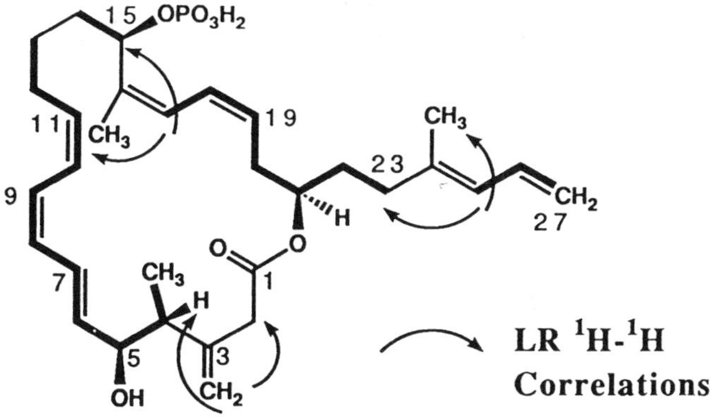

(14.8)

from ^{1}H–^{1}H decoupling experiments of the dephosphorylated derivative in CDCl$_3$ spiked with CD$_3$OD. Long-range correlations between the two vinyl methyls at C16 and C24 with H17 and H25, respectively, and the critical long-range couplings between H17 and H15 and between H25 and the allylic methylene protons at C23 established the contiguous chain Me–C4–C27. In addition, allylic coupling between the exocyclic methylene protons and the C4 methine and C2 methylene protons allowed placement of the remaining ester carbonyl group at C1, thereby completing the C1–C27 chain of oxydifficidin with ring closure at C21.

Dipolar Coupled (NOE) Methods.   As noted previously and for the sake of rigor, NOE experiments, which give indications of distance-dependent, "through-space" dipolar interactions, should be mainly reserved for providing stereochemical and conformational details. It is outside the scope of this review to deal with this comprehensive subject and the reader is referred to the recent text by Neuhaus and Williamson (1989) for a complete discussion of the various techniques available and their advantages and disadvantages. The manifestation of NOEs in many cases is intimately dependent on conformation, and unless substantial details of a structure are known, their interpretation can be fraught with uncertainty and ambiguity. For example, with the exception of the methylene protons, the strongest NOE cross peaks in the phase-sensitive NOESY (PS-NOESY) spectrum of variecolin (14.5) between H6 $\longrightarrow$ H10 and H7 $\longrightarrow$ H12 (Figure 14–2), are completely conformation dependent and provide conflicting information in terms of J connectivity during the early stages of the structure determination of variecolin (Hensens et al. 1991b).

Occasions where the NOE can be used to gain structural information include largely planar structures, sequencing of peptides and carbohydrates, and in situations generally, where a choice between structural possibilities has to be made. For example, NOE experiments were critical in determining the structure of the radical scavenger benthocyanin A (14.9) (Shin-ya et al. 1991). ^{1}H and ^{13}C NMR data suggested the presence of a 1,2,3-trisubstituted benzene, a phenyl group, a geranyl residue, and two isolated sp^{2} protons

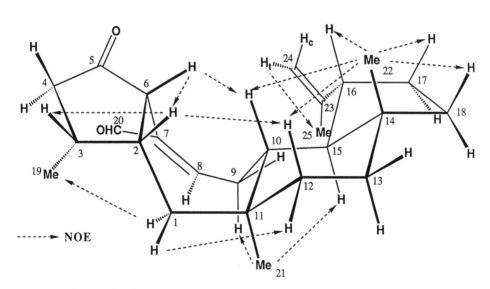

FIGURE 14–2 Conformation and relative stereochemistry of variecolin (14.5), indicating NOEs (––→) from phase-sensitive NOESY experiments. Reprinted with permission from O.D. Hensens, D. Zink, and J.M. Williamson, et al. (1991b) "Variecolin, a Sesterterpenoid of Novel Skeleton from *Aspergillus variecolor*," *J. Org. Chem.* 56(10), 3399–3403. Copyright 1991 American Chemical Society.

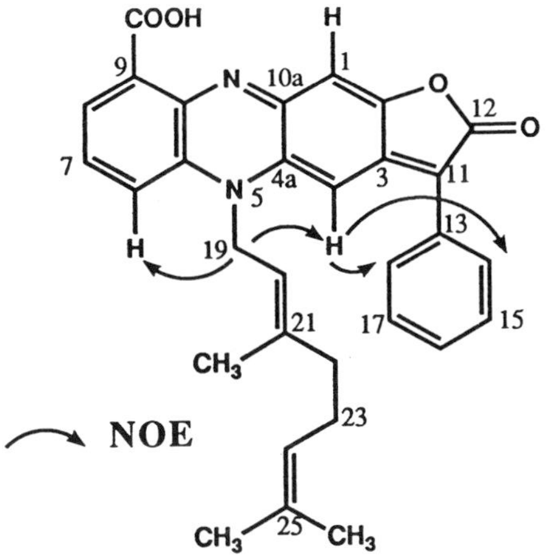

(14.9)

where the geranyl residue was linked to a nitrogen atom. Structure (14.9) was one possibility suggested on the basis of HMBC experiments but could not be distinguished from another alternative of having the COOH group at C6. Irradiation of the C19 methylene protons of the geranyl group resulted in signal intensity increases of H6 and H4, while H4 in turn showed NOE to the *ortho* protons of the phenyl ring consistent with structure (14.9). Another possibility (14.18) was ruled out on further interpretation of the HMBC data (see Section 14.2.1.2).

NOEs have also been used in establishing the positions of ether linkages in polyethers. For example on the basis of COSY and one-bond $^1H-^{13}C$ HETCOR data, three partial structures for the polyether macrolide goniodomin A were deduced (Murakami et al. 1988) and sequenced as shown in Figure 14–3A by using correlation via the long-range $^1H-^{13}C$ coupling experiment (COLOC) (see Section 14.2.1.2). On the basis of the NOEs between H2 and H6, H7 and H15, and 9–CH$_3$ and H15 (Figure 14–3B), the tetrahydrofuran rings C2 to C6, C7 to C11, and C11 to C15 were proposed. The remaining oxygen must connect to C21 and C24 where, because of the *trans* relationship of the H21 and H24 protons, no NOE was observed. It will be appreciated that in the absence of the sequencing COLOC data, the NOEs, in particular between H7 and H15 and between H15 and the 9–CH$_3$, do not readily lead to such a conclusion because they are conformation dependent. This methodology was initially developed during the elegant structure elucidations of the complex polyether toxic principles for yessotoxin (Murata et al. 1987), brevetoxin A (Pawlak et al. 1987), and pamamycin-607 (Kondo et al. 1987). Application of the technique proved essential for the structure determinations of the nitrogenous macrolide prorocentrolide (Figure 14–4) (Torigoe et al. 1988) and the complex polyether ciguatoxin (Murata et al. 1989). This was particularly critical in the latter case where long-range $^1H-^{13}C$ HMBC experiments were precluded because of the small, 0.35-mg sample size.

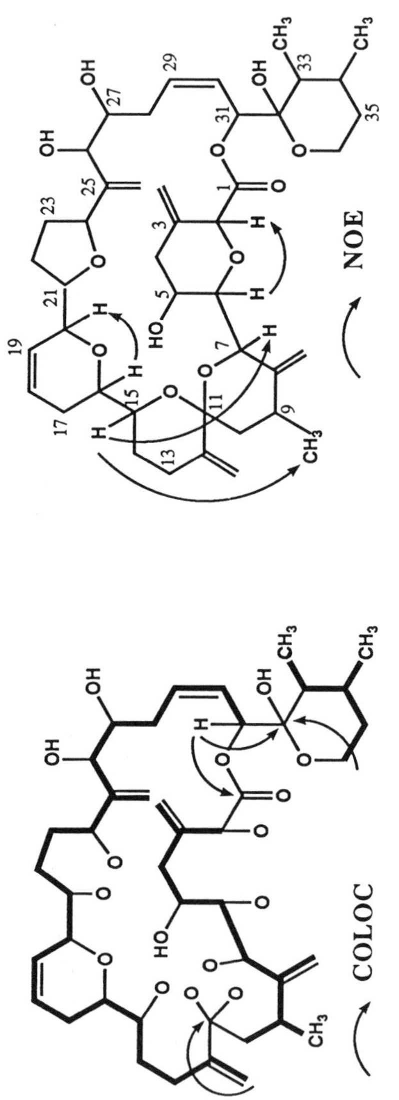

FIGURE 14–3 Construction of skeleton of goniodomin A using COLOC (see A) and NOE (see B) experiments (Murakami et al. 1988).

14.2.1.2 ^{1}H–^{13}C Long-Range Connectivities. Analysis of long-range ^{1}H–^{13}C correlation spectra is most beneficial when the chemist is armed with some knowledge of the magnitude of the coupling. This particularly holds true when the coupling mechanism may be expected to deviate from the usual 2- to 3-bond pathway. The reader is referred to the general reviews of Hansen (1981) and Marshall (1983) as well as an earlier review by Ewing (1975) dealing only with $^{2}J_{CH}$ couplings. Although couplings in aliphatic systems over about 5 to 6 Hz are usually vicinal and obey a Karplus-type dependence, $^{2}J_{CH}$ couplings can be appreciably larger in olefins. In aromatics, $^{3}J_{CH}$ is usually larger than $^{2}J_{CH}$ and can be readily distinguished on this basis. Couplings over four or more bonds are usually restricted to highly conjugated systems (Hansen 1981; Kunz et al. 1984; Koole et al. 1984; Benassi et al. 1987; Williamson et al. 1989; Kuhnt et al. 1990; Sawada et al. 1990) although they have been observed in simple aromatics (0.7 to 1.9 Hz) and α,β-unsaturated carbonyl derivatives (Hansen 1981).

Powerful as this widely applicable methodology of ^{13}C-detected and especially ^{1}H-detected ^{1}H–^{13}C long-range correlation experiments has become, it nevertheless has its shortcomings. By implication, this technique depends on each carbon being strategically located usually 2- or 3-bonds away from a proton in the molecule. This is not always the case and is often complicated in situations where a number of quaternary carbons or heteroatoms are contiguous or where the couplings into quaternaries are very small. In the latter case, depending on the size of the coupling, techniques such as HMBC or selective-INEPT (see as follows) may provide the necessary information because they can be fine-tuned for small couplings. Otherwise, the chemist often relies heavily on chemical shift arguments and comparison with appropriate model compounds. Given that the INADEQUATE experiment lacks sensitivity, it is exactly in these situations, however, that the technique would potentially be most useful.

The other obvious ambiguity arises if coupling beyond the usual 2- to 3-bond range is observed. These situations may increase as more sensitive and optimized methods for unravelling small couplings are developed.

The need to distinguish between 2- and 3-bond correlations is not always as serious as would at first appear. In many instances, when all long-range correlations are considered, including COSY and relayed ^{1}H–^{1}H connectivity data, a self-consistent interpretation of the data in terms of a unique structure often results. This is particularly facilitated by the presence of methyl groups of polypropionate- or isoprenoid-derived classes of natural products such as terpenoids, macrolides, and polyethers where assignments of 2- and 3-bond couplings involving the methyl protons can be made unequivocally (Seto et al. 1987, 1988a, 1988b).

Many of the examples in this section were chosen to highlight both the successes and deficiencies of the various techniques, in order to alert the chemist to potential pitfalls and ambiguities of interpretation.

Two-Dimensional ^{13}C Detection Methods. The long-range version of the HETCOR experiment introduced by Freeman and Morris (1978) provided the first opportunity to apply long-range ^{1}H–^{13}C correlations in a general sense to structure determination. In principle, the experiment allows the simultaneous detection of all long-range ^{1}H–^{13}C couplings in a molecule. From our random survey of natural product structure determinations since 1985, this experiment still enjoys great popularity as do the Kessler-Griesinger (1984) COLOC and less so the Reynolds et al. (1985) XCORFE (X-nucleus correlation with fixed evolution time) experiments, although all three are being challenged by the recent introduction of the Bax-Summers inverse ^{1}H detected HMBC experiment with its dramatic improvement in sensitivity as previously noted. The 1D long-range selective proton decoupling (LSPD) experiment of Seto et al. (1978) and the increasingly popular selective-INEPT experiment of Bax (1984) have also been extensively used.

In general, the HETCOR experiment has been optimized for $^{n}J_{CH}$ values in the range of 7 to 12 Hz (Martin and Zektzer 1988a) and our studies have shown that sample size often dictates the choice of the delay. For small samples we have obtained good results when the experiment is optimized for values near 10 Hz. For larger sample sizes a choice of smaller values (5 to 7 Hz) proved useful for detecting correlations characterized by smaller couplings. In most cases, therefore, it is necessary to offset optimization of defocusing and refocusing delays Δ_1 and Δ_2 against relaxation losses of signal due to the longer delays. The situation is complicated by the fact that one-bond modulation of the long-range response occurs when magnetization is being transferred to a protonated carbon. This is markedly influenced by the degree of protonation of the carbon, the multiplicity of the carbon due to long-range interactions, and the size of the couplings. Under certain circumstances it was shown both theoretically and experimentally that optimization even for 10 Hz can lead to negligible signal intensity. It is therefore critical to suppress or eliminate these modulation effects and this has been accomplished in part by employing BIRD pulses or low-pass J-filters (Martin and Zektzer 1988b; Krishnamurthy and Casida 1988a, 1988c; Atta-ur-Rahman et al. 1988; Tchouankeu et al. 1990). The modified pulse sequence was effective in corroborating the structure of the furanoid bisnorditerpenoid malabarolide (Atta-ur-Rahman et al. 1988) and in the assembly of partial fragments for the novel heptanortriterpenoid limonoid entilin A and its acetate B (Tchouankeu et al. 1990).

Moreover, modulation of long-range responses can also occur via other long-range couplings to the carbon. Krishnamurthy and Casida (1988b) have provided guidelines for choosing optimal delays as a function of J and Δ_1 values by using a modified pulse sequence with a BIRD pulse midway through the defocusing delay, Δ_1. There may, however, be an advantage in performing the experiment without broad-band decoupling during acquisition. Despite a factor of 2 loss in sensitivity, Krishnamurthy and Nunlist (1988) have shown that the standard heteronuclear shift correlation pulse sequence, modified with a one-step low-pass J filter before evolution and a BIRD pulse midway through Δ_2 and retaining ^1H coupling in the F_2 dimension, revealed the presence of critical long-range correlations in the spectrum of rotenone by comparison with its decoupled version. It was argued that, provided sufficient quantities are available, it is more advantageous to sacrifice sensitivity and remove the refocusing delay Δ_2 in this manner than to miss correlations altogether due to modulation effects.

It is not always clear from the literature what particular variant of the LR-HETCOR pulse sequence was used in a particular structure determination, but we speculate that the Bax-Morris (1981) modification of the original Morris-Freeman experiment predominates for most of the following applications. Significant quantities (~20 mg) are usually required but the power of the method in constructing a total structure from partial fragments can be clearly seen in recent examples, which include the acylpeptide antibiotics cepafungins (Terui et al. 1990), the polyketide heterocycle mycothiazole (Crews et al. 1988), the related marine sponge antifungals mycalamide A (Perry et al. 1988) and onnamide A (Sakemi et al. 1988), the pentacyclic aromatic alkaloids plakinidine A and B (Inman et al. 1990), the diterpene adenanthin (Xu et al. 1987), two diterpenes having the trinervitene skeleton (Goh et al. 1988), and the macrolides rhizoxin (Iwasaki et al. 1984), gabonolides A and B (Achenbach et al. 1985), and prorocentrolide (see Figure 14–4) (Torigoe et al. 1988).

The structure determination of the macrocycle polyether prorocentrolide (see Figure 14–4) is typical (Torigoe et al. 1988). ^1H–^1H COSY and ^1H–^{13}C one-bond HETCOR experiments in dimethyl sulfoxide DMSO-d_6 established the partial sequences as indicated by bold lines in Figure 14–4. Long-range ^1H–^1H connectivities across quaternary carbons were also used in their construction, particularly the allylic couplings between the exocyclic protons H51 and the protons at C8 and C10, H28/H30, H40/H42, and between the homo-allylic protons at C44 and C49 (see Figure 14–4A). LR-HETCOR experiments allowed the remaining linkages to be completed as indicated by the arrows in Figure 14–4B, in particular the construction of the aza-decalin ring system. The positions of the OH groups were established by ^2H-isotope induced shifts of their attached carbons (see Section 14.1.1), which allowed the location of the three ether linkages in B, from a consideration of the degree of unsaturation and number of rings from the molecular formula. These ether linkages and the geometry of the double bonds were firmly established from PS-NOESY data (see Section 14.2.1.1).

Attempts to circumvent relaxation-dependent signal loss and optimization difficulties in the Morris-Freeman experiment resulting from the length of fixed defocusing and refocusing delays Δ_1 and Δ_2, in combination with the incremented evolution time duration, has led to the development of several experiments employing constant evolution times. Of these, the Kessler-Griesinger (1984) COLOC and more recent Reynolds et al. (1985) XCORFE experiments are most noteworthy. Optimization procedures for the COLOC

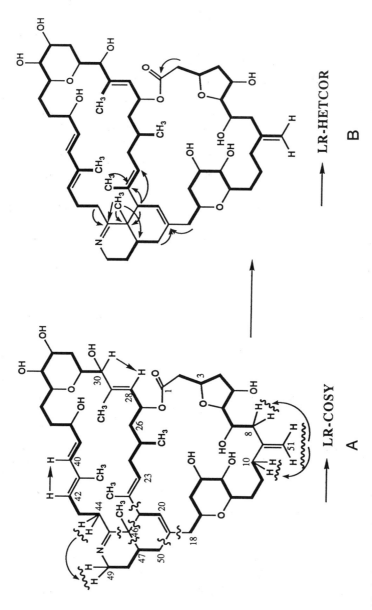

FIGURE 14–4 Construction of prorocentrolide skeleton from LR-COSY (see A) and LR-HETCOR (see B) experiments (Torigoe et al. 1988).

experiment have been described by Kessler et al. (1984) and improved by Perpick-Dumont et al. (1987) using a separate optimization sequence CINCH-C (calibrated indirect carbon hydrogen correlation spectra-COLOC). However, as Martin and Zektzer (1988b) have pointed out, a faster approach may be simply to repeat the experiment with different settings since optimization of all polarization transfers from ^1H to ^{13}C cannot be achieved with one experiment. The Δ_1 delay is incorporated into the evolution period and the experiment may be optimized as long as Δ_1 is longer than half the evolution time t_1, i.e.,

$$\Delta_1 > NI \times 1/2SW = t_1/2$$

where NI is the number of increments and 1/2SW is the t_1 evolution time increment in F_1. The expression does place restrictions on optimization of the experiment, especially for long-range couplings if high digitization in F_1 (^{13}C axis) is to be achieved simultaneously with a restricted chemical shift range. This has not restricted its application to natural products but it is not yet clear whether it has a sensitivity advantage over the Morris-Freeman experiment. Examples of the COLOC sequence that show its utility in constructing a full structure from partial fragments include a prodigiosin antibiotic (Laatsch et al. 1991), the aldose reductase inhibitor WF-2421 (Nishikawa et al. 1991), the polycyclic guanidine alkaloid ptilomycalin A (Kashman et al. 1989), and the diterpenoids rotalin A and B (Corriero et al. 1989), spongiolactone (Mayol et al. 1987), the fuscosides A through D (Shin and Fenical 1991), and several members of the cubitane and cembrane classes (Shin and Fenical 1991). In order to avoid one-bond modulation of the long-range response, as experienced with both the COLOC and the Morris-Freeman experiments, Krishnamurthy and Casida (1987) devised the COLOC-selective pulse sequence (COLOC-S), which combines the advantages of both pulse sequences. Several long-range couplings absent in the COLOC experiment were uncovered with the modified sequence when applied to two insecticides, ryanine and 1*R-cis*-phenothrin (Krishnamurthy and Casida 1987).

The Reynolds et al. (1985) XCORFE experiment is a variant of the COLOC pulse sequence and is similarly optimized (Perpick-Dumont et al. 1987). Several versions exist and the experiment is unique in that it is the only long-range ^1H–^{13}C correlation experiment that discriminates between $^2J_{CH}$ and $^3J_{CH}$ correlations via vicinal ^1H–^1H couplings. The $^3J_{CH}$ responses will appear as singlets in F_1 (^1H axis) whereas $^2J_{CH}$ responses are observed as doublets provided that sufficient digitization is available in F_1 without appreciable losses of signal due to spin-spin relaxation dependent homogeneity broadening (T_2^*). Care must be exercised, however, since lack of splitting of the signal may also reflect a near zero vicinal coupling. Discrimination on this basis applies only to protonated carbons because traces through quaternary carbons appear as singlets for both 2- and 3-bond couplings. This is indeed unfortunate because distinction between $^2J_{CH}$ and $^3J_{CH}$ involving protonated carbons, though useful, can often be made in conjunction with ^1H–^1H connectivity data, whereas there is no such recourse when quaternary carbons are involved.

Surprisingly, the XCORFE pulse sequence has had limited application in natural products but has been very useful where employed. Some applications include the phenolic glycoside acteoside (Numata et al. 1987), a novel tricyclic spiroketal, muamvatin (Roll et al. 1986), the diterpenoid alkaloid tatsirine (Zhang et al. 1990), the indole alkaloid aspidocarpine and cyanogen bromide reaction product (McLean et al. 1987), methyl 10,11-dihydroxy-3,7,11-trimethyl-2,6-dodecadienoate (Jacobs et al. 1987), triterpenoids (Reynolds et al. 1986), and a novel tetranortriterpenoid, guyanin (Jacobs et al. 1986; McLean et al. 1988).

Recently, the value of the FLOCK sequence was demonstrated in the structure elucidation of several cembrane and pseudopterane diterpenoid metabolites (Chan et al. 1991). Of the 68 possible 2- or 3-bond correlations in tobagolide (14.10), 57 were observed, which was sufficient to unequivocally establish the structure. The experiment demonstrated total suppression of cross peaks from one-bond connectivities, which can complicate assignments in complex molecules. The FLOCK sequence is a modification of the Morris-Freeman pulse sequence and incorporates three BIRD pulses. This gives the experiment significantly improved sensitivity and suppression of one-bond cross peaks over XCORFE and COLOC-S pulse sequences but retains some ability to distinguish between 2- and 3-bond couplings (Reynolds et al. 1989).

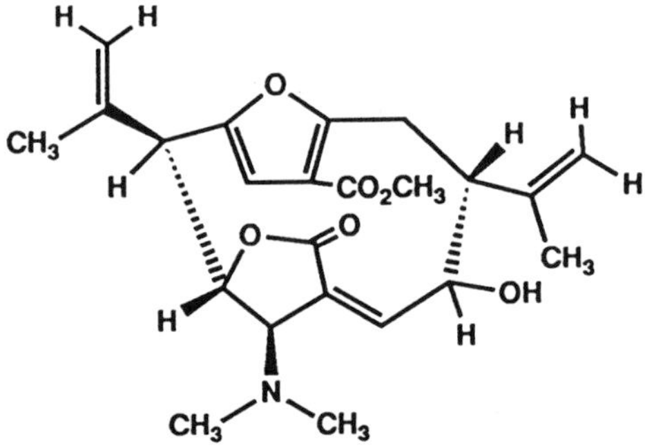

(14.10)

Two-Dimensional ^1H Detected Methods. As previously mentioned, the long-range HMBC version of the inverse ^1H detected HMQC experiment represents a dramatic improvement in sensitivity over ^{13}C detected experiments. The pulse sequence devised by Bax and Summers (1986) lacks the BIRD pulse of the HMQC sequence and incorporates two delays. The first Δ_1 ($= 1/(2^1J_{CH})$) interval serves as a low-pass J filter where typically $1/(2^1J_{CH}) = 140$ Hz followed by a $\Delta_2 = (1/(2^nJ_{CH}))$ delay. After Δ_2, long-range multiple quantum coherence is created by the application of a 90° ^{13}C pulse. The experiment can thus be fine-tuned for the particular coupling involved and is particularly useful for detecting small couplings. It is therefore beneficial, if time permits, to carry out the experiment with two or three settings for Δ_2

optimized for couplings of 4, 7, and 10 Hz, using an interpulse delay 1.5 times the longest proton T_1. A larger value of 8 to 10 Hz is usually optimal if time is limited. In case of extensive overlap in either the ^1H or ^{13}C dimension, the experiment is often run in two or more solvent systems. For best results in nulling the proton signals not coupled to ^{13}C nuclei, a T_1 experiment should be performed on each sample. The same considerations of digitization in the F_1 (^{13}C axis) and F_2 (^1H axis) dimensions apply as for the HMQC experiment (see Section 14.1.2.2). In our experience (G.L. Helms and O.D. Hensens, unpublished observations), long-range ^1H–^{13}C correlations can now be routinely obtained by ^1H detected experiments at 500 MHz on 1 to 2 mg of sample and even submilligram quantities for natural products of small to moderate size.

While writing this chapter, a comprehensive, general review of indirect ^1H detection methods, with extensive applications to natural products, has appeared (Martin and Crouch 1991) to which the reader is referred for further details. In addition to the more detailed examples illustrated below, recent applications that demonstrate the power of the technique in assembling partial fragments into a full structure include the spirolactone paecilospirone (Hirota et al. 1991), the sesquiterpene lactone illicinolide A (Fukuyama et al. 1990b), the sesquiterpene-neolignan clovanemagnolol (Fukuyama et al. 1990a), the diterpene aldehyde halitunal (Koehn et al. 1991), the antimicrobial diterpene oryzalic acid A (Kono et al. 1991), the carbazole antiostatins (Mo et al. 1990), the phosphate δ-lactones phospholine (14.2) (Ozasa et al. 1989) and phoslactomycins (Fushimi et al. 1989), the polyhydroxylactone discodermolide (Gunasekera et al. 1990, 1991), and the novel tetramic acid lydicamycin (Hayakawa et al. 1991).

An example in which the HMBC experiment demonstrated dramatic improvement over the LR-HETCOR experiment in our hands, was observed in determining the sequence of the C5$_a$ antagonist L-156,602 (14.4) (Hensens et al. 1991a). With widely varying experimental conditions of solvent (pyridine-d$_5$, CD$_3$CN, CD$_3$COOD, DMSO-d$_6$), concentration (5 to 45 mg/0.5 ml) and temperature (20 to 65°C), few correlations were observed for the backbone of the hexadepsipeptide in the LR-HETCOR experiment optimized for 8 to 10 Hz. We attribute this in part, to exchange broadening processes of the backbone in contrast to segmental motion of the C$_{14}$ side-chain. The sharp ^{13}C resonances of the latter moiety and observation of a sufficient number of long-range ^1H–^{13}C correlations in the LR-HETCOR spectrum, allowed elucidation of its novel tetrahydropyranylpropionic acid (THPP) structure, despite extensive overlap of the ^1H NMR resonances in the upfield region of the spectrum. Optimum results were obtained in pyridine-d$_5$ at 45 and 65°C. Optimizing for 8 Hz, HMBC experiments on only 4 mg of peptide in CD$_3$CN or pyridine-d$_5$ furnished sufficient correlations at 500 MHz (G.L. Helms and O.D. Hensens, unpublished observations), with signal to noise (S/N) to spare, to establish the sequence unequivocally as indicated in Figure 14–5. This confirmed the previous x-ray crystallographic results (Hensens et al. 1991a).

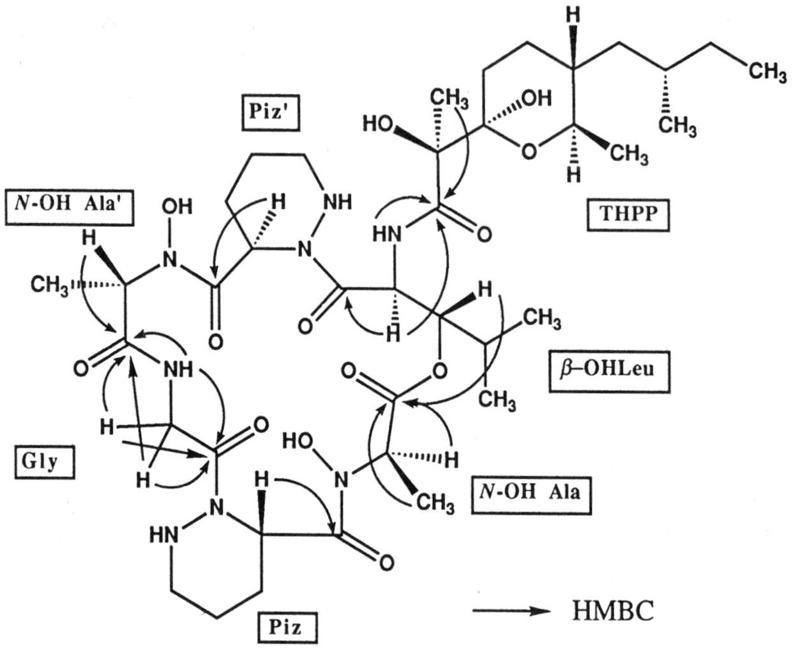

FIGURE 14–5 Sequence determination of the hexadepsipeptide L-156,602 (14.4) using the HMBC technique (O.D. Hensens and G.L. Helms, unpublished observations).

The strategy was similarly demonstrated for the related hexadepsipeptides variapeptin and citropeptin (Nakagawa et al. 1990).

The fascinating structure elucidation of the polythiazole-containing bicyclic peptide cyclothiazomycin (14.11) by Seto and colleagues (Aoki et al. 1991a, 1991b), is a case in point where the apparent lack of critical couplings, because of their small magnitude or relaxation effects, precluded the usefulness of the HMBC technique in the final assembly of the five partial fragments A through E (Figure 14–6) into the complete sequence. Here we have an example where the NOESY experiment was carefully used to obtain both sequence and stereochemical/conformational information in conjunction with acid hydrolysis experiments. The latter gave 1 mol each of Gly, L-Thr, L-Pro, and L-Asp, a minimum of 3 mol of L-Cys and two thiazole-containing fragments (14.12) and (14.13), where (14.12) was identified as saramycetic acid. Sequencing the fragments A $--\rightarrow$ D, C $--\rightarrow$ D, C $--\rightarrow$ E, and A $--\rightarrow$ E followed readily from the NOEs shown in Figure 14–7. Further arguments involving the structures (14.12) and (14.13) (see Figure 14–6) required that fragment D was linked to fragment B through a bond between terminal A and B. Consequently, fragment B was connected to fragment E through a bond between C and D, suggesting the bicyclic structure (14.11). However,

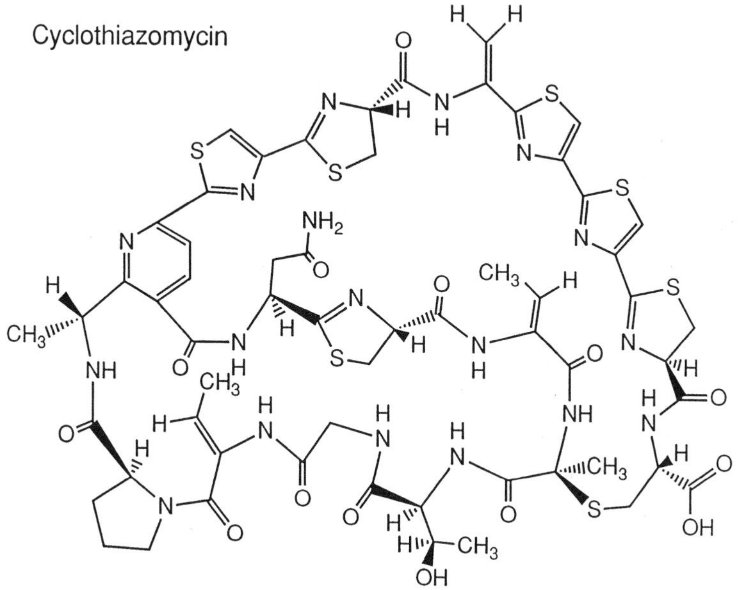

Cyclothiazomycin

(14.11)

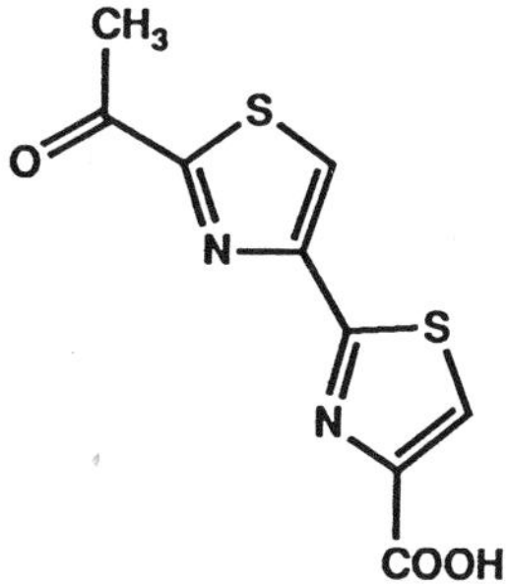

(14.12)

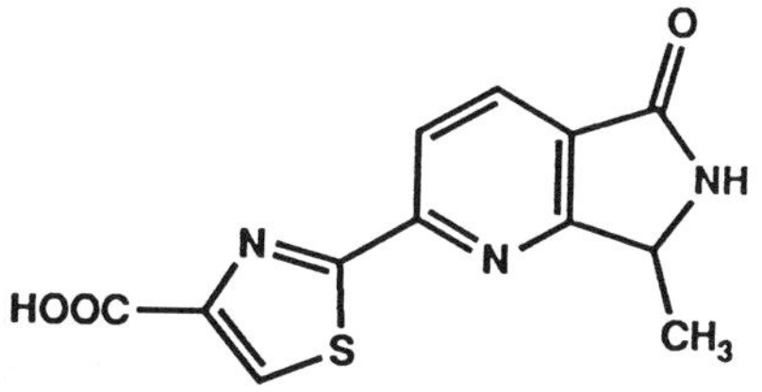

(14.13)

the power of the HMBC technique was clearly demonstrated in the construction of the fully elaborated partial structures A through E where correlations are shown by the arrows in Figure 14–6. The results were obtained from experiments in two solvent systems, namely, ^{13}C-depleted dimethyl formamide DMF-d$_7$ and CD$_3$OH–H$_2$O (9:1). All correlations are either through 2- or 3-bonds with the exception of the 4-bond CH3 $--\rightarrow$ CO correlation in the 2-aminobutenoic acid residue of fragment C, not unexpected from previous studies of αβ-unsaturated carbonyl derivatives (Hansen 1981).

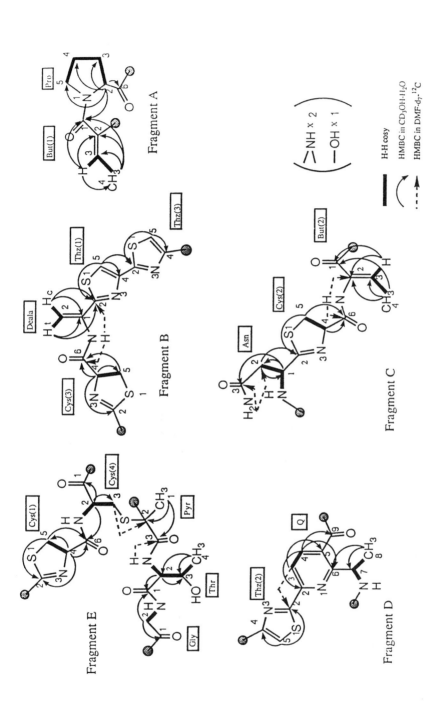

FIGURE 14-6 Construction of partial structures A through E of cyclothiazomycin (14.11) using $^1H-^1H$ COSY (----) and HMBC (—→) experiments. Reprinted with permission from *Tetrahedron Lett.* 32(2), 217–220, M. Aoki, T. Ohtsuka, Y. Itezono, et al. (1991a), "Structure of Cyclothiazomycin, a Unique Polythiazole-Containing Peptide with Renin Inhibitory Activity. Part 1. Chemistry and Partial Structures of Cyclothiazomycin," copyright 1991, Pergamon Press plc.

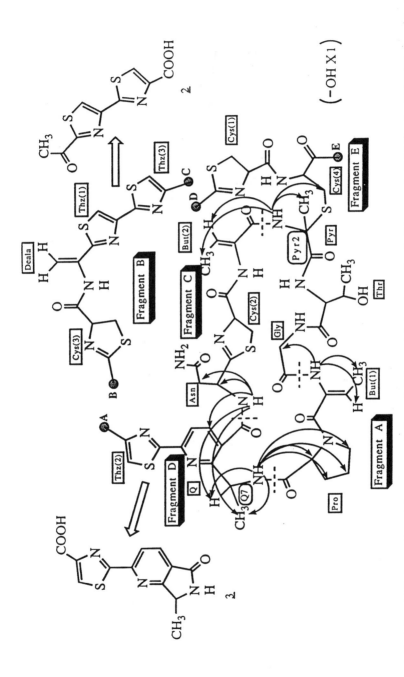

FIGURE 14-7 Partial linkage of partial structures A through E into total structure of cyclothiazomycin (14.11) using NOESY experiments in CD₃OD–H₂O (9:1). Reprinted with permission from *Tetrahedron Lett.* 32(2), 221–224, M. Aoki, T. Ohtsuka, Y. Itezono, et al. (1991b), "Structure of Cyclothiazomycin, a Unique Polythiazole-Containing Peptide with Renin Inhibitory Activity. Part 2. Total Structure," copyright 1991, Pergamon Press plc.

Whereas all NH protons could be assigned in $CD_3OH–H_2O$, few 2-bond correlations into their adjacent carbons, including carbonyls, were observed. On the other hand, only five NH protons were assigned (But(2), Deala, Cys(4), Thr, and Asn(NH$_2$)) in DMF-d$_7$ whereas the remainder showed no correlation peaks because of broadening (H. Seto, personal communication). This accounts for the fact that the linkage of fragments Q $--\rightarrow$ A, A $--\rightarrow$ E, E $--\rightarrow$ C, and C $--\rightarrow$ Q could not be completed through their NH protons. It is also interesting to note that results from experiments optimized for both small (5 and 6 Hz) and large couplings (8 and 10 Hz) (H. Seto, personal communication), allowed Thz(1) to be linked to Thz(3) in fragment B through the 3-bond coupling H5 $--\rightarrow$ C2. However, the corresponding correlations between Thz(2) and L-Cys(3) and between Thz(3) and L-Cys(1), were difficult to detect above the noise level in both solvents. This would otherwise have allowed completion of the attachment of fragment B to the bicyclic portion through Q and E.

The intricate structure determination of the major antitumor agent ecteinascidin 743 (14.14) from the Caribbean tunicate *Ecteinascidia turbinata* by two different groups, illustrates that in situations characterized by a lack of

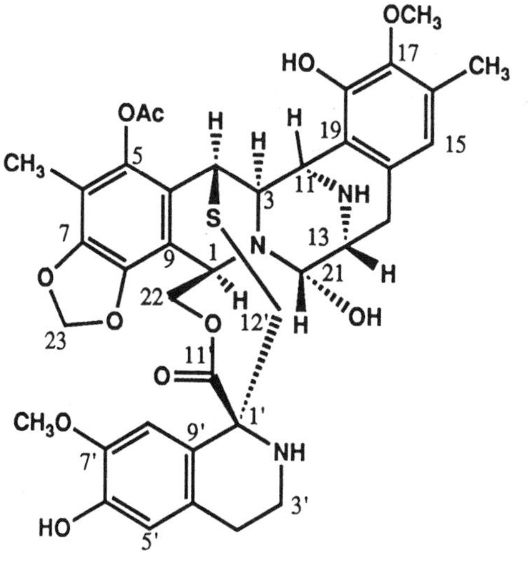

(14.14)

critical HMBC correlations, heavy reliance on chemical shift arguments may be necessary for the assembly of partial fragments into a total structure (Rinehart et al. 1990; Wright et al. 1990). The technique was essential in constructing the aromatic units A, B, and C (Figure 14–8) and clearly demonstrated the overlap of A and B. No protons or carbons in unit C were correlated with those in unit A or B or with the remaining and unassigned CH_2 and sulfide (S) units. The attachment of the ester carbonyl to C1' and spanning C1' of unit C and C4 of unit B with the CH_2–S moiety was therefore primarily accomplished on the basis of chemical shift arguments of C1', C11',

C12′, and C4 (see Figure 14–8). Both papers give detailed discussion on this point as well as consideration of alternative structures. The relative stereochemistry follows readily from ROESY experiments and is the same as in safracins and saframycins (Rinehart et al. 1990). The previously proposed structure in (14.15) (Wright et al. 1990) is chemically reasonable based on the apparent molecular ion M^+-H_2O from positive-ion FABMS measurements, although no confirmatory HMBC correlations between H21 and carbons of fragment C or between H3′ab and C21 were observed. Under negative-ion FABMS conditions however, the Illinois group had observed a molecular ion at m/z 760, consistent with (14.14). Both structures may be thought of as the addition products of water or the isoquinoline 2′-NH amine to the insipient $\Delta^{2(21)}$ imine function, respectively.

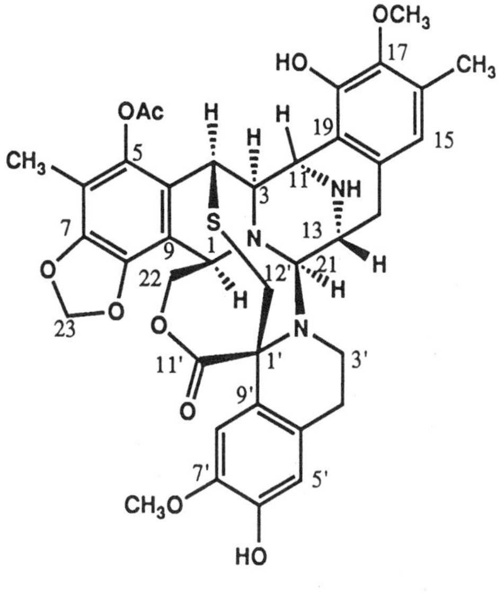

(14.15)

One-Dimensional ^{13}C *Detection Methods.* In spite of the high information content of 2D-heteronuclear shift correlation experiments and dramatic sensitivity gains in recent years, it is apparent from the literature that there are still occasions, dictated primarily by the type of instrumentation available, where only a few connectivities are required and where time-saving, 1D methods are useful. One such time-honored method is the LSPD (long-range selective proton decoupling) technique devised by Seto and colleagues (Seto et al. 1978; Takeuchi et al. 1977) before the advent of 2D-NMR techniques. The technique continues to be used for obtaining long-range ^1H-^{13}C correlations in natural products despite its lack of sensitivity. LSPD has an advantage over some recent methods in that the experiment does not require sophisticated instrumentation for its implementation. The experiment involves weak single-frequency ^1H irradiation where the decoupler power, usually 7 to 10 mgauss, is significantly less than the 0.4 to 0.5 gauss typically

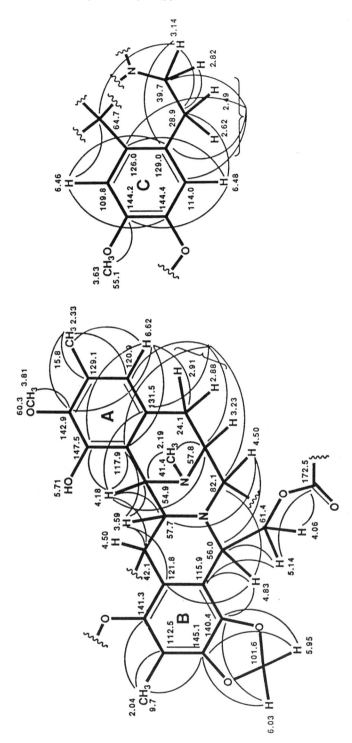

FIGURE 14–8 Construction of partial fragments of ecteinascidin 743 (14.14) from LR ¹H–¹³C correlation experiments (HMBC). Reprinted with permission from K.L. Rinehart, T.G. Holt, N.L. Fregeau, et al. (1990), "Ecteinascidins 729, 743, 759A, and 770: Potent Antitumor Agents from the Caribbean tunicate *Ecteinascidia turbinata*," *J. Org. Chem.* 55(15), 4512–4515. Copyright 1990 American Chemical Society.

required for collapse of the one-bond coupling. The technique, which may be regarded as the forerunner of more recent selective excitation experiments like SINEPT (selective-INEPT), involves the recording of a fully coupled ^{13}C spectrum and notes the simplification or collapse of long-range multiplet structure on selective irradiation of a particular proton. The technique has had considerable success in solving natural product structure problems, and some of its more recent applications include two bis-bibenzyl macrocycle marchantins (Tori et al. 1985), the isotetracenone antibiotics kerriamycin A, B, and C (Hayakawa et al. 1985), several benzonaphthacene quinone antibiotics benanomicins (Gomi et al. 1988, Kondo et al. 1990), the 1,8-diazaanthraquinones diazaquinomycins A and B (Omura et al. 1983), the novel lactone antibiotic kazusamycin B (Funaishi et al. 1987), and the antifungal phosphate phosphomycin C (Tomiya et al. 1990).

The need for a more sensitive 1D approach has been satisfactorily met by the introduction of the selective INEPT experiment of Bax that can optimally give a ^{13}C signal enhancement approaching a factor of four (Bax 1984; Bax et al. 1984). The experiment is variably referred to as SINEPT (Lin and Cordell 1986) or INAPT (Bax et al. 1985) and should be distinguished from the pulse sequence previously coined selective INEPT by Seto and co-workers (Furihata et al. 1983), which detects one-bond couplings. The Bax experiment is a variant of the refocused INEPT pulse sequence using a soft proton pulse ($\gamma H_2 \sim 20$ Hz) that affects predominantly the magnetization of a preselected proton. The same considerations for optimizing the defocusing Δ_1 and refocusing Δ_2 delays as in the LR-HETCOR experiment apply. The individual spectra only show those carbons that are long-range coupled to the particular proton that is irradiated.

This technique that in general requires only a few milligrams of sample, has been widely exploited by Cordell and co-workers and has recently been reviewed (Blasko and Cordell 1989; Cordell 1988). Recent applications to natural products include the coumarinolignans daphneticin, propacin, and aquillochin (Lin and Cordell 1986), the naphthoquinone larreantin (Luo et al. 1988), the depsidones auranticins A and B (Poch and Gloer 1991), a novel flavan-3-ol (Kim and Kinghorn 1987), several indole alkaloids (TePaske et al. 1989, 1990; Lin et al. 1989; Gogoll and Plobeck 1990), oxacanthin alkaloids (Pettit et al. 1988), the sesquiterpenoid eupatorenone (Ananvoranich et al. 1989), the diterpenoids tatsirine (Zhang et al. 1990) and prionitin (Blasko et al. 1988), the norditerpenoid xestenone (Northcote and Andersen 1988), the clerodane diterpenoids salvifolin and tilifodiolide (Rodriguez-Hahn et al. 1990), and the novel triterpenoid galactosides, pouosides A-E (Ksebati et al. 1988).

The technique was used recently to distinguish between two candidate structures (14.16) and (14.17) for the triterpenoid glycoside xestovanin A, proposed mainly on COSY and NOE data (Northcote and Andersen 1989). In two experiments optimized for a $^{n}J_{CH}$ of 7 Hz, irradiation of the secondary methyl group resulted in strong polarization transfer to the tertiary carbinol

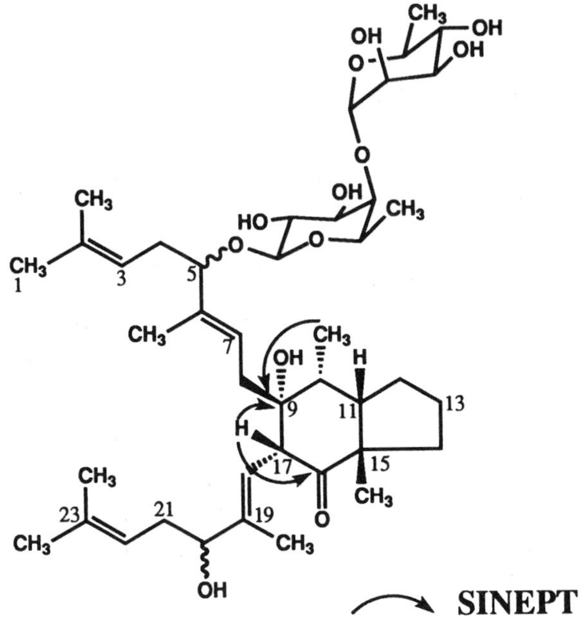

SINEPT   (14.16)

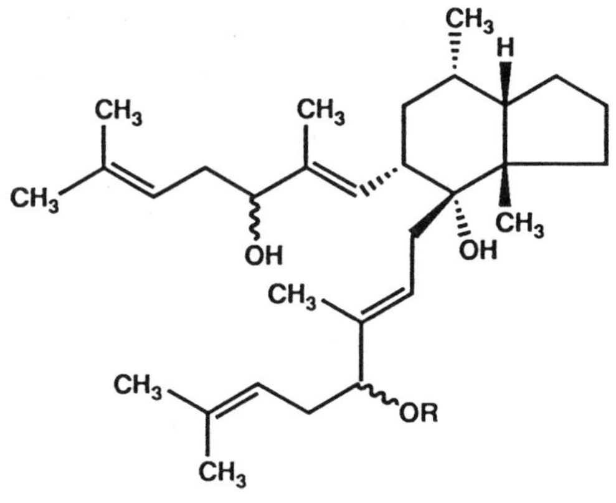

(14.17)

carbon (C9) whereas H17 showed correlations to both C9 and the ketone carbon (C16), clearly in favor of structure (14.16). An interesting example where the SINEPT experiment was successful over the HMBC experiment was recently reported during the structure determination of the ecteinascidins, discussed above (Wright et al. 1990; Rinehart et al. 1990). The presence of the hemiaminal linkage in ecteinascidin 743 (14.14) was readily confirmed by its rapid reaction with methanol to give the 21-OMe derivative. However, long-range correlation between the methoxyl protons and C21 could not be demonstrated in the HMBC experiment even when the position was labeled with ^{13}C enriched methanol (Wright et al. 1990). Correlation was, however, readily observed in the SINEPT experiment.

When suitably optimized for small couplings, the selective INEPT experiment is capable of detecting much weaker long-range couplings routinely as demonstrated in the structure determinations of the carbazole alkaloid tubingensin B (Figure 14–9A) (TePaske et al. 1989) and the unusual, polycyclic bis-ketal preussomerin A (Figure 14–9B) (Weber et al. 1990). In tubingensin B (see Figure 14–9A), two correlations between *para* positions of the indole ring were observed whereas in the bis-ketal (Figure 14–9B), an unusually large number of 4-bond correlations and one 5-bond correlation between H1 and C7 were detected, which complicated the structure elucidation. The 4-bond coupling pathways between H1/C4 and H1′/C4′ involve both sp^2 and sp^3 carbons and are reminiscent of the appreciable homoallylic ^{1}H–^{1}H couplings in 1,4-cyclohexadienes of dihydroaromatic systems (Rabideau 1978). Despite the large number of 51 long-range ^{1}H–^{13}C correlations observed, none were detected between the top and bottom half of the molecule and necessitated the connection of the two units through three oxygens in the only reasonable way possible as shown (see Figure 14–9B). The proposed structure including stereochemistry was confirmed by x-ray diffraction analysis (Weber et al. 1990). Four-bond ^{1}H–^{13}C couplings can be larger in heterocyclic systems (Riand and Chenon 1981; Chimichi et al. 1982; Denisov et al. 1988) and may therefore be detectable in less optimized experiments. Care should thus be exercised in eliminating possible alternative structures in aromatic systems where such correlations are possible. The 4-bond coupling between H11 and C4a in the alternative structure (14.18) considered for benthocyanin A (14.19) (see Section 14.2.1.1), was therefore not compelling evidence for its exclusion as much as the absence of correlations between H1 and the lactone carbonyl and between H4 and any of the phenyl carbons in (14.19) (Shin-ya et al. 1991).

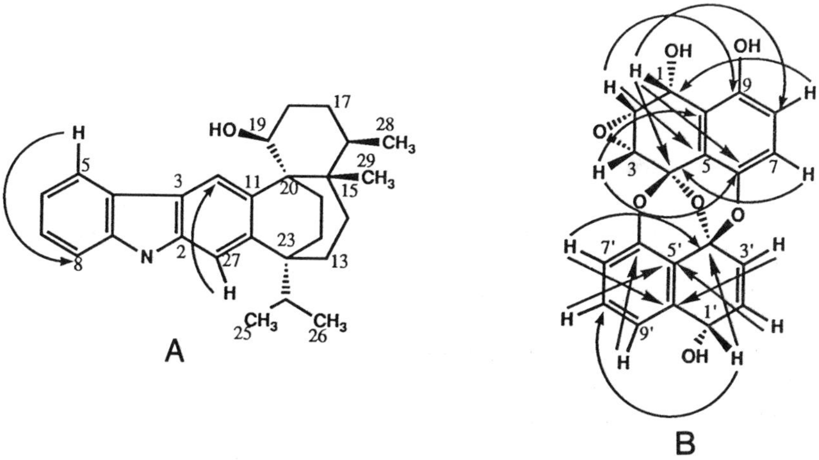

FIGURE 14–9 Four-bond ^{1}H–^{13}C couplings (SINEPT) in tubingensin (see A) (TePaske et al. 1989) and preussomerin A (see B) (Weber et al. 1990).

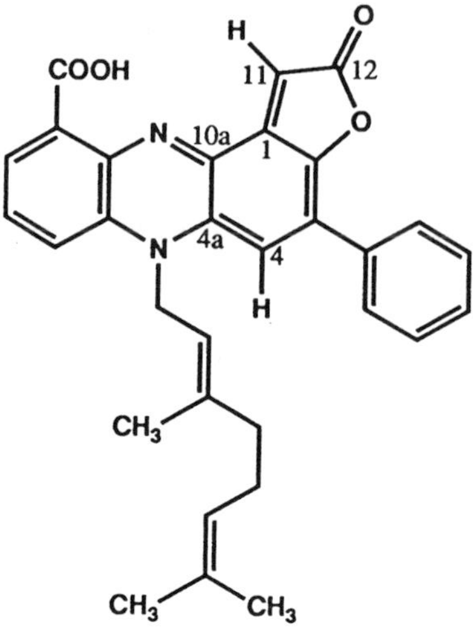

(14.18)

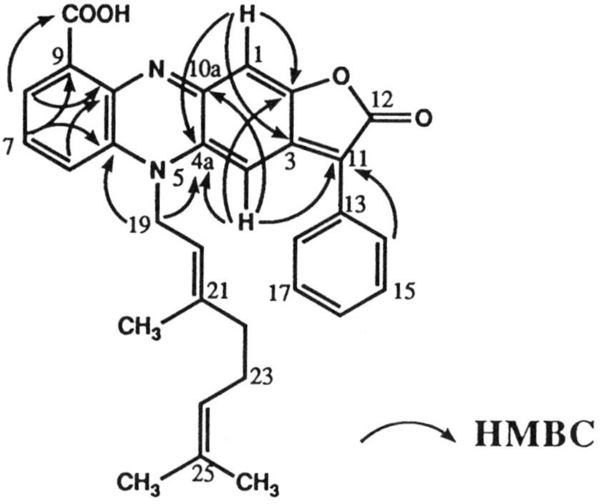

HMBC

(14.19)

14.3 CONCLUSION AND FUTURE TRENDS

The advent of NMR spectrometers with waveform generators will undoubtedly create a surge of renewed interest in selective excitation (Warren and Silver 1988; Geen and Freeman 1991; McDonald and Warren 1991). Hence several 1D analogs of both 2D homonuclear and heteronuclear correlation experiments have already been reported, offering considerable improvement in terms of time, resolution and sensitivity for more specific types of information (Bauer et al. 1984; Davis and Bax 1985; Subramanian and Bax 1987; Kessler et al. 1986, 1988b, 1989, 1990a; Davis 1989; Hall and Norwood 1990; Tellier et al. 1990; Xu et al. 1990; Wollborn et al. 1990; Martin and Crouch 1991). Three-dimensional or four-dimensional NMR methodologies, which have only recently played a significant role in establishing the 3D structures of macromolecules, have not as yet been applied to structure determination of natural products as far as we are aware. This is probably due to a number of factors, not the least being the considerable amounts of time required to obtain adequate digitization in the two labeling time domains (t_1 and t_2) in addition to the conventional acquisition time domain (t_3), for a 3D NMR experiment to be superior to a series of carefully selected 1D and 2D experiments. It is potentially here that the development of 1D or 2D analogs of the 3D experiment, by limiting sampling in one or two of the dimensions by selective excitation (Griesinger et al. 1987; Freeman et al. 1988; Friedrich et al. 1988; Vuister et al. 1988; Boudot et al. 1990; Homans 1990; Kessler et al. 1990b; Majumbar and Hosur 1990; Mueller et al. 1991; Nuzillard and Massiot 1991), may have application to complex natural products structure determination.

It is probably fair to say that given a choice, one needs a strong reason from a sensitivity standpoint not to use indirect detection methods for establishing ^1H–^{13}C correlations, especially the long-range HMBC version. Since the acquisition time domain F_2 now advantageously contains the less dispersed proton shifts, sufficient digitization in F_1 can usually be overcome by increasing the number of increments, t_1, by suitable folding of the ^{13}C sweep width (Bax et al. 1986; Summers et al. 1986), linear prediction in F_1 (Zhu and Bax, 1990), or by selective or semiselective excitation (Berger 1989; Bermel et al. 1989; Davis 1989; Crouch and Martin 1991; Martin and Crouch 1991). Very recent ^1H detected editing schemes (Davis 1990; Kessler et al. 1989, 1990b) and relayed experiments such as HMQC-COSY, HMQC-TOCSY, and HMQC-NOESY/ROESY (not treated in this chapter), also appear to have great promise (Martin and Crouch 1991). Of the ^{13}C detected experiments, only the selective INEPT experiment has demonstrated its utility with limited sample quantities. Because of the extreme usefulness of the HMBC experiment, it is highly recommended that for optimal results, several experiments be run under different experimental conditions (e.g., solvent or temperature), optimized for different couplings and that these experimental conditions be reported. Many of the publications we have reviewed do not mention the coupling for which the experiment was optimized. The tremendous success of this methodology so far has largely depended on detection of correlations characterized by reasonable coupling constants (probably >2 Hz), which has restricted discussion largely to 2- or 3-bond correlations. But as the detection of small couplings of less than 1 Hz improves, it is essential to have some notion of the magnitude of the coupling constant involved, in order to make structural sense out of assigning a 2- or 3-bond coupling and, importantly, to provide grounds for entertaining the possibility of a correlation of 4-bonds or more. This applies in particular to aromatic, dihydroaromatic, heterocyclic, and conjugated systems where 2- and 4-bond coupling constants often fall in the same range and where such possible ambiguity has already been demonstrated by the selective INEPT technique (TePaske et al. 1989; Weber et al. 1990). The development of sensitive techniques to facilitate the measurement of long-range ^1H–^{13}C couplings for specific situations in structure determination as well as in conformational analysis, is therefore highly desirable and remains a fertile area for research in natural products (Pratum 1991; Pratum et al. 1988; Bermel et al. 1989; Titman et al. 1989, 1991; Kessler et al. 1990a; Crouch and Martin 1991; Martin and Crouch 1991; Poppe and Halbeek 1991a, 1991b; Richardson et al. 1991) and macromolecules alike (Edison et al. 1991; Zuiderweg and Fesik 1991).

There is some reason to voice concern over the manner in which some structure elucidations of nontrivial natural products are reported. Because of space limitations, the need to be brief or for other reasons, editorial policies of some journals have not always been in the best interests of the field. There is no reason to believe that many of the natural product structures reported in the literature at large are in error, but the lack of discussion of possible alternative interpretations of the data and the systematic and methodical "weeding-out" of the correct solution is somewhat disturbing. Evidence is often presented "in support of" or "consistent with" a proposed structure, which is not the same as an unequivocal structure proof. As pointed out in this chapter, every ^1H–^1H and ^1H–^{13}C connectivity experiment has conceptionally, as well as in practice, its drawbacks and ambiguities, and despite the many advances that have been made, empirical chemical shift and coupling constant arguments that dictate substituent electronegativity, orientation, and bond strain, are still an integral part of the structure elucidation process. Hence, despite the apparent unequivocal manner in which the data are often presented, especially a structure depicted with many arrows representing these connectivities, it is incumbent upon authors to discuss possible alternative solutions consistent with the data and for journal editors and their referees to allow such discussion to be reported. Moreover, computer methods (Gray 1986) should be applied more widely to ensure that all possible structures are considered, especially for systems characterized by few connectivities and alternative carbon type assignments. In practice, this applies more to polycyclic and bridged ring-systems traditionally found, for example, in terpenoids, polyethers, and alkaloids than to largely linear systems including peptides and macrolides.

Despite remarkable additions to the structural chemist's arsenal of spectroscopic techniques during the last decade, nature has responded in kind and continues to provide increasingly larger, more structurally and stereochemically complex and biosynthetically intriguing structures that, without recourse to degradative studies or x-ray analysis, present unique challenges to the spectroscopist. The fascinating structures of such natural products as esperamicin A_1 (14.20), representative of the enediyne antitumor antibiotics esperamicins and calicheamicins (Golik et al. 1978a, 1987b; Lee et al. 1987a, 1987b), and the novel macrolides FK-506 (14.21) (Tanaka et al. 1987) and N-acetylsporaviridins (14.22) (R_1 = H/OH, R_2 = OH/NHAc. R_3 = Me/Et) (Kimura et al. 1987a, 1987b), are only a few recent examples. Moreover,

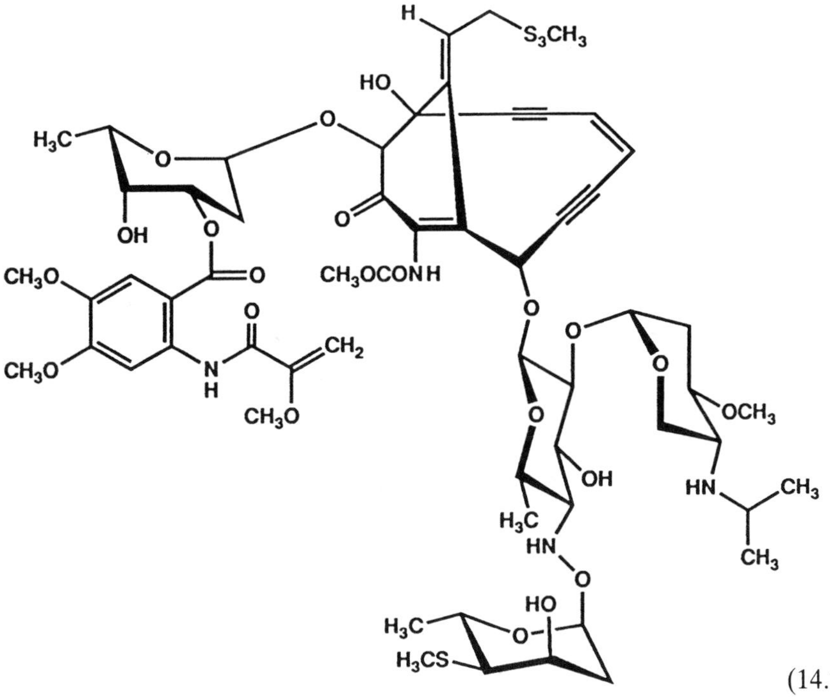

(14.20)

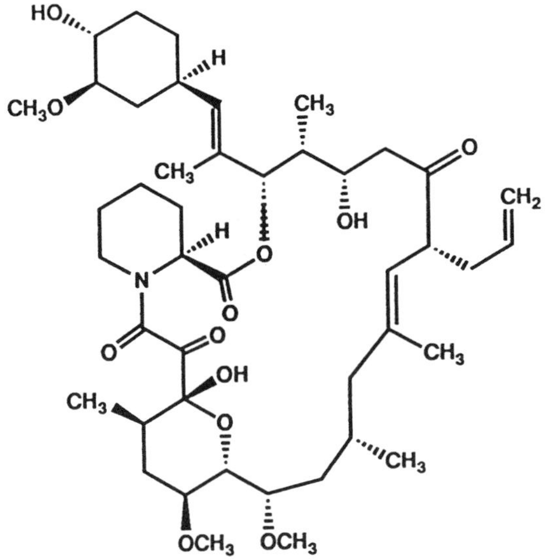

(14.21)

structures need not be large to be spectroscopically challenging, as exemplified by the tetracyclic polyketal (14.23) (Bittner et al. 1987) and the triterpene dilactone pseudolarolide E (14.24) (Chen et al. 1990), and demonstrated in the structure revision reports of sporol (14.25) (Ziegler et al. 1988) and aza-

(14.22)

(14.23)

(14.24)

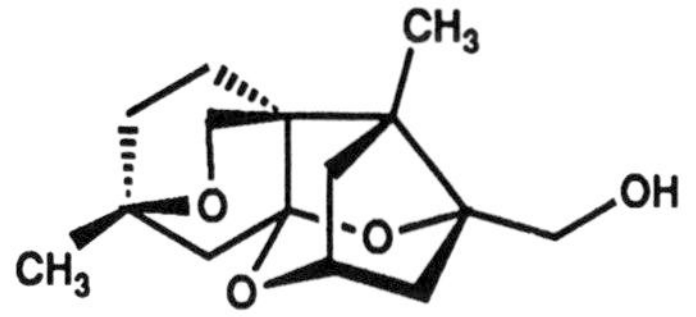

(14.25)

dirachtin (14.26) (Kraus et al. 1985, 1987; Bilton et al. 1987; Turner et al. 1987). It is therefore evident that the days of the structural natural products chemist are not yet numbered and that further advances in spectroscopic methodology are essential to meet nature's formidable challenges.

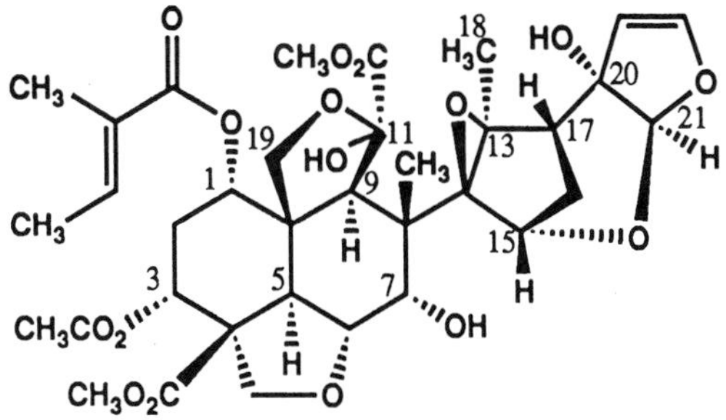

(14.26)

REFERENCES

Achenbach, H., Muhlenfeld, A., Fauth, U., and Zahner, H. (1985) *Tetrahedron Lett.* 26(50), 6167–6170.

Aljancic-Solaja, I., Milosavljevic, S., Djermanovic, M., Stefanic, M., and Macura, S. (1988) *Magn. Reson. Chem.* 26, 725–728.

Ananvoranich, S., Likhitwitayawuid, K., Ruangrungsi, N., Blasko, G., and Cordell, G.A. (1989) *J. Org. Chem.* 54(9), 2253–2255.

Anderegg, R.J. (1988) *Mass Spectrom. Rev.* 7, 395–424.

Andersen, N.H., Eaton, H.L., Nguyen, K.T., et al. (1988) *Biochemistry* 27(8), 2782–2790.

Aoki, M., Ohtsuka, T., Itezono, Y., et al. (1991a) *Tetrahedron Lett.* 32(2), 217–220.

Aoki, M., Ohtsuka, T., Itezono, Y., et al. (1991b) *Tetrahedron Lett.* 32(2), 221–224.

Atta-ur-Rahman (1986) *Nuclear Magnetic Resonance,* Springer-Verlag, New York.

Atta-ur-Rahman, Ahmad, S., Rycroft, D.S., et al. (1988) *Tetrahedron Lett.* 29(34), 4241–4244.

Atta-ur-Rahman (1989) *One and Two Dimensional NMR Spectroscopy,* Elsevier, Amsterdam.

Aue, W.P., Bartholdi, E., and Ernst, R.R. (1976) *J. Chem. Phys.* 64(5), 2229–2246.

Baker, G.H., Brown, P.J., Dorgan, R.J.J., and Everett, J.R. (1989) *J. Chem. Soc. Perkin Trans. II*, 1073–1079.

Barber, M., Bordoli, R.S., Sedgwick, R.D., and Tyler, A.N. (1981) *Nature* 293, 270–275.

Barnekow, D.E., and Cardellina, II, J.H. (1989) *Tetrahedron Lett.* 30(28), 3629–3632.

Batterham, T.J. (1973) *NMR Spectra of Simple Heterocycles,* Wiley and Sons, New York.

Bauer, C., Freeman, R., Frenkiel, T., Keeler, J., and Shaka, A.J. (1984) *J. Magn. Reson.* 58, 442–457.

Bax, A. (1982) *Two-Dimensional Nuclear Magnetic Resonance in Liquids,* Delft University Press, D. Reidel Publishing Co., Dordrecht, Holland.

Bax, A. (1984) *J. Magn. Reson.* 57, 314–318.

Bax, A. (1989) *Annu. Rev. Biochem.* 58, 223–256.

Bax, A., Aszalos, A., Dinya, Z., and Sudo, K. (1986) *J. Am. Chem. Soc.* 108(25), 8056–8063.

Bax, A., and Davis, D.G. (1985) *J. Magn. Reson.* 65, 355–360.

Bax, A., and Drobny, G. (1985) *J. Magn. Reson.* 61, 306–320.

Bax, A., Egan, W., and Kovac, P. (1984) *J. Carbohydr. Chem.* 3(4), 593–611.

Bax, A., Ferretti, J.A., Nashed, N., and Jerina, D.M. (1985) *J. Org. Chem.* 50(17), 3029–3034.

Bax, A., Freeman, R., and Frenkiel, T.A. (1981a) *J. Am. Chem. Soc.* 103(8), 2102–2104.

Bax, A., Freeman, R., Frenkiel, T.A., and Levitt, M.H. (1981b) *J. Magn. Reson.* 43, 478–483.

Bax, A., Freeman, R., and Kempsell, S.P. (1980a) *J. Am. Chem. Soc.* 102, 4851–4852.

Bax, A., and Lerner, L. (1986) *Science* 232, 960–967.

Bax, A., Mehlkopf, T., Smidt, J., and Freeman, R. (1980b) *J. Magn. Reson.* 41, 502–506.

Bax, A., and Morris, G.A. (1981) *J. Magn. Reson.* 42, 501–505.

Bax, A., and Subramanian, S. (1986) *J. Magn. Reson.* 67, 565–569.

Bax, A., and Summers, M.F. (1986) *J. Am. Chem. Soc.* 108, 2093–2094.

Benassi, R., Folli, U., Mucci, A., et al. (1987) *Magn. Reson. Chem.* 25, 804–810.

Bendall, M.R., Pegg, D.T., and Doddrell, D.M. (1982a) *J. Chem. Soc. Chem. Commun.,* pp. 872–874.

Bendall, M.R., Pegg, D.T., Doddrell, D.M., Johns, S.R.,and Willing, R.I. (1982b) *J. Chem. Soc. Chem. Commun.,* pp. 1138–1140.

Benn, R., and Günther, H. (1983) *Angew. Chem. Int. Ed. Engl.* 22, 350–380.

Berger, S. (1989) *J. Magn. Reson.* 81, 561–564.

Bermel, W., Wagner, K., and Griesinger, C. (1989) *J. Magn. Reson.* 83, 223–232.

Bhacca, N.S., and Williams, D.H. (1964) *Applications of NMR Spectroscopy in Organic Chemistry,* Holden-Day, San Francisco.

Bilton, J.N., Broughton, H.B., Jones, P.S., et al. (1987) *Tetrahedron* 43(12), 2805–2815.

Bittner, M., Gonzalez, F., Valdebenito, H., et al. (1987) *Tetrahedron Lett.* 28(35), 4031–4032.

Blasko, G., and Cordell, G.A. (1989) in *Studies in Natural Products Chemistry. Structure Elucidation,* vol. 5 (Atta-ur-Rahman, ed.), pp. 3–67, Elsevier, Amsterdam.

Blasko, G., Lin, L.-Z., and Cordell, G.A. (1988) *J. Org. Chem.* 53(26), 6113–6115.

Blumich, R. (1984) *J. Magn. Reson.* 60, 122–124.

Bock, K., and Lemieux, R.U. (1982) *Carbohydrate Res.* 100, 63–74.

Bodenhausen, G., and Freeman, R. (1977) *J. Magn. Reson.* 28, 471–476.

Bothner-By, A.A. (1965) in *Advances in Magnetic Resonance,* vol. 1 (Mooney, E.F., ed.), pp. 195–316, Academic Press, New York.

Boudot, D., Roumestand, C., Toma, F., and Canet, D. (1990) *J. Magn. Reson.* 90, 221–227.

Bovey, F.A. (1969) *Nuclear Magnetic Resonance Spectroscopy,* pp. 350–373, Academic Press, New York.

Boyd, J., Dobson, C.M., and Redfield, C. (1985) *J. Magn. Reson.* 62, 543–550.

Braunschweiler, L., and Ernst, R.R. (1983) *J. Magn. Reson.* 53, 521–528.

Brewster, A.I.R., and Hruby, V.J. (1973) *Proc. Nat. Acad. Sci. USA* 70(12), 3806–3809.

Brey, W.S. (1988) *Pulse Methods in 1D and 2D Liquid-Phase NMR,* Academic Press, New York.

Brill, G.M., McAlpine, J.B., Whittern, D.N., and Buko, A.M. (1990) *J. Antibiotics* 43(3), 229–237.

Brugel, W. (1979) *Handbook of NMR Spectral Parameters,* vols. 1–3, Heyden, London.

Burg, R.W., Hensens, O.D., Liesch, J.M., Cole, L.J., and Hernandez, S. (1989) U.K. patent no. 2,219,296 (Oct. 3).

Burg, R.W., Hensens, O.D., Liesch, J.M., Hernandez, S., and Cole, L.J. (1991) U.S. patent no. 5,028,537 (July 2).

Cardellina, II, J.H., Marner, F.J., and Moore, R.E. (1979) *Science* 204, 193–195.

Cavanagh, J., Chazin, W.J., and Rance, M. (1990) *J. Magn. Reson.* 87, 110–131.

Cavanagh, J., Hunter, C.A., Jones, D.N.M., Keeler, J., and Sanders, J.K.M. (1988) *Magn. Reson. Chem.* 26, 867–875.

Chan, W.R., Tinto, W.F., Laydoo, R.S., et al. (1991) *J. Org. Chem.* 56(5), 1773–1776.

Chen, G-F., Li, Z-L., Chen, K., et al. (1990) *J. Chem. Soc. Chem. Commun.,* pp. 1113–1114.

Chimichi, S., Tedeschi, P., Camparini, A., and Ponticelli, F. (1982) *Org. Magn. Chem.* 20(3), 141–143.

Christofides, J.C., and Davies, D.B. (1982) *J. Chem. Soc. Chem. Commun.,* pp. 560–562.

Christofides, J.C., and Davies, D.B. (1983a) *J. Chem. Soc. Chem. Commun.,* pp. 324–326.

Christofides, J.C., and Davies, D.B. (1983b) *J. Am. Chem. Soc.* 105(15), 5099–5105.

Cookson, R.C., Crabb, T.A., Frankel, J.J., and Hudec, J. (1966) *Tetrahedron* suppl. 7, 355–390.

Cordell, G.A. (1988) *Kor. J. Pharmacogn.* 19(3), 153–169.

Corriero, G., Madaio, A., Mayol, L., Piccialli, V., and Sica, D. (1989) *Tetrahedron* 45(1), 277–288.

Crews, P., Kakou, Y., and Quinoa, E. (1988) *J. Am. Chem. Soc.* 110(13), 4365–4368.

Crews, P., Naylor, S., Myers, B.L., Loo, J., and Manes, L.V. (1985) *Magn. Reson. Chem.* 23(8), 684–687.

Croasmun, W.R., and Carlson, R.M.K. (1987) *Two-Dimensional NMR Spectroscopy. Applications for Chemists and Biochemists,* VCH Publishers, New York.

Crouch, R.C., and Martin, G.E. (1991) *J. Magn. Reson.* 92, 189–194.

Dabrowski, J., Ejchart, A., Kordowicz, M., and Hanfland, P. (1987) *Magn. Reson. Chem.* 25, 338–346.

Davis, D.G., and Bax, A. (1985) *J. Am. Chem. Soc.* 107, 2820–2821.

Davis, D.G. (1989) *J. Magn. Reson.* 83, 212–218.

Davis, D.G. (1990) *J. Magn. Reson.* 90, 589–596.

Denisov, A.Y., Krivopalov, V.P., Mamatyuk, V.I., and Mamaev, V.P. (1988) *Magn. Reson. Chem.* 26, 42–46.

Derome, A.E. (1987) *Modern NMR Techniques for Chemistry Research, Organic Chemistry Series,* vol. 6, Pergamon Press, Oxford.

Derome, A.E. (1989) *Nat. Prod. Rep.* 6(2), 111–141.

Derome, A.E., and Williamson, M.P. (1990) *J. Magn. Reson.* 88, 177–185.

Doddrell, D.M., Pegg, D.T., and Bendall, M.R. (1982) *J. Magn. Reson.* 48, 323–327.

Doig, A.J., Williams, D.H., Oelrichs, P.B., and Baczynskyj, L. (1990) *J. Chem. Soc. Perkin Trans. I,* 2317–2321.

Edison, A.S., Westler, W.M., and Markley, J.L. (1991) *J. Magn. Reson.* 92, 434–438.

Eich, G., Bodenhausen, G., and Ernst, R.R. (1982) *J. Am. Chem. Soc.* 104, 3731–3732.

Ernst, R.R. (1966) *J. Chem. Phys.* 45(10), 3845–3861.

Evans, S.V., Hayman, A.R., Fellows, L.E., et al. (1985) *Tetrahedron Lett.* 26(11), 1465–1468.

Ewing, D.F. (1975) in *Annual Reports in NMR Spectroscopy,* vol. 6A (Mooney, E.F., ed.), pp. 389–437, Academic Press, New York.

Farrar, T.C. (1987a) *Anal. Chem.* 59(10), 679A–690A.

Farrar, T.C. (1987b) *Anal. Chem.* 59(11), 749A–761A.

Feeney, J., Partington, P., and Roberts, G.C.K. (1974) *J. Magn. Reson.* 13, 268–274.

Ferrari, P., Edwards, D., Gallo, G.G., and Selva, E. (1990) *J. Antibiotics* 43(11), 1359–1366.

Fesik, S.W., Perun, T.J., and Thomas, A.M. (1985) *Magn. Reson. Chem.* 23(8), 645–648.

Findlay, J.A., and Radics, L. (1980) *Can. J. Chem.* 58, 579–590.

Freeman, R., Friedrich, J., and Wu, X.-L. (1988) *J. Magn. Reson.* 79, 561–567.

Freeman, R., and Morris, G.A. (1978) *J. Chem. Soc. Chem. Commun.,* pp. 684–686.

Friedrich, J., Davies, S., and Freeman, R. (1988) *J. Magn. Reson.* 80, 168–175.

Fukuyama, Y., Otoshi, Y., and Kodama, M. (1990a) *Tetrahedron Lett.* 31(31), 4477–4480.

Fukuyama, Y., Shida, N., Kodama, M., Kido, M., and Nagasawa, M. (1990b) *Tetrahedron Lett.* 31(39), 5621–5622.

Funaishi, K., Kawamura, K., Sugiura, Y., et al. (1987) *J. Antibiotics* 40(6), 778–785.

Furihata, K., Seto, H., Ohuchi, M., and Otake, N. (1983) *Org. Magn. Reson.* 21(10), 624–627.

Fusetani, N., Nagata, H., Hirota, H., and Tsuyuki, T. (1989) *Tetrahedron Lett.* 30(50), 7079–7082.

Fushimi, S., Furihata, K., and Seto, H. (1989) *J. Antibiotics* 42(7), 1026–1036.

Gagnaire, D., and Vincendon, M. (1977) *J. Chem. Soc. Chem. Commun.,* pp. 509–510.

Geen, H., and Freeman, R. (1991) *J. Magn. Reson.* 93, 93–141.

Glickson, J.D., Dadok, J., and Marshall, G.R. (1974) *Biochemistry* 13(1), 11–14.

Goetz, M.A., McCormick, P.A., Monaghan, R.L., et al. (1985) *J. Antibiotics* 38(2), 161–168.

Gogoll, A., and Plobeck, N.A. (1990) *Magn. Reson. Chem.* 28, 635–641.

Goh, S.H., Chuah, C.H., Beloeil, J.C., and Morellet, N. (1988) *Tetrahedron Lett.* 29(1), 113–116.

Golik, J., Clardy, J., Dubay, G., et al. (1987a) *J. Am. Chem. Soc.* 109(11), 3461–3462.

Golik, J., Dubay, G., Groenewald, G., Kawaguchi, H., et al. (1987b) *J. Am. Chem. Soc.* 109(11), 3462–3464.

Gomi, S., Sezaki, M., and Kondo, S., et al. (1988) *J. Antibiotics* 41(8), 1019–1028.

Gray, N.A.B. (1986) *Computer-Assisted Structure Elucidation,* John Wiley & Sons, New York.

Griesinger, C., Sorensen, O.W., and Ernst, R.R. (1987) *J. Am. Chem. Soc.* 109, 7227–7228.

Gunasekera, S.P., Gunasekera, M., Longley, R.E., and Schulte, G.K. (1990) *J. Org. Chem.* 55(16), 4912–4915.

Gunasekera, S.P., Gunasekera, M., Longley, R.E., and Schulte, G.K. (1991) *J. Org. Chem.* 56(3), 1346.

Hall, L.D., and Norwood, T.J. (1987a) *J. Magn. Reson.* 74, 171–176.

Hall, L.D., and Norwood, T.J. (1987b) *J. Magn. Reson.* 74, 406–423.

Hall, L.D., and Norwood, T.J. (1990) *J. Magn. Reson.* 87, 331–345.

Hansen, P.E. (1981) in *Progress in NMR Spectroscopy,* vol. 14 (Elmsley, J.W., Feeney, J., and Sutcliffe, L.H., eds.), pp. 175–296, Pergamon Press, Oxford, England.

Hansen, P.E. (1983) in *Annual Reports in NMR Spectroscopy,* vol. 15 (Webb, G.A., ed.), pp. 105–234, Academic Press, New York.

Hansen, P.E. (1986) *Magn. Reson. Chem.* 24, 903–910.

Hansen, P.E. (1988) in *Progress in NMR Spectroscopy,* vol. 20 (Elmley, J.W., Feeney, J., and Sutcliffe, L., eds.), pp. 207–255, Pergamon Press, Oxford, England.

Hanstock, C.C., and Lown, J.W. (1984) *J. Magn. Reson.* 58, 167–172.

Hayakawa, Y., Furihata, K., Seto, H., and Otake, N. (1985) *Tetrahedron Lett.* 26(29), 3475–3478.

Hayakawa, Y., Kanamaru, N., Morisaki, N., Seto, H., and Furihata, K. (1991) *Tetrahedron Lett.* 32(2), 213–216.

Hayashi, F., Nagashima, K., Terui, Y., et al. (1990) *J. Antibiotics* 43(11), 1421–1430.

Helms, G.L., Moore, R.E., Niemczura, W.P., and Patterson, G.M.L. (1988) *J. Org. Chem.* 53(6), 1298–1307.

Hensens, O.D., and Albers-Schonberg, G. (1978) *Tetrahedron Lett.* 9(39), 3649–3652.

Hensens, O.D., and Albers-Schonberg, G. (1983a) *J. Antibiotics* 36(7), 799–813.

Hensens, O.D., and Albers-Schonberg, G. (1983b) *J. Antibiotics* 36(7), 814–831.

Hensens, O.D., and Albers-Schonberg, G. (1983c) *J. Antibiotics* 36(7), 832–845.

Hensens, O.D., Borris, R.P., Koupal, L.R., et al. (1991a) *J. Antibiotics* 44(2), 249–254.

Hensens, O.D., Monaghan, R.L., Huang, L., et al. (1983) *J. Am. Chem. Soc.* 105(11), 3672–3679.

Hensens, O.D., Zink, D., and Williamson, J.M., et al. (1991b) *J. Org. Chem.* 56(10), 3399–3403.

Hirota, H., Matsunaga, S., and Fusetani, N. (1990) *Tetrahedron Lett.* 31(29), 4163–4164.

Hirota, A., Nakagawa, M., and Hirota, H. (1991) *Agric. Biol. Chem.* 55(4), 1187–1188.

Hochlowski, J.E., Swanson, S.J., Ranfranz, L.M., et al. (1987) *J. Antibiotics* 40(5), 575–588.

Hoffman, R.E., and Davies, D.B. (1988) *Magn. Reson. Chem.* 26, 425–429.

Homans, S.W. (1990) *J. Magn. Reson.* 90, 557–560.

Homans, S.W., Dwek, R.A., Fernandes, D.L., and Rademacher, T.W. (1984) *Proc. Natl. Acad. Sci. USA* 81, 6286–6289.

Homans, S.W., Dwek, R.A., and Rademacher, T.W. (1987) *Biochemistry* 26, 6571–6578.

Hore, P.J., Scheek, R.M., Volbeda, A., Kaptein, R., and Van Boom, J.H. (1982) *J. Magn. Reson.* 50, 328–334.

Hosur, R.V., Govil, G., and Miles, H.T. (1988) *Magn. Reson. Chem.* 26, 927–944.

Hughes, D.W. (1988) *Magn. Reson. Chem.* 26, 214–223.

Inman, W.D., O'Neill-Johnson, M., and Crews, P. (1990) *J. Am. Chem. Soc.* 112(1), 1–4.

Ishida, T., Ohishi, H., Inoue, M., et al. (1989) *J. Org. Chem.* 54(22), 5337–5343.

Ishitsuka, M., Kusumi, T., Kakisawa, H., Kaya, K., and Watanabe, M.M. (1990) *J. Am. Chem. Soc.* 112(22), 8180–8182.

Isogai, A., Fukuchi, N., Yamashita, S., Suyama, K., and Suzuki, A. (1990) *Tetrahedron Lett.* 31(5), 695–698.

Iwasaki, S., Kobayashi, H., Furukawa, J., et al. (1984) *J. Antibiotics* 37(4), 354–362.

Jackman, L.M., and Sternhell, S. (1969) *Applications of Nuclear Magnetic Resonance in Organic Chemistry*, 2nd ed., Pergamon Press, Oxford, England.

Jacobs, H., Ramdayal, F., Reynolds, W.F., et al. (1986) *Tetrahedron Lett.* 27(13), 1453–1456.

Jacobs, H., Ramadayal, F., McLean, S., and McLean, S. (1987) *J. Nat. Prod.* 50(3), 507–509.

Kaptein, R., Boelens, R., Scheek, R.M., and Van Gunsteren, W.F. (1988) *Biochemistry* 27(15), 5390–5395.

Kashman, Y., Hirsh, S., McConnell, O.J., et al. (1989) *J. Am. Chem. Soc.* 111, 8925–8926.

Kato, Y., Fusetani, N., Matsunaga, S., and Hashimoto, K. (1986) *J. Am. Chem. Soc.* 108, 2780–2781.

Kay, L.E., Jones, P-J., and Prestegard, J.H. (1987) *J. Magn. Reson.* 72, 392–396.

Keeler, J., and Neuhaus, D. (1985) *J. Magn. Reson.* 63, 454–472.

Kessler, H., Gehrke, M., and Griesinger, C. (1988a) *Angew. Chem. Int. Ed. Engl.* 27, 490–536.

Kessler, H., Gemmecker, G., Haase, B., and Steuernagel, S. (1988b) *Magn. Reson. Chem.* 26, 919–926.

Kessler, H., Griesinger, C., Zarbock, J., and Loosli, H.R. (1984) *J. Magn. Reson.* 57, 331–336.

Kessler, H., Oschkinat, H., and Griesinger, C. (1986) *J. Magn. Reson.* 70, 106–133.

Kessler, H., Schmieder, P., Kock, M., and Kurz, M. (1990a) *J. Magn. Reson.* 88, 615–618.

Kessler, H., Schmieder, P., and Kurz, M. (1989) *J. Magn. Reson.* 85, 400–405.

Kessler, H., Schmieder, P., and Oschkinat, H. (1990b) *J. Am. Chem. Soc.* 112, 8599–8600.

Kettingring, J.K., Malabara, A., Vekey, K., and Cavalleri, B. (1990) *J. Antibiotics* 43(9), 1082–1088.

Kim, J., and Kinghorn, A.D. (1987) *Tetrahedron Lett.* 28(32), 3655–3658.

Kimura, I., Ota, Y., Kimura, R., et al. (1987a) *Tetrahedron Lett.* 28(17), 1917–1920.

Kimura, I., Ota, Y., Kimura, R., et al. (1987b) *Tetrahedron Lett.* 28(17), 1921–1924.

Kobayashi, J., Cheng, J., Ohta, T., et al. (1988) *J. Org. Chem.* 53(26), 6147–6150.

Koehn, F.E., Gunasekera, S.P., Niel, D.N., and Cross, S.S. (1991) *Tetrahedron Lett.* 32(2), 169–172.

Komoroski, R.A., Gregg, E.C., Shockcor, J.P., and Geckle, J.M. (1986) *Magn. Reson. Chem.* 24, 534–543.

Kondo, S., Gomi, S., Uotani, K., Inouye, S., and Takeuchi, T. (1990) *J. Antibiotics* 44(2), 123–129.

Kondo, S., Yasui, K., Katayama, M., et al. (1987) *Tetrahedron Lett.* 28(47), 5861–5864.

Kono, Y., Uzawa, J., Kobayashi, K., et al. (1991) *Agric. Biol. Chem.* 55(3), 803–811.

Koole, N.J., De Bie, M.J.A., and Hansen, P.E. (1984) *Org. Magn. Chem.* 22(3), 146–162.

Kraus, W., Bokel, M., Bruhn, A., et al. (1987) *Tetrahedron* 43(12), 2817–2830.

Kraus, W., Bokel, M., Klenk, A., and Pohn, H. (1985) *Tetrahedron Lett.* 26(52), 6435–6438.

Krishnamurthy, V.V., and Casida, J.E. (1987) *Magn. Reson. Chem.* 25, 837–842.

Krishnamurthy, V.V., and Casida, J.E. (1988a) *Magn. Reson. Chem.* 26, 362–367.

Krishnamurthy, V.V., and Casida, J.E. (1988b) *Magn. Reson. Chem.* 26, 367–372.

Krishnamurthy, V.V., and Casida, J.E. (1988c) *Magn. Reson. Chem.* 26, 980–989.

Krishnamurthy, V.V., and Nunlist, R. (1988) *J. Magn. Reson.* 80, 280–295.

Ksebati, M.B., Schmitz, F.J., and Gunasekera, S.P. (1988) *J. Org. Chem.* 53(17), 3917–3921.

Kuhnt, D., Anke, T., Besl, H., et al. (1990) *J. Antibiotics* 43(11), 1413–1420.

Kunz, R.W., Bilinski, V., Von Philipsborn, W., et al. (1984) *Org. Reson. Chem.* 22(5), 349–351.

Kusumi, T., Ooi, T., and Watanabe, M.M. (1987) *Tetrahedron Lett.* 28(40), 4695–4698.

Laatsch, H., Kellner, M., and Weyland, H. (1991) *J. Antibiotics* 44(2), 187–191.

Ladner, H.K., Led, J.J., and Grant, D.M. (1975) *J. Magn. Reson.* 20, 530–534.

Le Cocq, C., and Lallemand, J-Y. (1981) *J. Chem. Soc. Chem. Commun.*, pp. 150–152.

Lee, M.D., Dunne, T.S., Chang, C.C., et al. (1987b) *J. Am. Chem. Soc.* 109(11) 3466–3468.

Lee, M.D., Dunne, T.S., Siegel, M.M., et al. (1987a) *J. Am. Chem. Soc.* 109(11), 3464–3466.

Limat, D., Wimperis, S., and Bodenhausen, G. (1988) *J. Magn. Reson.* 79, 197–205.

Lin, L.-J., and Cordell, G.A. (1986) *J. Chem. Soc. Chem. Commun.*, 377–379.

Lin, L.-Z., Cordell, G.A., Ni, C.-Z., and Clardy, J. (1989) *J. Org. Chem.* 54(13), 3199–3202.

Luo, Z., Meksuriyen, D., Erdelmeier, C.A.J., Fong, H.H.S., and Cordell, G.A. (1988) *J. Org. Chem.* 53(10), 2183–2185.

Macura, S., Kumar, N.G., and Brown, L.R. (1984) *J. Magn. Reson.* 60, 99–105.

Madsen, J.C., Bildsoe, H., Jacobsen, H.J., and Sorensen, O.W. (1986) *J. Magn. Reson.* 67, 243–257.

Majumdar, A., and Hosur, R.V. (1990) *J. Magn. Reson.* 90, 597–599.

Marchand, A.P. (1982) *Stereochemical Applications of NMR Studies in Rigid Bicyclic Systems, Methods of Stereochemical Analysis,* vol. 1, Verlag Chemie, Deerfield Beach, FL.

Mareci, T.H. (1988) in *Pulse Methods in 1D and 2D Liquid-Phase NMR* (Brey, W.S., ed.), pp. 259–341, Academic Press, New York.

Mareci, T.H., and Freeman, R. (1983) *J. Magn. Reson.* 51, 531–535.

Marion, D., and Wüthrich, K. (1983) *Biochem. Biophys. Res. Commun.* 113(3), 967–974.

Marshall, J.L. (1983) *Carbon-Carbon and Carbon-Proton NMR Couplings,* pp. 11–64, Verlag Chemie International, FL.

Martin, G.E., and Crouch, R.C. (1991) *J. Nat. Prod.* 54(1), 1–70.

Martin, G.E., and Zektzer, A.S. (1988a) *Two-Dimensional NMR Methods for Establishing Molecular Connectivity,* VCH Publishers, New York.

Martin, G.E., and Zektzer, A.S. (1988b) *Magn. Reson. Chem.* 26, 631–652.

Matsumoto, M., Kawamura, Y., Yoshimura, Y., et al. (1990) *J. Antibiotics* 43(7), 739–747.

Mayol, L., Piccialli, V., and Sica, D. (1987) *Tetrahedron Lett.* 28(31), 3601–3604.

McDonald, S., and Warren, W.S. (1991) *Concepts Magn. Reson.* 3(2), 55–81.

McLean, S., Perpick-Dumont, M., Reynolds, W.F., et al. (1988) *J. Am. Chem. Soc.* 110, 5339–5344.

McLean, S., Reynolds, W.F., Zhu, X. (1987) *Can. J. Chem.* 65, 200–204.

Mierke, D.F., Schmieder, P., and Kessler, H. (1991) *Helv. Chim. Acta* 74, 1027–1047.

Mo, C-J., Shin-ya, K., Furihata, K., et al. (1990) *J. Antibiotics* 43(10), 1337–1340.

Moore, R.E., Bornemann, V., Niemczura, W.P., et al. (1989) *J. Am. Chem. Soc.* 111, 6128–6132.

Morales-Rios, M.S., and Joseph-Nathan, P. (1989) *Magn. Reson. Chem.* 27, 75–80.

Morris, G.A. (1986) *Magn. Reson. Chem.* 24, 371–403.

Morris, G.A., and Richards, M.S. (1985) *Magn. Reson. Chem.* 23(8), 676–683.

Muller, L. (1984) *J. Magn. Reson.* 59, 326–331.

Mueller, N., Di Bari, L., and Bodenhausen, G. (1991) *J. Magn. Reson.* 94, 73–81.

Murakami, M., Makabe, K., Yamaguchi, K., et al. (1988) *Tetrahedron Lett.* 29(10), 1149–1152.

Murata, M., Kumagai, M., Lee, J.S., and Yasumoto, T. (1987) *Tetrahedron Lett.* 28(47), 5869–5872.

Murata, M., Legrand, A.M., Ishibashi, Y., and Yasumoto, T. (1989) *J. Am. Chem. Soc.* 111(24), 8929–8931.

Nagayama, K. (1986) *J. Magn. Reson.* 66, 240–249.

Nakagawa, M., Hayakawa, Y., Furihata, K., and Seto, H. (1990) *J. Antibiotics* 43(5), 477–484.

Nakanishi, K., Goto, T., Ito, S., Natori, S., and Nozoe, S. (1974) *Natural Products Chemistry,* vol. 1, Academic Press, New York.

Nakanishi, K., Goto, T., Ito, S., Natori, S., and Nozoe, S. (1975) *Natural Products Chemistry,* vol. 2, Academic Press, New York.

Nakashima, Y., Yoshikawa, Y., Mitani, F., and Sasaki, K. (1986) *J. Magn. Reson.* 69, 162–164.

Neuhaus, D., and Williamson, M.P. (1989) *The Nuclear Overhauser Effect in Structural and Conformational Analysis,* VCH Publishers, New York.

Newmark, R.A., and Hill, J.R. (1980) *Org. Magn. Reson.* 13(1), 40–44.

Nishikawa, M., Tsurumi, Y., Murai, H., et al. (1991) *J. Antibiotics* 44(2), 130–135.

Northcote, P.T., and Andersen, R.J. (1988) *Tetrahedron Lett.* 29(35), 4357–4360.

Northcote, P.T., and Andersen, R.J. (1989) *J. Am. Chem. Soc.* 111(16), 6276–6280.

Numata, A., Pettit, G.R., Nabae, M., et al. (1987) *Agric. Biol. Chem.* 51(4), 1199–1201.

Nuzillard, J.M., and Massiot, G. (1991) *J. Magn. Reson.* 91, 380–385.

Omura, S., Nakagawa, A., Aoyama, H., Hinotozama, K., and Sano, H. (1983) *Tetrahedron Lett.* 24(34), 4643–4646.

Oppenheimer, N.J., and James, T.L. (1989a) *Nuclear Magnetic Resonance. Part A Spectral Techniques and Dynamics, Methods in Enzymology,* vol. 176, Academic Press, New York.

Oppenheimer, N.J., and James, T.L. (1989b) *Nuclear Magnetic Resonance. Part B Structure and Mechanism, Methods in Enzymology,* vol. 177, Academic Press, New York.

Otter, A., and Kotovych, G. (1986) *J. Magn. Reson.* 69, 187–190.

Otter, A., Liu, X., and Kotovych, G. (1990) *J. Magn. Reson.* 86, 657–662.

Otting, G., and Wüthrich, K. (1986) *J. Magn. Reson.* 66, 359–363.

Otoguro, K., Nakagawa, A., and Omura, S. (1988) *J. Antibiotics* 40(2), 250–252.

Ozasa, T., Tanaka, K., Sasamata, M., et al. (1989) *J. Antibiotics* 42(9), 1339–1343.

Patt, S.L., and Shoolery, J.N. (1982) *J. Magn. Reson.* 46, 535–539.

Pawlak, J., Tempesta, M.S., Golik, J., et al. (1987) *J. Am. Chem. Soc.* 109(4), 1144–1150.

Pegg, D.T., and Bendall, M.R. (1983) *J. Magn. Reson.* 55, 114–127.

Pelczer, I. (1991) *J. Am. Chem. Soc.* 113(8), 3211–3212.

Perpick-Dumont, M., Enriquez, R.G., McLean, S., Puzzuoli, F.V., and Reynolds, W.F. (1987) *J. Magn. Reson.* 75, 414–426.

Perry, N.B., Blunt, J.W., Munro, M.H.G., and Pannell, L.K. (1988) *J. Am. Chem. Soc.* 110(14), 4850–4851.

Pettit, G.R., Kamano, Y., Herald, C.L., et al. (1989) *J. Am. Chem. Soc.* 111, 5015–5017.

Pettit, G.R., Singh, S.B., Goswami, A., and Nieman, R.A. (1988) *Tetrahedron* 44(11), 3349–3354.

Pfeffer, P.E., Valentine, K.M., and Parrish, F.W. (1979) *J. Am. Chem. Soc.* 101(5), 1265–1274.

Piantini, U., Sorensen, O.W., and Ernst, R.R. (1982) *J. Am. Chem. Soc.* 104, 6800–6801.

Platzer, N., Goasdoue, N., and Davoust, D. (1987) *Magn. Reson. Chem.* 25, 311–319.

Poch, G.K., and Gloer, J.B. (1991) *J. Nat. Prod.* 54(1), 213–217.

Poole, C.F. (1977) in *Handbook of Derivatives for Chromatography,* (Blau, K., and King, G.S., eds.), pp. 152–200, Heyden, London.

Poppe, L., and Van Halbeek, H. (1991a) *J. Magn. Reson.* 92, 636–641.

Poppe, L., and Van Halbeek, H. (1991b) *J. Magn. Reson.* 93, 214–217.

Pouzard, G., Sukumar, S., and Hall, L.D. (1981) *J. Am. Chem. Soc.* 103(14), 4209–4215.

Pratum, T.K. (1991) *J. Magn. Reson.* 93, 642–647.

Pratum, T.K., Hammen, P.K., and Andersen, N.H. (1988) *J. Magn. Reson.* 78, 376–381.

Prestegard, J.H. (1988) in *Pulse Methods in 1D and 2D Liquid-Phase NMR,* (Brey, W.S., ed.), pp. 435–488, Academic Press, New York.

Rabenstein, D., and Nakashima, T. (1979) *Anal. Chem.* 51, 14651A–14657A.

Rabenstein, D.L., and Sayer, T.L. (1976) *J. Magn. Reson.* 24, 27–39.

Rabidau, P.W. (1978) *Accts. Chem. Res.* 11(4), 141–147.

Radeglia, R., and Porzel, A. (1989) *J. Magn. Reson.* 81, 530–537.

Rance, M., Sorensen, O.W., Bodenhausen, G., et al. (1983) *Biochem. Biophys. Res. Commun.* 117(2), 479–485.

Rance, M., Chazin, W.J., Dalvit, C., and Wright, P.E. (1989) in *Nuclear Magnetic Resonance, Part A Spectral Techniques and Dynamics, Methods in Enzymology*, vol. 176 (Oppenheimer, N.J., and James, T.L., eds.), pp. 114–134, Academic Press, New York.

Rance, M., Sorenson, O.W., Leupin, W., et al. (1985) *J. Magn. Reson.* 61, 67–80.

Reich, H.J., Jautelat, M., Messe, M.T., Weigert, F.J., and Roberts, J.D. (1969) *J. Am. Chem. Soc.* 91(26), 7445–7454.

Reuben, J. (1984) *J. Am. Chem. Soc.* 106(21), 6180–6186.

Reuben, J. (1985a) *J. Am. Chem. Soc.* 107(5), 1433–1435.

Reuben, J. (1985b) *J. Am. Chem. Soc.* 107(6), 1747–1755.

Reuben, J. (1985c) *J. Am. Chem. Soc.* 107(6), 1756–1759.

Reuben, J. (1986a) *J. Am. Chem. Soc.* 108, 1082–1083.

Reuben, J. (1986b) *J. Am. Chem. Soc.* 108(8), 1735–1738.

Reuben, J. (1987) *J. Am. Chem. Soc.* 109(2), 316–321.

Reynolds, W.F., Hughes, R.W., Perpick-Dumont, M., and Enriques, R.G. (1985) *J. Magn. Reson.* 63, 413–417.

Reynolds, W.F., McLean, S., Perpick-Dumont, M., and Enriques, R.G. (1989) *Magn. Reson. Chem.* 27, 162–169.

Reynolds, W.F., McLean, S., Poplawski, J., et al. (1986) *Tetrahedron* 42(13), 3419–3428.

Riand, J., and Chenon, M.-T. (1981) *Org. Magn. Reson.* 15(1), 18–21.

Richardson, J.M., Titman, J.J., and Keeler, J. (1991) *J. Magn. Reson.* 93, 533–553.

Rinehart, K.L., Holt, T.G., Fregeau, N.L., et al. (1990) *J. Org. Chem.* 55, 4512–4515.

Rodriguez-Hahn, L., O'Reilly, R., Esquivel, B., et al. (1990) *J. Org. Chem.* 55(11), 3522–3525.

Roll, D.M., Biskupiak, J.E., Mayne, C.L., and Ireland, C.M. (1986) *J. Am. Chem. Soc.* 108(21), 6680–6682.

Roy, R., Vishwakarma, R.A., Varma, N., and Tandon, J.S. (1990) *Tetrahedron Lett.* 31(24), 3467–3470.

Sadler, I.H. (1988) *Nat. Prod. Rep.* 5(2), 101–127.

Sakai, S., Aimi, N., Yamaguchi, K., et al. (1984) *Chem. Pharm. Bull.* 32, 354–357.

Sakemi, S., Ichiba, T., Kohmoto, S., Saucy, G., and Higa, T. (1988) *J. Am. Chem. Soc.* 110(14), 4851–4853.

Sanders, J.K.M., and Hunter, B.K. (1987) *Modern NMR Spectroscopy*, Oxford University Press, Oxford.

Sasaki, S. (1985) *Handbook of Proton-NMR Spectra and Data*, vols. 1–5, Academic Press, New York.

Sasaki, S. (1986) *Handbook of Proton-NMR Spectra and Data*, vols. 6–10, Academic Press, New York.

Sasaki, S. (1987) *Handbook of Proton-NMR Spectra and Data*, Index to vols. 1–10, Academic Press, New York.

Sawada, Y., Tsuno, T., Yamamoto, H., et al. (1990) *J. Antibiotics* 43(11), 1367–1374.

Schneider, H-J., Buchheit, U., Becker, N., Schmidt, G., and Siehl, U. (1985) *J. Am. Chem. Soc.* 107, 7027–7039.

Seto, H., Furihata, K., Guangyi, X., Xiong, C., and Deji, P. (1988a) *Agric. Biol. Chem.* 52(7), 1797–1801.

Seto, H., Furihata, K., and Ohuchi, M. (1988b) *J. Antibiotics* 41(8), 1158–1160.

Seto, H., Furihata, K., Saeki, K., et al. (1987) *Tetrahedron Lett.* 28(29), 3357–3360.

Seto, H., Sasaki, T., Yonehara, H., and Uzawa, J. (1978) *Tetrahedron Lett.* 19(10), 923–926.

Shin, J., and Fenical, W. (1991) *J. Org. Chem.* 56(9), 3153–3158.

Shin-ya, K., Furihata, K., Hayakawa, Y., et al. (1991) *Tetrahedron Lett.* 32(7), 943–946.

Shoolery, J.N. (1984) *J. Nat. Prod.* 47(2), 226–259.

Skelton, N.J., Williams, D.H., Monday, R.A., and Ruddock, J.C. (1990) *J. Org. Chem.* 55(12), 3718–3723.

Skelton, N.J., Williams, D.H., Rance, M.J., and Ruddock, J.C. (1991) *J. Am. Chem. Soc.* 113(10), 3757–3765.

Sorensen, O.W., and Jakobsen, H.J. (1988) in *Pulse Methods in 1D and 2D Liquid-Phase NMR* (Brey, S., ed.), pp. 149–258, Academic Press, New York.

Sowinski, P., Gariboldi, P., Czerwinski, A., and Borowski, E. (1989) *J. Antibiotics* 42(11), 1631–1638.

States, D.J., Haberkorn, R.A., and Ruben, D.J. (1982) *J. Magn. Reson.* 48, 286–292.

Steffens, J.C., Roark, J.L., Lynn, D.G., and Riopel, J.L. (1983) *J. Am. Chem. Soc.* 105(6), 1669–1671.

Sternhell, S. (1969) *Quart. Rev.* 23(2), 236–270.

Subramanian, S., and Bax, A. (1987) *J. Magn. Reson.* 71, 325–330.

Summers, M.F., Marzilli, L.G., and Bax, A. (1986) *J. Am. Chem. Soc.* 108(15), 4285–4294.

Szymanski, H.A., and Yelin, R.E. (1968) *NMR Handbook,* Plenum, New York.

Takeuchi, S., Uzawa, J., Seto, H., and Yonehara, H. (1977) *Tetrahedron Lett.* 18(34), 2943–2946.

Tanaka, H., Kuroda, A., Marusawa, H., et al. (1987) *J. Am. Chem. Soc.* 109(16), 5031–5033.

Tchouankeu, J.C., Tsamo, E., Sondengam, B.L., Connolly, J.D., and Rycroft, D.S. (1990) *Tetrahedron Lett.* 31(31), 4505–4508.

TePaske, M.R., Gloer, J.B., Wicklow, W.T., and Dowd, P.F. (1989) *Tetrahedron Lett.* 30(44), 5965–5968.

Terui, Y., Nishikawa, J., Hinoo, H., Kato, T., and Shoji, J. (1990) *J. Antibiotics* 43(7), 788–795.

Titman, J.J., Foote, J., Jarvis, J., Keeler, J., and Neuhaus, D. (1991) *J. Chem. Soc. Chem. Commun.,* pp. 419–421.

Titman, J.J., Neuhaus, D., and Keeler, J. (1989) *J. Magn. Reson.* 85, 111–131.

Tomer, K.B. (1989) *Mass Spec. Rev.* 8, 445–482.

Tomiya, T., Uramoto, M., and Isono, K. (1990) *J. Antibiotics* 43(1), 118–121.

Tori, M., Toyota, M., Harrison, L.J., Takikawa, K., and Asakawa, Y. (1985) *Tetrahedron Lett.* 26(39), 4735–4738.

Torigoe, K., Murata, M., Yasumoto, T., and Iwashita, T. (1988) *J. Am. Chem. Soc.* 110, 7876–7877.

Traficante, D.D. (1990a) *Concepts Magn. Reson.* 2(3), 151–167.

Traficante, D.D. (1990b) *Concepts Magn. Reson.* 2(4), 181–195.

Turner, C.J., Tempesta, M.S., Taylor, R.B., et al. (1987) *Tetrahedron* 43(12), 2789–2803.

Von Dreele, P.H., Rae, I.D., and Scheraga, H.A. (1972) *Biochemistry* 17(6), 956–961.

Vuister, G.W., Boelens, R., and Kaptein, R. (1988) *J. Magn. Reson.* 80, 176–185.

Waltho, J.P., Williams, D.H., Selva, E., and Ferrari, P. (1987) *J. Chem. Soc. Perkin Trans.* 1, 2103–2107.

Warren, W.S., and Silver, M.S. (1988) in *Advances in Magnetic Resonance*, vol. 12 (Waugh, J.S., ed.), pp. 247–384, Academic Press, New York.

Weber, H.A., Baenziger, N.C., and Gloer, J.B. (1990) *J. Am. Chem. Soc.* 112(18), 6718–6719.

Wemmer, D.E. (1989) *Concepts Magn. Reson.* 1(2), 59–72.

Werner, G., Hagenmaier, H., Albert, K., et al. (1983) *Tetrahedron Lett.* 24, 5193–5196.

Werner, G., Hagenmaier, H., Drautz, H., et al. (1984) *J. Antibiotics* 37, 110–117.

Williamson, D.S., Smith, R.A., Nagel, D.L., and Cohen, S.M. (1989) *J. Magn. Reson.* 82, 605–612.

Wilson, K.E., Flor, J.E., Schwartz, R.E., et al. (1987) *J. Antibiotics* 40(12), 1682–1691.

Wise, W.B., Pfeffer, P.E., and Kovac, P. (1984) *J. Carbohyd. Chem.* 3(4), 513–524.

Wolborn, U., Domke, T., and Leibfritz, D. (1990) *J. Magn. Reson.* 90, 575–579.

Wright, A.E., Forleo, D.A., Gunawardana, G.P., et al. (1990) *J. Org. Chem.* 55(15), 4508–4512.

Wüthrich, K. (1986) *NMR of Proteins and Nucleic Acids,* John Wiley & Sons, New York.

Wüthrich, K. (1989) *Acc. Chem. Res.* 22(1), 36–44.

Wüthrich, K. (1990) *J. Biol. Chem.* 265(36), 22059–22062.

Wyss, D.F., Lahm, H-W., Manneberg, M., and Labhardt, A.M. (1991) *J. Antibiotics* 44(2), 172–180.

Xu, Y., Sun, H., Wang, D., et al. (1987) *Tetrahedron Lett.* 28(5), 499–502.

Xu, P., Wu, X-L., and Freeman, R. (1990) *J. Magn. Reson.* 89, 198–204.

Yamaguchi, K. (1970) *Spectral Data of Natural Products,* vol. 1, Elsevier, Amsterdam.

Yashiro, M., Yano, S., and Yoshikawa, S. (1986) *J. Am. Chem. Soc.* 108(5), 1096–1097.

Zabriskie, T.M., Foster, M.P., Stout, T.J., Clardy, J., and Ireland, C.M. (1990) *J. Am. Chem. Soc.* 112(22), 8080–8084.

Zabriskie, T.M., Mayne, C.L., and Ireland, C.M. (1988) *J. Am. Chem. Soc.* 110, 7919–7920.

Zhang, X., Snyder, J.K., Joshi, B.S., Glinsky, J.A., and Pelletier, S.W. (1990) *Heterocycles* 31(10), 1879–1888.

Zhu, G., and Bax, A. (1990) *J. Magn. Reson.* 90, 405–410.

Ziegler, F.E., Nangia, A., and Tempesta, M.S. (1988) *Tetrahedron Lett.* 29(14), 1665–1668.

Zuiderweg, E.R.P., and Fesik, S.W. (1991) *J. Magn. Reson.* 93, 653–658.